Routledge International Handbook of Social and Environmental Change

Today, the risks associated with global environmental change and the dangers of extreme climatic and geological events remind us of humanity's dependence on favourable environmental conditions. Our relationships with the landscapes and ecologies that we are a part of, the plants and animals that we share them with, and the natural resources that we extract, lie at the heart of contemporary social and political debates. It is no longer possible to understand key social scientific concerns without at the same time also understanding contemporary patterns of ecosystem change.

The *Routledge International Handbook of Social and Environmental Change* reviews the major ways in which social scientists are conceptualizing more integrated perspectives on society and nature, from the global to local levels. The chapters in this volume, by international experts from a variety of disciplines, explore the challenges, contradictions and consequences of social-ecological change, along with the uncertainties and governance dilemmas they create. The contributions are organized around the themes of:

- climate change, energy and adaptation
- urban environmental change and governance
- risk, uncertainty and social learning
- (re)assembling social-ecological systems

With case studies from sectors across both developed and developing worlds, the *Handbook* illustrates the interconnectedness of ecosystem health, natural resource condition, livelihood security, social justice and development. It will be of interest to students and scholars across the social and natural sciences, as well as to those interested and engaged in environmental policy at all levels.

Stewart Lockie is Professor of Sociology and Director of The Cairns Institute at James Cook University, Australia.

David A. Sonnenfeld is Professor of Sociology and Environmental Policy at the State University of New York College of Environmental Science and Forestry (SUNY-ESF).

Dana R. Fisher is Associate Professor of Sociology and Director of the Program for Society and the Environment at the University of Maryland.

Routledge International Handbook of Social and Environmental Change

*Edited by Stewart Lockie,
David A. Sonnenfeld and
Dana R. Fisher*

First published 2014
by Routledge
2 Park Square, Milton Park, Abingdon, Oxon OX14 4RN

and by Routledge
711 Third Avenue, New York, NY 10017

Routledge is an imprint of the Taylor & Francis Group, an informa business

© 2014 selection and editorial material Stewart Lockie, David A. Sonnenfeld and Dana R. Fisher; individual chapters, the contributors

The right of the editors to be identified as the authors of the editorial material, and of the authors for their individual chapters, has been asserted in accordance with sections 77 and 78 of the Copyright, Designs and Patents Act 1988.

All rights reserved. No part of this book may be reprinted or reproduced or utilized in any form or by any electronic, mechanical, or other means, now known or hereafter invented, including photocopying and recording, or in any information storage or retrieval system, without permission in writing from the publishers.

Trademark notice: Product or corporate names may be trademarks or registered trademarks, and are used only for identification and explanation without intent to infringe.

British Library Cataloguing-Publication Data
A catalogue record for this book is available from the British Library

Library of Congress Cataloging in Publication Data
Routledge international handbook of social and environmental change / edited by Stewart Lockie, David A. Sonnenfeld and Dana R. Fisher.
  pages cm
Includes bibliographical references and index.
ISBN 978-0-415-78279-1 (hardback) — ISBN 978-0-203-81455-0 (ebook) 1. Social change–Environmental aspects. 2. Social change. 3. Global environmental change–Social aspects. 4. Climate change. I. Lockie, Stewart. II. Sonnenfeld, David Allan. III. Fisher, Dana, 1971–HM856.R68 2013
303.4—dc23
2013011534

ISBN: 978-0-415-78279-1 (hbk)
ISBN: 978-0-203-81455-0 (ebk)

Typeset in Bembo
by Cenveo Publisher Services

Printed and bound in the United States of America by Publishers Graphics, LLC on sustainably sourced paper.

Contents

List of illustrations — viii
Notes on contributors — ix
Acknowledgements — xv

1   Socio-ecological transformations and the social sciences — 1
    Stewart Lockie, David A. Sonnenfeld and Dana R. Fisher

PART I
Challenges, contradictions and consequences of global socio-ecological change — 13

2   Ecological modernization theory: taking stock, moving forward — 15
    Arthur P.J. Mol, Gert Spaargaren and David A. Sonnenfeld

3   The emergence of new world-systems perspectives on global environmental change — 31
    Andrew K. Jorgenson and Jennifer E. Givens

4   China's economic growth and environmental protection: approaching a 'win–win' situation? A discussion of ecological modernization theory — 45
    Dayong Hong, Chenyang Xiao and Stewart Lockie

5   Eco-imperialism and environmental justice — 58
    Anja Nygren

6   Neoliberalism by design: changing modalities of market-based environmental governance — 70
    Stewart Lockie

7   Dilemmas for standardizers of sustainable consumption — 81
    Magnus Boström and Mikael Klintman

Contents

PART II
Climate change, energy and adaptation   93

8   Climate, scenario-building and governance: comprehending the temporalities of social-ecological change   95
Stewart Lockie

9   From Rio to Copenhagen: multilateral agreements, disagreements and situated actions   106
Chukwumerije Okereke and Sally Tyldesley

10   Marriage on the rocks: sociology's counsel for our struggling energy–society relationships   118
Debra J. Davidson

11   Sustainability as social practice: new perspectives on the theory and policies of reducing energy consumption   133
Harold Wilhite

12   Environmental migration: nature, society and population movement   142
Anthony Oliver-Smith

PART III
Urban environmental change, governance and adaptation   155

13   Climate change and urban governance: a new politics?   157
Harriet Bulkeley

14   Recovering the city level in the global environmental struggle: going beyond carbon trading   170
Saskia Sassen

15   Hybrid arrangements within the environmental state   179
Dana R. Fisher and Erika S. Svendsen

16   The new mobilities paradigm and sustainable transport: finding synergies and creating new methods   190
Rachel Aldred

PART IV
Risk, uncertainty and social learning   205

17   Towards a socio-ecological foundation for environmental risk research   207
Ortwin Renn

18 Uncertainty and claims of uncertainty as impediments to risk management 221
    Raymond Murphy

19 Transboundary risk governance: co-constructing environmental issues and political solutions 231
    Rolf Lidskog

20 The role of professionals in managing technological hazards: the Montara blowout 241
    Jan Hayes

21 Social learning to cope with global environmental change and unsustainability 253
    J. David Tàbara

PART V
(Re)assembling social-ecological systems   267

22 The social-ecological co-constitution of nature through ecological restoration: experimentally coping with inevitable ignorance and surprise 269
    Matthias Gross

23 Biological invasions as cause and consequence of 'our' changing world: social and environmental paradoxes 280
    Cécilia Claeys

24 Biological resources, knowledge and property 292
    Luigi Pellizzoni

25 Disassembling and reassembling socionatural networks: integrated natural resource management in the Great Bear Rainforest 307
    Justin Page

26 Land use tensions for the development of renewable sources of energy 319
    Giorgio Osti

Index   331

Illustrations

Tables

6.1	Typology of market-based instruments	75
12.1	Estimates of people displaced by the effects of environmental change	148
13.1	The ICLEI Cities for Climate Protection milestone methodology	159
15.1	Distribution of hybrid arrangements	183
16.1	Contrasting approaches to transport planning	197
20.1	Summary of well control barriers	245
21.1	Interpretations of environmental impairment	261
22.1	Types of ignorance	271

Figure

26.1	Framework of renewable energy source development in rural areas	323

Contributors

Rachel Aldred is a sociologist based in the Westminster University Department of Planning and Transport. She has written extensively on cycling, having led the ESRC Cycling Cultures research project (2010–11). Current projects include the ESRC seminar series, Modelling on the Move, and the ESRC Secondary Data Analysis project, Changing Commutes.

Magnus Boström is Associate Professor in Sociology and environmental lecturer at Södertörn University, Sweden. His research and teaching interest generally concern politics, governance, participation, communication, organization and responsibility in relation to transnational environmental issues. He has studied how political, regulatory, organizational and discursive circumstances shape green consumerism and organized activism. His recent publications include the books *Eco-standards, Product Labelling and Green Consumerism* (Palgrave, 2008), co-authored with Mikael Klintman, and *Transnational Multi-Stakeholder Standardization: Organizing Fragile Non-State Authority* (Edward Elgar, 2010), co-authored with Kristina Tamm Hallström. He has also written several articles on these topics.

Harriet Bulkeley is Professor at the Department of Geography, Durham University. Her research interests are in the nature and politics of environmental governance and focus on climate change, energy and urban sustainability. Her recent books include *Cities and Climate Change: Critical Introductions to Urbanism and the City* (Routledge, 2013), *Governing Climate Change* (with Peter Newell, Routledge, 2010), and *Cities and Low Carbon Transitions* (with Vanesa Castan-Broto, Mike Hodson and Simon Marvin, eds, Routledge, 2011).

Cécilia Claeys, Associate Professor at Aix-Marseille University, is a member of the LPED (Laboratoire Population Environnement Développement). An environmental sociologist, her research explores public policies, public perceptions and socio-technical controversies in the context of several topics including protected natural area creation, use and management, urban and suburban nature, invasive species management, and so on. Dr Claeys has developed interdisciplinary programmes drawing together the social and natural sciences, realising original field studies, exploring the epistemological heritages of the disciplines and developing theoretical debates between anthropocentrism and biocentrism.

Debra J. Davidson is Associate Professor of Environmental Sociology in the Department of Resource Economics and Environmental Sociology at the University of Alberta, with teaching and research interests in the social dimensions of energy and climate change. She is co-author of *Challenging Legitimacy at the Precipice of Energy Calamity* (2011, Springer), and recent articles have appeared in *Science*, *British Journal of Sociology* and *Society and Natural Resources*. She is currently a

Contributors

lead author of Chapter 26 of the Intergovernmental Panel on Climate Change's Fifth Assessment Report, Working Group II.

Dana R. Fisher is Associate Professor of Sociology and Director of the Program for Society and the Environment at the University of Maryland. Her research focuses on environmental policy, civic participation and activism more broadly. She has written extensively on climate politics in the USA and comparatively across nations, including *National Governance and the Global Climate Change Regime* (Rowman and Littlefield, 2004). Fisher is currently finishing up two grants from the US National Science Foundation: Comparing Climate Change Policy Networks (COMPON) project (co-investigator) and Understanding the Dynamic Connections Among Stewardship, Land Cover, and Ecosystem Services in New York City's Urban Forest (lead investigator). With support from the US Forest Service, she is expanding her work on urban stewardship to compare New York City to other cities, including Philadelphia.

Jennifer E. Givens is a Ph.D. student in Sociology at the University of Utah. Her interests include environmental sociology, global inequality and international development. Her published work appears in such venues as *Social Science Research*, *Organization and Environment*, *Environment and Behavior* and *Nature and Culture*.

Matthias Gross is Professor of Environmental Sociology at Helmholtz Centre for Environmental Research–UFZ in Leipzig and, by joint appointment, at the University of Jena, Germany. His current research focuses on the sociology of non-knowledge, experimental implementation strategies and the sociology of tapping geothermal energy. Co-founder and editor of the journal *Nature and Culture*, his most recent monograph is *Ignorance and Surprise: Science, Society, and Ecological Design* (MIT Press, 2010).

Jan Hayes has thirty years' experience in safety and risk management. Her current activities cover academia, consulting and regulation. She holds a Senior Research Fellow appointment at the Australian National University, where she is Program Leader for the social science research activities of the Energy Pipelines Co-operative Research Centre. Her research interests include decision making, safety in design, professionalism and learning. In addition, she consults part time with a small group of clients on safety performance improvement projects. Dr Hayes is a member of the Advisory Board of Australia's National Offshore Petroleum Safety and Environmental Management Authority.

Dayong Hong is Professor of Sociology at Renmin University of China, the Director of the Institute of Environmental Sociology, RUC, and also Chair of the Environmental Sociology Committee of the Chinese Sociological Association. His research focuses on the sociology of development, environmental sociology and social policy. His recent scholarly publications include *Social Change and Environmental Problems* (2001, Capital Normal University Press), *Social Assistance in Transitional China* (2004, Liaoning Education Press), *The Growing Non-Governmental Forces for Environmental Protection in China* (with Chuanjin Tao Mei; 2007, Shizheng Feng), and *The Social Bases of an Environmentally Friendly Society* (with Chengyang Xiao, 2012; Renmin University of China Press).

Andrew K. Jorgenson is Professor and Director of Graduate Studies in Sociology at the University of Utah. The primary area of his research is coupled human and natural systems, with a focus on the political economy and human ecology of global environmental change. His secondary areas of research include the macro-level causes of public health outcomes in developing nations, the globalization of environmental concern and the political economy of development. His recent work appears in such venues as *American Journal of Sociology, Social Problems, Social Forces, Social Science Research, Global Environmental Politics, Ecological Economics, Urban Studies* and *PLoS ONE*.

Mikael Klintman is Professor of Sociology at Lund University, Sweden. His research addresses environmental and urban issues, science and technology studies, public understandings of science, ethical and political consumption, social movements and globalization, policy analysis, and democracy and participation. Recent books include *Citizen-Consumers and Evolutionary Theory: Reducing Environmental Harm Through our Social Motivation* (Palgrave, 2012) and *Eco-Standards, Product Labelling and Green Consumerism* (with Magnus Böstrum, Palgrave, 2008).

Rolf Lidskog is Professor of Sociology at the Center for Urban and Regional Studies, Örebro University, Sweden. His research interests include environmental policy and politics at the international and national levels, especially the role of expertise in environmental governance. He has studied a variety of environmental policy areas: air pollution, climate change, biodiversity, nuclear waste and hazardous waste. He is co-author (with Linda Soneryd and Ylva Uggla) of the book *Transboundary Risk Governance* (Earthscan, 2009) and co-editor of *Governing the Air: The Dynamics of Science, Policy, and Citizen Interaction* (MIT Press, 2011).

Stewart Lockie is Professor of Sociology and Director of The Cairns Institute at James Cook University, Australia. He is also President of the International Sociological Association's Research Committee on Environment and Society, a member of the International Council for Science's Committee on Scientific Planning and Review and a Fellow of the Academy of Social Sciences in Australia. Professor Lockie is an environmental sociologist with research interests in food production and consumption networks, environmental governance, natural resource management and risk governance. Recent publications include *Risk and Social Theory in Environmental Management* (co-edited with Thomas Measham; CSIRO Publishing, 2012) and *Agriculture, Biodiversity and Markets: Livelihoods and Agroecology in Comparative Perspective* (co-edited with David Carpenter; Earthscan, 2010).

Arthur P.J. Mol is Professor in Environmental Policy at both Wageningen University, the Netherlands, and Renmin University of China. He is also Director of the Wageningen School of Social Sciences, and co-editor of the journal *Environmental Politics*. His main fields of interests and publications are in globalization, social theory and the environment, information and transparency, environmental governance, ecological modernization, and sustainable production and consumption. His latest books are *Environmental Reform in the Information Age* (2008, Cambridge University Press) and *The Ecological Modernisation Reader* (edited with David A. Sonnenfeld and Gert Spaargaren, 2009, Routledge).

Raymond Murphy is Emeritus Professor, and former Chair, of the Department of Sociology at the University of Ottawa, Canada. He is past-president of the Environment and Society Research Committee of the International Sociological Association. Professor Murphy is author of *Sociological Theories of Education* (McGraw-Hill Ryerson, 1979); *Social Closure* (Oxford University Press, 1988), which was translated into Japanese; *Rationality and Nature* (Westview, 1994), which was translated into Korean; *Sociology and Nature* (Westview, 1997), named by the journal *Choice* as one of the Outstanding Academic Books published in the United States in 1997; and *Leadership in Disaster: Learning for a Future with Global Climate Change* (McGill-Queen's University Press, 2009).

Contributors

Anja Nygren is an Associate Professor of Development Studies and Adjunct Professor of Environmental Policy at the University of Helsinki, Finland. She has long-term research experience in Costa Rica, Nicaragua, Honduras and Mexico. Anja's research interests include environmental governance, political ecology, urban ethnography, risks and vulnerabilities, forests and society, certifications and fair trade, corporate responsibility, environmental justice and social movements. Anja is currently a leader of a research project, New Forms of Environmental Governance: Managing the Risks and Vulnerabilities in Southern Cities, funded by the Academy of Finland.

Chukwumerije Okereke is a Reader in Environment and Development at the School of Human and Environmental Sciences, University of Reading, UK. Before joining Reading, he was a Senior Research Fellow and Head, Climate and Development Centre at the Smith School of Enterprise and the Environment, University of Oxford. He remains a Visiting Fellow at the Smith School and Oxford University's Environmental Change Institute (ECI). His research focuses on the links between global climate governance systems and international development.

Anthony Oliver-Smith is Professor Emeritus of Anthropology at the University of Florida. He has undertaken anthropological research on issues relating to disasters and involuntary resettlement in Peru, Honduras, India, Brazil, Jamaica, Mexico, Japan and the USA since the 1970s. His work on disasters has focused on issues of post-disaster social organization, including class/race/ethnicity/gender based patterns of differential aid distribution, social consensus and conflict, grief and mourning issues and social mobilization of community-based reconstruction efforts. He has served on the Social Sciences Committee of the Earthquake Engineering Research Institute and is a member of *La Red de Estudios Sociales en Prevención de Desastres en America Latina* (The Network for Social Studies on Disaster Prevention in Latin America).

Giorgio Osti is a rural sociologist and Associate Professor in the Department of Political and Social Sciences, University of Trieste, Italy. He is interested in socio-spatial relationships and in the working of reciprocity in place. He has been involved in research concerning local development in fragile areas and environmental issues like waste management and energy transition. Recently he has published 'The moral basis of a forward society: Relations and forms of localism in Italy' (*Local Economy*, March 1, 2013).

Justin Page is an environmental sociologist with over ten years of experience examining the social dimensions of natural resource management. Dr Page has completed rigorous analyses of land use planning, conservation, public acceptability, community resilience, social capital, Aboriginal consultation, and environmental justice in the forestry, aquaculture, mining and fishing sectors. Currently an environmental consultant working on the social impacts of large-scale natural resource projects, Dr Page continues to contribute to our understanding of natural resource management policy, Aboriginal relations, and the social causes and consequences of environmental change.

Luigi Pellizzoni is Associate Professor in Environmental Sociology at the Department of Political and Social Sciences, University of Trieste. His research interests intersect two main areas: environment, technoscience and social change; and participation, conflict and transformation of the ways of governing. On these topics he has published some books and several articles (*Global Environmental Change, Environmental Politics, European Journal of Social Theory, Theory Culture and Society*, etc.). In recent years he has especially focused on new mobilizations and the impact of new and emerging technosciences, with their regulatory arrangements, on the society-nature relationship. He is co-editor, with Marja Ylönen, of *Neoliberalism and Technoscience: Critical Assessments* (Ashgate, 2012).

Ortwin Renn is Professor and Chair of Environmental Sociology and Technology Assessment at Stuttgart University, Germany. He directs the Center for Interdisciplinary Risk and Innovation Studies at the University of Stuttgart (ZIRIUS) and the non-profit company DIALOGIK, a research institute for the investigation of communication and participation processes in environmental policy making. He also serves as Adjunct Professor for Integrated Risk Analysis at Stavanger University, Norway, and as Affiliate Professor at Beijing Normal University.

Saskia Sassen is the Robert S. Lynd Professor of Sociology and co-Chair of the Committee on Global Thought, Columbia University (www.saskiasassen.com). Recent books are *Territory, Authority, Rights: From Medieval to Global Assemblages* (Princeton University Press, 2008), *A Sociology of Globalization* (W.W. Norton, 2007), and the fourth edition of *Cities in a World Economy* (Sage, 2011). *The Global City* came out in a new edition in 2001. Her books are translated into over twenty languages. She is currently working on *When Territory Exits Existing Frameworks* (Harvard University Press). She contributes regularly to www.OpenDemocracy.net and www.HuffingtonPost.com.

David A. Sonnenfeld is Professor of Sociology and Environmental Policy at the State University of New York's College of Environmental Science and Forestry (SUNY ESF), in Syracuse, USA; and Research Associate, Environmental Policy Group, Wageningen University, the Netherlands. His research interests include environmental reform in newly industrializing countries, popular participation in environmental reform and technological environmental innovation. Among his most recent publications are *Food, Globalization and Sustainability* (with Peter Oosterveer, Earthscan, 2012) and *The Ecological Modernisation Reader: Environmental Reform in Theory and Practice* (co-edited with Arthur P.J. Mol and Gert Spaargaren, Routledge, 2009).

Gert Spaargaren is a Senior Researcher and Professor of Environmental Policy for Sustainable Lifestyles and Consumption at Wageningen University. His main research interests and publications are in the field of environmental sociology, sustainable consumption and behaviour and the globalization of environmental reform. His latest books are *The Ecological Modernisation Reader* (edited with Arthur P.J. Mol and David A. Sonnenfeld, Routledge, 2009) and *Food in a Sustainable World; Transitions in the Consumption, Retail and Production of Food* (edited with Peter Oosterveer and Anne Loeber, Routledge, 2011).

Erika S. Svendsen is a Research Social Scientist with the US Forest Service. Her work includes understanding the spatial, temporal and political aspects of environmental stewardship, focusing on governance and social well-being. Dr Svendsen is based in New York City, where she is the Forest Service representative to the NYC Urban Field Station. Founded in partnership with the New York City Department of Parks and Recreation, the Field Station's mission is to improve quality of life in urban areas by conducting and supporting research about social-ecological systems and natural resource management.

J. David Tàbara works at the Global Climate Forum and is Associate Senior Researcher at the Institute of Environmental Sciences and Technology at the Autonomous University of Barcelona. He is a member of the European Sustainability Science Group and a member of the Board of the

Contributors

Research Committee on Environment and Society of the International Sociological Association. Recently he has contributed to the following books: *Making Climate Change Work for Us* (Hulme et al., CUP, 2010), *Reframing the Problem of Climate Change* (Jaeger et al., Earthscan, 2011) and *European Research for Sustainable Development.* (Jaeger et al., Springer, 2011).

Sally Tyldesley is a Policy Adviser working within the Sustainability theme at the Royal Society's Science Policy Centre. Before joining the Royal Society, Sally was a research assistant at the Smith School of Enterprise and the Environment, University of Oxford. Her work focuses on climate science and policy, climate resilient development and energy.

Harold Wilhite is Professor of Social Anthropology and Research Director at the University of Oslo's Centre for Development and Environment. His main research interests have been associated with theorizing consumption in countries of both the North and South and developing innovative policy instruments promoting sustainable consumption. He has published widely on consumption, development and societal change based on ethnographic field studies in North America, Latin America, Japan, Norway and India. Over the past five years he has been Academic Director for the University of Oslo Interfaculty programme on sustainable energy and environmental change (MILEN).

Chenyang Xiao is Assistant Professor of Sociology at American University, USA. His research focuses on public beliefs of and attitudes towards environmental problems such as global warming, as well as general environmental values and worldviews, often with an international scope, and public attitudes towards science and technology.

Acknowledgements

Undertaking a project such as this is not possible without the encouragement and support of numerous people, over a period of many months.

Gerhard Boomgaarden, Senior Publisher at Routledge, approached us with the idea for a compilation on the broad topic of environment and society, and was generous with his encouragement and advice as we refined and brought this idea to life. His colleague, Emily Briggs, Editorial Assistant at Routledge, has done a wonderful job of keeping us on track.

Contributors to the book have been uniformly easy to deal with. We thank them for their patience and responsiveness as we have worked through the processes of reviewing and editing their chapters.

We all were saddened by the passing in 2010 of William R. Freudenberg, Professor of Environmental Studies at the University of California, Santa Barbara. One of the intended contributors to this volume, Bill's work on the social causes of environmental degradation and disasters continue to be extremely influential, as is evident in its use by others whose work does appear here.

Lastly, we would like to thank all of our colleagues in the International Sociological Association's Research Committee on Environment Society (RC24), whose tireless and deeply thoughtful scholarship on societal and environmental change in many places around the globe deeply inspires us, giving us hope for a greener planet for generations to come.

1
Socio-ecological transformations and the social sciences

Stewart Lockie, David A. Sonnenfeld and Dana R. Fisher

Our relationships with the landscapes and ecologies that we are a part of, the plants and animals that we share them with, and the natural resources that we extract, lie at the heart of contemporary social and political debates. Mitigation of anthropogenic climate change is among a handful of issues that dominate the world's political imagination early in the third millennium of the common era. Risks associated with global environmental change combine with dangers of extreme climatic and geological events to remind us of humanity's dependence on favourable environmental conditions. In every one of the first ten years of the century, natural disasters affected some 200–300 million people and caused around US $100 billion in damage (Armstrong et al. 2011). Such costs increased fourfold in 2011, following the Tōhoku earthquake and tsunami off the east coast of Japan, as well as a spate of (comparatively smaller) disasters in other advanced economies including the USA, Australia and Aotearoa/New Zealand (Ferris and Petz 2012).

The technological infrastructure of industrial society has not isolated us from environmental threats; it has simply changed the nature and distribution of vulnerability. Numerous social movements seek to strengthen our appreciation of the relationships between people, health, livelihoods, dignity, safety, environmental quality and access to the planet's ecosystems and resources. Environmental and climate justice groups highlight the disproportionate exposure to pollution, resource degradation and climate risk borne by the poor and marginalized. Nanotechnologies, genetic engineering, pesticides, indigenous land rights, ecological footprints, food security and unfettered global networks and material flows sit alongside traditional interests in habitat conservation and nuclear non-proliferation in the campaign portfolios of mainstream environmental NGOs. The major issues of our times, to paraphrase Latour (1993), blend politics, biology, economics, ecology, chemistry, ethics, engineering and so on, with no respect for distinctions between society, nature, technology and economics.

Even still, human–environment interactions do not receive the attention or response that they deserve. Countries' failure to meet targets established at the 1992 Rio Earth Summit for slowing the rate of species extinction by 2010 attracted little media or political attention. The secretariat of the Convention on Biological Diversity merely announced new targets, more attuned to the underlying causes of biodiversity degradation (Perrings et al. 2011). This decision is despite suggestions that biodiversity loss already may have passed critical thresholds of irreversibility, and despite what we know about the contributions that biodiversity makes to ecosystem adaptability and resilience in the face of other processes of environmental change such as global warming (Rockström et al. 2009). Whether inadequate governmental responses are due to continued faith in the project of modernity, vested political and economic interests, lack of institutional capacity or other causes, it is clear that we have entered a new era of approaches to addressing today's environmental and natural resource challenges. Environmental politics can no longer be simply about preserving 'pristine' nature in isolated parks and reserves. Natural resource management can no longer be focused exclusively on soil conservation and sustainable forest yields. Sustainable development – contested though it may be – presupposes the interdependence of social and environmental well-being across space and through time.

Such parallel shifts in our material (inter)relations with the planet and in political consciousness mirror a new consensus across the social sciences that the environment is within our domains of interest and expertise. Biophysical nature is not something to be ignored by social scientists. Nor does it lie somehow outside of society. The (socio-)natural environment is now everyday social science fodder; a staple of undergraduate and graduate programmes in anthropology, sociology, human geography etc.; the foci of numerous sub- and cross-disciplinary professional organizations; and increasingly visible in mainstream social science literature and meetings. To many, the relevance of the social sciences to understanding and dealing with environmental issues is self-evident. Human beings are, and have been for some time, major agents of global environmental change and – to the extent that this change is threatening or undesirable – human patterns of organization, production and consumption must be reformed and human beliefs, attitudes and dispositions must be reoriented. At the same time, the influence of environmental change on traditional foci of social scientific concern such as inequality, political conflict, social movement mobilization and so on, is increasingly apparent. The negative consequences of environmental change are unequally distributed, as are exposures to risks associated with natural disasters and hazardous industries. Political conflicts abound over how environmental matters should be prioritized relative to other values and concerns, and uncertainties inherent to scientific understanding of global environmental change (climate change, in particular) have become lightning rods for anti-environmental movements.

Several of these dynamics may be incorporated readily within the dominant theoretical and methodological approaches of the social sciences. Environmental attitudes, politics, inequalities, behaviours and movements are no less amenable to description, measurement and/or comparison using accepted concepts and social research methods than are, for example, financial inequalities, class politics or gender relations. Each of these topics allows us to leave questions of environmental change to the natural sciences and to concentrate our efforts on determining the causes and consequences of human-induced environmental change. Other human–environment dynamics, however, require us to reconsider the adequacy of existing theory and method. Take, for example, the politics of knowledge about global climate change. It is a basic assumption of the social sciences that all knowledge is, in some way, socially constructed; a reflection of institutional processes, political values, existing theory, etc. Given this point, it is important to map competing knowledge claims. However, it is also necessary to participate in the generation of new socio-ecological knowledge and inform transformations in policy and practice in relation to the ways people and environments interact.

With these two related but different goals in mind, the following section of this chapter introduces some of the ways in which social scientists are attempting to conceptualize more integrated or co-evolutionary perspectives on society and nature. By way of providing an overview and introduction to the themes of this volume, the chapter then explores several frontiers of contemporary environmental social science. These perspectives include theories of global socio-ecological change; climate change, energy and adaptation; urban environmental change and governance; risk, uncertainty and social learning; and the (re)assembly of vital social-ecological systems.

Conceptualizing social-ecological change

It is no longer possible (if indeed it ever were) to understand broad patterns of social inequality and a number of other key social scientific concerns without understanding patterns of ecosystem change. Social theory that is environmentally blind is as potentially problematic as social theory that is gender-, race- or class-blind. It still makes sense to talk about the 'social' (the ways in which people interact and organize), and many or even most 'social problems' have little or nothing to do with 'nature' or 'the environment'. What makes less sense is talking about the natural as a residual domain of everything that is left over when people are taken out, and vice versa. Some branches of social theory have grasped this notion better than others and social scientists have a long history of debating the best ways to incorporate human interaction with the biophysical environment within social theory. Numerous terms have been coined to capture the interpenetration or inseparability of the social and the natural – co-evolution, co-constitution, methodological symmetry, social metabolism and so on. Similarly, a variety of methodological approaches have been developed to integrate this basic insight into social-ecological research.

How, then, are we to conceptualize relationships between 'the social' and 'the natural', 'society' and 'environment', and what roles do such conceptualizations imply for the social sciences? One of the more common ways to organize the various theoretical approaches to environment and society has been to assimilate them into two broad categories: those deemed 'constructivist' in orientation and those deemed 'materialist'. The social construction of reality refers to the ways in which we humans make sense of and in turn shape our social and natural environments through language, symbols, culture and so on. It follows, according to constructivists, that 'there is no socially unmediated position from which to apprehend material reality' (Lockie 2004a: 30). In its more extreme idealist and postmodernist guises (often referred to as the 'strong programme' in social construction), constructivism argues that there is in fact 'no reality whatsoever outside the symbolic world-building activities of humans and no way of knowing about that reality that is, in principle, any better or worse than any other way of knowing' (Lockie 2004a: 30). Materialist perspectives, conversely, challenge the primacy in such arguments accorded to ideas, symbols and beliefs. They accept the realist proposition that phenomena in both the social and natural worlds exist independently of humans' perception of them and thus argue the importance of scientific inquiry as a rational and systematic means to comprehend reality. Some variants of materialism go further in arguing that key, historically situated dimensions of social organization such as the capitalist mode of production largely determine people's ideological dispositions (environmental values, for example, reflecting particular class locations).

The distinction between constructivist and materialist perspectives is useful but, as a means of categorizing major approaches to eco-social theory, has its limitations. In practice, few environmental social scientists ascribe to a strong constructivist position, while even the most materialist in orientation would acknowledge that the ways in which people understand and communicate environmental change play a recursive role in those very same processes of transformation. This point is exemplified in Ortwin Renn's contribution to this volume (Chapter 17). Renn argues that acknowledging the role of social construction in knowledge creation is crucial to bringing our understanding of social and ecological change into better alignment with the realities of these processes. There is a lot at stake here. Inadequate knowledge undermines our capacity to identify and deal with environmental problems. But this issue is not simply about collecting more data and correcting misconceptions. Human values and aspirations (not to mention routines and ways of life) are deeply implicated in the definition and prioritization of environmental problems, and in the goals of environmental research and management interventions. Failure to recognize this fact potentially amplifies pressure on ecosystems by misdirecting policy and research.

While Renn argues that it is analytically useful to maintain clear distinctions between 'society', 'nature' and 'the environment' (the latter as those aspects of nature that have been transformed through interaction with humans), others argue for a more inclusive notion of the social in which people are joined by non-human species, ecosystem processes, technologies and so on in diverse networks, or collectives, that enable, ultimately, the expression of human agency. In developing this argument, several contributors to this book draw on actor–network theory (ANT), a perspective developed primarily through sociological studies of science and technology (see Law and Hassard 1999). ANT adopts what may be called a 'radically relational' approach to the theorization of key concepts in the social sciences. Power, agency, structure, subjectivity, micro- and macro-levels of organization, even the social itself, are all conceived as network effects; that is, as the generative outcomes of relationships between entities within a network. Social scientists are required not to define concepts like agency or power in advance, let alone attribute them to particular actors or treat them as self-evident explanatory categories. Rather, the actor–network perspective requires social scientists to approach power, agency, etc. as phenomena that may take a variety of forms and that therefore require description and explanation in specific collective contexts (Callon 1995; Latour 1999). The admittance of non-humans to these collectives is not intended to discount uniquely human traits such as language, consciousness or intentional action and resistance, but to highlight the ways in which non-human objects, beings and processes both enable and confound the attempts of humans to enact their own projects (Lockie 2004b).

The 'relational materialism' of ANT (see Law 1999) has certainly attracted criticism, not least due to suspicion that its decentring of human agency may lead to a discounting of power and domination in the production of social and environmental damage. We will not review this debate in detail here (see Lockie 2004b). However, we will point to the very useful contributions that actor–network studies have made to our understanding of how knowledge, techniques and technologies are implicated in the constitution and extension of power (a theme we return to below, in relation to the assembling and reassembling of social-ecological systems). We will also point to the implications of relational materialism for our understanding of scale. A network perspective suggests that there is no change of scale in the social domain between the micro/actor and the macro/structural. According to Latour (1999), macro/structural phenomena are neither the sum of localized interactions nor the context for those interactions. They are better conceived, he argues, as attempts to *sum up* and thence to influence 'interactions through various kinds of devices, inscriptions, forms and formulae, into a very local, very practical, very tiny locus' (Latour 1999: 17). Several contributions to this book problematize spatial and/or temporal scale in a way consistent with the idea that scale ought to be conceived as a potentially unstable network effect.

Categorizing theoretical perspectives on environment and society according to a constructivist–materialist dualism not only is somewhat forced; it also understates diversity and debate among theorists who might otherwise be positioned on the same side of this dualism. An excellent example of such diversity is provided by proponents of, variously, ecological modernization and ecological Marxism, both of which are predominantly materialist in orientation. These and other key theoretical perspectives on global socio-ecological change are presented in Part I of this volume.

Challenges, contradictions and consequences of global socio-ecological change

The contributions to Part I of this volume address the conceptual and policy challenges associated with the global acceleration of markets' social inequality, and anthropogenic environmental change early in the third millennium. Written from a variety of perspectives and empirical points of reference, the chapters all take up the seriousness of the challenges; the necessity of drawing from social scientific theories, methods and evidence to discern important patterns and dynamics in those changes; and the fashioning of effective policy instruments to encourage change in more socially and ecologically benign or even beneficial directions. Underlying each of the six chapters is a critical realist approach to the thoughtful integration of theory, evidence and policy application. Altogether, these six contributions give evidence of the significant breadth of issues, variety of approaches and depth of critical engagement and debate being utilized in contemporary environmental social science inquiry on the challenges, contradictions and consequences of global socio-ecological change.

Arthur Mol, Gert Spaargaren and David Sonnenfeld (Chapter 2) take stock of scholarship on processes of *ecological modernization* in advanced and developing economies around the world. Reviewing the accomplishments of three decades of such studies, they discuss key debates involving this socio-environmental approach and outline a research agenda for the study of institutional environmental transformation. For the further development of social theory of (global) environmental change, they argue, three advances are critical: extending the geographical scope of environmental social science to include more places around the world; increasing analysis of transboundary social and environmental flows, and their regulation and reform; and examining the vital cultural dimensions of change necessary for greater sustainability.

In Chapter 3, Andrew Jorgenson and Jennifer Givens explore two frontier areas within quantitative, cross-national, *world-systems* analysis of global social and environmental change: *ecologically unequal exchange theory*, which 'considers how international trade' allows more powerful developed nations to externalize at least some of their environmental impacts to less-developed countries' (p. 40); and studies of *foreign investment dependence and environmental impact*, exploring the relationship between 'increased levels of foreign investment within less-developed countries … [and] dirtier forms of extraction and production … environmental harms, and … costs to human beings' (ibid.).

Dayong Hong, Chenyang Xiao and Stewart Lockie (Chapter 4) use the dramatic *rise of China* as a global economic power and its attendant significance for global environmental change to explore limitations of Western socio-environmental theory. Chronicling both China's extraordinary economic accomplishments and the daunting environmental challenges now facing the Chinese people and government, the authors push ecological modernization theory, in particular, to: address more fully the unevenness of socio-environmental development and impacts at multiple levels, thus designing and carrying out studies at global, regional, national and local scales; take into greater account the diversity of social, political and economic forms between and within various countries around the world; pay more attention to problems of social inequality; examine more closely both bottom-up and top-down approaches to environmental regulation and reform; and, finally, place more emphasis on the study of 'humans' lifestyles, consuming behaviours and ideologies' (p. 55).

In Chapter 5, Anja Nygren, argues that *environmental justice* perspectives have much to offer to a critical understanding of inequality and power on a global scale, including with respect to environmental policymaking. Aiming to develop more sophisticated understandings of 'justice and equity', she argues for greater diversification of social theory of environment – especially in relation to 'the variety of environmental mobilizations taking place in different parts of the world' (p. 58). After Newell (2008), Nygren suggests that 'environmental justice is only one among many organizing principles [in the Global South], including indigenous rights, food security, human rights and democracy' (p. 59). Observing that 'in recent years, environmental justice movements have become increasingly transnational' (p. 64), she finds dangerous signs of *eco-imperialism* at work: 'the forceful imposition of Northern environmental views on the Global South ... [e.g.] Indians are essentialized as peoples of simplicity and environmental wisdom, while the non-Indian peasants are portrayed as rootless, corrupted, and lacking in environmental knowledge' (pp. 59, 65).

Stewart Lockie (Chapter 6) examines the limitations of market-based environmental policy instruments in effectively addressing global environmental issues, with particular reference to climate change. He argues that

> it is not just any environment, any nature or any natural resource that is incorporated within neoliberal regimes of governance. It is natures that can be defined through clearly delineated property rights. Natures that can be reduced to simple and transferable common values. Natures that comply with the requirement of markets for well understood and calculable risks. Natures that are not, therefore, characterized by excessive uncertainty, complexity, or the possibility of discontinuity and threshold effects.
>
> (p. 79)

Magnus Boström and Mikael Klintman (Chapter 7) focus on the 'inherent contradictions and limitations' of eco-labelling and certification efforts as strategies to advance sustainable consumption. Drawing on case studies at multiple levels of scale, they explore four dilemmas for the design and utilization of such market-oriented environmental policy instruments, whether: certification standards should be formulated in an inclusive or exclusive manner; such schemes should be broad or narrow in scope; certification criteria should be universal or adapted to particular local ecologies and cultures; and sustainability standards should be presented to consumers in a simplified form.

Climate change, energy and adaptation

Part II of the Handbook presents an array of perspectives on issues related to the social aspects of climate change. In particular, the collection of chapters presented here contextualizes the current state of knowledge regarding climate governance (Chapter 8), the climate regime and multilateral agreements more generally (Chapter 9), the relationship between energy and society (Chapters 10 and 11) and the ways people are adapting to climate change through 'environmental migration' (Chapter 12). This section highlights the fact that the appropriate scales – spatial, temporal and social – at which to address global environmental change are not obvious. Moreover, the policy and management responses are also not evident. As a result, responses to these challenges have required considerable innovation and numerous tensions and contradictions remain.

In Chapter 8, Stewart Lockie looks to policymaking around the issue of climate change to discuss challenges to environmental governance. He focuses his discussion on 'the temporalities embedded in climate governance through the use of climate modelling and scenario-building' (p. xxx). In other words, by exploring the ways that future predictions are integrated into climate politics through specific scientific techniques, the author begins to unpack the ways in which governments and others attempt to bring the future into the present as an object of policy and action. Through this chapter, the author illustrates 'the partiality of any knowledge that we bring to bear on environmental problems and the associated need to maintain a constructively critical stance towards our knowledge, the conceptual frameworks with which we organize it, and the policy responses it informs' (p. 104).

Chapter 9 looks more closely at the international climate change regime, tracing its development from the Earth Summit in Rio de Janeiro in 1992 to the Copenhagen round of climate negotiations (COP15) in 2009. Chukwumerije Okereke and Sally Tyldesley focus their attention on the innovations in governance that have emerged within the regime while it was responding to and managing what they call 'complexities, political imperatives and competing interests' (p. 106). The authors highlight four themes: the quest for justice and equality within the regime; challenges to determining agency, scale and authority; identifying the most appropriate policy mechanism and institutions for addressing climate change; and recognizing the connections between mitigation and adaptation.

One of the major factors contributing to anthropogenic climate change is, of course, energy consumption. In her chapter on energy–society relationships (Chapter 10), Debra Davidson points to the enormous political, economic and cultural influence of the fossil fuel industry and the manner in which this industry has turned its resources to campaigning against climate action. The chapter outlines 'three major drivers in the energy–society relationship: the escalation of fossil fuel dependency's social disruption; the escalation of fossil fuel dependency's environmental disruption; and macro-social transition prospects' (p. 119). Davidson notes that the social sciences have the capacity to contribute significantly to understanding the energy–society relationship and the chapter concludes by noting promising trends in research around this issue. However, the author also points out 'the magnitude of the work that remains' to move forward in our understanding of the complex relationship between energy and society (p. 129). Here, Davidson specifically notes the need to move beyond the structure–agency distinction to focus on a more integrated approach.

In Chapter 11, Harold Wilhite presents social practice theory as a means of drawing 'together perspectives on social, cultural and material contributions to consumption' (p. 133). The framework of social practice theory leads to the identification of practical strategies to influence energy consumption practice. In light of the politics of energy discussed by Davidson, we might also note that: the legitimacy of interventions is potentially undermined by mistrust in institutions promoting change or regulating behaviour; participation and involvement has the capacity to increase trust; and 'choice editing' may strike a balance between the democratization of socio-technical systems and the many life pressures we cope with through our daily routines.

Not all embodied practice is the practice of the archetypal Western consumer. The poor of the Global South often are cast as victims in debates over climate change mitigation and adaptation – in some cases emerging as climate refugees. Accordingly, it is important to examine livelihood strategies including those incorporating mobility. In Chapter 12, Anthony Oliver-Smith specifically discusses environmental migration. He points out that 'environmental, social, economic and political forces combined to increase the risk of uprooting for many vulnerable populations in exposed regions' (p. 142). Given the complexity of the forces that contribute to environmental migration, the author notes that it is extremely difficult to estimate numbers when it comes to environmental migration. Nonetheless, he concludes that, given the scientific predictions of environmental change in the not-too-distant future, 'we must be prepared for significant increases in the role environmental factors will play in displacement and migration in the relatively near future' (p. 151).

Stewart Lockie, David A. Sonnenfeld and Dana R. Fisher

Urban environmental change, governance and adaptation

As Okereke and Tyldesley point out in Chapter 9 of this volume, the nation state is not the sole jurisdictional unit for climate governance. Nor does the nation state necessarily emerge unchanged from engagement with climate governance. In Part III of this Handbook, the city is analysed as a site for socio-ecological change. Two unique perspectives on cities and climate change are included. The section begins with Chapter 13 by Harriet Bulkeley, who focuses her attention on the emergence of urban regimes of climate governance. The chapter argues that 'we have witnessed a shift from a period of *municipal voluntarism* to one that could be described as *strategic urbanism*' (p. 157). Bulkeley reviews a number of what she calls 'governance innovations' or 'climate change experiments'. She concludes with discussions about how these examples of strategic urbanism present challenges to cities in the Global South. Discussion focuses on how to integrate a 'more radical sense of urban economic development with the climate change agenda' (p. 167).

Next, in Chapter 14, Saskia Sassen makes a case for bringing the city level into global climate regimes. She claims that this expansion to multi-scalar global governance 'might be one of the most effective ways of achieving protection without national protectionism' (p. 170). By looking at cities, the urgency of global environmental change is rendered more visible and concrete. Moreover, Sassen notes that it is much easier to see the interactions between policy instruments and scientific knowledge at the city level. A focus at this scale will counteract what she calls 'the excessive weight of markets (i.e. carbon trading) as a means to address the environmental crisis' (p. 176).

Following these works, Dana Fisher and Erika Svendsen dig deeply into one aspect of urban environmental governance. Chapter 15 explores how the environmental state has expanded in recent years to include innovative partnerships across different social actors, or what the authors call 'hybrid arrangements'. Using data collected in New York City, they assess the types of social arrangements that are emerging in cities. Due to resource constraints facing cities around the world, these types of arrangements are becoming increasingly common as actors and organizations that historically did not collaborate begin to work together to yield 'highly creative and innovative forms of environmental governance seem to be emerging from a growing social network of actors and organizations' (p. 186).

In Chapter 16, Rachel Aldred highlights the role of transportation, exploring how the 'new mobilities paradigm' can contribute to our understanding of sustainable transport. Although cities are not the specific focus of the chapter, transportation and mobility within cities emerge as a clear theme. According to Aldred, this paradigm has usefully drawn social scientists' attention to commuting and other forms of movement as meaningful arenas of human experience. Relatively neglected, however, are the movement of 'things' (freight), and the social and ecological consequences of particular mobility practices. The chapter concludes, therefore, by arguing for more cross-fertilization between the theoretical innovation of the mobilities paradigm and the explicitly normative agenda of sustainable transport research.

Risk, uncertainty and social learning

Many recent innovations in governance may be characterized as attempts to transform incalculable uncertainty into manageable risk, at a variety of spatial and temporal scales. At the same time, innovations in governance have aimed to address sometimes competing political imperatives such as between (re)asserted national sovereignty and perceived local, community or even transboundary resource access rights. Contemporary institutions, both public and private, have sophisticated systems for dealing with potentially adverse risk events, the probability of which can be predicted with a reasonable degree of certainty. While risk calculation and management have been subject to sustained critique by social scientists, today's institutions may be worse at dealing with uncertainty than they are with risk. Large-scale socio-environmental change further challenges risk calculation and management through the magnification of uncertainty.

The contributions to Part IV of this volume examine risk management, governance and social learning with respect to socio-ecological change. The five chapters take strikingly different approaches to the analysis of risk, uncertainty and social learning for sustainability in the contemporary era. Together, they underscore the conceptual, empirical and behavioural complexities involved, as well as the value of continued theoretical and empirical explorations into this rich field.

In the first selection, Ortwin Renn (Chapter 17) calls for us to '[find] the most appropriate balance between too little and too much precaution', advocating for the adoption of a 'moderate realistic' socio-ecological approach to assessing environmental risk. For him, such an approach is necessarily problem-oriented, has a 'close connection' between theory and practice, utilizes multiple disciplines and methods and involves the direct participation of affected groups and individuals (pp. 213–217). The ultimate goal, he argues, 'is to strive for sustainable development through human intervention for all concerned', while (and, in part, through) coming 'to terms with environmental hazards' (p. 217).

Raymond Murphy (Chapter 18) takes up the critical role of uncertainty in the management of risk associated with socio-ecological challenges as diverse as major earthquakes, global climate change, extreme weather events and peak oil. The danger, from his perspective, is one of social, political and institutional paralysis over claims and counter-claims related to scientific uncertainty about risk. He suggests, on the one hand, that 'there is no certainty that technological solutions will appear when needed' (pp. 228–9). He argues, on the other hand, that to move forward it is essential to distinguish between 'total and partial uncertainty, between ignorance and valuable indicative knowledge' (p. 230).

The dynamic, interdependent character of science, nature and institutions in risk governance is the subject of Rolf Lidskog's (Chapter 19) contribution. International regulation of air pollution in Europe through the Convention on Long-Range Transboundary Air Pollution provides a fascinating case study. Since the initial adoption of this Convention in 1979, tremendous changes have occurred: in the biophysical nature of the problem; in the science and technique for measuring, modelling and understanding these changes; and in the constitution and functioning of political institutions responsible for regulating them. According to Lidskog, these changes were 'potentially unforeseeable' (p. 238). For him, the key is that policy networks and scientific models must be updated regularly to keep pace with such changes, or risk obsolescence or irrelevance.

Jan Hayes (Chapter 20) explores the important and conflicting roles of technical professionals and senior management in a complex series of events and institutional decision making that led up to a major blowout in the Montara oil field off Australia's northwest coast in 2009. For her, risk is imbued with power and conflict within organizational and institutional settings. It is not simply knowledge – but also its use, the hierarchical roles within organizations, and the expedient, but faulty priorities that can override sound technical reasoning – that give her pause. She sees an enhanced 'leadership role' for technical professionals as a potential remedy; she is concerned, however, that industry may be moving even further away from what is needed in this regard (p. 250).

The final selection in Part IV is written by David Tàbara (Chapter 21). It focuses on the social learning involved with coping with global environmental change. Using a series of 'vignettes', he reflects on 'the complexity of global environmental change, sustainability and social learning' (p. 254). For him, the key is 'the restoration or re-creation of social-ecological feedback' so that humans can 'learn from their actions on global and local ecosystems' (p. 262). Among the new competencies required for greater sustainability, he argues, are the 'aptitude to translate complexity among different audiences, capacity to integrate different sources of knowledge, and the abilities to lead and create sustainability partnerships and to bridge organizations across different scales and domains of action' (p. 255). Tàbara argues that nothing short of 're-creating our identities ... and setting limits to our actions' is required for sustainability.

(Re)assembling social-ecological systems

Contributors to Part V of this book take up the relational materialism advocated by actor–network theorists, exploring ways in which 'landscapes', 'natures' and even 'species' are constructed through politicized interactions of people, organisms and ecosystem processes. In the first of these chapters, Matthias Gross (Chapter 22) examines ecological restoration practices – deliberate attempts to rehabilitate landscapes degraded or contaminated by industrial activity. Gross's chapter illustrates well the theoretical point that 'society' and 'nature' cannot be separated. Not only are restoration landscapes products of human and non-human energies; the very idea of 'restoration' evokes complex questions about both the desirable end-point of ecosystem interventions (the 'nature' to be constructed) and the feasibility of strategies to reach this end-point (the extent to which 'nature' cooperates in its planned transformation). For Gross, restoration is less a reinstatement of the natural order than an informed journey into the unknown, a process of social-ecological *experimentation*. Emphasizing the centrality of ignorance and surprise, he argues that, when the unexpected occurs, these characteristics call for flexible institutional arrangements and a focus on social learning, rather than 'finger pointing'.

Plants and animals, embedded in ecosystems, often transgress roles and boundaries that we as humans assign to them. As Cécilia Claeys (Chapter 23) reminds us, such transgressions are not always examples of the unexpected; they are examples also of the politics of nature and of environmental transformations. The concept of 'biological invasions' has been developed to describe the ways in which species movements and population dynamics affect biodiversity and other ecosystem values. On the one hand, the transformation of 'alien' species into 'weeds' and 'pests' presents a major, and costly, challenge for conservation and economic activities such as agriculture, fisheries and forestry. On the other hand, as Claeys argues, the classification of living organisms into a host of ostensibly conflicting categories – 'indigenous' versus 'alien', 'naturalized' versus 'invasive', etc. – obscures competition among humans to define desirable ecological futures (the end-point, as it were, of ecosystem interventions) and the place of various species within them.

In a further exploration of the politics of classification, Luigi Pellizzoni (Chapter 24) examines gene technologies and the governance regimes developed to regulate their application. Such regimes define the 'novel species' created by horizontal gene transfer as forms of private (or corporate) intellectual property protected by patents, licensing arrangements, and so on. Intended to protect the rights of those who invest in genetic research and development, this regime has been critiqued for privatizing public assets, discouraging genuine innovation and uncritically assuming the superiority of market-based mechanisms for the transfer of rights. In his intervention, Pellizzoni argues that the establishment of markets in genetic material reconfigures nature into a repository of genetic resources, the value of which is determined by intellectual property arrangements rather than the countless other ways in which living organisms contribute to ecosystem functioning and human well-being. His point is not that using intellectual property arrangements to regulate gene technologies is necessarily wrong. Rather, it is that within the neoliberal regime of governance underlying such arrangements there seems little room to debate 'the virtues and drawbacks of property rights-based regulation' (p. 302).

It would be a mistake, of course, to regard this re-designation of nature as a repository of genetic resources as complete. The last two chapters in Part V thus return to the landscape scale and examine processes of classification and contestation through which natures, communities and economic interests are constructed at broader spatial scales. Justin Page (Chapter 25) takes what may appear self-evident – the importance of 'integrated natural resource management' to Canada's temperate coastal rainforests – and explores the lengthy process through which such approaches have been achieved. Integrated resource management requires, he argues, far more than bringing together relevant stakeholders in pursuit of common objectives. What was to become known as British Columbia's 'Great Bear Rainforest' was not simply there, waiting to be discovered and managed. It was created through two decades of research, conflict, alliance building, compromise and negotiation between environmentalists, governments, First Nations, forestry companies, their customers and forest-dependent communities. Conceiving forest ecosystems as hybrid entities – neither 'pristine nature' nor fully exploitable economic resources – is straightforward enough. Accommodating the interests, values and rights of such diverse stakeholder groups required profound transformation in the identities and interests of institutions, interest groups, ecosystems and, for that matter, grizzly bears.

In Chapter 26, Giorgio Osti explores the social, economic and infrastructural transformations necessary to achieve the seemingly self-evident need for expanded renewable energy generation. While it may be tempting to dismiss resistance to renewable energy as an expression of 'not in my backyard' attitudes, Osti demonstrates that the technological choices associated with renewable energy have highly variable implications for rural land use and, as a consequence, for rural economies, labour markets and community identities. Wind farms, photovoltaic panels, biogas plants, hydroelectric facilities, etc. have different spatial requirements and impacts. The resolution of land use conflicts, he argues, is often determined by rural peoples' political and economic resources. Where the latter are limited, rural landscapes are more readily colonized and transformed into 'energy landscapes' through the imposition of infrastructure incompatible with traditional rural land uses (i.e. agriculture and forestry). By contrast, where communities are organized and have the necessary social resources, opportunities exist for landscape transformations that enhance local livelihoods and vitality.

Conclusion

Today's environmental crises are of a scale and magnitude that demand the continued development and transformation of the social sciences. As argued above, many of the questions posed by contemporary processes of socio-ecological change call for the application of well-established theories and methodologies; others call for innovation and experimentation. Irrespective of whether we apply traditional or novel approaches to social research, there is a strong need for social scientists to move from the margins to the core of environmental research. Global environmental change requires social scientific contributions that go beyond identifying 'barriers' to popular acceptance of scientific findings or policy instruments. Environmental change calls for the sustained interrogation of governance regimes and their implicit (often naive) sociological assumptions. It calls, as well, for analysis of how impacts are distributed (and by whom), and the ways in which those impacts interact with other dimensions of disadvantage to reinforce or transform inequalities. And it calls social scientists to take leadership roles within multidisciplinary efforts to comprehend and reshape in more propitious terms our socio-ecological futures on Planet Earth.

Inevitably, the issue of global climate change looms large throughout this book. At the same time, the book deliberately avoids treating climate change in isolation from other processes of social and ecological change. In contrast with works that take a narrowly prescriptive look at how to avoid undesirable ecological change, this collection highlights the many uncertainties, ambiguities and value conflicts associated with socio-ecological change and the ways they are understood and addressed. This volume, thus, has several aims:

- to present multidisciplinary social science perspectives on the co-evolution of human societies and the natural environment;
- to identify and analyse key governance regimes and adaptation strategies for social and environmental change;
- to evaluate the ability of social science perspectives on social and environmental change to contribute to more just and sustainable socio-ecological relationships;
- to highlight research questions and analytical approaches that are underrepresented in the research literature and/or policy circles; and, finally,
- to argue the case for research and policy agendas that take a more sophisticated and holistic approach to the combined social and ecological dimensions of environmental change.

In pursuit of these aims, as is always the case, gaps remain in our coverage. We encourage and invite readers not simply to point these out, but to fill them in and, in so doing, contribute to the further development of more robust social-ecological futures for all.

References

Armstrong, S., Curtis, M., Kent, R., Maxwell, D., Mousseau, F., Pearce, F., Sadler, K., Tamminga, P. and Tansey, G. (2011) *World Disasters Report: Focus on Hunger and Malnutrition*. Geneva: International Federation of Red Cross and Red Crescent Societies.

Ferris, E. and Petz, D. (2012) *The Year That Shook the Rich: A Review of Natural Disasters in 2011*. London: Brookings Institution.

Latour, B. (1993) *We Have Never Been Modern*. Cambridge, MA: Harvard University Press.

Law, J. and Hassard, J. (eds) (1999) *Actor Network Theory and After*. Oxford, UK: Blackwell.

Lockie, S. (2004a) 'Social Nature: The Environmental Challenge to Mainstream Social Theory', pp. 26–42 in R. White (ed.), *Controversies in Environmental Sociology*. London: Cambridge University Press.

Lockie, S. (2004b) 'Collective Agency, Non-Human Causality and Environmental Social Movements: A Case Study of the Australian "Landcare Movement"', *Journal of Sociology*, 40, 1, pp. 41–57.

Newell, P. (2008) 'Contesting Trade Politics in the Americas: The Politics of Environmental Justice', pp. 49–74 in D.V. Carruthers (ed.), *Environmental Justice in Latin America: Problems, Promise, and Practice*. Cambridge, MA: MIT Press.

Perrings, C., Naeem, S., Ahrestani, F.S., Bunker, D.E., Burkill, P., Canziani, G., Elmqvist, T., Fuhrman, J.A., Jaksic, F.M., Kawabata, Z., Kinzig, A., Mace, G.M., Mooney, H., Prieur-Richard, A.-H., Tschirhart, J. and Weisser, W. (2011) 'Ecosystem Services, Targets, and Indicators for the Conservation and Sustainable use of Biodiversity', *Frontiers in Ecology and the Environment*, 9, pp. 512–520.

Rockström, J., Steffen, W., Noone, K., Persson, A., Chapin, S., Lambin, E., Lenton, T., Scheffer, M., Folke, C., Joachim Schellnhuber, H., Nykvist, B., de Wit. C., Hughes, T., van der Leeuw, S., Rodhe, H., Sörlin, S., Snyder, P., Costanza, R., Svedin, U., Falkenmark, M., Karlberg, L., Corell, R., Fabry, V., Hansen, J., Walker, B., Liverman, D., Richardson, K., Crutzen, P. and Foley, J., (2009) 'A Safe Operating Space for Humanity', *Nature*, 461, pp. 472–475.

Part I
Challenges, contradictions and consequences of global socio-ecological change

2

Ecological modernization theory
Taking stock, moving forward[1]

Arthur P.J. Mol, Gert Spaargaren and David A. Sonnenfeld

With the rebirth of environmental concern among social scientists in the 1960s and 1970s, scholars initially were preoccupied with explaining environmental devastation. Their central concern was how human behaviour, capitalist institutions, a culture of mass consumption, failing governments and states, and industrial and technological developments, among others, contributed to the ongoing deterioration of the physical environment. In the 1970s and 1980s, with environmental problems manifesting themselves on a worldwide scale, there were many reasons to look for explanations of the widening and deepening environmental crises. The result was an expansive literature – both theoretical and empirical in nature – on the main causes of continued environmental deterioration. Various disciplines and schools of thought emphasized different structural, institutional and behavioural traits as the fundamental origins and causes of the environmental crisis.

Beginning in the 1980s and maturing in the 1990s, attention in environmental sociology and environmental politics started to widen, focusing not only on environmental deterioration as the variable to be explained, but also on environmental reform. This led to what sociologist Frederick Buttel (2003) labelled the sociology or social sciences of environmental reform. Strongly driven by empirical and ideological developments in the European environmental movement, by practices and institutional developments in some 'environmental frontrunner states' and developments in private companies, some European social scientists began reorienting their focus from explaining ongoing environmental devastation towards understanding processes and outcomes of environmental reform. Later, and sometimes less strongly, this new environmental social science agenda was followed by US and other non-European scholars and policy analysts. By the turn of the millennium, this focus on understanding and explaining environmental reform had become mainstream, not so much instead of, but rather as a complement to, studies on environmental deterioration.

Within the 'social sciences of environmental reform', *ecological modernization* stands out as one of the strongest, well-known, most used and widely cited and constantly debated concepts. The notion of ecological modernization may be defined as *the social scientific interpretation of environmental reform processes at multiple scales in the contemporary world*. As a still young but growing body of scholarship, ecological modernization studies reflect on how various institutions and social actors attempt to integrate environmental concerns into their everyday functioning, development and relations with others and the natural world.

From the launching of the term by Martin Jänicke and Joseph Huber around 1980, and its explicit foundation into social theory by Arthur Mol and Gert Spaargaren around 1990, ecological modernization has been applied around the world in empirical studies, has been at the forefront in theoretical debates, and even has been used by politicians to frame environmental reform programmes in, among other countries, Germany, the Netherlands, the UK and China. There is now wide interest and research in ecological modernization throughout the world, including Asia (Japan, Korea, Vietnam, Thailand, Malaysia and especially China), North America, Australia and New Zealand, Latin America (Argentina, Peru, Chile and especially Brazil), as well as elsewhere on the wider European continent (including Russia). After three decades of scholarship, quite a number of volumes have been published on ecological modernization (see Barrett 2005; Mol and Sonnenfeld 2000; Mol et al. 2009; Young 2001).

In this chapter, we aim to take stock of the vital and still-developing body of literature on ecological modernization and suggest directions for continued advancement of this school of thought. What have been the key accomplishments of ecological modernization studies to date? The critical debates involving ecological modernization theory until now? What should the research agenda be for ecological modernization studies in the new millennium? These questions are given centre stage in this chapter. First, we draw up the balance of three decades of ecological modernization theorizing and empirical studies, assessing its achievements in terms of its key academic contributions and societal impacts. Second, we examine the various debates and criticisms that ecological modernization has been engaged in and sometimes has triggered in its still relatively short history. Third, and finally, these assessments are drawn upon to suggest elements of a research agenda for future ecological modernization scholarship.

Taking stock of ecological modernization studies

When ecological modernization research started to gain firm ground in the environmental social sciences from the mid-1980s onwards, scholars in environmental sociology, political sciences, human geography and related fields still were solidly dominated by frames and traditions for investigating and explaining *environmental crises*. Such scholars mainly focused their investigations on 'the roots of the environmental crisis', as David Pepper (1984) so adequately summarized. And, of course, good reasons existed for such a preoccupation. After a period of rapid environmental capacity building in OECD countries in the first half of the 1970s, environmental policymaking and implementation – especially at the national level – stagnated, and the failure of governments to adequately address contemporary environmental problems was widely perceived as a dominant trend (Jänicke 1986). Moreover, most private firms were reluctant to take environmental responsibilities on board. Only continuous and strong (societal and state) pressure resulted in what were still often minimal environmental improvements. Finally, the environmental movement in many countries was internally focused and divided over strategy. After the heydays of the early 1970s and their radicalization at the end of that decade, environmental movements struggled between grassroots radicalism and professional lobbyism (see Gottlieb 1993), and were confronted with less favourable (and sometimes even hostile; e.g. UK and US) administrations.[2] It should not come as a surprise, therefore, that the social, political, economic and environmental conditions of the time were reflected in environmental social scientists preoccupation with analysing the poor environmental records of modern societies and institutions.

Although rooted in the environmental conditions of the time, the overall critical and pessimistic outlook of environmental social scientists in the late 1970s and early 1980s was shaped as well by factors internal to the environmental social science disciplines. While ecological economists established relationships with policymakers at an early stage in the development of their discipline, in the 1970s and 1980s environmental sociologists, political scientists and human geographers were internally divided, inwardly directed and very much involved in (neo-)Marxist debates over the roots of environmental degradation. The structuralist argument that genuine environmental improvements are impossible while the main institutions of capitalism remain in place had (and continues to have) a strong resonance in the environmental social sciences (Dickens 1991; Dobson 1990; Pepper 1984, 1999; Schnaiberg 1980).

Against this background, ecological modernization scholars made important contributions to social theory through the development of a systematic theory of institutional environmental reform; the introduction of a variety of theoretical innovations on the relation between society and the natural environment; the elaboration of new approaches in environmental policy and practice; and contributions to the 'globalization' of general social theory.

A new emphasis: analysing and co-constructing environmental reform

The first, and arguably most important, contribution of ecological modernization research has been to open up, diversify and provide a systematic theoretical framework for integrating social science scholarship and policy perspectives on the ways in which contemporary societies interact and deal with their biophysical environments. Interpreting, explaining and theorizing the social processes and dynamics of environmental reform were key scientific innovations resulting in the initiation of a new field of study complete with its own research agenda, themes and concepts. Previously, there had been fragmented (especially policy, business and economic) studies of successful environmental policymaking, of the development of environmental technologies and of proactive firm behaviour – to name a few key ecological modernization subjects. With its emergence, the ecological modernization frame provided a basis for bringing these disparate studies together into a more-or-less-coherent body of knowledge that has had an enduring impact on environment-oriented studies across the social sciences.

The endeavour of advancing a scientific and policy framework for environmental reform resulted at times in strong academic reactions and criticisms. Most scholars would accept the fact that individual, ad hoc cases of successful environmental reform can exist even in an overall destructive late-modern, capitalist–industrialist society. However, arguing that these exceptions might exemplify or foreshadow something of a general (green) rule with respect to the fragmented but ongoing institutionalization of environmental concerns is a different story. When the dominant view in the discipline refers to the impossibility of major environmental improvements under conditions of modernity, an eco-modernist perspective that used the mounting case studies of environmental reform to argue that late-modern capitalist–industrialist societies could, in the end, be environmentally reformed was bound to yield major criticisms. Hence, ecological modernization – whether framed as an academic social theory or as a political programme for politicians and environmental NGOs – remains under strong debate and accusations of various kinds (see next section).

With hindsight, it can be concluded that ecological modernization scholars were timely in formulating a new perspective that has become mainstream in policymaking circles and well accepted in the social sciences. What started to develop was a new branch of scholarship: the social sciences of environmental reform. Theories and methods of environmental change were developed in close interaction with and reflecting upon major developments in the environmental discourse during the late 1980s and early to mid-1990s. Among the major events and dynamics were the publication of the Brundtland report (WCED 1987); the wave of international environmental treaties being negotiated and signed; the successful UNCED conference in Rio in 1992; the formation of environmental ministries, laws and policies in most developing countries from the late 1980s onwards (see Sonnenfeld and Mol 2006); the worldwide spread of the awareness of environmental risks among major parts of the population; and a resurgence in environmental activism (see Sonnenfeld 2002).

From the mid-1980s, ecological modernization scholarship helped to promote new societal roles and orientations for the environmental social sciences while contributing significantly to a broad reworking of the theoretical landscape in sociology, political science, human geography, business studies and other fields. This reorientation has resulted in, among other things, a (positive) re-evaluation of modernization theories, especially in their 'reflexive forms' as suggested by Anthony Giddens (1990) and Ulrich Beck (Beck et al. 1994), and a diminishing (or less exclusive) influence of neo-Marxism in the environmental social sciences (Mol 2006). Building upon the reflexive modernization theories of Giddens and Beck, ecological modernization scholarship was instrumental for social scientists and policymakers when trying to move beyond the 1980s debate over whether capitalism (Marxism) *or* industrialism/technology (industrial society theory) should be regarded as the most important driver of environmental degradation (Mol 1995).

Conceptual and theoretical innovations

A second major contribution of ecological modernization scholarship is to be found in the introduction of a variety of innovative concepts, theoretical notions and major research themes into social theory. They include, amongst others: the notion of emerging processes of ecological rationalization (akin to, but different from, Weber's idea of institutional rationalization); the notion of political modernization catalysed by civic and institutional environmental response and interaction; the incorporation of market dynamics, market actors and market-based instruments into environmental policymaking and practice. We shortly elaborate on the three innovations in what follows.

At the heart of the theory is the understanding of ecological modernization as a process of differentiation – and 'emancipation' – of an ecological sphere and the concomitant articulation of an independent ecological rationality (see Dryzek 1987; Mol 1995; Spaargaren and Mol 1992). This conceptual move brings a number of different developments under one common denominator and makes room for the environment in more general social theories.[3]

Another major conceptual innovation is to be found in the introduction of, and the ensuing debate on, the concept of 'political modernization' (see Jänicke 1993; Mez and Weidner 1997; van Tatenhove and Pieter 2003; van Tatenhove et al. 2000). Political modernization refers to the renovation and reinvention of state environmental policies and politics in order to make environmental reform better adapted to the new conditions of late-modern societies. The debate on political modernization within environmental politics can be seen as an early formulation of themes and basic ideas of environmental governance. Moreover, the concept of political modernization connected ideas on innovative governance in a direct and explicit way with (the management of) environmental change. The notion of political modernization already included ideas of multiple actors and multi-level governance and made room for various modes of steering and policymaking applied by different actors outside the framework of national environmental governments. As a further specific example of conceptual innovation in this respect, we can refer to the notion of 'environmental capacity' as it has been developed within the ecological modernization school of thought by the group around Martin Jänicke in Germany (e.g. Jänicke 1995).

In developing their ideas on political modernization and environmental governance, ecological modernization scholars have been innovative in allowing economic categories and concepts to enter theories of environmental reform. Of course, this emphasis on the importance of economic/market-based concepts and schemes for environmental policymaking – in particular, for technology policies and for the management of production–consumption chains and networks – was not unique for ecological modernization scholars. In this respect, a number of ideas were borrowed from environmental and ecological economics in particular. From the 1980s onwards, environmental/ecological economists contributed significantly to the development of (eco)economic valuation models, criteria and theories which were used to 'internalize externalities'. When applied by ecological modernization scholars, this process was referred to as the 'economizing of ecology' and the 'ecologizing of the economy' (Huber 1982).

Nowadays, environmental governance scholars from various perspectives acknowledge the crucial role of economic actors and instruments in environmental policymaking. In an early phase of the debate on economic instruments in environmental policy, ecological modernization scholars made significant contributions to the literature in two respects. First, they brought many of the market and monetary dynamics – such as eco-taxes, environmental auditing, corporate environmental management, green consumerism, valuing environmental goods, annual environmental reports, environmental insurances, green niche markets, green branding and eco-labelling, – together in a coherent broader framework. Second, ecological modernization scholars interpreted these market and monetary developments in terms of a redefinition of the role of states and markets in environmental reform, moving away from the basic idea of a monopoly of state authority on the protection of public goods. With the help of this broader framework of new relationships between private firms, states and civil society actors and organizations, it became possible to move beyond narrow economic, neoliberal frameworks for understanding the role of privatization, marketization and liberalization in environmental politics while, at the same time, being able to understand better the new roles in environmental governance assigned to environmental movements and to citizen-consumers (see Sonnenfeld 2002; Spaargaren and Mol 2008).

Next to these primary theoretical contributions, ecological modernization scholars have also been recognized for a number of models, concepts and approaches that, while not exclusive to ecological modernization research, have been developed in collaboration with colleagues in other fields. This includes: the application and elaboration within environmental political sciences of discourse analyses (Hajer 1995, 1996; Weale 1992); major contributions to the field of technology studies and to environmental technologies in particular (Huber 1982, 1989, 2004; Sonnenfeld 2002); the elaboration and extensive empirical use of the triad network model (Mol 1995); the application of the core concepts of risk and trust, especially in the context of the globalization of production and consumption (Cohen 1997; Oosterveer 2007); the elaboration of the social practices model, especially in the field of sustainable consumption (Spaargaren 2003, 2011); the first outline of an environmental sociology of networks and flows (Spaargaren et al. 2006); and the analyses of the emerging role of what is now labelled informational governance (Heinonen et al. 2001; Kleindorfer and Orts 1999; Mol 2008; van den Burg 2006).

Contributions to environmental politics and management

While the core of ecological modernization studies remains academic in outlook, most scholars in this tradition have engaged themselves with applied, policy-relevant studies as well, moving beyond 'ivory tower' criticism of contemporary developments. Through their applied research, they have engaged with environmental politics and practices; they have joined discussions and planning sessions with environmental NGOs on their position, strategy, alliances and priorities (see Smith et al. 2006); and they have involved themselves with business (associations) in designing pro-active strategies and in exploring niche market developments. We would thus argue that the third category of major contributions of ecological modernization is its substantial contributions to studies of environmental politics and management. These contributions are the direct result of the explicit and sustained focus on major innovations in environmental policy and practice. The fact that governmental administrations, political parties, as well as environmental movements, have used the notion of ecological modernization to refer to their main aims and strategies is indicative of the 'practical' proliferation of ecological modernization ideas. At the same time, it gives evidence of the fact that in today's reflexively modern world, academic concepts are spiralling in and out of environmental governance practices more frequently and at an ever-faster pace. A range of empirical studies on concrete processes of ecological modernization have found – to varying degrees – their way into environmental policies and management. Hence, ecological modernization studies very often give witness of the mutual influence between theoretical models and concepts on the one hand, and empirical developments and practices in society on the other. This fact has resulted in the frequent intermingling of the often-quoted two sides of ecological modernization: the academic/analytical theory and the normative/prescriptive/policy-oriented model of environmental change.

Globalization and environment

A fourth key contribution of ecological modernization scholars is their encouragement of and significant contributions to globalization theory and research in the environmental social sciences. While, in the twenty-first century, more and more social science research moves beyond the 'nation-state container' (Beck 2005), this was not true in the 1980s and most of the 1990s. Several factors can explain the early internationalization of ecological modernization research and the explicit comparative perspective used to study processes of environmental change. For one, the object of study meant that environmental studies were leading in internationalization within the social sciences. Cross-border problems and – especially in the 1990s – growing international efforts and coordination for solving environmental problems triggered this international and global outlook. More specifically for ecological modernization, scholars' need to learn from (successful) environmental reform practices and developments in different countries spurred comparative research both within Europe and between different regions in the world economy. The fact that ecological modernization emerged in Europe contributed to an early international outlook of ecological modernization. Hence, comparative and international studies are well represented in the ecological modernization research tradition (see, for some early examples, Andersen 1994; Jänicke 1990; Liefferink et al. 1993; Weale 1992), and ecological modernization scholars have contributed significantly to international, comparative and global social science research on the environment.

Ecological modernization scholars have also been productive in bringing the environment into globalization studies through a number of books and special issues (see Mol 2001; Sonnenfeld 2008; Sonnenfeld and Mol 2006, 2011; Spaargaren et al. 2000, 2006). These studies helped develop a more balanced interpretation of how globalization dynamics affect the environment, while contributing as well to the analysis and conceptual understanding of globalization with respect to environmental change.

Taken together, the four contributions to social theory make a positive balance for ecological modernization as a recently established school of thought within the environmental social sciences. Individually, many of these achievements are not unique or exclusive to ecological modernization studies. Taken as a whole, however, they represent the distinct approach, coherent perspective and active research programme of ecological modernization theory. In the course of three decades, steadily accumulating scholarship in this school of thought managed to contribute to an increased commitment across the social sciences with understanding and facilitating social and institutional change related to the environment, natural resources and climate change worldwide.

Critical debates on ecological modernization

From its inception, ecological modernization theory has met considerable opposition. The critiques have become more frequent and fierce, however, with its growing recognition, popularity and acceptance, together with its widening geographical scope of application and its profound policy impact. The fiercest debates relate to its positive commitment to environmental reform under conditions of modernity. As scientific researchers and policy analysts of institutional and cultural environmental reform worldwide, ecological modernization scholars have been characterized as being theoretically mistaken (Blühdorn 2000), Eurocentric (Mol 1995), politically naive (Hobson 2002), empirically wrong (Weinberg et al. 2000), blind to issues related to consumption (Carolan 2004) and social inequality (Harvey 1996), 'colonized by the economic and cultural system' (Jamison 2001: 4), and even 'cursed with an unflappable sense of technological optimism' (Hannigan 2006: 26). For those active in social scientific disciplines and fields of study, which have been preoccupied for decades now with explaining the sources and continuity of environmental crises and deterioration, one should perhaps not be surprised when meeting views of this kind. Instead, well-formulated scientific critiques can and must be used to develop a more deeply nuanced understanding of key assumptions and notions. Debates on ecological modernization theory over the last two decades have been summarized and reviewed in various publications (cf. Carolan 2004; Christoff 1997; Mol 1995; Mol and Spaargaren 2000, 2004; York and Rosa 2003). Here, we briefly address critiques that have been of strategic relevance to ecological modernization scholarship. We start by discussing some of the more well-known earlier perspectives that have been reflexively addressed and incorporated into more recent formulations of ecological modernization theory. Second, we review a number of critical views that are difficult or impossible to address within the basic framework and understandings of ecological modernization theory.

Critiques that have been addressed over the course of time

In response to perspectives and formulations common in early ecological modernization studies, critics raised a number of objections or limitations of this new approach. These included arguments about ecological modernization theory's shortcomings with respect to technological determinism, its focus on production processes and consequent neglect of practices of consumption, its lack of analyses of social inequality and power and its Eurocentric outlook. Since then, critiques on these topics have been acknowledged, the theoretical approach has been revised and strengthened and new studies have been undertaken. Comments on the Eurocentric outlook, for example, have resulted in studies of ecological modernization outside of northwestern Europe; for instance on the North American continent and Australia/New Zealand, but also in Asia and Latin America (Mol et al. 2009).

Criticisms of ecological modernization theory's technocratic outlook and the ring of technological determinism attached to earlier formulations have resulted in a refinement of ecological modernization perspectives with respect to the role of technology in bringing about social change and environmental reform. This approach is complicated, as well, by a growing diversity of perspectives *within* ecological modernization theory reflecting, for example, different evaluations by ecological modernization theorists of (environmental) technologies as driving forces for environmental change.[4] Such intellectual diversity notwithstanding, most ecological modernization studies within environmental sociology, political science and human geography today have become sensitive and reflexive with respect to the role of technology in environmental change. Using the work of Beck (1986, 1992) on science and risk, Giddens (1990, 1991) on trust and abstract systems, and responding to ideas from actor–network theory (Latour 1993; Urry 2000) and science and technology studies (Geels 2005; Schot 1992; Shove 2003), ecological modernization theorists today have made strong contributions to a more reflexive stance on the use and role of environmental technologies in environmental policy. Mol et al. (2009) provide recent documentation of the changing role of technology in ecological modernization studies.

The critique of early ecological modernization theory's relative neglect of social inequality and issues of power has become a focus of more recent scholarship as well. The inclusion of such themes as inequality and green trade (Oosterveer 2007), green consumption as a 'Western' phenomenon (Spaargaren and van Koppen 2010), power and inequality related to environmental and informational flows (Mol 2008), the differential effects of stringent environmental policies and the unequal distribution of environmental risks (Smith et al. 2006) all bear witness to the active involvement of ecological modernization scholars with the themes of power and inequality, especially at global, regional and international levels of analysis. Against this background, we would argue that contemporary criticisms of ecological modernization theory focusing on the issues discussed above are seriously out of date.

Lasting controversies

Several critiques of the ecological modernization perspective find their origin in radically different paradigms and approaches to the theme of late modernity, social change and environmental sustainability. Because these approaches have fundamentally different starting positions, assumptions and worldviews, it is far from easy to incorporate them into the ecological modernization framework, or vice versa. As a result, some of the controversies and debates with respect to these frameworks, and the theory itself, will endure with little near-term prospect of reconciliation, agreement or synthesis. Three rival social theories of the environment fall into this category: those rooted in neo-Marxism, radical or deep ecology and structural human ecology/neo-Malthusianism.

As noted above, neo-Marxist perspectives on contemporary societies and the natural environment were dominant in some parts of the world in the late twentieth century. These embrace several strands including the 'treadmill of production' perspective by Allan Schnaiberg and colleagues (Pellow et al. 2000; Schnaiberg et al 2002; Weinberg et al. 2000); environmental sociologists and others working within the world-systems theory perspective, identified with the online *Journal of World-Systems Theory*; the eco-socialist perspective of James O'Connor, Michael Goldman, Patrick Bond and others associated with the journal *Capitalism, Nature, Socialism*; and the more structural Marxist perspective of John Bellamy Foster and others associated with *Monthly Review*.

All emphasize and prioritize the fundamental continuity of the (global) capitalist order, which disallows meaningful and enduring structural environmental reform in contemporary, market-oriented societies. Whether in the form of the treadmill of production (Gould et al. 2008), the second contradiction of capital (O'Connor 1997) or any other conceptualization, the fundamental criticism remains essentially the same: environmental conditions continue to deteriorate everywhere to the point of global and local crises; enduring, effective structural environmental reform is impossible in (increasingly globalized) capitalist societies. The basic notion of ecological modernization processes aimed at 'repairing one of the crucial design faults of modernity' is held to be theoretically impossible.

Such debates are sometimes heated, but can also be productive. The enduring confrontation between neo-Marxism and ecological modernization theory has been fruitful in clarifying the fundamental differences between the perspectives (cf. the contributions in Harvey 1996; Mol and Buttel 2002). The interrelation between environmental and social exploitation and degradation has been a key thread in such debates. Basic differences notwithstanding, periodic attempts have been made to find common ground and enjoin theoretical and empirical challenges (see Fisher 2002; Mol 2011; Spaargaren et al. 2006). At the same time, some empirical studies have tried to use both perspectives in a complementary rather than mutually exclusive way (Lang 2002; Smith et al. 2006; Wilson 2002).

Other scholars inspired by a variety of radical or deep ecology values and informed by discursive, neo-institutional theories of political change are sceptical of what they see as the reformist agenda of ecological modernization. They view ecological modernization theory as an elite-centred approach to reform resulting in 'light-green', superficial forms of social and environmental change. Against the 'pragmatic' outlook of ecological modernization theorists, such radical and deep ecologists argue instead for more fundamental, or 'dark-green', forms of institutional and especially bottom-up political change in order to bring contemporary societies beyond the political structures of late modernity and into more ecologically sustainable social relations and institutional configurations. Whether eco-feminist, socialist, postmodernist or anarchist, the alternatives put forward by such scholars keep (considerable) distance between the ideal, arguably utopian, green futures on the one hand and what in principle can, or has already been, realized in terms of environmental reform so far on the other. Among others, Andrew Dobson (1990), John Barry (1999), Robyn Eckersley (1992, 2004) and John Dryzek (1987) in political science, and Tom Princen (Princen et al. 2002), Mikko Jalas (2006) and Kerstin Hobson (2002) in the field of consumption studies, all represent mild or strong versions of 'deep-green'/discursive thinking as suggested here.

Radical ecologists' positions continue to evolve, importantly. Christoff (1997), for example, has attempted to build constructively on the controversy described here by distinguishing different (green) shades of ecological modernization. He argues that the Beck- and Giddens-inspired variants of (reflexive) ecological modernization could be further developed into 'strong versions' of ecological modernization theory. Mol and Spaargaren (2000) aimed to contribute to this debate, as well, distinguishing between 'green radicalism' positions on the one hand and 'socio-economic radicalism' on the other, thereby creating conceptual space for a debate on discrete forms of radicalism within eco-modernization perspectives. Despite these efforts, radical or deep-ecology-inspired schools of thought do not easily mix and match with ecological modernization ideas. In addition to disagreement on the desired pace and scope of environmental change, the perceived 'anthropocentric outlook' (Eckersley 1992) of many ecological modernization studies also seems to contribute to a continued divide.

Structural human ecologists inspired, in part, by neo-Malthusian notions of overpopulation and absolute natural limits have aimed to quantify cross-national environmental impacts and mathematically relate them to a variety of anthropogenic drivers. Eugene Rosa, Thomas Dietz and Richard York are the most visible representatives of this stream of thinking (York and Rosa 2003; York et al. 2003). Inspired as well by Wackernagel and Rees' (1998) 'ecological footprint' analyses, and natural-science-based historical interpretations of human–society metabolism (Fisher-Kowalski 1997), structural human ecologists have concluded that, due to increased affluence and growing population, environmental impacts are only increasing as eco-technological development cannot keep pace with the former two causes. From this, they challenge ecological modernization theory since 'historical patterns of modernization and economic development have clearly led to increased pressure on the global environment' rather than to effective environmental reforms (York et al. 2003: 44–45). Structural human ecology/neo-Malthusian perspectives diverge significantly from ecological modernization theory in that the former are highly abstract rather than richly particular; are structurally deterministic rather than reflexive and change-oriented; and are profoundly pessimistic rather than opening up windows to institutional and cultural environmental change.

All three 'competing' schools of social theory criticize ecological modernization studies as: (1) being one-sided in focusing only on environmental reform; (2) utilizing non-representative case studies rather than cross-national, statistical analyses based on large data-sets; (3) not addressing the basic, structural drivers behind environmental degradation; and, for that reason, (4) being overly optimistic/naive about the potential for environmental change and sustainable development. When related back to the basic starting points and premises characteristic of these schools of thought, the points raised in debate with ecological modernization are usually internally logical and coherent. Since disagreements tend to go back to fundamental assumptions – on science, its role in society, the relationship between theoretical and empirical work, and the present state of the world – we expect such controversies to be long-lasting.

While ecological modernization scholars acknowledge that environmental deterioration continues forcefully and widely, at the same time they find much evidence of significant environmental reform around the world. Among the paradigmatic assumptions of ecological modernization is the contention that scientific efforts to identify, analyse, understand and design new, more environmentally friendly and sustainable socio-technical systems, institutions, policy arrangements and social relations are not only of key academic importance in themselves, but are central also to the identification and understanding of structural, anthropogenic drivers of environmental decay.

Moving forward: future directions

Ecological modernization has moved from a peripheral position in both the environmental social sciences and the general social sciences to an acknowledged school of thought in this new millennium. Yet much work remains to be done in order to understand the extent to which environmental interests are included in all kinds of economic, cultural and political practices and institutional developments, and the success of these, at different levels, geographies and time frames. The need for a wide variety of theoretically informed empirical studies – quantitative, qualitative, comparative, longitudinal, etc.[5] – remains high and will be an important part of the future research profile of ecological modernization. The last section of this chapter focuses on three areas within the broad scope of ecological modernization theory that we believe are ripe for further study: extending the geography of ecological modernization studies; studying the environmental reform of global flows; and focusing on cultural dimensions of ecological modernization.

The extended geographical scope of ecological modernization studies

As mentioned above, there has been progress with respect to moving away from the Eurocentrism contained in the first generation of ecological modernization studies. However, the non-OECD countries under ecological modernization study to date mainly represent the rapidly emerging economies of central and eastern Europe, East and Southeast Asia and, to a lesser degree, Latin America. Over the last decade, however, the relevance of ecological modernization for countries and regions with low or negative growth rates and with thin and fragmented connections with the world network society has been taken up as a pressing and theoretically challenging theme. Research on critically important environmental infrastructures in rapidly urbanizing sub-Saharan Africa, for example, has been organized around the key concept of modernized mixtures[6] (cf. Hegger 2007; Spaargaren et al. 2005). In the African context, modernized mixtures refer to an ecological modernization strategy that is sensitive to and adapted for the specific circumstances of societies with fragmented urban infrastructures and ill-functioning institutions and (health and sanitary) practices. These new studies are expected to deepen ecological modernization theory's appreciation of the effects of North–South relations and inequalities on the opportunities, limitations and particular forms of ecological modernization in less developed countries.

Environmental reform of global flows

Hyperglobalization and the emergence of a global network society are arguably key processes that fundamentally change the face of the world – and of the earth. Hence, one of the key challenges for the development and relevance of ecological modernization theory lies in understanding such processes of hyperglobalization and the network society in relation to environmental reform. In working on the new dynamics of change in global modernity, Mol (2001) has addressed the ecological modernization of the global economy; and Sonnenfeld (2002), Huber (2004), Angel and Rock (2005) and others have addressed industrial and technological environmental transformation on a global scale. More recently, and fundamentally, Spaargaren et al. (2006) have related ecological modernization theory to the new sociology of networks and flows (Urry 2000) and, as such, have opened up a new field for research on environmental reform. Instead of the conventional notion of place-based environmental reform, the emphasis shifts to environmental reform in the 'space of flows', related to globalized, deterritorialized and denationalized mobilities and flows. This shift has started to inspire a range of empirical studies on the environmental reform of all kinds of global environmental issues and transnational flows. For instance, Presas (2005) uses this conceptual framework for investigating the sustainable construction of transnational buildings in global cities; Oosterveer and colleagues (Bush and Oosterveer 2007; Oosterveer 2007; Oosterveer et al. 2007) apply the framework to global food production and consumption; Mol (2007, 2010) investigates environmental reform of global biofuel networks and of mega events such as the Olympics; and van Koppen (2006) uses the conceptual framework for exploring biodiversity flows. The so-called environmental sociology of networks and flows raises a number of questions, challenges and insights for environmental reform studies in the twenty-first century (as reported by Rau 2010; Spaargaren et al. 2006), and much work lies ahead of both theoretical and empirical significance. It can be expected that such an elaboration of ecological modernization – with the help of key concepts of networks, scapes, hybrids and flows – will result in new insights into the dynamics of environmental change under conditions of hyperglobalization, and to the development of a new generation of governance approaches to affect those dynamics in more positive (and environmental) directions at the global and local scales.

The cultural dimensions of ecological modernization

When trying to address environmental flows in the context of the global network society, ecological modernization scholars are confronted with the increased intermingling of social and ecological subsystems that so far have been analytically separated in (ecological) modernization theory. From its roots in systems theory, states, markets and civil society were conceived of as independent spheres, each interacting in a specific way with each other and with the emerging ecological sphere. This point brings us back to where it all started in ecological modernization: the differentiation of an independent ecological sphere or subsystem in late-modern societies in the last decades of the second millennium. But, after the conceptual emancipation of the ecological sphere, there has to follow a re-embedding of the ecological sphere in society by reconnecting the ecological sphere to the spheres of market, state and civil society. And, while the relationship with economic and political rationalities has already been discussed and explored in some depth, the theoretical and practically oriented anchoring of ecological rationalities in the socio-cultural sphere of civil society remains a huge task yet to be fully taken up. What images of the good, sustainable life do we have to offer to lay people, to concerned citizen-consumers, the deprived, householders, youngsters, middle classes in transitional economies, inhabitants of slums, etc.? What is at stake here is the need to develop the cultural dimension of ecological modernization in much greater detail. Especially in the field of consumption studies, this (re)connecting of ecological rationalities to everyday life has been taken up as a challenging task (Spaargaren 2011). Again, most of these studies tend to be confined to OECD countries, and it is unknown as yet what relevance they will have for analysing the lifeworlds and lifestyles of the citizen-consumers in the fast-emerging middle classes of the upcoming economies of China, India, Brazil, Russia and elsewhere.

Epilogue

This chapter has given a broad overview of the accomplishments of three decades of ecological modernization studies, key debates involving ecological modernization theory during this period, and elements of a research agenda for future studies in this area. In the early years of the second decade of the third millennium, the environmental challenges faced by humankind are both better understood, and socially and politically as daunting, as ever (see Sonnenfeld and Mol 2011). Ecological modernization scholarship's challenge is to provide the conceptual and change-oriented frameworks, and empirical examples and evidence from around the world, to enable scholars, policymakers and citizens to understand, design and implement institutional and social arrangements that address those environmental challenges, albeit in ways suitable to local contexts. As all that happens in a rapidly changing world and with accumulating insights and knowledge, ecological modernization ideas will continue to evolve dynamically over time, deepening their scientific basis while increasing their salience for policymaking.

Notes

1 This chapter is based partially on Spaargaren, Mol and Sonnenfeld, 'Ecological Modernisation: Assessment, Critical Debates and Future Directions', pp. 501–520 in Mol, Sonnenfeld and Spaargaren (eds), *The Ecological Modernisation Reader* (Routledge 2009).
2 The early 1980s were marked by the rise of global neoliberalism, led by US President Ronald Reagan and UK Prime Minister Margaret Thatcher, and strong corporate and governmental backlashes against the environmental regulations of the 1970s and the social movements that had supported them.
3 The absence of environmental themes in most of the major social theories had been a worry among environmental sociologists from the very early days, and resulted in the formulation of the so-called HEP–NEP dichotomy: the Human Exemptionalist Paradigm (HEP) of the mother discipline and the New Ecological Paradigm (NEP) of environmental sociology (cf. Catton and Dunlap 1978a, 1978b; Mol 2006). Instead of building upon the HEP–NEP debate, ecological modernization scholars used the European sociological tradition of system theory (Luhmann, Habermas) and discourse analysis (Hajer) as their main starting points.
4 It may be suggested, for example, that Huber's (2004) study on environmental technologies retains the technocentric orientation of his earlier work.

5 Ecological modernization analyses do seem to move further into quantitative studies (e.g. Buitenzorgy and Mol 2011; Choy 2007; Sonnenfeld and Mol 2006).
6 Mixed modernities or 'modernized mixtures' refer to socio-technical configurations of infrastructures in which features of different (modern) systems have been deliberately and reflexively reconstructed to deal with dynamic social, economic and environmental contexts and challenges.

References

Andersen, M.S. (1994) *Governance by Green Taxes. Making Pollution Prevention Pay*. Manchester: Manchester University Press.
Angel, D.P. and Rock, M.T. (2005) *Industrial Transformation in the Developing World*. Oxford, UK: Oxford University Press.
Barrett, B. (ed.) (2005) *Ecological Modernization in Japan*. London: Routledge.
Barry J. (1999) *Rethinking Green Politics*. London: Sage.
Beck, U. (1986) *Risikogesellschaft. Auf dem Weg in eine andere Moderne*. Frankfurt: Suhrkamp.
Beck, U. (1992) 'From Industrial Society to the Risk Society: Questions of Survival, Social Structure and Ecological Enlightenment', *Theory, Culture and Society*, 9, pp. 97–123.
Beck, U. (2005) *Power in the Global Age. A New Global Political Economy*. Cambridge: Polity.
Beck, U., Giddens, A. and Lash, S. (1994) *Reflexive Modernization, Politics, Tradition and Aesthetics in the Modern Social Order*. Cambridge, UK: Polity Press.
Blühdorn, I. (2000) 'Ecological Modernization and Post-Ecologist Politics', pp. 209–228 in G. Spaargaren, A.P.J. Mol and F. Buttel (eds), *Environment and Global Modernity*. London: Sage.
Buitenzorgy, M. and Mol, A.P.J. (2011) 'Does Democracy Lead to A Better Environment? Deforestation and the Democratic Transition Peak', *Environmental and Resource Economics*, 48, 1, pp. 59–70.
Bush, S.R. and Oosterveer, P. (2007) 'The Missing Link: Intersecting Governance and Trade in the Space of Place and the Space of Flows', *Sociologia Ruralis*, 47, 4, pp. 384–399.
Buttel F.H. (2003) 'Environmental Sociology and the Explanation of Environmental Reform', *Organization and Environment*, 16, 3, pp. 306–344.
Carolan, M. (2004) 'Ecological Modernization: What about Consumption?', *Society and Natural Resources*, 17, 3, pp. 247–260.
Catton, W.R. and Dunlap, R.E. (1978a) 'Environmental Sociology: A New Paradigm', *American Sociologist*, 13, pp. 41–49.
Catton, W.R., and Dunlap, R.E. (1978b) 'Paradigms, Theories, and the Primacy of the HEP–NEP Distinction', *American Sociologist*, 13, pp. 256–259.
Choy E.A. (2007) 'A Quantitative Methodology to Test Ecological Modernization Theory in the Malaysian Context, Ph.D. thesis, Wageningen University.
Christoff, P. (1997) 'Ecological Modernization, Ecological Modernities', *Environmental Politics*, 5, 3, pp. 476–500.
Cohen, M. (1997) 'Risk Society and Ecological Modernization: Alternative Visions for Postindustrial Nations', *Futures*, 29, 2, pp. 105–119.
Dickens, P. (1991) *Society and Nature. Towards a Green Social Theory*. New York, NY: Harvester Wheatsheaf.
Dobson, A. (1990) *Green Political Thought*. London: Unwin Hyman.
Dryzek, J.S. (1987) *Rational Ecology. Environment and Political Economy*, Oxford/New York, NY: Blackwell.
Eckersley, R. (1992) *Environmentalism and Political Theory: Toward an Ecocentric Approach*. London: UCL Press.
Eckersley, R. (2004) *The Green State: Rethinking Democracy and Sovereignty*. Cambridge, MA: MIT Press.
Fisher, D. (2002) 'From the Treadmill of Production to Ecological Modernization? Applying a Habermasian Framework to Society–Environment Relationships', pp. 53–64 in A.P.J. Mol and F.H. Buttel (eds), *The Environmental State under Pressure*. Amsterdam: Elsevier.
Fisher-Kowalski, M. (1997) 'Society's Metabolism: On the Childhood and Adolescence of a Rising Conceptual Star', pp. 119–137 in M. Redclift and G. Woodgate (eds), *The International Handbook of Environmental Sociology*. Cheltenham, UK: Edward Elgar.
Geels, F.W. (2005) *Technological Transitions and System Innovation: A Co-Evolutionary and Socio-Technical Analysis*. Cheltenham, UK: Edward Elgar.
Giddens, A. (1990) *The Consequences of Modernity*. Cambridge, UK: Polity Press.
Giddens, A. (1991) *Modernity and Self-Identity. Self and Society in the Late Modern Age*. Cambridge, UK: Polity Press.

Gottlieb, R. (1993) *Forcing the Spring: The Transformation of the American Environmental Movement*. Washington, DC: Island Press.

Gould, K.A., Pellow, D.N. and Schnaiberg, A. (2008) *The Treadmill of Production: Injustice and Unsustainability in the Global Economy*. Boulder, CO: Paradigm Publishers.

Hajer, M.A. (1995) *The Politics of Environmental Discourse: Ecological Modernization and the Regulation of Acid Rain*. Oxford, UK: Oxford University Press.

Hajer, M.A. (1996) 'Ecological Modernization as Cultural Politics', pp. 246–268 in S. Lash et al. (eds), *Risk, Environment and Modernity: Towards a New Ecology*. London: Sage.

Hannigan, J. (2006) *Environmental Sociology: A Social Constructionist Perspective*. London/New York, NY: Routledge.

Harvey, D. (1996) *Justice, Nature and the Geography of Difference*, Malden, MA: Blackwell.

Hegger, D. (2007) 'Greening Sanitary Systems, An End-User Perspective', dissertation, Wageningen University.

Heinonen, S., Jokinen, P. and Kaivo-oja, J. (2001) 'The Ecological Transparency of the Information Society', *Futures*, 33, pp. 319–337.

Hobson K. (2002) 'Competing Discourses of Sustainable Consumption: Does the "Rationalisation of Lifestyles" Make Sense?' *Environmental Politics*, 11, 2, pp. 95–120.

Huber, J. (1982) *Die verlorene Unschuld der Ökologie. Neue Technologien und superindustrielle Entwicklung*. Frankfurt am Main: Fisher.

Huber, J. (1989) *Technikbilder. Weltanschaulich Weichenstellungen der Technik- und Umweltpolitik*. Opladen: Westdeutcher Verlag.

Huber, J. (2004) *New Technologies and Environmental Innovation*. Cheltenham, UK: Edward Elgar.

Jalas, M. (2006) 'Sustainable Consumption Innovations: Instrumentalization and Integration of Emergent Patterns of Everyday Life', pp. 461–478 in M. Munch Andersen and A. Tukker (eds), *Perspectives on Radical Changes to Sustainable Consumption and Production (SCP). Proceedings of the workshop of the SCORE-network.* Copenhagen, Denmark, 20–21 April 2006. Available at http://www.risoe.dtu.dk/rispubl/art/2006_117_proceedings.pdf.

Jamison A. (2001) 'Environmentalism in an Entrepreneurial Age: Reflections on the Greening of Industry Network', *Journal of Environmental Policy and Planning*, 3, pp. 1–13.

Jänicke, M. (1986) *Staatsversagen. Die Ohnmacht der Politik in der Industriegesellschaft*. München: Piper.

Jänicke, M. (1990) 'Erfolgsbedingungen von Umweltpolitik im internationalen Vergleich', *Zeitschrift für Umweltpolitik und Umweltrecht*, 3, pp. 213–232.

Jänicke, M. (1993) 'Über Ökologische und Politische Modernisierungen', *Zeitschrift für Umweltpolitik und Umweltrecht*, 2, pp. 159–175.

Jänicke, M. (1995) *The Political System's Capacity for Environmental Policy*, Berlin: Freie Universität.

Kleindorfer, P.R. and Orts, E.W. (1999) 'Informational Regulation of Environmental Risks', *Risk Analysis*, 18, pp. 155–170.

Lang, G. (2002) 'Deforestation, Floods and State Reactions in China and Thailand', pp. 195–220 in A.P.J. Mol and F.H. Buttel (eds), *The Environmental State under Pressure*. Amsterdam/New York, NY: Elsevier.

Latour, B. (1993) *We Have Never Been Modern*. Cambridge MA: Harvard University Press.

Liefferink, J.D., Lowe, P.D. and Mol, A.P.J. (eds) (1993) *European Integration and Environmental Policy*. London/New York, NY: Belhaven Press.

Mez, L., and Weidner, H. (1997) (eds) *Umweltpolitik und Staatsversagen. Perspektiven und Grenzen der Umweltpolitikanalyse*. Berlin: Ed Sigma.

Mol, A.P.J. (1995) *The Refinement of Production. Ecological Modernization Theory and the Chemical Industry*, Utrecht: Jan van Arkel/International Books.

Mol, A.P.J. (2001) *Globalization and Environmental Reform: The Ecological Modernization of the Global Economy*. Cambridge, MA/London: MIT Press.

Mol, A.P.J. (2006) 'From Environmental Sociologies to Environmental Sociology? A Comparison of US and European Environmental Sociology', *Organization & Environment*, 19, 1, pp. 5–27.

Mol, A.P.J. (2007) 'Boundless Biofuels? Between Environmental Sustainability and Vulnerability', *Sociologia Ruralis*, 47, 4, pp. 297–315.

Mol, A.P.J. (2008) *Environmental Reform in the Information Age. The Contours of Informational Governance*, Cambridge, UK/New York, NY: Cambridge University Press.

Mol, A.P.J. (2010) 'Sustainability as Global Attractor: The Greening of the 2008 Beijing Olympics', *Global Networks*, 10, 4, pp. 510–528.

Mol, A.P.J. (2011) 'China's Ascent and Africa's Environment', *Global Environmental Change*, 21, 3, pp. 785–794.

Mol, A.P.J. and Buttel, F.H. (eds) (2002) *The Environmental State Under Pressure*. Amsterdam: JAI.

Mol, A.P.J. and Sonnenfeld, D.A. (eds) (2000) *Ecological Modernisation Around the World. Perspectives and Critical Debates*. London and Portland: Frank Cass/Routledge.

Mol, A.P.J., and Spaargaren, G. (2000) 'Ecological Modernisation Theory in Debate: A Review', *Environmental Politics*. 9, 1, pp. 17–49.

Mol, A.P.J. and Spaargaren, G. (2004) 'Ecological Modernization and Consumption: A Reply', *Society and Natural Resources*, 17, pp. 261–265.

O'Connor, J. (1997) *Natural Causes: Essays in Ecological Marxism*. New York, NY: Guilford.

Oosterveer, P. (2007) *Global Governance of Food Production and Consumption*. Cheltenham, UK: Edward Elgar.

Oosterveer P., Guivant, J.S. and Spaargaren, G. (2007) 'Shopping for Green Food in Globalizing Supermarkets: Sustainability at the Consumption Junction', pp. 411–428 in J. Pretty, A. Ball, T. Benton, J. Guivant, D. Lee, D. Orr, M. Pfeffer and H. Ward (eds), *Sage Handbook on Environment and Society*. London: Sage.

Pellow, D.N., Weinberg, A.S. and Schnaiberg, A. (2000) 'Putting Ecological Modernization to the Test: Accounting for Recycling's Promises and Performance', *Environmental Politics*, 9, 1, pp. 109–137.

Pepper, D. (1984) *The Roots of Modern Environmentalism*. London: Croom Helm.

Pepper, D. (1999) 'Ecological Modernization or the "Ideal Model" of Sustainable Development? Questions Prompted at Europe's Periphery', *Environmental Politics*, 8, 4, pp. 1–34.

Presas, L.M. (2005) *Transnational Buildings in Local Environments*. Aldershot, UK: Ashgate.

Princen, T., Maniates, M. and Conca, K. (eds) (2002) *Confronting Consumption*. Cambridge, MA: MIT Press.

Rau, H. (2010) '(Im)mobility and Environment–Society Relations: Arguments for and Against the "Mobilisation" of Environmental Sociology', pp. 237–254 in M. Gross and H. Heinrichs (eds), *Environmental Sociology: European Perspectives and Interdisciplinary Challenges*. Dordrecht: Springer.

Schnaiberg, A. (1980) *The Environment: From Surplus to Scarcity*. Oxford, UK: Oxford University Press.

Schnaiberg, A., Weinberg, A.S. and Pellow, D.N. (2002) 'The Treadmill of Production and the Environmental State', pp.15–32 in A.P.J. Mol and F.H. Buttel (eds), *The Environmental State Under Pressure*. Amsterdam: JAI.

Schot, J. (1992) 'Constructive Technology Assessment and Technology Dynamics: The Case of Clean Technologies', *Science, Technology and Human Values*, 17, 1, pp. 36–56.

Shove, E. (2003) *Comfort, Cleanliness and Convenience: The Social Organisation of Normality*. New York, NY: Berg.

Smith, T., Sonnenfeld, D.A. and Pellow, D.N. (eds.)(2006) *Challenging the Chip: Labor Rights and Environmental Justice in the Global Electronics Industry*. Philadelphia: Temple University Press.

Sonnenfeld, D.A. (2002) 'Social Movements and Ecological Modernization: The Transformation of Pulp and Paper Manufacturing', *Development and Change*, 33, 1, pp. 1–27.

Sonnenfeld, D.A. (ed.) (2008) 'Symposium on "Globalisation and Environmental Governance: Is Another World Possible?"' *Global Environmental Change*, 18, p. 3.

Sonnenfeld, D.A. and Mol, A.P.J. (2006) 'Environmental Reform in Asia: Comparisons, Challenges, Next Steps', *Journal of Environment and Development*, 15, 2, pp. 112–137.

Sonnenfeld, D.A. and Mol, A.P.J. (eds) (2011) 'Symposium on "Social Theory and the Environment in the New World (Dis)Order"', *Global Environmental Change*, 21, p. 3.

Spaargaren, G. (2003) 'Sustainable Consumption: A Theoretical and Environmental Policy Perspective', *Society and Natural Resources*, 16, pp. 1–15.

Spaargaren, G. (2011) 'Theories of Practices: Agency, Technology, and Culture. Exploring the Relevance of Practice Theories for the Governance of Sustainable Consumption Practices in the New World-Order', *Global Environmental Change*, 21, 3, pp. 813–822.

Spaargaren, G. and Mol, A.P.J. (1992) 'Sociology, Environment and Modernity: Ecological Modernization as a Theory of Social Change', *Society and Natural Resources*, 5, 4, pp. 323–344.

Spaargaren, G. and Mol, A.P.J. (2008) 'Greening Global Consumption: Redefining Politics and Authority', *Global Environmental Change*, 18, 3, pp. 350–359.

Spaargaren, G., Mol, A.P.J. and Buttel, F.H. (eds) (2000). *Environment and Global Modernity*. London: Sage.

Spaargaren, G., Mol, A.P.J. and Buttel, F.H. (eds) (2006) *Governing Environmental Flows: Global Challenges to Social theory*. Cambridge, MA/London: MIT Press.

Spaargaren, G., Mol, A.P.J. and Sonnenfeld, D.A. (eds) (2009) *The Ecological Modernisation Reader: Environmental Reform in Theory and Practice*. London: Routledge.

Spaargaren, G., Oosterveer, P., van Buuren J. and Mol, A.P.J. (2005) *Mixed Modernities: Toward Viable Urban Environmental Infrastructure Development in East Africa*. Position Paper, Wageningen University.

Spaargaren, G. and van Koppen, C.S.A. (2010) 'Provider Strategies and the Greening of Consumption Practices: Exploring the Role of Companies in Sustainable Consumption', pp. 81–100 in H. Lange and L. Meier (eds), *The New Middle Classes: Globalizing Lifestyles, Consumerism and Environmental Concern*. London/New York, NY: Springer.

Urry, J. (2000) *Sociology Beyond Society*. London: Routledge.

van den Burg, S.W.K. (2006) 'Governance Through Information: Environmental Monitoring from a Citizen-Consumer Perspective', Ph.D. dissertation, Wageningen University.

van Koppen, C.S.A. (2006) 'Governing Nature? On the Global Complexity of Biodiversity Conservation', pp. 187–220 in G. Spaargaren, A.P.J. Mol and F.H. Buttel (eds), *Governing Environmental Flows. Global Challenges for Social Theory*. Cambridge, MA: MIT Press.

van Tatenhove, J., Arts, B. and Leroy, P. (eds) (2000) *Political Modernization and the Environment: The Renewal of Policy Arrangements*. Dordrecht: Kluwer.

van Tatenhove, J. and Pieter, L. (2003) 'Environment and Participation in a Context of Political Modernization', *Environmental Values*, 12, 2, pp. 155–174.

Wackernagel, M., and Rees, W. (1998) *Our Ecological Footprint: Reducing Human Impact on Earth*. Gabriola Island, BC, Canada: New Society Publishers.

WCED (World Commission on Environment and Development (1987) *Our Common Future*. Oxford, UK: Oxford University Press.

Weale, A. (1992) *The New Politics of the Environment*, Manchester, UK: Manchester University Press.

Weinberg, A.S., Pellow, D.N. and Schnaiberg, A. (2000) *Urban Recycling and the Search for Sustainable Community Development*. Princeton, NJ: Princeton University Press.

Wilson, D.C. (2002) 'The Global in the Local: The Environmental State and the Management of the Nile Perch Fishery on Lake Victoria', pp. 171–192 in A.P.J. Mol and F.H. Buttel (eds), *The Environmental State Under Pressure*. Amsterdam: JAI.

York, R., and Rosa, E.A. (2003) 'Key Challenges to Ecological Modernization Theory', *Organization and Environment*, 16, 3, pp. 273–287.

York, R., Rosa, E.A. and Dietz, T. (2003) 'A Rift in Modernity? Assessing the Anthropogenic Sources of Global Climate Change with the STIRPAT Model', *International Journal of Sociology and Social Policy*, 23, 10, pp. 31–51.

Young, S. (ed.) (2001) *The Emergence of Ecological Modernization*. London: Routledge.

3

The emergence of new world-systems perspectives on global environmental change[1]

Andrew K. Jorgenson and Jennifer E. Givens

The world is 'full' and 'unequal' (Milanovic 2005). This is scarcely a novel assertion and yet the implications remain fundamentally underappreciated. With roughly seven billion people and counting, and the consumption of energy and other natural resources at historically unprecedented levels, human societies are pushing up against the limits of global ecological systems. That the world remains vastly unequal in the wake of such pressures foreshadows the challenges facing the development prospects of the poorer countries – and it is an indication of the links between power, disparate economic development and patterns of global environmental change. World-systems analysis and emergent perspectives derived from it provide useful analytical tools for better understanding such complex society–nature relationships in their various manifestations (e.g. Goldfrank et al. 1999; Hornborg et al. 2007; Jorgenson and Kick 2006; Podobnik 2006; Roberts and Grimes 2002).

Broadly speaking, world-systems analysis (e.g. Chase-Dunn 1998; Wallerstein 1974) argues that a country's domestic social and environmental conditions must be understood within the context of the entire capitalist world economy, which is characterized by relational processes of unequal development that generate and reproduce a core–periphery hierarchy. The central concept here is that a country's economic development, levels of domestic income inequality, resource use, environmental harms and overall human well-being are not entirely a function of its own internal processes. Rather, they are at least partly a function of that country's location in a global economic hierarchy, an international division of labour comprised of three zones: a core, semi-periphery and periphery. The core of the world economy, which consists of what are commonly referred to as developed countries, is characterized by capital-intensive production and high-wage labour. Labour-intensive production and low-wage labour are located in the periphery. Semi-peripheral nations hold relatively intermediate positions in the stratified interstate system, possessing a combination of both core-like and periphery-like characteristics. For analytical and empirical reasons, peripheral and semi-peripheral nations are commonly combined into the category of less-developed countries or developing countries.

Early quantitative, comparative work on the environment from a world-systems perspective largely focused on how resource consumption and environmental degradation patterns tightly correlate with the stratified interstate system (e.g. Burns et al. 1994, 1997; Goldfrank et al. 1999;

Kick et al. 1996). That scholarship was foundational in that it provided clear, robust evidence of such society–nature patterns within the core–periphery hierarchy.[2] However, that work was limited by the lack of theoretical and empirical attention paid to how political–economic conditions, relations and processes within the contemporary world served as key mechanisms leading to the well-documented patterns of global environmental inequalities within the world-system.

In recent years, two strands of theory and cross-national quantitative research have emerged to help reconcile limitations of earlier world-systems scholarship on the environment and to advance our collective understanding of the human dimensions of global environmental change. These two emergent areas of macrosociological work, commonly referred to as the 'ecologically unequal exchange' and 'foreign direct investment (FDI) dependence and the environment' traditions, are the primary focus of this chapter. In the next section, we review the theory of ecologically unequal exchange and summarize related research that employs relational measures and model estimation techniques to test its fundamental propositions. Next, we review and summarize findings for the FDI dependence and the environment tradition. In a brief conclusion, key points are summarized, with an emphasis on the two emergent traditions.

Ecologically unequal exchange

Recognition of the substantive, structured ecological relations between countries is increasingly articulated through the theory of ecologically unequal exchange, a perspective that describes the unequal material exchange relations and consequent ecological interdependencies within the capitalist world economy, all of which are fundamentally tied to wide disparities in socio-economic development and power embedded within the world-system (e.g. Hornborg 1998a, 1998b, 2009; Jorgenson 2006a; Jorgenson and Clark 2009, 2011; Rice 2007a; Roberts and Parks 2007, 2009). 'Unequal exchange' can be broadly defined as the assertion of asymmetrical power relationships between more-developed and less-developed countries, wherein the former gain disproportionate advantages at the expense of the latter through patterns of trade as well as other structural relationships. The assertion of unequal exchange relations diverges from neoclassical economic thought by inquiring into the historical power relations shaping present comparative advantages rather than taking present comparative advantage as a given.

In turn, 'ecologically unequal exchange' refers to the environmentally damaging withdrawal of energy and other natural resource assets from and the externalization of environmentally damaging production and disposal activities within less-developed countries. It constitutes the obtainment of 'natural capital' (stocks of natural resources that yield important goods and services) and the usurpation of 'sink-capacity' (waste assimilation properties of ecological systems in a manner enlarging the domestic carrying capacity of more powerful developed countries) to the detriment of developing countries. It is, therefore, focused upon the manner and degree to which less-developed countries tend to fulfil a role in the world-system as a tap for the raw materials and a sink for the waste products of industrialized (and post-industrial) countries, thereby underwriting the disproportionate production–consumption–accumulation processes of more-developed countries.

The unequal geographical distribution of energy and other natural resources suggests that trade among nations is a necessity and can contribute to more efficient and productive utilization from a global perspective. And yet in the modern world economy many export-oriented, less-developed countries remain mired in poverty, having failed to exhibit the vertical and horizontal economic diversification and growth that should follow temporally from specialization in their comparative advantages (Mahutga 2006). A conundrum, moreover, underlies the juxtaposition between those countries exhibiting the greatest consumption of natural resources and those characterized by the greatest degradation or loss of natural resource assets: nations with the highest levels of natural resource consumption, principally the most industrialized and post-industrial countries, are typically characterized by the lowest domestic levels of environmental degradation (e.g. deforestation). In turn, the most intense natural resource degradation processes frequently beset the poorest countries in the world, those exhibiting minimal natural resource consumption demand. This inconsistency is referred to as the 'consumption/environmental degradation paradox' (e.g. Jorgenson 2003).

Bunker (1984, 1985) and Bunker and Ciccantell (2005) have crafted a body of comparative-historical work that is particularly responsive to the ecologically unequal connections forged through world-system processes. They illustrate the crucial role that reliable access to cheap natural resources has played in fuelling the rise of hegemonic powers within a given historical era. From their perspective, orthodox theories of development have insufficiently recognized the fundamental differences between the internal dynamics and logic of accumulation of extractive and productive economies, respectively. It is not extraction of natural resources and energy per se that promotes ecologically unequal exchange but the socio-organizational consequences this tends to produce between and within exporting and importing nations. The historical interactions between modes of extraction and production create path-dependent dynamics shaping the historical development trajectories of differentially situated countries.

Ecologically unequal exchange, therefore, is contingent upon differential cross-national social organization and accelerated production–consumption–accumulation linkages in the industrialized countries – facilitating the ability of state and private capital interests to determine global demand for natural resources (Bunker 1985; Hornborg 2011). Their capacity to control demand ensures that core interests engage in the substantive decisions regarding global export activity, and subjects less-developed countries to ever-changing market demands (Bunker 1985; Bunker and Ciccantell 2005). Local populations, social organization, infrastructure and ecosystems within peripheral extractive regions are often disrupted in the face of the malleable needs of core countries. Extractive regions failing to conform to core interests are likely to be subject to declining terms of trade or abandoned entirely in lieu of other, more accommodating locations. Differential cross-national social power, in turn, is based upon historically contingent exchange relationships forged through the ability to control asymmetrical flows of environmental resources and risks (Hornborg 2001).

These forms of transnational and international processes are also part and parcel of what McMichael (2008) refers to as the 'globalization project'. Pointing to a shift that began in the late 1970s and early 1980s, his analysis focuses on export-oriented production and the attraction of foreign direct investment as two related means that less-developed nations attempt to employ to stimulate economic development. Through formal and informal mechanisms, global institutions such as the World Bank and International Monetary Fund increasingly encourage such activities, which they promote as critically important for less-developed countries to establish positions in the increasingly integrated world economy (e.g. Babb 2005, 2009). In turn, the wealthier nations, by increasing consumption of manufactured products, agricultural goods and extracted materials, exacerbate environmental harms within developing countries. These constitute some of the key mechanisms underlying the ecologically unequal exchanges between developed and less-developed countries (e.g. Hornborg 1998a; Jorgenson 2006a; Rice 2007b).

The use of quantitative comparative methods to assess key assertions of ecologically unequal exchange theory is quite challenging.[3] As implied by the theory, the vertical flow of exports from lesser-developed nations to relatively more-developed nations is a structural mechanism through which ecologically unequal consequences are born and maintained. On the flipside, it also matters where human-caused waste is generated, and where it is ultimately sent for disposal.

Over the past few years, a number of sociologists have attempted to develop and employ appropriate measures in comparative international studies to test propositions derived from ecologically unequal exchange theory. Early on, Jorgenson (2004) designed a measure – referred to as 'weighted export flows' – that quantifies the relative extent to which a nation's exports are sent to more-developed countries. This weighted index, which involves the use of relational data (export flows between sending and receiving nations) and attributional data (levels of development of receiving nations), includes all primary sector and secondary sector exports. These data were used initially in a study of deforestation from 1990 to 2000, which concluded that the vertical flow of all exports contributes to forest degradation in less-developed countries, net of demographic and political–economic factors, including levels of exports and classic trade dependence measures (Jorgenson 2006a). In a related study, Jorgenson and Rice (2005) employ the same weighted index in a cross-sectional analysis of the per capita ecological footprints of less-developed countries. Results indicate that the vertical flow of exports suppresses consumption-based environmental demand within less-developed countries, many of which consume natural resources well below globally sustainable thresholds. Shandra et al. (2009b) use Jorgenson's (2006a) weighted export flows measure as one of multiple predictors in an analysis of industrial organic water pollution in developing nations. Net of other political–economic and human-ecological factors, they find that less-developed countries with relatively greater levels of exports sent to more economically developed nations exhibit higher per capita levels of industrial organic water pollutants emitted per day in the year 2000.

Rice (2007b) conducts a cross-sectional analysis of the per capita ecological footprints of nations for a sample of both developed and less-developed countries. To test key tenants of ecologically unequal exchange theory, Rice employs a measure that identifies the percentage of a given country's exports that are sent to core nations. While such a variable differs from the weighted export flows measures used by others, the general logic of the variable and its suitability for testing hypotheses is consistent with the former, and the two measures tend to be highly correlated (see Jorgenson 2012). Most notably, Rice (2007b) creates interactions between his trade measure and dummy variables for nations in differing income quartiles based on the World Bank's country income classifications (i.e. low, lower-middle, upper-middle and high income). Results of cross-sectional models that include the interactions suggest that low and lower-middle income countries characterized by a greater proportion of exports to the core countries exhibit lower per capita footprints relative to nations that are upper-middle and high income. These results are consistent with propositions of ecologically unequal exchange theory and with world-systems logic in general, and highlight non-trivial differences in such structural relationships for nations at varying levels of development.

Combined, the above studies suggest that the resource consumption/environmental degradation paradox is, to some extent, two sides of the same coin in the context of ecologically unequal relationships between more-developed and less-developed countries. However, these studies are limited in particular ways. Jorgenson's (2006a) deforestation analysis lacks temporal depth, and the use of a measure for all exports is problematic since research on forest degradation often emphasizes the relevance of trade in primary sector goods (e.g. Rudel 2005). Jorgenson and Rice (2005) conduct a cross-sectional analysis, which is indeed limited, yet the use of the weighted export flows measure for all commodity types is less problematic in this study since the dependent variable is the comprehensive per capita ecological footprint. The same cross-sectional limitations apply to Rice (2007b).

In part to address weaknesses in prior empirical work, Jorgenson et al. (2010a) employ a weighted flows measure for only primary sector goods in analyses of deforestation in less-developed countries from 1990 to 2005. Findings reveal a strong association between forest degradation and the vertical flow of primary sector exports (see also Austin 2010 and Jorgenson 2010). Likewise, Shandra et al. (2009a) employ a weighted exports flow measure for primary sector commodities in an analysis of threatened mammals in less-developed countries in 2005. Their export measure is similar to that used in Rice (2007a) in that it quantifies the percentage of a sending nation's primary sector exports that goes to OECD countries, most of which are core nations. Consistent with the theory of ecologically unequal exchange, the results of their negative binomial regression model estimates indicate that numbers of threatened mammals in poor nations are positively associated with flows of primary sector exports to rich nations.

To help resolve the temporal limitations of prior cross-sectional studies, Jorgenson (2009b) employs more rigorous methods in panel analyses of the vertical flow of exports and the per capita ecological footprints of less-developed countries from 1975 to 2000. Results confirmed a negative association between the overall consumption-based environmental demands per person in developing countries and the flow of total exports from those countries to relatively more-developed nations. The association increased in magnitude over the entire twenty-five-year period, suggesting that these relationships became more ecologically unequal through time. Jorgenson et al. (2009) employ a similar approach, employing a weighted export flows measure for only primary sector commodities in panel analyses of deforestation and a refined, primary sector-oriented ecological footprint – known as the cropland, grazing land and timber footprint (CGT footprint) – for a sample of less-developed countries from 1970 to 2000. Like the preceding studies, their export flows measure is weighted by the levels of economic development of receiving countries. Jorgenson et al. (2009) estimate fixed and random effects models for both outcomes, allowing for more rigorous hypothesis testing. Consistent with prior research, they found that increases in the vertical flow of primary sector exports contributes to increases in forest degradation and concomitant suppression of the CGT footprint. The results hold, net of various controls and across both types of panel model estimations.

In perhaps one of the most thorough cross-national analyses in the ecologically unequal exchange tradition to date, Jorgenson and Clark (2009) integrate the tradition with two contemporary theories in environmental sociology: the treadmills of production and destruction, respectively. Treadmill of production theory focuses on how an economic system driven by endless growth, on an ever larger scale, generates widespread ecological degradation (Gould et al. 2008). Treadmill of destruction theory suggests that the military has its own expansionary dynamics, which involve significant environmental and ecological costs (Hooks and Smith 2005; Jorgenson et al. 2010a). Jorgenson and Clark (2009) argue that the ecologically unequal exchange perspective intersects with both treadmill orientations. The treadmill of production propels the world economy toward constant expansion, demanding more and more resources to meet its insatiable appetite, especially in the articulated consumer markets of developed countries. Similarly, in the interests of national security, technological innovation, political power and geopolitical influence, the treadmill of destruction facilitates the increased consumption of resources by the nations' militaries and their supporting sectors. As suggested by world-systems scholars, increased military strength enhances access to the natural resources and sink capacity of less-powerful, underdeveloped nations (e.g. Chase-Dunn 1998).

Jorgenson and Clark (2009) argue that the populations of more-developed *and* militarily powerful countries are positioned advantageously in the contemporary world economy, and thus more likely to secure and maintain favourable terms of trade allowing for greater access to the natural resources and sink capacity of bioproductive areas within less-developed countries. These advantageous positions facilitate the externalization of environmental costs of resource extraction and consumption to less-developed countries and help create conditions where more-developed countries and those with more powerful militaries are able to overutilize global environmental space. The misappropriation of global environmental space suppresses resource consumption opportunities for the populations of many less-developed countries. Given the structure and acceleration of both the treadmill of destruction and treadmill of production, it is quite likely that the consequences of these processes for less-developed countries are more pronounced than for more-developed countries, and that they increase through time.

To test their arguments and assess the extent to which these perspectives intersect in meaningful and empirically valid ways, Jorgenson and Clark (2009) create and employ two export flows measures. One is weighted by the levels of economic development of receiving countries, the other by military expenditures per soldier of receiving countries. The two export flows measures are treated as predictors in panel analyses of the ecological footprints of nations from 1975 to 2000. Most notably, the results of their panel model estimates indicate that countries with relatively higher levels of exports sent to economically developed and militarily powerful nations experience suppressed consumption levels, and these effects – that are independent of one another – are especially pronounced and increasingly so for the less-developed countries, many of which consume resources well below globally sustainable thresholds. In other words, both forms of structural relationships between nations have become increasingly unequal in ecological contexts.

All of the scholarship discussed so far in this chapter focuses on how the structure of international trade in ecologically unequal contexts contributes to increased likelihood of environmental degradation in less-developed countries as well as the suppression of resource consumption for domestic populations in developing nations, often well below globally sustainable limits. The latter also contributes to the well-being of these populations, which underscores the complexity of resource use and human health associations (e.g. Rice 2008). At the same time, it is argued that ecologically unequal exchange relationships between nations are likely to contribute to an increase in the production and dumping of various forms of environmental waste in less-developed countries. Such waste is the result of the 'off-shoring' of environmentally intensive manufacturing and the externalizing of the post-consumption disposal costs associated with manufactured goods for the articulated markets in more-developed core countries. The focus here is on the international treatment of less-developed countries as 'sinks' for waste. While such unequal structural relationships and consequences are central to the theory of ecologically unequal exchange, comparative international research on these topics is limited. Three notable exceptions in sociology are Roberts and Parks (2007), Stretesky and Lynch (2009) and Jorgenson (2012), all of which focus on how ecologically unequal relationships contribute to anthropogenic carbon dioxide emissions. Given the scientific consensus on the role such emissions play in global climate change, these studies are far more than academic exercises.

Roberts and Parks (2007) estimate cross-sectional models of total emissions, per capita emissions, emissions per unit of production and cumulative emissions per capita for a large sample of nations. The first three outcomes are measured in 1999, while the fourth measure is a cumulative score for the 1950 to 1999 period. Net of multiple controls, they find that nations with a greater reliance on the export of manufactured goods have higher levels of all four types of emissions. Even though Roberts and Parks (2007) do not include relational measures of trade, their results are generally consistent with the arguments of ecologically unequal exchange theory. Stretesky and Lynch (2009) take a different approach that includes relational measures. Their cross-national panel analyses for the 1984 to 2004 period assess the extent to which a reliance on exports to the United States relative to other nations contributes to growth in carbon dioxide emissions. Results indicate that relatively higher levels of reliance on exports to the United States do contribute to growth in per capita carbon emissions. This study highlights the importance of focusing on specific trade relationships and their potential ecologically unequal consequences. Though Stretesky and Lynch do not situate their research in an unequal exchange theoretical framework, they nonetheless make notable methodological and substantive contributions to that literature. Jorgenson (2012) conducts a longitudinal analysis of per capita carbon dioxide emissions for a large sample of developed and less-developed nations for the 1960 to 2005 period. Due to data availability challenges, he employs the same trade measure as Rice (2007b) to evaluate the propositions of ecologically unequal exchange theory. He finds that the effect of a country's proportion of exports to developed nations on per capita emissions is positive and much larger in magnitude for less-developed countries than for developed countries. Furthermore, and of particular note, in models for a sample restricted to less-developed countries, Jorgenson finds the magnitude of the effect of the vertical flow of exports on per capita emissions increases through time, providing evidence of increasingly ecologically unequal relationships between developed and less-developed countries. In the context of greenhouse gas emissions (Jorgenson 2012) and resource consumption levels (Jorgenson and Clark 2009), relationships between more-developed and less-developed nations appear to have become increasingly ecologically unequal.

The theory of ecologically unequal exchange has arguably become central in certain areas of academic discourse about society–nature relationships in international contexts, and the perspective has deep analytical roots in world-systems analysis. Comparative-historical work illustrates the long-term systemic processes that create and maintain conditions permitting such exchanges to occur between nations. Through quantitative studies, scholars using this approach have developed and employed measurements to test certain propositions of the theory, particularly those concerned with the environmental consequences of the upward, vertical flow of exports from less-powerful, less-developed nations to the more powerful and developed nations of the Global North. Further studies will refine such approaches and apply them to additional questions. Of particular need is for scholars to design measures that allow for testing propositions in comparative international analyses of the other side of the ecologically unequal exchange coin: the extent to which the flow of hazardous materials and other forms of waste from the Global North to the Global South impacts the environment and well-being of human populations in the latter (see Pellow 2007). In line with Jorgenson and Clark (2009), future research on ecologically unequal exchange relationships also needs to consider forms of power other than economic and, like Rice (2008) suggests, scholars working in this tradition would do well to pay closer attention to the human well-being consequences of such international and global dynamics.

This chapter turns now to a discussion of foreign investment dependence and the environment, a second emerging area of socio-environmental scholarship with deep roots in the world-systems tradition.

Foreign investment dependence and the environment

During recent decades, many less-developed countries experienced a deepening of foreign debt, resulting in austerity measures developed by global governance and finance institutions, in turn characteristics of the aforementioned 'globalization project' (McMichael 2008). These austerity measures, such as structural adjustment programmes, encourage the governments of indebted countries to create more favourable conditions for foreign investors and transnational corporations. At the heart of the austerity measures is the assumption that attracting foreign capital will stimulate economic development, thus assisting in debt repayment and increasing the overall well-being of domestic populations. It is often suggested (e.g. OECD 1999) that the longer-term benefits of foreign investment will outweigh any short-term environmental harms and human well-being costs, and that attracting foreign investment might eventually lead to more environmentally-friendly forms of extraction and production in different sectors through technology transfers and spillover effects. It is anticipated, further, that investing transnational firms may be more likely to install and employ environmentally friendly technologies than their counterparts in host countries. Often referred to as the 'pollution halo hypothesis', this proposition is commonplace in economics (e.g. Birdsall and Wheeler 1993; Cole et al. 2008). Empirical support for the pollution halo hypothesis is lacking, however, especially in macro-comparative contexts (see Hoffman et al. 2005; Letchumanan and Kodama 2000; Perkins and Neumayer 2009).

Partly in an effort to attract foreign investment and transnational enterprises, many less-developed countries implement relaxed labour laws and tax reductions as well as exemptions to environmental regulations designed to protect the natural environment from activities in different sectors of the economy (Leonard 1988; McMichael 2008).[4] The real or perceived threat of capital flight could be viewed as an additional incentive for less-developed countries to offer regulatory concessions to foreign capital. Further, prior research shows that some less-developed countries are less likely than many developed countries to ratify international environmental treaties, many of which deal explicitly with extractive and productive activities that are of direct relevance for transnational corporations (Roberts and Parks 2007). At least partly resulting from these unfolding political–economic processes, in general, the relative presence of foreign investment stocks for all economic sectors combined within less-developed countries increased substantially during recent decades (Chase-Dunn and Jorgenson 2007), and prior research shows that international debt and the implementation of structural adjustment programmes in less-developed countries do contribute to increases in inward foreign direct investment (Shandra et al. 2003).

With the above conditions in mind, and contrary to the pollution halo hypothesis, some world-systems influenced scholars have argued that a large proportion of secondary sector foreign direct investment in less-developed countries finances highly polluting and environmentally unfriendly manufacturing processes and facilities, much of which are outsourced from developed countries (e.g. Grimes and Kentor 2003; Jorgenson 2007; Jorgenson et al. 2007). Transnational manufacturing firms often experience economic benefits from this form of environmental cost-shifting since ecologically harmful manufacturing methods tend to include relatively outdated and inexpensive mechanization processes and equipment. This also allows transnational firms to further distance themselves in the public eye from the environmental impacts of their activities.[5] Moreover, the transportation vehicles owned and operated by foreign-owned manufacturing centres in less-developed countries for the movement of inputs, outputs and labour are often outdated, energy-inefficient and thus more polluting (Jorgenson et al. 2007). What is more, the 'on-the-ground' transportation infrastructure of many less-developed countries tends to be poorly maintained relative to developed countries. For example, roadways are less likely to be paved on a regular basis, and rail systems are more likely to be spotty in different areas. Such conditions can lead to the increased use of fossil fuels by transnational firms for the transportation of raw materials, manufactured goods, and labour (Grimes and Kentor 2003). In line with these arguments, comparative international research indicates that within less-developed countries, manufacturing sector FDI positively affects growth in carbon dioxide emissions (e.g. Grimes and Kentor 2003; Jorgenson 2009a; York 2008) as well as other greenhouse gases (Dick and Jorgenson 2010; Jorgenson 2006b) and industrial organic water pollution (Jorgenson 2009b). These results hold regardless if the emissions or pollution outcome are measured by scale, intensity or per unit of production.

In a related vein, scholars in this tradition have empirically demonstrated that foreign investment in the primary sector (i.e. agriculture, forestry, mining) commonly finances agricultural activities, forestry operations and extractive enterprises that contribute to the degradation of forested areas in less-developed countries (e.g. Jorgenson 2008; Jorgenson et al. 2011). As agricultural enterprises are integrated into the world economy, especially those owned by transnational firms, the scale and intensity of their production tends to increase dramatically (Harper and Le Beau 2003). To augment production, forest areas are cleared in a variety of ways, including the burning of biomass and the use of tractors as well as other types of machinery (Altieri 2000). Forest areas are also cleared for export-oriented livestock operations, many of which are controlled by foreign capital (Burns et al. 1994; Jorgenson and Birkholz 2010). Forms of capital-intensive agriculture can deplete the soil of nutrients, which leads to further expansion and concomitant deforestation (Magdoff et al. 2000).

Many less-developed countries, especially those with relatively larger forest areas, are prime locations for logging operations, and indebted countries are often encouraged to utilize their natural resources, including forested areas, as a form of comparative advantage to attract foreign capital (McMichael 2008). Thus, like agriculture, forestry in general and logging operations in particular have gradually become transnationally organized and globally distributed. The extraction of minerals and other raw materials are the starting points for most global production systems, and transnational firms are key actors in these primary sector activities (Bunker and Ciccantell 2005). Mining activities are carried out in a series of stages, each of which are potentially detrimental to forested areas (Rudel 2005).

Comparative international research in this arena also links primary sector FDI in less-developed countries to growth in the use of synthetic pesticides and fertilizers in agricultural production (e.g. Jorgenson and Kuykendall 2008; see also Frey 1995). In the contemporary world economy, much of the production of agricultural goods is globally distributed and largely controlled by transnational corporations headquartered in developed countries (McMichael 2008). Harriett Friedmann (1990) refers to the rising oligopolistic control of global food production as the 'world food order'. As farming systems in less-developed countries are integrated into the world economy, often through the influence and control of transnational corporations and foreign capital, crop rotation and recycling of organic matter is more likely to be replaced by the high-intensity use of pesticides and synthetic fertilizers (Altieri 2000).[6] Transnational corporations investing in or directly operating capital-intensive agriculture within less-developed countries are principal customers for pesticides and fertilizers, some of which are banned in developed countries, but provide potential markets for their producers in locales with fewer environmental protection barriers (Magdoff et al. 2000). What is more, in a treadmill-like fashion, the use of such agrochemicals often increases through time as more are needed to maintain or increase crop yields, and longitudinal research shows a strong correlation in less-developed countries between increases in primary sector FDI and heightened use of both pesticides and synthetic fertilizers (Jorgenson and Kuykendall 2008).

Overall, research results generally supports the world-systems-influenced propositions that increased FDI in primary and secondary sectors contributes to a concomitant increase in environmental problems in less-developed, investment-dependent nations. Considering the globalization and outsourcing of service industries, future research needs to examine the potential environmental impacts of tertiary sector FDI (e.g. tourism, telecommunications and financial and legal services) in less-developed countries as well. Further, emerging studies in the world society tradition within sociology and its sister disciplines suggest that various civil society and institutional factors, such as environmental international non-governmental organizations and national environmental ministries, are capable of mitigating – to some extent – the environmental impacts of foreign direct investment in less-developed countries (e.g. Jorgenson 2009c; Jorgenson et al. 2011; see also Schofer and Hironaka 2005). Future research on the environmental impacts of any form of FDI would do well to consider such institutional and civil society factors and their mitigating capabilities.

Conclusion

This chapter has highlighted two emergent lines of comparative analysis on society–nature relationships, both of which are influenced by foundational world-systems and environment works of earlier decades. First, ecologically unequal exchange theory and related research considers how particular aspects of international trade – especially the vertical flow of exports – allow more powerful developed nations to externalize at least some of their environmental impacts to less-developed nations, leading to increases in various forms of environmental harms in the latter, including deforestation, loss of biodiversity, industrial water pollution and greenhouse gas emissions. Resource consumption levels of the less-developed countries often remain suppressed well below globally sustainable thresholds, with significant public health implications. Evidence indicates increasingly unequal relationships through time. Second, scholarship on foreign investment dependence and the environment empirically demonstrates that increased levels of foreign investment within less-developed countries, rather than leading to the spread of cleaner technologies and production processes in the secondary and primary sectors, often are linked to dirtier forms of extraction and production, various forms of environmental harms and subsequent costs to human well-being. Theoretical explanations for these relationships are complex, involving various historical and relational processes within the world-system.

While these two emergent areas of scholarship increase our understanding of human dimensions of global environmental change and address limitations of past studies on the relationships between the contemporary capitalist world-system and the environment, like all areas of social scientific inquiry, they have limitations as well. Most notably, work in the ecologically unequal exchange tradition has yet to adequately examine how the structure of international exchanges allows wealthier and more powerful nations to export hazardous waste to less-developed and less-powerful nations. Though much needed, in-depth case studies in this tradition are few and far between. The latter limitation applies as well to scholarship on foreign investment dependence and the environment. In addition, that area of study has yet to focus on possible environmental impacts of foreign direct investments in the services sector, an increasingly important area of investment in many less-developed countries. The two emerging traditions of scholarship on world-systems and the environment will be further enriched as they address these limitations in coming years.

Notes

1 Portions of this chapter also appear in Jorgenson and Rice (2012) and Jorgenson and Winitzky (2012).
2 A decade ago, Roberts and Grimes (2002) provided a thorough discussion of the need to study environmental and ecological disruptions from a world-systems perspective. Their essay reviews the foundational bodies of research that identified strong correlations between various environmental harms and world-system position.
3 Recent historical analyses (e.g. Clark and Foster 2009) and in-depth case studies (e.g. Hornborg 2006) focus on the underlying mechanisms that collectively shape and reinforce the broader structural relationships identified by the quantitative inquiries.
4 However, such exemptions are less common now than in past decades.
5 However, partly due to the pressures of social justice and environmental justice groups, major corporations such as Apple and Nike are increasingly being held accountable for the behaviours and practices of their facilities, contractors and subcontractors.

6 Of additional importance, the use of pesticides and fertilizers in agricultural production is linked to a variety of human health and environmental problems. See Jorgenson and Kuykendall (2008) for an extended discussion of the known public health and environmental problems associated with their use.

References

Altieri, M. (2000) 'Ecological Impacts of Industrial Agriculture and the Possibilities for Truly Sustainable Farming', pp.77–92 in F. Magdoff, J. Foster and F. Buttel (eds), *Hungry for Profit: The Agribusiness Threat to Farmers, Food, and the Environment*. New York, NY: Monthly Review Press.

Austin, K. (2010) 'The Hamburger Connection as Ecologically Unequal Exchange: A Cross-National Investigation of Beef Exports and Deforestation in Less-Developed Countries', *Rural Sociology*, 75, 2, pp. 270–299.

Babb, S. (2005) 'The Social Consequences of Structural Adjustment: Recent Evidence and Current Debates', *Annual Review of Sociology*, 31, pp. 199–222.

Babb, S. (2009) *Behind the Development Banks: Washington Politics, World Poverty, and the Wealth of Nations*. Chicago, IL: University of Chicago Press.

Birdsall, N. and Wheeler, D. (1993) 'Trade Policy and Industrial Pollution in Latin America: Where are the Pollution Havens?' *Journal of Environment and Development*, 2, 1, pp. 137–147.

Bunker, S.G. (1984) 'Modes of Extraction, Unequal Exchange, and the Progressive Underdevelopment of an Extreme Periphery: The Brazilian Amazon, 1600–1980', *American Journal of Sociology*, 89, 5, pp. 1017–1064.

Bunker, S.G. (1985) *Underdeveloping the Amazon: Extraction, Unequal Exchange, and the Failure of the Modern State*. Urbana, IL: University of Illinois Press.

Bunker, S.G. and Ciccantell, P.S. (2005) *Globalization and the Race for Resources*. Baltimore, MD: Johns Hopkins University Press.

Burns, T., Davis, B. and Kick, E. (1997) 'Position in the World-System and National Emissions of Greenhouse Gases', *Journal of World-Systems Research*, 3, 3, pp. 432–466.

Burns, T., Kick, E., Murray, D. and Murray, D. (1994) 'Demography, Development, and Deforestation in a World-System Perspective', *International Journal of Comparative Sociology*, 35, 3–4, pp. 221–239.

Chase-Dunn, C. (1998) *Global Formation: Structures of the World-Economy*. Lanham, MD: Rowman and Littlefield.

Chase-Dunn, C. and Jorgenson, A. (2007) 'Trajectories of Trade and Investment Globalization', pp. 165–184 in I. Rossi (ed.), *Frontiers of Globalization Research*. New York, NY: Springer.

Clark, B. and Foster, J.B. (2009) 'Ecological Imperialism and the Global Metabolic Rift: Unequal Exchange and the Guano/Nitrates Trade', *International Journal of Comparative Sociology*, 50, 3–4, pp. 311–334.

Cole, M., Elliott, R. and Strobl, E. (2008) 'The Environmental Performance of Firms: The Role of Foreign Ownership, Training, and Experience', *Ecological Economics*, 65, 3, pp. 538–546.

Dick, C. and Jorgenson, A. (2010) 'Sectoral Foreign Investment and Nitrous Oxide Emissions: A Quantitative Investigation', *Society and Natural Resources*, 23, 1, pp. 71–82.

Frey, R.S. (1995) 'The International Traffic in Pesticides', *Technological Forecasting and Social Change*, 50, 2, pp. 151–169.

Friedmann, H. (1990) 'The Origins of Third World Food Dependence', pp. 13–31 in B. Bernstein, B. Crow, M. McIntosh and C. Martin (eds), *The Food Question: Profits vs People*. New York, NY: Monthly Review Press.

Goldfrank, W., Goodman, D. and Szasz, A. (eds) (1999) *Ecology and the World-System*. Westport, CT: Greenwood Press.

Gould, K., Pellow, D. and Schnaiberg, A. (2008) *The Treadmill of Production: Injustice and Unsustainability in the Global Economy*. Boulder, CO: Paradigm Publishers.

Grimes, P. and Kentor, J. (2003) 'Exporting the Greenhouse: Foreign Capital Penetration and CO_2 Emissions 1980–1996', *Journal of World-Systems Research*, 9, 2, pp. 261–275.

Harper, C. and LeBeau, B. (2003) *Food, Society, and Environment*. Upper Saddle River, NJ: Prentice Hall.

Hoffman, R., Chew Sing, L., Ramasamy, B. and Yeung, M. (2005) 'FDI and Pollution: A Granger Causality Test Using Panel Data', *Journal of International Development*, 17, 3, pp. 311–317.

Hooks, G. and Smith, C. (2005) 'Treadmills of Production and Destruction: Threats to the Environment Posed by Militarism', *Organization and Environment*, 18, 1, pp. 19–37.

Hornborg, A. (1998a) 'Ecosystems and World Systems: Accumulation as an Ecological Process', *Journal of World-Systems Research*, 4, 2, pp. 169–177.

Hornborg, A. (1998b) 'Towards an Ecological Theory of Unequal Exchange: Articulating World System Theory and Ecological Economics', *Ecological Economics*, 25, 1, pp. 127–136.

Hornborg, A. (2001) *The Power of the Machine: Global Inequalities of Economy, Technology, and Environment*. New York, NY: AltaMira Press.

Hornborg, A. (2006) 'Footprints in the Cotton Fields: The Industrial Revolution as Time–Space Appropriation and Environmental Load Displacement', *Ecological Economics*, 59, 1, pp. 74–81.

Hornborg, A. (2009) 'Zero-Sum World: Challenges in Conceptualizing Environmental Load Displacement and Ecologically Unequal Exchange in the World-System', *International Journal of Comparative Sociology*, 50, 3–4, pp. 237–262.

Hornborg, A. (2011) *Global Ecology and Unequal Exchange: Fetishism in a Zero-Sum World*. New York, NY: Routledge.

Hornborg, A., McNeill, J.R. and Martinez-Alier, J. (eds)(2007) *Rethinking Environmental History: World-System History and Global Environmental Change*. Lanham, MD: Alta Mira Press.

Jorgenson, A.K. (2003) 'Consumption and Environmental Degradation: A Cross-National Analysis of the Ecological Footprint', *Social Problems*, 50, 3, pp. 374–394.

Jorgenson, A.K. (2004) 'Export Partner Dependence and Environmental Degradation, 1965–2000'. Ph.D. dissertation, Department of Sociology, University of California, Riverside.

Jorgenson, A.K. (2006a) 'Unequal Ecological Exchange and Environmental Degradation: A Theoretical Proposition and Cross-National Study of Deforestation, 1990–2000', *Rural Sociology*, 71, 4, pp. 685–712.

Jorgenson, A.K. (2006b) 'Global Warming and the Neglected Greenhouse Gas: A Cross-National Study of Methane Emissions Intensity, 1995', *Social Forces*, 84, 3, pp. 1777–1796.

Jorgenson, A.K. (2007) 'Does Foreign Investment Harm the Air We Breathe and the Water We Drink? A Cross-National Study of Carbon Dioxide Emissions and Organic Water Pollution in Less-Developed Countries, 1975–2000', *Organization and Environment*, 20, 2, pp. 137–156.

Jorgenson, A.K. (2008) 'Structural Integration and the Trees', *Sociological Quarterly*, 49, 3, pp. 503–527.

Jorgenson, A.K. (2009a) 'The Transnational Organization of Production, the Scale of Degradation and Ecoefficiency: A Study of Carbon Dioxide Emissions in Less-Developed Countries', *Human Ecology Review*, 16, 1, pp. 64–74.

Jorgenson, A.K. (2009b) 'The Sociology of Unequal Exchange in Ecological Context: A Panel Study of Lower-Income Countries, 1975–2000', *Sociological Forum*, 24, 1, pp. 22–46.

Jorgenson, A.K. (2009c) 'Foreign Direct Investment and the Environment, the Mitigating Influence of Institutional and Civil Society Factors, and Relationships between Industrial Pollution and Human Health: A Panel Study of Less-Developed Countries', *Organization and Environment*, 22, 2, pp. 135–157.

Jorgenson, A.K. (2010) 'World-Economic Integration, Supply Depots, and Environmental Degradation: A Study of Ecologically Unequal Exchange, Foreign Investment Dependence, and Deforestation in Less-Developed Countries', *Critical Sociology*, 36, 3, pp. 453–477.

Jorgenson, A.K. (2012) 'The Sociology of Ecologically Unequal Exchange and Carbon Dioxide Emissions, 1960–2005', *Social Science Research*, 41, pp. 242–252.

Jorgenson, A.K., Austin, K. and Dick, C. (2009) 'Ecologically Unequal Exchange and the Resource Consumption/Environmental Degradation Paradox: A Panel Study of Less-Developed Countries, 1970–2000', *International Journal of Comparative Sociology*, 50, 3–4, pp. 263–284.

Jorgenson, A.K. and Birkholz, R. (2010) 'Assessing the Causes of Anthropogenic Methane Emissions in Comparative Perspective, 1990–2005', *Ecological Economics*, 69, pp. 2634–2643.

Jorgenson, A.K. and Clark, B. (2009) 'The Economy, Military, and Ecologically Unequal Relationships in Comparative Perspective: A Panel Study of the Ecological Footprints of Nations, 1975–2000', *Social Problems*, 56, 4, pp. 621–646.

Jorgenson, A.K. and Clark, B. (2011) 'Societies Consuming Nature: A Panel Study of the Ecological Footprints of Nations, 1960–2003', *Social Science Research*, 40, 1, pp. 226–244.

Jorgenson, A.K. and Clark, B. (2012) 'Are the Economy and the Environment Decoupling? A Comparative International Study, 1960–2005', *American Journal of Sociology*, 118, 1, pp. 1–44.

Jorgenson, A.K., Clark, B. and Kentor, J. (2010a) 'Militarization and the Environment: A Panel Study of Carbon Dioxide Emissions and the Ecological Footprints of Nations, 1970–2000', *Global Environmental Politics*, 10, 1, pp. 7–29.

Jorgenson, A.K., Dick, C. and Austin, K. (2010b) 'The Vertical Flow of Primary Sector Exports and Deforestation in Less-Developed Countries: A Test of Ecologically Unequal Exchange Theory', *Society and Natural Resources*, 23, 9, pp. 888–897.

Jorgenson, A.K., Dick, C. and Mahutga, M. (2007) 'Foreign Investment Dependence and the Environment: An Ecostructural Approach', *Social Problems*, 54, 3, pp. 371–394.

Jorgenson, A.K., Dick, C. and Shandra, J. (2011) 'World Economy, World Society, and Environmental Harms in Less-Developed Countries', *Sociological Inquiry*, 81, 1, pp. 53–87.

Jorgenson, A.K. and Kick, E. (eds)(2006) *Globalization and the Environment*. Leiden: Brill.

Jorgenson, A.K. and Kuykendall, K. (2008) 'Globalization, Foreign Investment Dependence, and Agriculture Production', *Social Forces*, 87, 1, pp. 529–560.

Jorgenson, A.K. and Rice, J. (2005) 'Structural Dynamics of International Trade and Material Consumption: A Cross-National Study of the Ecological Footprint of Less-Developed Countries', *Journal of World-Systems Research*, 11, 1, pp. 57–77.

Jorgenson, A. K. and Rice, J. (2012) 'The Sociology of Ecologically Unequal Exchange in Comparative Perspective', pp. 431–439 in S. Babones and C. Chase-Dunn (eds), *Routledge Handbook of World-Systems Analysis*. London: Routledge.

Jorgenson, A. K. and Winitzky, J. (2012) 'The Environmental Impacts of Foreign Direct Investment in Less-Developed Countries', pp. 445–448 in S. Babones and C. Chase-Dunn (eds), *Handbook of World-Systems Analysis: Theory and Research*. London: Routledge.

Kick, E., Burns, T., Davis, B., Murray, D. and Murray, D. (1996) 'Impacts of Domestic Population Dynamics and Foreign Wood Trade on Deforestation: A World-System Perspective', *Journal of Developing Societies*, 12, 1, pp. 68–87.

Leonard, J. (1988) *Pollution and the Struggle for the World Product*. Cambridge, UK: Cambridge University Press.

Letchumanan, R. and Kodama, F. (2000) 'Reconciling the Conflict Between the "Pollution Haven Hypothesis and an Emerging Trajectory of International Technology Transfer', *Research Policy*, 29, 1, pp. 59–79.

McMichael, P. (2008) *Development and Social Change: A Global Perspective*. Thousand Oaks, CA: Pine Forge Press.

Mahutga, M. (2006) 'The Persistence of Structural Inequality? A Network Analysis of International Trade, 1965–2000', *Social Forces*, 84, 4, pp. 1863–1889.

Magdoff, J., Foster, J.B. and Buttel, F. (2000) *Hungry for Profit: The Agribusiness Threat to Farmers, Food, and the Environment*. New York, NY: Monthly Review Press.

Milanovic, B. (2005) *Worlds Apart: Measuring International and Global Inequality*. Princeton: Princeton University Press.

Organisation for Economic Co-operation and Development (OECD) (1999) *Foreign Direct Investment and the Environment*. Paris: OECD Publications.

Pellow, D. (2007) *Resisting Global Toxins: Transnational Movements for Environmental Justice*. Cambridge, MA: MIT Press.

Perkins, R. and Neumayer, E. (2009) 'Transnational Linkages and the Spillover of Environment-Efficiency into Developing Countries', *Global Environmental Change*, 19, 3, pp. 375–383.

Podobnik, B. (2006) *Global Energy Shifts: Fostering Sustainability in a Turbulent Age*. Philadelphia, PA: Temple University Press.

Rice, J. (2007a) 'Ecological Unequal Exchange: Consumption, Equity and Unsustainable Structural Relationships within the Global Economy', *International Journal of Comparative Sociology*, 48, 1, pp. 43–72.

Rice, J. (2007b) 'Ecological Unequal Exchange: International Trade and Uneven Utilization of Environmental Space in the World System', *Social Forces*, 85, 3, pp. 1369–1392.

Rice, J. (2008) 'Material Consumption and Social Well-Being within the Periphery of the World Economy: An Ecological Analysis of Maternal Mortality', *Social Science Research*, 37, 4, pp. 1292–1309.

Roberts, J.T. and Grimes, P. (2002) 'World-System Theory and the Environment: Toward a New Synthesis', pp. 167–196 in R. Dunlap, F. Buttel, P. Dickens and A. Gijswijt (eds), *Sociological Theory and the Environment: Classical Foundations, Contemporary Insights*. Lanham, MD: Rowman and Littlefield.

Roberts, J.T. and Parks, B. (2007) *A Climate of Injustice: Global Inequality, North–South Politics, and Climate Policy*. Cambridge, MA: MIT Press.

Roberts, J.T. and Parks, B. (2009) 'Ecologically Unequal Exchange, Ecological Debt, and Climate Justice: The History and Implications of Three Related Ideas for a New Social Movement', *International Journal of Comparative Sociology*, 50, 3–4, pp. 385–409.

Rudel, T. (2005) *Tropical Forests: Regional Paths of Destruction and Regeneration in the Late Twentieth Century*. New York, NY: Columbia University Press.

Schofer, E. and Hironaka, A. (2005) 'The Effects of World Society on Environmental Outcomes', *Social Forces*, 84, 1, pp. 25–47.

Shandra, J., Leckband, C., McKinney, L. and London, B. (2009a) 'Ecologically Unequal Exchange, World Polity, and Biodiversity Loss: A Cross-National Analysis of Threatened Mammals', *International Journal of Comparative Sociology*, 50, 3–4, pp. 285–310.

Shandra, J., Ross, R. and London, B. (2003) 'Global Capitalism and the Flow of Foreign Direct Investment to Non-Core Nations, 1980–1996', *International Journal of Comparative Sociology*, 44, 3, pp. 199–238.

Shandra, J., Shor, E. and London, B. (2009b) 'World Polity, Unequal Exchange, and Organic Water Pollution: A Cross-National Analysis of Less Developed Nations', *Human Ecology Review*, 16, 1, pp. 51–64.

Stretesky, P. and Lynch, M. (2009) 'A Cross-National Study of the Association Between Per Capita Carbon Dioxide Emissions and Exports to the United States', *Social Science Research*, 38, 1, pp. 239–250.

Wallerstein, I. (1974) *The Modern World System*, vol 1., *Capitalist Agriculture and the Origins of the European World-Economy in the Sixteenth Century*. New York: Academic Press.

York, R. (2008) 'De-Carbonization in the Former Soviet Republics, 1992–2000: The Ecological Consequences of De-Modernization', *Social Problems*, 55, 3, pp. 370–390.

4

China's economic growth and environmental protection

Approaching a 'win–win' situation? A discussion of ecological modernization theory[1]

Dayong Hong, Chenyang Xiao and Stewart Lockie

Modern China is one of the main engines of global economic growth. A three-decade-long trend of annual GDP growth of around 9 per cent was maintained through the global financial crisis 2007–2012 and residents' average incomes, consumption and savings continue to improve (NBS 2011). Thus, China's per capita GDP increased from 379 yuan per annum in 1978 to 29,920 yuan in 2010. Per capita disposable income for urban households grew from 343 to 19,109 yuan over the same period and for rural households from 134 to 5,919 yuan. This is reflected in consumption, which rose from 405 yuan per annum in 1978 to 15,907 yuan in 2010 for urban Chinese and from 138 to 4,455 yuan for rural Chinese. Household savings over this period increased by a factor of 1,439. Living standards have risen and many Chinese, positively, have been lifted out of poverty (NBS 2011). However, questions remain as to the ecological and public health costs of such rapid and sustained economic growth. Even as growth has slowed slightly in coastal centres such as Beijing and Shanghai in response to government efforts to control inflation and unemployment, and achieve more balanced income distribution, etc., economic growth in the vast central and western regions of China has accelerated.

The Chinese government, like many others, has adopted the language of 'sustainable development' and 'sustainable growth' to reconcile economic growth with environmental protection. Legislative and institutional changes in support of sustainability are outlined in more detail below. According to the *2010 State of the Environment Report in China*, some significant progress has been made. For example, reductions in total national emissions of chemical oxygen demand and sulphur dioxide of 12.5 and 14.3 per cent were achieved between 2005 and 2010, exceeding goals established in *The Eleventh Five-Year Plan For National Economic and Social Development of the People's Republic China* (MEP 2011a). In 2010, 69 per cent of urban respondents and 58 per cent of rural respondents reported that they were 'satisfied' or 'somewhat satisfied' with their surrounding environment (9.8 and 10.3 per cent more than in 2009) (MEP 2011a). There is some evidence to suggest, therefore, that China's economic growth and environmental protection are approaching a 'win–win' situation. In this chapter, we present a detailed analysis of this win–win trend and identify what we believe are hidden threats that need special attention. We also examine the relevance and implications of ecological modernization theory (EMT) – a product of Western environmental sociology – for analysing and guiding environmental reform in China.

Chinese environmental governance steadily strengthening

Some parallels can be drawn between growth in the Chinese economy and growth in instruments and institutions of environmental governance. Dutch scholar Arthur Mol (2010) identifies four elements in the creation and transformation of Chinese environmental governance. These include: the political modernization of the 'environmental state'; the role of economic actors and market dynamics; the emerging civil-society institutions; and processes of international integration. Mol (2010) concedes that, while growing, the contribution of civil society organizations to reform of environmental governance is not yet significant. However, the consequence of global economic integration, he argues, is increasing international scrutiny of environmental standards imposed, under the centralized state system of China, by government. We will focus here, therefore, on the role of the Chinese government in expanding environmental governance through policy and planning, legislation, institutional development and expenditure.

When opening of the Chinese economy commenced at the end of 1970s, it was underpinned by the ideology that 'only development counts'. However, the Chinese government soon issued a number of statements linking the modernization of China with appropriate use of natural resources, pollution control and ecological protection. In 1992, the Chinese government issued *Report of Environment and Development in People's Republic of China*, outlining its basic stand on sustainable development and, in 1994, *The White Book on China's Agenda in the 21st Century – China's Population, Environment and Development in the 21st Century*, establishing the strategic framework for sustainable development. The Fifth Planetary Session of the 14th CPC Central Committee 1995 further emphasized the need for population control, resource preservation and economic growth that is coordinated with the environment and natural resources to form a benign cycle. In practice, implementation of the sustainable development strategy was characterized by a bias towards economic growth.

Following a review of progress in 2002, a new *Program of Action for Sustainable Development in China in the Early 21st Century* was released in 2003. Also released in 2003, the *Decision of the Central Committee of the Communist Party of China on Some Issues Concerning the Improvement of the Socialist Market Economy* set out a new conceptual framework summarized as the 'Scientific Viewpoint of Development'. This viewpoint rejects the prevailing thinking on development that puts materials first but neglects people. Instead, it seeks to embed a 'people first' principle, adjusting the mode of economic growth to recognize both that development depends on people and that the benefits of development ought be shared by all members of society. The Scientific Viewpoint of Development emphasizes comprehensiveness, coordination, sustainability and the balancing of five key 'aspects' of development. These include balancing urban and rural development, balancing development among regions, balancing economic and social development, balancing development of man and nature and balancing domestic development with the opening of China to the outside world.

The National People's Congress and its Chairmen's Council have enacted nine environmental protection laws and fifteen natural resource protection laws since 1949 (MEP 1995–2011; SCIO 1996, 2006). Since 1996, the National Congress has passed or updated a series of laws on water pollution control, protection of the oceanic environment, air pollution control, noise pollution control, solid waste pollution control, environmental impact assessment, and radioactive pollution treatment, as well as other environmental protection-related laws and legislations on water, clean production, renewable energy, agriculture, grassland and livestock. By the end of 2010, China had passed 61 national and a total of nearly 400 provincial and local environmental protection bills, as well as more than 1,000 local administrative regulations. China now has a comprehensive environmental legal system, which has the *Constitution of People's Republic of China* as the base, *Environmental Protection Act of People's Republic of China* as the main body, and specialized national laws of environmental protection, related national laws of natural resources, administrative regulations and rules of environmental protection, as well as local environmental protection regulations, as the content. China has also established national and local systems of standards for environmental protection. National environmental protection standards include national environmental quality standards, national pollution emission (control) standards, national environmental standard sample standards, and others. Local environmental protection standards include local environmental quality standards and local pollution emission standards. China has also established a series of assessment goals, or targets, which are set out in the eleventh (2006–2010) and twelfth (2011–2015) five-year key plans for national economy and social development.[2]

Starting from scratch in the 1970s, China has developed a comprehensive system of environmental protection institutions from the national level all the way through to local townships. In 2008, the National Bureau of Environmental Protection was upgraded to become the Ministry of Environmental Protection of China, a part of the National Council. By 2010, there were a total of 12,849 environmental protection agencies nationwide, including 43 national, 371 provincial, 1,937 regional-municipal, 8,606 county-level, and 1,892 township-level units, with a total of 193,900 personnel. From 1996 to 2010, the number of environmental protection agencies grew 3.8 per cent annually, and the number of personnel increased 7.4 per cent annually, which is close to China's annual economic growth rate (MEP 1995–2011). More importantly, the Chinese government has continuously increased expenditure on environmental pollution control. Based on data compiled from multiple years of *National Statistical Communiqué on the Environment and NBS's Report* (MEP 1995–2011; NBS 2008), in 1981, such government expenditure was 2.5 billion yuan; accounting for only 0.51 per cent of the year's GDP. By 2010, expenditure on environmental protection had risen to 665.5 billion yuan; accounting for 1.67 per cent of GDP. Of this total, urban environmental infrastructure construction took 422.4 billion yuan, industrial pollution source control cost 39.7 billion yuan and water conservation features embedded in main construction projects used the remaining 203.3 billion yuan.

According to a number of indicators, China has shown signs of improvement in environmental conditions. As mentioned at the beginning of this chapter, emissions of both chemical oxygen demand and sulphur dioxide have fallen (MEP 2011a). Further, although waste water discharge has increased, ammonia-nitrogen emissions have decreased significantly since 2006. Nationwide, 73 per cent of municipal waste water was treated in 2010, up nearly 10 percentage points from 2009. In 2010, 95 per cent of industrial waste water discharge met environmental protection standards, an increase of 1 percentage point from 2009. Regarding air pollution, smog emissions peaked in 2005 at nearly 12 million metric tonnes and have been steadily decreasing since to just over 8 million metric tonnes in 2010. Emissions of industrial dust have halved from a peak of 9 million metric tonnes in 2005. Also, in 2010, 92 per cent of industrial sulphur dioxide emissions, 91 per cent of industrial fume emissions, 91 per cent of industrial dust emissions met standards and 88 per cent of industrial ammonia-nitrogen emissions met environmental protection standards. In terms of solid waste control, the national total amount of industrial solid waste increased steadily from 887 million tonnes in 2001 to 2.4 billion tonnes in 2010. However, the amount of multi-purpose use of solid waste also increased significantly, reaching 1.6 billion tonnes in 2010, a 17 per cent increase on the previous year (MEP 2011b).

Technological advancement, economic restructuring, and the strengthening of energy-conserving and emission-reduction policies have seen China's economic transformation accompanied by steadily decreasing energy consumption per GDP unit. In 1990, for every 100 million yuan of GDP, energy consumption was nearly 53,000 standard tonnes of coal. By 2010, this had decreased to only 8,000 standard tonnes of coal (NBS 2011). The government-issued *2010 State of the Environment Report in China* thus strikes an optimistic tone:

> Environmental protection has made the important transition from awareness to practice, and has entered the main line, primary battle field, and grand stage of economic and social developments. The task of pollution reduction has been accomplished beyond expectation, environmental quality is steadily improving; it appears that the whole society's environmental protection awareness has generally strengthened.
>
> *(MEP 2011a: Overview)*

Under the guidance of the Scientific Viewpoint of Development, the Chinese government now walks a strategic path that attempts to combine five aspects of development into a coherent whole: economic, political, cultural, social and ecological civilization. Such a path will not only strengthen the win–win interplay of economic growth and environmental protection, but also promote environmental protection as an independent social value that both parallels other social values and exerts its own normative effect on other social values. As such, it seems that in China there is evidence for the emergence of an 'ecological modernization' process as depicted by some Western scholars. Before assessing arguments for the ecological modernization of China in more detail, we will critically evaluate evidence for the 'greening' of the Chinese economy.

Threats to win–win outcomes for environment and economy

We believe that China ought to be congratulated for progress towards a win–win relationship between economic growth and environmental protection. However, we also believe that the trend towards a win–win relationship ought to be examined critically. Closer examination, we argue, shows this trend to be fragile and indistinct, and the path to a more robust relationship between economy and environment to be littered with obstacles. The need to do more, further, is recognized by the Chinese people. The 2010 Chinese General Social Survey (CGSS) found that 69 per cent of respondents believed environmental problems in China are serious. Only 38 per cent agreed that central government was doing its best and making progress towards environmental protection. An even lower percentage (27 per cent) believed local governments were making progress. We must therefore examine carefully those data that suggest an improvement in the environment.

First, the reliability of official environmental protection data needs qualification. China's environmental statistics are built upon environmental monitoring work from different levels of monitoring stations. Each level of stations is organized and coordinated by the main environmental protection agencies at the same level, and only receives professional guidance from higher levels of stations. The operational budgets of these stations are part of the local governmental budget at the same level. As a result, all data published by environmental monitoring stations are filtered by local governments. Such a system creates a possibility that monitoring stations may succumb to local interest and manipulate data. Moreover, most basic monitoring work is carried out by monitoring stations at the municipal and county levels. These stations (especially in regions that are relatively underdeveloped) tend to have limited personnel, equipment, working space and operational budgets, weakening their capacity to monitor environmental quality and pollution sources for control purposes in their corresponding regions. Monitoring is consequently lacking in both depth and breadth, rendering the accuracy and comprehensiveness of data questionable.

Second, the standards adopted in environmental monitoring have a direct effect on the reporting of environmental quality. For instance, raising or lowering the index of permissible pollutant density in a given area could create the impression of improving or deteriorating environmental quality. For example, using the current standard for air quality, the *2010 State of the Environment Report in China* report concluded that 'air quality in cities is good nationwide and is better than last year' (MEP 2011a: Air). However, if the new standard of Particulate Matter 2.5 ($PM_{2.5}$) to be implemented in 2016 were applied, according to MEP Vice Minister Xiaoqing Wu in a news conference, then two-thirds of all cities nationwide would not meet requirements for air quality (China News Center 2012).

Third, scale effects and the distribution of pollution, etc. across space are not necessarily captured by aggregate data. For example, at the national level, China's air and water quality may have improved. However, this apparent improvement is concentrated in eastern urban areas, while air and water quality has deteriorated in mid and western rural areas. Investment in environmental protection is clearly aimed at cities and industrial centres via urban environmental infrastructure, industrial pollution source control, etc. Nevertheless, the deterioration of rural environments is attracting more attention. The *First National General Survey of Pollution Sources* (2010) and *2010 State of the Environment Report in China* (MEP 2011a) acknowledge that rural pollution sources are becoming relatively more important. These include both non-point sources of pollution, such as livestock, and point sources, such as mining and industrial facilities. As pollution controls tighten in urban areas, there has been a trend to relocate polluting activities to rural areas with less effective controls (MEP 2011a). Sun (2012: 4) argues that:

> Due to historical shortage of investment in rural environmental protection, there are 40,000 townships and 600,000 administrative villages that have virtually nothing on pollution control.

Fourth, environmental monitoring is focused on a narrow range of biophysical indicators (air, water, solid waste, etc.) and little consideration is given to other issues implicated in the material flows of industrial production and the consumer society they support. Material flows play a key role in the interplay of the natural environment and modern society, raising concerns among Chinese people about a plethora of issues such as genetically modified crops, food safety, medicine, building standards, indoor environments, household consumption. However, assessment and debate over risks associated with such material flows is not yet considered in government's environmental monitoring and reporting. We believe that environmental sociologists have a contribution to make here in examining and highlighting the materialization of human living and its environmental impact.

In addition to doubts about official environmental data, we also call for caution in assessing changes to China's mode of economic development. Although the national government has taken steps to slow economic growth and improve the quality of the economy, this is resisted in various parts. Many provincial units in China had economic growth in 2011 higher than the national average of 9 per cent; some had it higher than 13 per cent. This is fuelled both by inter-regional economic inequalities within China and by international competition and imperatives to maintain relatively rapid growth. In China, we can thus observe a duality. On the one hand, the developed eastern regions are beginning to seek better economic quality and to actively slow their growth rates in order to achieve it. On the other, the less developed central and western regions are still pursuing the maximum possible growth rate. Such a duality may give an impression that China's overall economy is slowing down and improving its quality, but this does not mean that there has been a fundamental change in the mode of economic development across China.

China advocates a path to a green and circular economy via technological advancements and improved management so as to reduce energy and resource consumption in the long run. As pointed out above, China's energy consumption per GDP unit has been steadily declining over the past twenty years. This has made a substantial contribution to the reduction of relative greenhouse gas emissions. However, total energy consumption (and thus total emissions) have risen rapidly, from 987 million tonnes of standard coal in 1990 to 3.3 billion tonnes in 2010 (NBS 2011). The consumption of other resources, and hence impacts on the environment, have been similarly expanding. In other words, even though China's economy has improved its efficiency, the ever-expanding size and total resource consumption has not reduced pressure on the environment and is not likely to, any time soon. An alternate possibility, in theory, would be economic growth in less material and energy intensive industries such as the services sector. To date, expansion of the relative contribution made by services to China's GDP (up from 24 per cent in 1978 to 43 per cent in 2010) has been at the cost of agriculture (down from 28 per cent to 10 per cent) (NBS 2011). The share contributed by industrial production has remained relatively steady at around 45 per cent. Slowing industrial growth – despite its contribution to pollution – would be very difficult for China. Industrial production is a major contributor to GDP and to employment. Adjustments to the national economic structure will lead to regional adjustments and redistributions. In fact, the slowing of economic growth in the east is less the result of technological and economic transformations than it is the result of polluting industries relocating to middle and western rural areas in pursuit of less environmental regulation and cheaper labour. For the foreseeable future, therefore, it seems likely that environmental pressures will continue to build and we will see an even more materialized Chinese society.

The Chinese government, further, faces a number of structural limitations in relation to its ability to contain environmental pressures (Hong 2008). The first of these relates to its double functions of promoting economic growth and environmental protection. Despite official discourses of balanced and sustainable development, China has low per capita incomes and wealth when compared to Western developed nations. The 'dream of a strong country' ideology is extremely powerful and there is tremendous pressure on the Chinese government to maintain economic growth and to expand employment opportunities. Official interpretation of the Scientific Viewpoint of Development sets 'development' as the top priority, yet China's development has created conspicuous environmental pollution. This pollution, along with a range of resource constraints, now threatens continued growth of the Chinese economy while the international community is increasingly concerned about the impacts China's development on the global environment (e.g. Mol 2011). The Chinese government must therefore be both a 'developmentalist government' and an 'environmental government', as China's leaders have realized. However, in reality, it is very difficult to play both roles well, due to technological, institutional, political and other domestic and international social constraints. To date, compromise has been achieved by favouring economy at one time and place, and favouring environmental protection in another time and place (as evident in the urban-to-rural relocation of polluting industries). Obviously, such a strategy will only have limited effects on promoting substantive, comprehensive and system-wide environmental governance.

The next structural limitation stems from differentiation between the Chinese central government and local governments. Chinese government is not a coherent and solid institutional whole, as many believe, and there are multiple conflicting interests within and between levels of governments. Practically speaking, the central government's control over local government has gradually weakened since reform and opening up began in the late 1970s. Reform has given local governments more flexibility and incentives to promote development, in the process stimulating competition among local governments to attract and retain development. This has been one driving force for China's rapid economic growth. In fact, some local governments have been directly engaged in coalitions with business enterprises to pursue economic interests. Relatively speaking, China's central government plays a more distant role and focuses on longer-term development for the entire nation. Because of this, China's central government is more active in promoting environmental protection and sustainable development, initiates more environmental work and is perceived more positively by the public in this respect. However, the institutional arrangements for environmental protection place substantive responsibility for much environmental regulation on the shoulder of local governments. Local environmental protection agencies report to local governments, which are generally more concerned with economic development, and the central government's environmental protection policies are rarely, as a consequence, fully implemented or enforced. Local governments often do little more than pay lip service to environmental policies and regulations, finding 'innovative' ways to interpret these so as to avoid conflict with economic goals. As the common saying goes, 'There are policies from the top, then there are counter-policies from the bottom.'

Increasingly tight global economic and political connections, along with domestic pressures for political reform, expose the Chinese government to a growing number of constraints. The Chinese government has signed many bilateral and multilateral agreements regarding environmental protection and showed willingness to honour these agreements. At the same time, environmental awareness among the public in China is strengthening and a meaningful civil society is coming into being, which also increases expectations on government for social and environmental reform (Hong 2008). The Chinese government has taken a number of steps to encourage greater public participation in environmental protection. In 2006, for example, the *Act of Public Participation in Environmental Impact Assessment* provided details on how the public can obtain relevant information and then formally take part in environmental impact assessment processes. In 2008, an even more important regulation was issued, *The Act of Environmental Information Disclosure (trial)*, which mandates that government needs to disclose environmental information to the public. Despite these pressures and responses, the Chinese government also expresses its disagreement with some international treaties for the sake of maintaining sovereignty, upholding the right to economic development, and to selectiveness and flexibility in the actual practices of government. Reform of the overall political system of China is slow and the public has limited opportunity and capacity to directly participate in environmental protection and to effectively check government's activities. The Chinese government still has great freedom, which tends to detract from its environmental protection responsibility and goals. It is especially important to note that, in this market economy, some environmental protection agencies and their personnel even use the name of environmental protection to seek personal benefit via illegal actions such as demanding bribery, damaging the image both of government and of environmental protection.

Ecological modernization and China

On the surface, the pursuit of a win–win relationship between environmental protection and economic development in China has some synergy with the concept of ecological modernization as proposed by environmental sociologists. Ecological modernization theory (EMT) asserts that not only is environmental sustainability potentially compatible with industrialization, technological advancement and economic growth, but that the latter can, in fact, become important factors and mechanisms to promote environmental sustainability. EMT argues, in other words, that one of the solutions to environmental problems generated by industrialization is a kind of 'super-industrialization' (as opposed to deindustrialization) based on ecological principles such as recycling, energy conservation or substitution of hazardous materials (Simonis 1989; Spaargaren and Mol 1992). Overcoming environmental crises, from this perspective, is not about a retreat from modernization but about a new kind of modernization. German sociologist Joseph Huber (one of the founders of EMT) argues that there are three stages in the development of an industrial society: (1) industrial breakthrough; (2) the establishment of an industrial society; and (3) the ecological transformation of the industrial system via superindustrialization. The third stage is made possible by the invention and implementation of new technologies (Huber 1985). Whether or not this is inevitable, as implied by Huber, is a moot point. Nonetheless, the focus of ecological modernization theory has become the growing prominence of ecological perspectives in modern society (Mol 1995).

EMT can be used both analytically and normatively. Analytically, EMT focuses on the characteristics and processes of adjustment that modern industrial society makes, based on ecological principles, to production, consumption, national practices and political discourses. In this sense, it is an empirical sociology of environmental reform. Normatively, EMT promotes a particular approach to environmental governance based on superindustrialization, the coordination of ecology and economy and the self-interest of business enterprises able to reduce costs arising from waste and environmental degradation, maintain access to natural resource, and build markets for new green products and services (Mol 1995).

Mol and Sonnenfeld (2000: 6–7) identify five key social and institutional transformations associated with the process of ecological modernization. These five transformations have been the core of much scholarship analysing the ecological modernization process and stage of a nation or region. They include:

1. the changing role of science and technology, which are more valued for their role in curing and preventing environmental problems;
2. the increasing importance of market dynamics and economic agents in environmental reform;
3. the move from top-down decision making by nation states to more decentralized, flexible and consensual styles of environmental governance and the provision of more opportunities for non-state actors to participate in environmental governance;
4. modifications in the position, role and ideology of social movements, which are increasingly institutionalized and involved in public and private decision making regarding environmental reforms; and
5. the emergence of new ideologies, such that complete neglect of the environment and the fundamental counter-positioning of economic and environmental interests are no longer accepted as legitimate positions.

Prior to the mid-1990s, ecological modernization research was conducted primarily in a few Western European nations that shared several common characteristics in terms of their economy, society and politics (such as welfare states, advanced industrial technologies, state-coordinated market economies and popular environmental awareness). A persistent question has been whether EMT can thus be generalized. Different nations have their own political cultures, governments and social institutions. The form and degree of institutionalization of environmental problems thus varies, leading to different processes and results of environmental reform. Davidson and Frickel (2004: 477) argue that:

> For every empirical study supportive of the potential for ecological modernization, there are now a number of empirical analyses that raise numerous caveats regarding the propensity for industry actors to undergo the 'greening' process of their own accord, particularly when we move beyond the advanced countries of Western Europe.

This highlights two persistent sticking points in the literature critical of EMT (e.g. Buttel 2000): first, whether enthusiasm for ecological modernization (the normative approach) may obscure the manipulation by polluters of incremental reform (i.e. 'greenwashing'); second, whether the particular transformations identified by ecological modernization theorists represent the only ways in which economy and environment may be brought together. Proponents insist that EMT is a dynamic and constantly evolving perspective that can usefully be applied to various economic, cultural, political and geographical settings around the world. Mol and Sonnenfeld (2000: 11) argue:

> the approach and tools of Ecological Modernization Theory are useful for social scientific analysis and policy formation, even where all conditions for development of ecologically modern institutions do not yet exist. At the same time, some processes of ecological modernization are global (even while others are not), and thus this body of theory remains at least partially relevant around the world.

As the largest developing nation in the world, China's rapid economic growth and accompanying environmental problems naturally attract attention from foreign scholars, including ecological modernization theorists (e.g. Liu and Diamond 2005; Mol 2010; Mol and Carter 2006). Mol and Carter (2006) highlight a range of environmental reforms in China that seem fairly consistent with EMT. The environmental state has become significantly more powerful; political leaders have a clearer awareness of environmental crises and have issued specific promises regarding environmental governance; market dynamics increasingly reflect the full price of natural resources and certain environmental externalities; the legal system for environmental protection has been strengthened; investment for environmental pollution control is rapidly increasing; there are more opportunities and space for the public to participate and cooperate with the government; and there are more effective regulations that mandate environmental information disclosure. Lei Zhang et al. (2007) directly apply the analytical framework of ecological modernization to China's efforts to reduce the consumption of natural resources and energy in economic development via improved efficiency, to strongly encourage the development of renewable energy industry in order to reduce pollution, and to develop a recycling economy. Applications of a similar analytical perspective are increasingly common among Chinese scholars (see CCMR 2008; PTCMS 2007).

We do not question the importance of understanding processes and implications of environmental reform. In the remainder of this section we raise several questions concerning the utility of ecological modernization theory as it has been developed in Europe to the analysis of environmental reform in China, and we will suggest corresponding ways in which EMT might develop to better inform such analysis.

The first question concerns the units of analysis in EMT. Empirically, it is possible to investigate ecological modernization at the level of one enterprise, one industry, one administrative unit, one region, one nation, etc. Further, it is possible to study these units at one point in time, over years or over decades. Regardless of the units chosen, care must be taken in drawing conclusions about ecological modernization more generally. Improvement, for example, in a local environment does not necessarily mean that environmental quality is improving at a larger scale. Conversely, improvement at a regional or national scale does not mean that all local environments are improving. Urban–rural pollution transfers in China illustrate that what appears to be ecological modernization at the national and some regional levels may be at the cost of environmental deterioration elsewhere. This is not unique to China. The apparent dematerialization of some wealthy countries is based in large part on the displacement of production to developing countries in the Asia-Pacific region (Schandl et al. 2011). At the same time, many of the gains in resource-use efficiency within wealthy countries have been offset by rising levels of material consumption. As a consequence of these two trends, global resource-use efficiency over the last decade has actually declined (Schandl et al. 2011). In an era of deep globalization, we must take a systemic and global environmental perspective to the question of ecological modernization. The so-called ecological modernization of some regions (both within and outside China) has not occurred in isolation from other regions. On the basis of material and waste transfers we suggest, further, that this apparent ecological modernization is most likely temporary and superficial and will not necessarily lead to global ecological modernization.

The second question concerns the normative inclination of ecological modernization. Analytical constructs developed through EMT studies are often applied as indexes with which to evaluate changes in, for example, technology, market, state, social movements. EMT scholars develop these indexes based on empirical data for Western Europe where associations have been found between, for example, relatively deregulated markets, democratic institutions and environmental protection. Therefore, the more in common a particular national context has with Western Europe, economically and politically, the more ecologically modernized it appears to be. This ignores (or at least neglects) the possibility of multiple paths and models of ecological modernization (in the broad sense of reconciling economic and environmental objectives), and it especially misperceives the effects of government and market. As stated above, environmental reforms in China have been driven largely by the central government. It cannot be discounted, therefore, that further decentralizing or weakening the nation state will similarly weaken environmental governance. China's environmental reforms, further, have been based on upholding socialist institutions. This political framework has not been adequately dealt with by EMT scholars. Certainly, EMT theorists argue that capitalism is 'neither … an essential precondition for, nor … the key obstruction to, stringent or radical environmental reform' (Mol and Spaargaren 2000: 23). Nevertheless, the centrality of market dynamics and economic agents in EMT largely presupposes free market capitalism and theorists appear to favour free market capitalism for its ability to self-evolve and to 'be accountable'. We argue here it should not be assumed that indexes or policy prescriptions derived from the economic and political transformations of capitalist economies in Western Europe will uncover or promote those transformations of most importance to China. Care must be taken to avoid turning the normative thrust of EMT into an unreflexive extension of Eurocentrism.

The third question involves the relationship between social equity and ecological modernization. EMT emphasizes the positive environmental effects of social institutional transformations associated with modernity but pays inadequate attention to the consequences of modernity: consequences including technological risks and wealth-based social stratification and environmental injustice. Stratification persists at the group, regional, national and global levels. It is a frequent precipitating factor for environmental deterioration and a prominent barrier to environmental reform. If the benefit of development cannot be shared by all people, then such development cannot be sustainable. Disparities in wealth and access to resources at multiple levels fuel developmental competition and increase pressure on the environment. Entering the twenty-first century, the Chinese government began to put significant effort into promoting social development, social justice and equity such that the benefits of development can be shared by all people. Although we do not yet have direct evidence to show that these efforts have assisted in the implementation of environmental reforms, we believe that they will, in the long run, be proven a necessary precondition for meaningful ecological modernization. Looking beyond China, this raises important questions for EMT theorists concerning the negative consequences of modernity and mechanisms to promote global social fairness.

The fourth question concerns social conflict and its implications for ecological modernization. Looking at existing practices, we believe that environmental reform is far from the smooth self-evolution of modern social institutions implied by EMT. Nor have discourses of sustainability, corporate responsibility, green capitalism etc. rendered alternative ideological frames unacceptable or illegitimate in the political sphere. On the contrary, environmental values and standards are frequently seen as subservient to, or in competition with, other values and standards. Possibilities for environmental reform remain embedded in specific political and economic structures and constrained by a complicated web of interests and power relations. Specific reforms face conflict and resistance that go beyond merely different interpretations of how best to pursue the goal of sustainability. In China, for example, there are a great number of medium and small businesses, as well as some industries, that are not ready for ecological modernization, but which provide employment and livelihoods for many people. These businesses and industries are concentrated in the rapidly growing mid- and west regions. If ecological modernization is forced onto these businesses, passive or even active resistance is inevitable. We argue, therefore, that to implement ecological modernization in China nationwide would most likely create a new green versus non-green dualist social structure that will divide people, industries and regions. China will not become a homogeneous ecologically modernized society but an even more unequal society. Globally, we also see resistance to the perceived forcing of environmental reform on developing countries and concern that agreements on, for example, greenhouse gas emissions strengthen the existing advantages enjoyed by developed nations in the North.

The final question concerns extension of the concept 'environment' in EMT. Early articulations of ecological modernization were concerned with protection of the 'external environment' from human activities by mimicking ecological processes in human systems. Modern industrial production, however, has transformed human society's material environments into various material flows embedded in social systems. In other words, the external environment has been internalized as a key component of human society. Modern society is highly materialized and flows of material and energy have thus become crucial mediums between human society and what was once known as the external environment. EMT, therefore, needs to expand the concepts of 'environment' and 'environmental reform' such that the whole materialization process of human life and the 'internalized' material world receive due attention. The importance of such a task is acknowledged by EMT scholars such as Mol (2010: 71), who now advocates a refocusing of EMT research as an 'environmental sociology of networks and flows' to recognize that globalization, in short, will change the nature of environmental reform (see also Mol and Spaargaren 2006). However, we argue that it cannot be assumed that this change will be positive. Nor can be assumed that the institutional and political differences between nation states will become inconsequential. It is critical to begin examination of the deepening overmaterialization of human society (an overmaterialization that, to date, technological advances and environmental reforms have failed to reverse) and to take a holistic view of humans' lifestyles, consuming behaviours and ideologies.

Conclusion

China's pursuit of socialist modernization seems to have generated a win–win trend regarding economic growth and environmental protection. The Chinese government has engaged in deep reflection over its developmental ideology, and is actively reforming it, with important effects. However, there are numerous grounds for pessimism over environmental reform both within China and globally. Before either endorsing or dismissing Chinese attempts to harmonize economic and environmental management, more research is needed; specifically, research that might inform processes of what we might broadly call ecological modernization without imposing Eurocentric normative or political agendas. The goals of environmental protection, social equity and poverty alleviation are fundamental to the construction of an ecological civilization in China. Theoretically, the case of China can be linked to ecological modernization theory as developed within Western environmental sociology and there is much to be learned in China from the experience of environmental reform elsewhere. However, we call for more in-depth reflections on EMT based on the analysis of China and caution against simply transferring research findings from Europe to observe, analyse and plan China's environmental reform.

Notes

1 This chapter presents initial results of the 2008 key project, 'Environmental Sociological Research on Ecological Civilization', funded by National Social Science Foundation of China (08ASH001), and held by Dayong Hong.
2 The eleventh five-year plan aimed to keep the total national population under 1.36 billion, maintain the total area of arable land at 130 million hectares, reduce energy consumption per GDP unit by 20 per cent and water consumption per unit of industrial added value by 30 per cent, increase the rate of multipurpose use of industrial solid waste to 60 per cent, reduce the emission of main pollutants (chemical oxygen demand and sulphur dioxide) by 10 per cent and increase forest coverage to 20 per cent. These goals have proved fairly effective in regulating governmental and main economic activities. The twelfth five-year plan aimed, in addition, to increase the index of water efficiency of agricultural irrigation to 0.53 and the proportion of non-fossil fuel in the primary energy consumption to 11.4 per cent, reduce energy consumption per GDP unit by 16 per cent and carbon dioxide emission per GDP unit by 17 per cent, reduce emissions of ammonia-nitrogen and NOx by 10 per cent, raise forest coverage to 21.66 per cent and increase forest volume by 600 million cubic metres (see *The Eleventh Five-Year Plan for National Economic and Social Development of the People's Republic of China* and *The Twelfth Five-Year Plan for National Economic and Social Development of the People's Republic of China*.

References

Buttel, F.H. (2000) 'Ecological Modernization as Social Theory', *Geoforum*, 31, 1, pp. 57–65.
Center for China's Modernization Research (CCMR), Chinese Academy of Sciences (eds) (2008) *China's Ecological Modernization Strategy*. Beijing: China Environmental Science Press.
China News Center (2012) 'MEP: Two-Thirds of Cities Cannot Meet Stricter Air Quality Standards', 2 March 2012. Available at www.chinanews.com/shipin/2012/03-02/news58335.shtml.
Davidson, D.J. and Frickel, S. (2004) 'Understanding Environmental Governance: A Critical Review', *Organization and Environment*, 17, 4, pp. 471–492.
The Eleventh Five-Year Plan for National Economic and Social Development of the People's Republic of China (2006). Available at www.gov.cn/gongbao/content/2006/content_268766.htm.
First National General Survey of Pollution Sources (2010) 10 February 2010. Available at http://cpsc.mep.gov.cn/pv_obj_cache/pv_obj_id_210AD4D88459E94855053045C030A77B90CE0400/filename/W020100225545523639910.pdf.
Hong, D. (2008) 'Proposing New Directions for Improving Environmental Governance in China', *Hunan Social Sciences*, 3, pp. 79–82.
Huber, J. (1985) *The Rainbow Society: Ecology and Social Policy*. Frankfurt: Fischer.

Liu, J. and Diamond, J. (2005) 'China's Environment in a Globalizing World', *Nature*, 435, 1179–1186.
Ministry of Environmental Protection, China (MEP) (1995–2011) *National Statistical Communiqué on the Environment*. Available at http://zls.mep.gov.cn/hjtj/qghjtjgb/.
Ministry of Environmental Protection, China (MEP) (2011a) *2010 State of the Environment Report in China*. Available at http://jcs.mep.gov.cn/hjzl/zkgb/2010zkgb/.
Ministry of Environmental Protection, China (MEP) (2011b) *2010 Annual Report of Environmental Statistics*. Available at http://zls.mep.gov.cn/hjtj/nb/2010tjnb/.
Mol, A.P.J. (1995) *The Refinement of Production: Ecological Modernisation Theory and the Chemical Industry*. Utrecht: Van Arkel/International Books.
Mol, A.P.J. (2010) 'Environmental Reform in Modernizing China', pp. 378–393 in M. Redclift and G. Woodgate (eds), *The International Handbook of Environmental Sociology, Second Edition*. Cheltenham, UK: Edward Elgar.
Mol, A.P.J. (2011) 'China's Ascent and Africa's Environment', *Global Environmental Change*, 21, 3, pp. 785–794.
Mol, A.P.J. and Carter, N.T. (2006) 'China's Environmental Governance in Transition', *Environmental Politics*, 15, 2, pp. 149–170.
Mol, A.P.J. and Sonnenfeld, D.A. (eds) (2000) *Ecological Modernisation around the World: Perspectives and Critical Debates*. London/Portland, OH: Frank Cass/Taylor and Francis.
Mol, A.P.J. and Spaargaren, G. (2000) 'Ecological Modernization Theory in Debate: A Review', *Environmental Politics*, 9, 1, pp. 17–49.
Mol, A.P.J. and Spaargaren, G. (2006) 'Toward a Sociology of Environmental Flows: A New Agenda for Twenty-First-Century Environmental Sociology', pp. 39–82 in G. Spaargaren, A.P.J. Mol and F.H. Buttel (eds), *Governing Environmental Flows: Global Challenges to Social Theory*. Cambridge, MA: MIT Press.
National Bureau of Statistics, China (NBS) (2011) *2011 China Statistical Yearbook*. Available at www.stats.gov.cn/tjsj/ndsj/2011/indexch.htm.
National Bureau of Statistics, China (NBS) (2008) *Environmental Protection Has Made Positive Progress: Report on China's Economic and Social Development During 30 Years Since Reform and Opening Up (No. XV)*. Available at www.stats.gov.cn/was40/gjtjj_detail.jsp?searchword=%BB%7%BE%B3%B1%A3%BB%A4&channelid=6697&record=294.
Project Team on China's Modernization Strategy (PTCMS) (2007) *Report of China's Modernization 2007: Research on Ecological Modernization*. Peking: Peking University Press.
Schandl, H., Alexander, K., Collins, K., Heyenga, S., Poldy, F., Turner, G., West, J., Keen, S., Bengtsson, M., Hotta, Y., Hayashi, S., Akenji, L., Mishra, A. and Chen, S. (2011) *Resource Efficiency: Economics and Outlook for Asia and the Pacific*. Bangkok: United Nations Environment Program.
Simonis, U. (1989) 'Ecological Modernization of Industrial Society: The Strategic Elements', *International Social Science Journal*, 121, pp. 347–361.
Spaargaren, G. and Mol, A.P.J. (1992) 'Sociology, Environment and Modernity: Ecological Modernization as a Theory of Social Change', *Society and Natural Resources*, 5, 323–344.
State Council Information Office, China (SCIO) (1996) *White Book on Environmental Protection in China*. Available at www.scio.gov.cn/zfbps/ndhf/1996/200905/t307976.htm.
State Council Information Office, China (SCIO) (2006) *White Book on Environmental Protection in China, 1996–2005*. Available at www.gov.cn/zwgk/2006-06/05/content_300288.htm.
Sun, X. (2012) 'Environment Generally Deteriorating in China's Rural Areas, 600,000 Villages without Pollution Control', *People's Daily*, 31 January 2012, p. 4.
The Eleventh Five-Year Plan for National Economic and Social Development of the People's Republic of China (2006). Available at www.gov.cn/gongbao/content/2006/content_268766.htm.
The Twelfth Five-Year Plan for National Economic and Social Development of the People's Republic of China (2011). Available at www.gov.cn/2011lh/content_1825838.htm.
Zhang, L., Mol, A.P.J. and Sonnenfeld, D.A. (2007) 'The Interpretation of Ecological Modernization in China', *Environment Politics*, 16, 4, pp. 659–668.

5
Eco-imperialism and environmental justice

Anja Nygren

In recent years, environmental justice has become an important framework for the analysis of diverse environmental conditions and a powerful catalyst for popular mobilization in different parts of the world. Growing numbers of activists, scholars and policymakers refer to the framework of environmental justice when trying to understand the multiple concerns and claims over environmental degradation, resource rights, food security, climate change, local environmental knowledge and other environmental issues (Carruthers 2008). In this rapidly evolving field, discourses related to eco-imperialism are gaining popularity, especially when exploring the disproportionate impact of environmental degradation on the lives of poor communities and ethnic minorities in the Global South.

This chapter aims to provide an overview of environmental justice as both a framework for the analysis of environmental concerns and a discourse for political action. The main focus will be on the multifaceted links between the critique of eco-imperialism and the struggles over environmental justice in the Global South. As Joan Martínez Alier (2009) remarks, environmental issues are not just a matter of academic inquiry, or a luxury of the rich. They are deeply woven into the everyday lives and livelihoods of people, and their appreciation and practice of justice. The first section explores environmental justice and eco-imperialism as conceptual issues. The second section provides a brief introduction to the theoretical discussions on environmental justice, and illustrates the ways this framework has been used as an approach for interpreting diverse environmental concerns. The third section provides examples of socio-environmental movements that have used eco-imperialism and environmental justice as a political discourse and a means of popular mobilization in the struggle for justice and equity. By drawing on lessons from different parts of the world, it illustrates the multiple ways in which environmental justice activists articulate issues of justice. The fourth section deals with the growing transnationalization of environmental justice movements and the diverse tactics adopted by them in different socio-political and cultural contexts. Insofar as the notions of eco-imperialism and environmental justice highlight questions of distribution, participation, conflict and equity, they provide a broad but helpful framework for understanding the variety of environmental mobilizations taking place in different parts of the world.

Conceptual issues

Both the academic and activist literature surrounding eco-imperialism and environmental justice utilize a number of conceptions of justice. In brief, 'environmental justice' refers to a socio-spatial distribution and recognition of environmental benefits and burdens within human populations, while 'ecological justice' focuses on the relationship between human beings and the rest of the natural world (Baxter 2005). This distinction is, however, not clear-cut. Several scholars have recently called for the formulation of a broader approach that would combine issues of environmental and ecological justice (Agyeman 2006; Scholsberg 2007). Environmental justice as a concept emerged in the United States in the early 1980s, when certain African American, Latino and Native American communities affected by industrial pollution began to protest against environmental racism (Bullard 2000; Bryant 2003). Many environmental justice scholars make particular reference to the civil rights activists of Warren County, North Carolina, who organized themselves to stop the dumping of soil contaminated with hazardous materials in areas with a high proportion of African Americans. This prompted the launch of the environmental justice movement, which adopted civil rights and social justice approaches to environmental concerns. Shortly thereafter, environmental activists began to recognize similar struggles around the world (Mohai et al. 2009).

Rather than considering the environmental justice movement as a US or a Northern phenomenon which then spread to the Global South, we should pay attention to the diverse claims for environmental justice that have unfolded in different parts of the world, along with the social context in which they are grounded (Carruthers 2008). As Newell (2008: 51) remarks, there is a long history of environmental justice struggles around the world; it is more that these struggles have not always been framed in terms of environmental justice by Western observers. This especially concerns many Southern struggles, in which environmental justice is only one organizing principle among many issues, including indigenous rights, food security, human rights and democracy. In recent years, increasingly visible environmental justice campaigns have emerged in different parts of the world. This phenomenon is partly the result of the publicity given to environmental issues in contemporary development discourses and policies. Another impetus has been the increase of environmental awareness and public advocacy of environmental justice among different actors in civil society.

The discussion of environmental justice has close links to the discourse of 'eco-imperialism', which refers to the forceful imposition of Northern environmental views on the Global South. The debate over eco-imperialism arose from criticism of historical explanations of European colonization of the rest of the world in environmental terms. A popular example of such views may be found in Alfred Crosby's (1986) *Ecological Imperialism: The Biological Expansion of Europe, 900–1900*, which explains European control over the so-called New World in terms of the introduction of plants, animals and diseases to settler colonies. Corresponding ideas characterize Jared Diamond's (2005) bestseller *Collapse: How Societies Choose to Fail or Succeed*, in which bio-geographical factors such as the shape of the continents and the distribution of domesticable plants and animals are considered as the leading factors for Western European domination of other societies from the sixteenth to nineteenth century. Although Diamond claims that Western imperialism did not arise because of racial or cultural superiority, his analyses have been criticized for undertones of eco-imperialism. According to Blumer (2006) and Demeritt (2005), among others, Diamond uses overpopulation and environmental carrying capacity as the main factors to validate his Eurocentric view of world history, while largely ignoring analysis of changes in political economy.

Recently, many Southern researcher-activists have argued that eco-imperialistic views of the environment place the well-being of nature above the well-being of human populations, particularly at the margins of the Global South. Ramachandra Guha (1990) has claimed that Northern conservationists value the protection of endangered species more than the well-being of local people. Indeed, thousands of local inhabitants have been removed from their traditional lands to make way for protected areas dedicated to the preservation of nature and recreation for Northern tourists (Neumann 1998). Such agendas overlook questions that are crucial for Southern environmental movements, such as the struggle for territorial rights and the defence of local livelihoods. As Tim Ingold (2000) and James Carrier (2004) note, for many indigenous peoples and Southern communities the environment is not a distinct object of protection separate from human engagement. In contrast, it is a sphere of life activity and a place where one dwells, works and makes a living. On this basis, many Southern environmental justice groups question how campaigning for nature protection can be carried out without taking into account issues of social (in)equality and (under)development. These groups reject the one-sided approach of wilderness protection and seek to integrate issues of environmental conservation with local livelihoods. At the same time, they criticize Northern conservationists' tendency to reinvent indigenous peoples as noble stewards of pristine nature (Conklin and Graham 1995). Corresponding issues have been raised concerning global patterns of resource extraction. According to Arturo Escobar (2001, 2008), the Northern capitalist conception of nature produces a view of the natural environment as a realm to be appropriated through commodification and control. By aiming to break down the conventional boundary between nature and culture, Escobar emphasizes multiple constructions of nature and diverse modes of knowing about the environment. He argues for embedding questions of environmental justice within a broader framework of social justice, where the issues of whose resources are being exploited, whose knowledge is being appropriated and what kinds of environmental and social costs local people have to bear in the name of a form of development from which they rarely benefit are the crucial points of examination.

What is common to both the critique of eco-imperialism and the struggle for environmental justice is the search for alternatives to Northern-driven development models. In his book *Eco-Imperialism: Green Power, Black Death*, Paul Driessen (2005) argues that present-day eco-imperialists are similar to European imperialists of the seventeenth century in that they attempt to keep developing countries poor for the benefit of the developed world. According to Driessen, radical environmental activists have become inflexible in their demands for environmental protection and insensitive to the needs of the billions of people who lack food, health care and other basic necessities. Driessen criticizes Northern environmental-development models for having caused poverty and suffering in the Global South and urges developing countries to generate sustainable strategies for endogenous development, without dependency on foreign aid.

Correspondingly, Joan Martínez Alier (2002, 2009) remarks that Northern economic accounting systems overlook indigenous concepts for valuing the environment, such as territorial rights, cultural attachment to the land, and other non-monetary values. The loss of people's place-based livelihoods and identities in Nigeria because of extensive oil extraction, or the destruction of the mangrove forests in Honduras to make way for export-driven shrimp production thus cannot be simplified to the issue of economic compensation (Schroeder 2000; Watts 2004). According to Martínez Alier (2009), when poor people campaign for environmental protection, it is not because they are professional environmentalists but because their livelihoods are threatened. Crucial to this kind of 'environmentalism of the poor' is the logic of a moral economy, which is incompatible with the profit-based extraction of oil, minerals, wood and agro-fuels practised by corporations at the commodity frontiers (Martínez Alier 2009: 1111).

Whether such an environmentalism of the poor exists, and whether it can counteract the depletion of natural resources at the hands of big business, is a moot point even among those criticizing eco-imperialism and arguing for global environmental justice. As Peet et al. (2010)

note, Northern and Southern views of nature are in themselves too heterogeneous to allow for strict categorizations. Most environmental justice scholars today agree that concepts of justice are hybrid and contested. It is not possible to consider every environmental conflict an issue of environmental justice, nor can every environmental justice issue be explained in terms of environmental conflict (Pellow and Brulle 2005; Lockie 2009). The initial focus on racism in the conceptualization of environmental justice has recently shifted towards more nuanced views of the relationship between race, class, ethnicity and other forms of social difference, including gender and intergenerational justice (Buckingham and Kulcur 2009).

Theoretical orientations

As the movements that organize around environmental justice articulate diverse notions of justice, it is little wonder that the theoretical approaches used to explain them are also pluralistic. Traditionally, the *distributional* aspect of justice has dominated environmental justice thinking. Scholars relying on the ideas of John Rawls (1971) have focused on the questions of who benefits from environmental resources and who bears the environmental and social costs (Dobson 1998). By examining the controversial use of land, such as the location of hazardous waste sites and polluting industrial facilities in socially marginalized areas, these studies have demonstrated that parcels of land are not simply empty fields awaiting human action. Rather, land is linked to specific interests and power relations (Bullard 2000). In recent years, the distributional aspect of environmental justice has expanded to better capture articulations of justice in everyday life. As Walker (2009) notes, environmental injustice does not arise simply from inequalities in the spatial distribution of risk; it is also a question of how they interact with unevenness in the social distribution of vulnerability and well-being.

Mark Pelling's (2003) study of urban flooding in Guyana provides an interesting example of how the spatial distribution of inequality interacts with social patterns concerning who is most vulnerable to impacts of flooding and how this vulnerability is produced for different people and places. Although risks might be similar, people do not experience or cope with them in the same way. This observation extends to the unevenness of psychological and social impacts. In the case of floods, a series of contributory factors to vulnerability need to be considered such as access to insurance, pre-existing health problems, availability of the resources needed to recover, and the effectiveness of authorities' responses to the emergency (Bickerstaff and Walker 2003). Pelling (2003) argues the majority of the urban poor in Southern cities live in areas (often informal or 'squatter' settlements with poor planning and services) in which they are exposed to a heightened level of environmental risk. The lack of political will to apply risk prevention strategies in 'shantytowns' stereotyped as the source of various social ills and health hazards creates a vicious circle of poor livelihood options, increasing vulnerability and further environmental degradation. In this light, the outcomes of injustice cannot be reduced to the issues of who lives in areas prone to flooding or how they came to live there; people's different experiences of vulnerability are also important.

To take account of this point, theorists have sought to complement the distributive approach with additional conceptual approaches to environmental justice. Researchers such as Iris Young (1990), Nancy Fraser (1997) and Axel Honneth (2001) have argued that justice must also address the processes that construct and legitimize practices of misdistribution, including both individual and social *recognition* (i.e. respect for the dignity and status of others). Here, what is central is not only the psychological component of recognition but also structural asymmetries and the social position of those less well off in distributional schemes (Fraser and Honneth 2003). In this respect, there is a rich body of research in anthropology, geography and development studies that questions eco-imperialist explanations of the causes and consequences of environmental problems in the Global South. Several scholars have demonstrated how dominant representations of Southern environmental problems, such as deforestation or soil erosion, construct hegemonic narratives of degradation with limited empirical evidence (Escobar 1995; Rocheleau et al. 1995; Forsyth 2003). Correspondingly, in a study of an Argentine shantytown with high levels of pollution, Auyero and Swistun (2009) demonstrate how state officials, lawyers and media reporters visiting the neighbourhood created a hegemonic narrative of a contaminated place and contaminated people. Admittedly, the residents were worried about the pollution. However, they were also preoccupied with many other matters, such as poverty, high levels of unemployment and lack of public security. At the core of such misrecognition are institutional processes that devalue some people and places in comparison to others. Once certain communities get 'associated with trash', they become a 'natural' target for further unwanted land due to the attitude that marginal people live in a disorderly way and do not care for their environment (Pellow 2002: 38). This attitude then further limits official efforts to address the environmental problems in such places (Leichenko and Solecki 2008; Hastings 2009).

Many contemporary justice theorists also highlight aspects of *procedural* justice, which has motivated initiatives for greater political participation and more authentic citizenship in issues such as environmental rights, occupational health and human rights (Schrader-Frezette 2002; Barnett and Low 2004). Conventionally, scholars arguing for procedural justice have asserted that those who are most affected by environmental decisions should have a particular right to have their voices heard. However, as Walker (2009) points out, this raises the question of how to define those who are most affected and how to decide the area for which compensations should be negotiated when the impact of environmental disasters arguably extend far beyond the immediate locality. The dilemma of who is included, and who is not, is complex – as illustrated by negotiations over rights to indigenous knowledge (Schroeder 2000). Recently, theorists of procedural justice have extended their views of procedural justice to include the interaction between people, ideas and perspectives across different institutions and sectors of society. The degree to which transparent interaction is genuinely achieved is the crucial test of whether procedural justice has been realized (Walker 2009).

Nancy Fraser (2009), David Scholsberg (2004, 2007), and Gordon Walker (2009) have been particularly influential in the conceptualization of justice as an integrated multi-dimensional phenomenon. To summarize, justice as distribution focuses on the socio-spatial distribution of environmental impacts and responsibilities. Justice as recognition emphasizes the ways in which people's status or merit is evaluated in comparison with others. Justice as procedure looks at inclusions and exclusions in environmental policies and decision making. Additionally, David Scholsberg and David Carruthers (2010) have recently introduced a capabilities framework to better capture the communal dimensions of environmental justice. Based on the theories of Martha Nussbaum and Amartya Sen (1992), the capabilities framework emphasizes the socially differentiated opportunities people have to take command of their lives and the well-being of their communities. Collective experiences of justice are crucial for many Southern communities, where the central question may not be how satisfied an individual is, or the level of resources she or he commands, but the capabilities of the community to renew itself and act collectively (Sen 2010). In short, the different approaches to justice demonstrate the multiple dimensions of the phenomenon, even if all these approaches might not be equally important on every occasion.

Movements and strategies

Analytical approaches to justice can demonstrate the ways in which environmental–social relations are characterized by dominance and contribute to consideration of the transformations available within different political contexts (Fraser 2009). However, as many environmental activists emphasize, issues of environmental injustice go beyond analytical thinking, as they have implications for people's everyday life and thus require direct engagement. The way to grapple with acute injustice is thus considered to be through political struggle and social mobilization, even if the distinction between analytical reflection and political practice is often blurred as the ideas move between activist, scholarly and policy circles. Nevertheless, as Amartya Sen (2010: vii) notes, an important impetus for mobilization for many activists is

> not the realization that the world falls short of being completely just – which few of us expect – but that there are clearly remediable injustices around us which we want to eliminate.

Environmental justice movements, especially in the Global South, are usually characterized by a plurality of actors and agendas. Depending on the case, there may be environmental and human rights activists, small producers and rural workers, academic scholars and urban popular movements, women's associations, labour unions and indigenous movements involved in the struggle for environmental justice. The way justice claims are made also varies. This is partly because, as Harvey states (1996: 6), 'different socio-ecological circumstances imply different approaches to the question of what is or is not just'; partly because acts of claim-making are strategic. Movements typically employ heterogeneous conceptions of justice and change their agendas according to the conditions. In environmental justice struggles against the mining industry in West Africa, for example, environmental issues, labour rights and social justice issues are interwoven, as it is a situation where hundreds of workers have died in accidents and of occupational illnesses, child labour is commonly used and the minerals mined in war zones are often sold to finance insurgencies.

Another common characteristic of Southern environmental justice struggles is the incorporation of current concerns over injustice into a broader historical pattern of eco-imperialistic resource exploitation. The expansion of Northern market actors into Southern peripheries often brings the interests of global capital into conflict with local communities (Newell 2008). This issue is illustrated in the Mapuche Indians' struggle against the construction of hydroelectric dams and the granting of concessions to mining and timber companies in Chile. As Schlosberg and Carruthers (2010) note, the Mapuche articulated their concerns over justice as a critique of a neoliberalist development model that favours transnational corporations over the well-being of the native communities. Their disgust was clearly expressed by one of the Mapuche leaders when he claimed: 'We don't want *your* progress to rub out *our* culture' (cited in Scholsberg and Carruthers 2010: 27). Crucial to the Mapuche's protest is the resistance to the opening up of new areas for resource exploitation by taking advantage of ignorance about indigenous resource rights. In addition, there is controversy over the appropriation of indigenous knowledge. Here, the Mapuche conceptualize Northern intellectual property rights as a form of colonialism which overrides collective rights to knowledge. Similar issues have been highlighted by Di Chiro (2007), who explores how indigenous activists in different parts of the world criticize the ongoing 'genetization' of environmental issues. They consider this eagerness to commodify 'life itself' as a continuation of centuries-old patterns of eco-imperialism.

The unequal power relations that mediate control over environmental resources in many parts of the world complicate verification of the disproportionate impact of environmentally harmful activities on socially vulnerable people. Polluting industries often claim that isolating the impact of their activities is difficult as the populations concerned are also affected by suboptimal living conditions such as poor housing, malnutrition, neighbourhood crime and psychological stress. The paucity of systematic environmental and public health data presents an additional constraint. Furthermore, industrial sectors are often reluctant to release information on their activities while governments may enforce environmental laws unevenly (Carruthers 2008). Another tactic employed by governments is to use consensus-forming techniques to shift politicized protests away from confrontation towards collaboration. In a situation where industries seek to avoid communities that are capable of mounting effective opposition, those communities with limited economic and political power become an easy target for interventions containing a high degree of uncertainty. This does not justify the argument that the poor are uninterested in environmental health and sustainability. According to Freidberg (2004), the small producers in Burkina Faso and Zambia who grow high-quality vegetables for European markets do have concerns about food safety and sustainable livelihoods. The question is more that in precarious living conditions they cannot afford healthy consumption practices themselves.

Support for environmental justice movements has traditionally been weak in many countries. Formal spaces for environmental policymaking are more accessible for civil society organizations that are willing to support official development programmes, while groups that question the appropriateness of governmental policies and advance environmental justice claims find themselves excluded from the official decision-making process. When they are frustrated in this way, many of these groups adopt strategies of protest and resistance to strengthen calls for environmental justice and to contest development interventions that lure transnational corporations to Southern peripheries with promises of abundant natural resources, cheap labour and minimal environmental regulation (Newell 2008). Campaigns against transnational oil companies' extraction activities on contested lands in Bolivia and Ecuador provide an illuminating example of such struggles (Perreault and Valdivia 2008). Another example is provided by protests against genetically modified crops in Brazil, India and South Africa. Many environmental justice groups, in alliance with small farmers, are protesting against the deterioration of local livelihoods and food security, in a situation where 25 per cent of the global food production is consolidated in the hands of ten multinational companies (Dicken 2007: 367–368; Scoones 2008).

In addition to calls for more equal distribution of environmental resources and political rights, many Southern environmental justice movements emphasize issues related to cultural identities, collective values and symbolic relationships with nature. This is clear, for example, in the environmental justice movement '13 Pueblos', in Morelos, Mexico. For years, the communities participating in this movement have fought against plans by the municipal government and a private company to construct sanitary landfills and dumping grounds in their territories (Risdell 2010). In this battle, struggles over material conditions and cultural meanings are tightly intertwined. Local residents articulate claims over environmental rights and ecological sustainability alongside calls for the protection of traditional ways of life, respect for sacred sites and the preservation of communal capabilities. At the same time, this struggle demonstrates that environmental justice movements are far from internally homogeneous. Diverse conceptions of justice and how that justice can be sought, together with complicated power relations, form part of the dynamics of many environmental justice movements. Categorical distinctions between local resistance and outside intervention are thus difficult to justify.

Localized images – globalized alliances

In recent years, environmental justice movements have become increasingly transnational. Interaction between different levels of activity, from local proximities to globalized arenas, is an important tactic used by many activists to gain wider attention to their plight (Roberts 2007; Borras et al. 2008). When the underprivileged in the Global South can demonstrate that they suffer from environmental hazards caused by the privileged of the North, the claims for justice sometimes gain particular weight (Gedicks 2009). In campaigns against the massive transfer of hazardous waste products from the North to Southern peripheries, activist groups have strategically presented the waste as the rubbish of the rich dumped in the backyards of the poor. Similarly in the electronics industry, where production workers are exposed to many harmful chemicals, environmental justice activists have campaigned against the high level of inequality in global networks of production, consumption and responsibility. In this way, they have succeeded in changing some corporations' environmental strategies and improving national and international legislation to address the worst excesses of electronic waste. Some activists have also been able to pressure transnational companies into recycling their electronics at the end of the products' lifecycle and to reduce the use of toxic inputs in their production processes (Smith et al. 2006).

Correspondingly, in negotiations over climate policies, environmental justice advocates have stressed that resilience to climate change is unequally distributed. This includes who causes the problem, who suffers most, who is expected to act and who has the resources to do so (Ikeme 2003; Roberts and Parks 2007). Many advocates also point out that global production and consumption networks do not respect territorial boundaries, and thus the most challenging environmental justice issues today incorporate transnational claims, which clearly demonstrates that even the remotest corners of the world are interconnected (Scholte 2005: 75–84).

In such fields of transcultural encounter, local forms of environmental consciousness mingle with global symbolic politics. Community identities and local environmental knowledge are often used as key symbols to promote campaigns where localized struggles over environmental justice are linked to transnational advocacy networks in order to gain more visibility and wider reach than a single movement could achieve on its own. The challenge in such alliances is often how to integrate the diverse interests involved in socially meaningful ways. According to Conklin and Graham (1995), alliances between Amazonian Indians and international environmentalists, for example, are often founded on the assertion that native peoples' environmental perceptions are consistent with Northern conservationist principles, where the Indians are represented as 'guardians of forest' and 'people dwelling in nature, according to nature'. Such images undermine both the complexity of the Indians' way of life and their priorities for environmental justice. In fact, what many native groups are seeking from such alliances is better recognition of their territorial rights, while environmentalists are looking for Indians to provide a human face for their biodiversity conservation agenda and legitimacy for their engagement in Southern environmental politics, which could otherwise be interpreted as a form of eco-imperialism. There is a risk that the Indians are accepted as useful partners in such networks only to the extent that they conform to Northern images of what constitutes an authentic conservationist. It is clear that more analysis is needed to better understand the promises and pitfalls offered by the transnationalization of environmental justice struggles.

In regard to environmental justice, both academic and public attention has focused often on those movements that have achieved media exposure or that have been successful in confronting the environmental threats affecting them. This is evident, for example, when comparing the attention paid to indigenous versus non-indigenous environmental struggles in the tropics. The way tropical forest-dwellers are often portrayed is based on a sharp dichotomy between those who are considered environmentally noble and those who are not. Indigenous peoples conforming to what is perceived as a traditional way of life are essentialized as peoples of simplicity and environmental wisdom. Non-indigenous peasants (or peasants who have simply lost or abandoned obvious markers of indigeneity) are portrayed as rootless, corrupted and lacking in environmental knowledge. Indigenous movements whose agendas for local cultural revitalization can be linked to global strategies for tropical conservation are often privileged over the struggles of poor peasants on the degraded agricultural frontiers. As Nygren (2004) shows, this selective attention limits the ability of non-indigenous forest-dwellers involved in land conflicts and suffering from political–economic marginalization and human rights violation to gain access to transnational advocacy networks and global media.

To fully understand the heterogeneity of environmental justice concerns, it is important to pay attention not only to those battles with visible protests and confrontations but also to fragmented concerns with latent tensions and hidden forms of resistance. Jordi Diez and Reyes Rodríguez (2008) draw important lessons from the everyday forms of environmental justice important to a Mexican community subjected to serious health hazards from a polluting chemical enterprise. Their study demonstrates how institutional and social barriers can seriously constrain open mobilization for environmental justice. However, when periods of uncertainty and abnormality do emerge, destabilizing hegemonic interpretations of justice and allowing previously subordinated interpretations to make their way into public debate, there is much to be learned (Fraser 2009). More attention is needed to the fact that the main target of environmental justice activism has conventionally been transnational companies likely to be vulnerable to attacks on their global public image and, therefore, to be influenced by transnational organizing. There is little evidence that small and medium-sized domestic firms automatically operate in more environmentally sustainable and socially just ways, and thus no reason they should be spared the analytical and political gaze of environmental justice (Newell 2008: 66–67).

David Schlosberg has explained the transnational fluidity of environmental justice movements by likening them to a rhizome, as such movements sprout underground in various directions and connect in ways that are not easily visible from above (Scholsberg 1999: 96, 120). Correspondingly, Tom Connor (2004) demonstrates how environmental justice groups and anti-sweatshop movements have been successful in persuading transnational corporations in the clothing and footwear industry to better respect environmental regulations and labour rights, through the strategies of scattered campaigns and mobile multi-scale networks. By dividing and merging, proliferating and contracting, without sharing allegiance to a particular organizational form or strategy of action (Keck and Sikkink 1998: 5), these movements have been able to raise effective campaigns without the risk of the entire movement becoming the target of oppression. In contrast to the 1990s, when most such campaigns originated in Northern organizations, today an increasing number of initiatives stem from Southern groups of justice activists, with logistic and financial support from sympathetic transnational networks.

Conclusion

This chapter has analysed eco-imperialism and environmental justice as frameworks for analytical interpretation and as catalysts for social movements and political mobilization in different parts of the world. The critique of Northern-driven environmental-development models commonly presented both in the environmental justice literature and in circles contesting eco-imperialistic ideologies makes these concepts particularly useful for interpreting the diversity of environmental justice concerns in the Global South. As a phenomenon, environmental justice has a long history. For decades, there have been myriad forms of environmental mobilization and struggle in different parts of the world, even if those struggles have not always been framed in terms of environmental justice. Over the last decade, there has been a significant broadening in the academic and public understanding of the ways in which environmental issues, political rights, cultural values and social justice matters are intertwined with environmental justice struggles, especially in the Global South.

Today, many scholars emphasize the fact that environmental justice is a complex concept which must be understood from the viewpoint of several interrelated strands – distribution, recognition, representation and capabilities – to gain a fuller understanding of the different perceptions, scopes and meanings involved in this phenomenon in different circumstances. Such multi-dimensional approaches to justice incorporate the concept of inequity in the distribution of environmental goods and bads, together with a range of issues concerning recognition, participation and community-based capabilities. The environmental justice framework acts as a catalyst for social mobilization and political struggle and challenges the conventional forms of environmental policymaking by highlighting environmental struggles as disputes over material resources and cultural meanings. The multiplicity of environmental justice concerns and claims in different environmental, socio-cultural and political–economic circumstances demonstrates the heterogeneity and hybridity of environmental justice struggles, especially in the Global South. While globalization has brought new threats and uncertainties to the livelihoods of the Southern poor, it has also prompted new forms of mobilization and organization in order to gain wider attention to local demands for justice. Numerous Southern environmental justice movements have established strategic alliances with transnational advocacy networks concerned with issues of global environmental justice.

References

Agyeman, J. (2006) *Sustainable Communities and the Challenge of Environmental Justice*. New York, NY: New York University Press.

Auyero, J. and Swistun, D. A. (2009) *Flammable: Environmental Suffering in an Argentine Shantytown*. Oxford, UK: Oxford University Press.

Barnett, C. and Low, M. (eds) (2004) *Spaces of Democracy: Geographical Perspectives on Citizenship, Participation and Representation*. London: Sage.

Baxter, B. (2005) *A Theory of Ecological Justice*. London: Routledge.

Bickerstaff, K. and Walker, G. (2003) 'The Place(s) of Matter: Matter out of Place – Public Understandings of Air Pollution', *Progress in Human Geography*, 27, 1, pp. 45–67.

Blumer, M. (2006) 'Review of *Collapse: How Societies Choose to Fail or Succeed* (J. Diamond)', *The Geographical Review*, 96, 3, pp. 519–521.

Borras, S.M., Jr, Edelman, M. and Kay, C. (2008) 'Transnational Agrarian Movements: Origins and Politics, Campaigns and Impact', *Journal of Agrarian Change*, 8, 2–3, pp. 169–204.

Bryant, B. (2003) 'History and Issues of the Environmental Justice Movement', pp. 3–24 in G.R. Visgilio and D.M. Whitelaw (eds), *Our Backyard: A Quest for Environmental Justice*. Lanham, ML: Rowman and Littlefield.

Buckingham, S. and Kulcur, R. (2009) 'Gendered Geographies of Environmental Justice', *Antipode*, 41, 4, pp. 659–683.

Bullard, R.D. (2000) *Dumping in Dixie: Race, Class, and Environmental Quality*. Boulder, CO: Westview Press.

Carrier, J.G. (2004) 'Introduction', pp. 1–29 in J. G. Carrier (ed.), *Confronting Environments: Local Understandings in a Globalizing World*. Walnut Creek, CA: AltaMira Press.

Carruthers, D.V. (2008) 'Introduction: Popular Environmentalism and Social Justice in Latin America', pp. 1–22 in D.V. Carruthers (ed.), *Environmental Justice in Latin America: Problems, Promise, and Practice*. Cambridge, MA: MIT Press.

Conklin, B. and Graham, L. (1995) 'The Shifting Middle Ground: Amazonian Indians and Eco-Politics', *American Anthropologist*, 97, 4, pp. 695–710.

Connor, T. (2004) 'Time to Scale up Cooperation? Trade Unions, NGOs and the International Anti-Sweatshop Movement', *Development in Practice*, 14, 1–2, pp. 61–70.

Crosby, A.W. (1986) *Ecological Imperialism: The Biological Expansion of Europe, 900–1900*. Cambridge, UK: Cambridge University Press.

Demeritt, D. (2005) 'Perspectives on Diamond's *Collapse: How Societies Choose to Fail or Succeed*', *Current Anthropology*, 46, pp. S92–S94.

Diamond, J. (2005) *Collapse: How Societies Choose to Fail or Succeed*. New York, NY: Penguin Books.

Di Chiro, G. (2007) 'Indigenous Peoples and Biocolonialism: Defining the "Science of Environmental Justice" in the Century of the Gene', pp. 251–284 in R. Sandler and P.C. Pezzullo (eds), *Environmental Justice and Environmentalism: The Social Justice Challenge to the Environmental Movement*. Cambridge, MA: MIT Press.

Dicken, P. (2007) *Global Shift: Mapping the Changing Contours of the World Economy*. New York, NY: Guilford Press.

Diez, J. and Rodríquez, R. (2008) 'Environmental Justice in Mexico: The Peñoles Case', pp. 161–179 in D.V. Carruthers (ed.), *Environmental Justice in Latin America: Problems, Promise and Practice*. Cambridge, MA: MIT Press.

Dobson, A. (1998) *Justice and the Environment: Theories of Distributive Justice*. Oxford: Oxford University Press.

Driessen, P. (2005) *Eco-Imperialism: Green Power, Black Death*. New Delhi: Academic Foundation.

Escobar, A. (1995) *Encountering Development: The Making and Unmaking of the Third World*. Princeton, NJ: Princeton University Press.

Escobar, A. (2001) 'Culture Sits in Places: Reflection on Globalism and Subaltern Strategies of Localization', *Political Geography*, 20, 2, pp. 139–174.

Escobar, A. (2008) *Territories of Difference: Place, Movements, Life, Redes*. Durham, NC: Duke University Press.

Forsyth, T. (2003) *Critical Political Ecology: The Politics of Environmental Science*. London: Routledge.

Fraser, N. (1997) *Justice Interruptus: Critical Reflections on the 'Postsocialist' Condition*. London: Routledge.

Fraser, N. (2009) *Scales of Justice: Reimagining Political Space in a Globalizing World*. New York, NY: Columbia University Press.

Fraser, N. and Honneth, A. (2003) *Redistribution and Recognition? A Political–Philosophical Exchange*. London: Verso.

Freidberg, S. (2004) *French Beans and Food Scares: Culture and Commerce in an Anxious Age*. Oxford, UK: Oxford University Press.

Gedicks, A. (2009) *Dirty Gold: Indigenous Alliances to End Global Resource Colonialism*. Cambridge, MA: South End.

Guha, R. (1990) *Environmentalism: A Global History*. New York, NY: Longman.

Harvey, D. (1996) *Justice, Nature and the Geography of Difference*. Oxford, UK: Blackwell.

Hastings, A. (2009) 'Planning, Anti-Planning and the Infrastructure Crisis Facing Metropolitan Lagos', *Urban Studies*, 43, 2, pp. 371–396.

Honneth, A. (2001) 'Recognition or Redistribution: Changing Perspectives on the Moral Order of Society', *Theory, Culture and Society*, 18, 2–3, pp. 43–55.

Ikeme, J. (2003) 'Equity, Environmental Justice and Sustainability: Incomplete Approaches in Climate Change Politics', *Global Environmental Change*, 13, 3, pp. 195–206.

Ingold, T. (2000) *The Perception of the Environment: Essays in Livelihood, Dwelling and Skill*. London: Routledge.

Keck, M. and Sikkink, K. (1998) *Activists Beyond Borders: Advocacy Networks in International Politics*. Ithaca, NY: Cornell University Press.

Leichenko, R. M. and Solecki, W. D. (2008) 'Consumption, Inequity, and Environmental Justice: The Making of New Metropolitan Landscapes in Developing Countries', *Society and Natural Resources*, 21, 7, pp. 611–624.

Lockie, S. (2009) 'Deliberation and Actor-Networks: The "Practical" Implications of Social Theory for the Assessment of Large Dams and Other Interventions', *Society and Natural Resources*, 20, 9, pp. 787–799.

Martínez Alier, J. (2002) *The Environmentalism of the Poor: A Study of Ecological Conflicts and Valuation*. Cheltenham, UK: Edward Elgar.

Martínez Alier, J. (2009) 'Socially Sustainable Economic De-Growth', *Development and Change*, 40, 6, pp. 1099–1119.

Mohai, P., Pellow, D. and Roberts, J.T. (2009) 'Environmental Justice', *Annual Review of Environment and Resources*, 34, 16, pp. 405–430.

Neumann, R.P. (1998) *Imposing Wilderness: Struggles over Livelihood and Nature Preservation in Africa*. Berkeley, CA: University of California Press.

Newell, P. (2008) 'Contesting Trade Politics in the Americas: The Politics of Environmental Justice', pp. 49–74 in D.V. Carruthers (ed.), *Environmental Justice in Latin America: Problems, Promise, and Practice*. Cambridge, MA: MIT Press.

Nussbaum, M. and Sen, A. (1992) *The Quality of Life*. Oxford, UK: Oxford University Press.

Nygren, A. (2004) 'Contested Lands and Incompatible Images: The Political Ecology of Struggles over Resources in Nicaragua's Indio-Maíz Reserve', *Society and Natural Resources*, 17, 3, pp. 189–205.

Peet, R., Robbins, P. and Watts, M. (2010) 'Global Nature', pp. 1–47 in R. Peet, P. Robbins and M. Watts (eds), *Global Political Ecology*. London: Routledge.

Pelling, M. (2003) 'Toward a Political Ecology of Urban Environmental Risk: The Case of Guyana', pp. 73–93 in K.S. Zimmerer and T.J. Bassett (eds), *Political Ecology: An Integrative Approach to Geography and Environment-Development Studies*. New York, NY: Guilford Press.

Pellow, D.N. (2002) *Garbage Wars: The Struggle for Environmental Justice in Chicago*. Cambridge, MA: MIT Press.

Pellow, D.N. and Brulle, R.J. (2005) 'Power, Justice, and the Environment: Toward Critical Environmental Justice Studies', pp. 1–22 in D.N. Pellow and R.J. Brulle (eds), *Power, Justice, and the Environment: A Critical Appraisal of the Environmental Justice Movements*. Cambridge, MA: MIT Press.

Perreault, T. and Valdivia, G. (2010). 'Hydrocarbons, Popular Protest and National Imaginaries: Ecuador and Bolivia in Comparative Context', *Geoforum*, 41, 5, pp. 689–699.

Rawls, J. (1971) *A Theory of Justice*. Cambridge, MA: Harvard University Press.

Risdell, N.M. (2010) '"¡No al relleno en Loma de Mejía!" Conflicto, injusticia ambiental y movilización en Morelos'. Paper presented at 'Congreso Nacional de Antropología Social y Etnología, Ciudad de México', Mexico, 22–24 September.

Roberts, J.T. (2007) 'Globalizing Environmental Justice: Trend and Imperative', pp. 285–308 in R. Sandler and P.C. Pezzullo (eds), *Environmental Justice and Environmentalism: The Social Justice Challenge to the Environmental Movement*. Cambridge, MA: MIT Press.

Roberts, J.T. and Parks, B.C. (2007) *A Climate of Injustice: Global Inequality, North–South Politics, and Climate Change*. Cambridge, MA: MIT Press.

Rocheleau, D., Steinberg, P.E. and Patricia A.B. (1995) 'Environment, Development, Crisis, and Crusade: Ukambani, Kenya, 1890–1990', *World Development*, 23, 6, pp. 1037–1051.

Scholsberg, D. (1999) *Environmental Justice and the New Pluralism: The Challenge of Difference for Environmentalism*. Oxford, UK: Oxford University Press.

Scholsberg, D. (2004) 'Reconceiving Environmental Justice: Global Movements and Political Theories', *Environmental Politics*, 13, 3, pp. 517–540.

Scholsberg, D. (2007) *Defining Environmental Justice: Theories, Movements and Nature*. Oxford, UK: Oxford University Press.

Scholsberg, D. and Carruthers, D. (2010) 'Indigenous Struggles, Environmental Justice, and Community Capabilities', *Global Environmental Politics*, 10, 4, pp. 12–35.

Scholte, J.A. (2005) *Globalization: A Critical Introduction, Second Edition*. New York, NY: Palgrave Macmillan.

Schrader-Frezette, K. (2002) *Environmental Justice: Creating Equality, Reclaiming Democracy*. Oxford, UK: Oxford University Press.

Schroeder, R. (2000) 'Beyond Distributive Justice: Environmental Justice and Resource Extraction', pp. 52–64 in C. Zerner (ed.), *People, Plants and Justice: The Politics of Nature Conservation*. New York, NY: Columbia University Press.

Scoones, I. (2008) 'Mobilizing against GM Crops in India, South Africa and Brazil', pp. 147–176 in S.M. Borras Jr, M. Edelman and C. Kay (eds), *Transnational Agrarian Movements Confronting Globalization*. Malden, MA: Wiley-Blackwell.

Sen, A. (2010) *The Idea of Justice*. London: Penguin Books.

Smith, T., Sonnenfeld, D.A. and Pellow, D.N. (eds) (2006) *Challenging the Chip: Labor Rights and Environmental Justice in the Global Electronics Industry*. Philadelphia, PA: Temple University Press.

Walker, G. (2009) 'Beyond Distribution and Proximity: Exploring the Multiple Spatialities of Environmental Justice', *Antipode*, 41, 4, pp. 614–636.

Watts, M. (2004) 'Resource Curse: Governmentality, Oil and Power in the Niger Delta, Nigeria', *Geopolitics*, 9, 1, pp. 50–80.

Young, I. (1990) *Justice and the Politics of Difference*. Princeton, NJ: Princeton University Press.

6
Neoliberalism by design
Changing modalities of market-based environmental governance

Stewart Lockie

Economist Nicholas Stern's advice to the British Government in 2008 that anthropogenic climate change is 'the greatest and widest-ranging market failure ever seen' (Stern 2007: i) was widely hailed as a seminal moment in the political and economic mainstreaming of environmental issues. In arguing that the benefits of strong and immediate action to mitigate against damaging climate change were likely to far outweigh any costs in relation to short-term economic growth, Stern seemed to be stating the obvious. Higher mean temperatures, shifting precipitation patterns, more frequent extreme weather events, sea level rise and the increased potential for abrupt large-scale shifts in climate states are all likely to impose significant costs and to damage long-term economic growth. Addressing the threat posed by climate change to the well-being of human communities therefore requires us to look beyond old debates over conservation versus development, wildlife versus people, etc. and to recognize that the health of the global economy depends on the health of global ecosystems. Stern's declaration did not, of course, close off debate over whether anthropogenic climate change is actually occurring or, to the extent that it is, whether it is economically rational to do anything about it. Nevertheless, this statement was emblematic of a growing global orthodoxy that environmental degradation, where it does occur, is best conceptualized as a form of market failure that is most effectively dealt with through the application of various forms of market-based policy instrument.

Market-based instruments (MBIs) take a variety of forms that are defined in more detail in following sections. For climate mitigation, according to Stern (2007), the key issue is to address the underlying market failure by putting a price on carbon. Such a price might be imposed through tradable emission permits, pollution taxes or regulatory restrictions on emissions. Whichever mechanism is chosen, the goal is to stop treating greenhouse gas (GHG) emissions as externalities and to absorb the cost of their mitigation in the production and exchange of goods and services. In his advice to the Australian Government in 2008, economist Ross Garnaut was more prescriptive. He argued that the most efficient way to price carbon was through the creation of a market for tradable emission permits and cautioned against 'compromising' emissions trading schemes with free permits, industry exemptions and/or price controls. 'Seemingly small compromises', according to Garnaut (2008: 321), would rapidly undermine the effectiveness of a 'well-designed' scheme. The use of non-market alternatives, he argued, would be even less effective, putting governments in the game of picking technological and sectoral 'winners' and reducing incentives for innovation across the economy. Stern and Garnaut are not unique in their advocacy of market-based policy instruments to abate GHG emissions. Such instruments are the cornerstone of international agreements such as the United Nations Framework Convention on Climate Change's Kyoto Protocol and its associated Joint Implementation and Clean Development Mechanisms. Nor is climate change the only domain of environmental policy in which market mechanisms have been embraced as the almost self-evidently superior policy option. Markets in pollution, biodiversity, water, soil conservation and so on are increasingly common from the multilateral to the national and sub-national scales. Across virtually every arena of environmental policy, market instruments are the new common sense.

This chapter will contextualize the current interest in MBIs in broader processes of social and economic reform that have reshaped environmental governance over the last few decades. It contends that technical arguments regarding the superiority of MBIs have not simply been discovered. They have been developed both as the most recent manifestation of peculiarly neoliberal ways of thinking about governance and as a response to the contradictions and limitations of other neoliberal experiments in environmental policy. The chapter will outline neoliberal approaches to environmental governance in more detail before going on to evaluate the assumptions and arguments underlying MBIs and the opportunity costs that are imposed by choosing these over other potential policy instruments. In doing so, it will argue that understanding the potential impacts, both positive and negative, of neoliberal governance requires us to recognize that the relationship between understanding of environmental problems and policy responses to those problems is neither linear nor unidirectional. The rationalities, or ways of thinking, that inform environmental governance shape recognition and comprehension of environmental problems and causes just as surely as they shape policy design. This leads to another question. If neoliberal rationalities do attempt to reconstitute the world in their own image, what contradictions and limitations are likely to emerge from the project of rethinking environmental degradation as a market failure amenable to action through the deployment of MBIs?

Market-based policy in context: neoliberalization and the environment

The concept of neoliberalism is used to denote a diverse range of governmental philosophies and programmes that rely in some way on the unitary logic of 'the market' (Lockie 2010). These include free-market ideologies, global realignments of economic and social interests, processes of market-led institutional and regulatory reform and bundles of preferred policy instruments embracing various forms of market discipline and the development of entrepreneurial capacities and subjectivities among citizens (Brenner et al. 2010: 183). Reflecting this diversity, neoliberalism has attracted the attention of a range of social theoretical perspectives (these are reviewed in detail by Brenner et al. 2010). Despite significant differences between these perspectives it is suggested here that two in particular are useful for the purpose of understanding market-based environmental governance. The first, drawing on critical political economy, examines processes, patterns and trajectories of market-oriented regulatory and institutional restructuring. *Neoliberalization* is identified in this literature as a flexible process of experimentation in economic and social reform that is simultaneously interconnected, contested, spatially uneven and locally specific (Brenner et al. 2010; see also Brenner and Theodore, 2002; McCarthy and Prudham, 2004). The second perspective, drawing on Foucault's (1991) lectures on governmentality, conceptualizes neoliberalism as a rationality which renders governance thinkable and actionable. *Neoliberal rationality* provides guidance on the framing of governance problems, what might be done about them, the boundaries of acceptable intervention and the actors through whom strategies may be applied (Dean 1999; Foucault 1991).

While neoliberal rationality may be seen logically to precede and inform processes of neoliberalization, in practice the relationship is more dynamic and recursive. It is not enough to provide guidance in advance of a policy decision. Government is a complex and 'congenitally failing' enterprise (Miller and Rose 1990: 10). To attain any kind of durability, rationalities of governance must be able to account for failure, resolve competing demands and adapt to new circumstances. According to Brenner et al. (2010: 190):

> Successive rounds of [neoliberal] restructuring are ... shaped by the conflicts, failures and contradictions associated with previous iterations of this 'layering' process, just as they reflect experimental policy borrowing, learning, inter-referentiality and co-evolutionary influences from both local and extra-jurisdictional sites.

In order to contextualize market-based environmental policy it is useful to identify three broad phases of neoliberalization; *rollback* and *rollout neoliberalization* as identified by Peck and Tickell (2002), and *designer neoliberalization*, which will be proposed here. Noting comments already made in this chapter concerning discontinuities and diversity in the neoliberal project, it is important to stress that these phases should not be treated as exhaustive, universally applicable nor necessarily sequential historical categories. They should be treated rather as attempts to sum up spatial and temporal patterns in the process of neoliberalization and thus to identify the relationships between variable local manifestations of neoliberalism and extra-local networks and processes (Peck and Tickell 2002: 380).

Rollback neoliberalization emerged in the 1980s as an aggressive rejection of Keynesian welfare state economics. Associated by most authors with the Thatcher (UK) and Reagan (US) administrations (but just as evident in numerous other national and supranational contexts), rollback neoliberalization was manifest in discursive attacks on government agencies and public servants, the dismantling of regulatory systems, the sale of state enterprises, the winding back and/or privatization of social welfare programmes, the deregulation of financial markets, and so on. Rollback neoliberalization was extremely uneven in the environmental sphere. Nevertheless, a variety of jurisdictions including the US and Canada saw the repeal of legislative controls over pollution, resource use, etc. (replacing these with voluntary standards, self-monitoring and industry codes of conduct) and the privatization or corporatization of statutory authorities responsible for the management of natural resources such as water. The capacity of environmental agencies either to regulate or to work proactively with resource-users was attacked through dramatic cuts to budgets and staffing combined with the imposition of regulations on the environmental agencies themselves (such as the need to conduct a full cost–benefit analysis of any proposed standards or regulations) (Prudham 2004).

Resistance to the destructive tendencies of rollback neoliberalization concentrated, according to McCarthy and Prudham (2004) around environmental concerns. By the 1990s, rollout neoliberalization saw renewed attention to institution-building and government intervention (Peck and Tickell 2002). This represented a deepening of neoliberalization rather than a return to Keynesianism as states sought creative new ways both to deal with the contradictions and consequences of rollback policies and to extend market discipline to new domains of social and environmental policy. This occurred in a variety of ways, including the promotion of partnerships between the public and private sectors, the mobilization of non-government groups in service delivery and the devolution of responsibilities to communities and individuals through various forms of self-help and capacity-building programmes. Devolutionist programmes included promotion of community-based natural resource management groups, business and natural resource planning, voluntary conservation covenants, etc. This phase of neoliberalization resonates with Foucault's (1991) analysis of neoliberal rationality, through which he contends individuals are constructed as behaviourally manipulable beings who might be equipped to respond rationally to changing environmental conditions. The market becomes the organizing principle behind the state and 'government itself becomes a sort of enterprise whose task it is to universalize competition and invent market-shaped systems of action for individuals, groups and institutions' (Lemke 2001: 197). According to Dean (1998), the results of this change can be seen in two broad and interrelated categories of governmental technique. *Technologies of agency* are used to enable and encourage self-regulating and self-managing behaviour among members of a population, while *technologies of performance* are used to monitor, inform and, where necessary, hold citizens accountable.

Technologies of agency are often accompanied by rhetorics of empowerment, partnership, mutual obligation and citizenship. Despite this, they are routinely characterized in the literature as a form of regulatory dumping through which states work around the contradictions of neoliberalism by assigning local institutions and actors responsibility to fill the voids of rollback neoliberalization without the requisite power or resources to do so (Brenner et al. 2010). However, to characterize all rollout policy as regulatory dumping assumes, rather problematically, that in the absence of public–private partnerships, devolution, etc. governments necessarily have the power and resources themselves to address environmental and social concerns. Further, it assumes that centralized state action (either through management interventions or through regulatory controls) is necessarily more appropriate and effective than its alternatives. This is not necessarily the case. Governments may have limited capacity to monitor and enforce regulations. The legitimacy of interventions may be contested. Other policy options may simply be more effective. Lockie and Higgins (2007) take this up in relation to a well-known experiment in community-based natural resource management: Australia's National Landcare Programme (NLP). The NLP was initiated in 1988 to improve resource management on primarily agricultural lands by encouraging landholders to join the self-help community Landcare groups. These groups were provided small amounts of funding to assist in group coordination and the establishment of local research, education and watershed planning projects. Over time they became delivery mechanisms for other programmes in farm and business planning. Critically, the NLP was supported by peak organizations representing both conservation and agricultural interests. This reflected the ability of this programme to reconcile a number of potentially competing political discourses. Landcare was seen as environmentally responsible and community-building. At the same time, it linked improved natural resource management with improved business management. Landcare avoided confrontation with landholders' perceived private property rights. It did not pay subsidies that could potentially distort trade, and it recognized that the causes and consequences of rural land degradation were so widely dispersed that no amount of government money, by itself, would ever 'fix' the problem. As a form of rollout neoliberalization, the NLP brought multiple policy goals together in a hybrid form of governance that sought environmental and social outcomes through the explicit promotion of economically rational business management; resource degradation being conceived as one of many risks which prudent landholders – equipped with appropriate managerial and entrepreneurial capacities – will accommodate in their normal business planning (Lockie and Higgins 2007).

What we see with Landcare and numerous other experiments in rollout neoliberalization is not the displacement of potentially competing rationalities (e.g. ecological or communicative rationality) but their incorporation within the underlying logic of neoliberalism. This incorporation has gone some way to resolve the legitimation problems facing more punitive rollback policies. By itself, however, discursive accommodation is not sufficient to guarantee the ongoing legitimacy of the neoliberalization process. Programmes that fail to achieve their goals stimulate a need for constant innovation and reinvention of the neoliberal project. Returning to the Landcare example, by the mid-1990s approximately one third of Australian farm businesses were directly involved in Landcare groups, with evidence that many more had changed resource management practices as an indirect consequence of community Landcare groups' activities (Lockie 2006). Yet criticism had also emerged that private landholders had insufficient capacity – in light of poor financial returns to agriculture – to undertake environmental works on a scale that would reverse landscape scale declines in resource condition. A flurry of investment in regional planning and funding for high priority works followed, but the fundamental source of concern remained unresolved and the NLP has progressively been stripped both of funding and of its status as the centrepiece of Australian agri-environmental policy.

Devolutionist policy measures based on capacity building, active citizenship and entrepreneurialism seek to address market failures allegedly evident in environmental degradation by encouraging individuals and businesses to internalize the costs of avoiding or repairing this damage. Recent years have seen a rising chorus of claims, however, that these suasive or educational measures are, at best, inferior to explicitly market-based incentives and, at worst, largely ineffective. As the next section will show, MBIs may, or may not, attempt actually to deal with the underlying causes of market failure. What they have in common is a view that, regardless of the problem to which they are directed, market-like mechanisms can be designed to target investment in an intrinsically more efficient and effective manner than alternative policy instruments. *Designer neoliberalization* may thus be conceptualized both as a variant on rollout neoliberalization with its emphasis on institution-building and proactive policy, and as a variant of rollback neoliberalization that delegitimates and dismantles many existing rollout institutions and programmes. Designer neoliberalization may also be described as a new pinnacle of economic triumphalism. Environmental degradation is conceptualized as the outcome of factors external to a properly functioning market. Where market solutions prove ineffective in addressing these factors, their own failure is rationalized as the outcome of 'design flaws' to be addressed through the application of more economic expertise (Lockie 2013). Markets are never seen as the problem but they are always seen as the solution (Lockie 2013; Muradian et al. 2010). The balance of this chapter will assess the veracity of this argument by identifying and assessing the often hidden assumptions of market-based environmental policy.

Neoliberalism by design: the logic of market-based environmental policy

Market-based instruments are generally classified into market friction, price-based and quantity-based measures (see Table 6.1). Market friction measures attempt to remove obstacles to the recognition of environmental costs and benefits through instruments such as standards and eco-labels. Price-based measures include both attempts to encourage existing markets to internalize environmental costs (i.e. market reform), through instruments such as eco-taxes, and attempts to create new markets through which to allocate payments for the provision of ecosystem services (i.e. market design) using instruments such as biodiversity auctions. Quantity-based measures work in tandem with price-based measures in order to meet specified targets for the provision of ecosystem services or the reduction of environmental harm.

The rationale for MBIs is elegant and, at face value, compelling. The long-term viability of all businesses depends on sustainable resource use. The costs of conservation, therefore, are a cost of production that in a properly functioning market would be internalized and passed on to consumers. However, inadequate understanding of the long-term impact of resource use practices, poorly defined property rights, and pricing of natural resource inputs below their full economic and environmental cost all encourage overutilization. Environmental degradation is

Table 6.1 Typology of market-based instruments

Classification	Market intervention	Examples	Suited to:
Market friction	Improving efficiency of existing markets by removing obstacles to recognition of ecosystem services	Standards, certification, eco-labelling, ethical investment schemes, capacity building	Outcomes that can be improved through reduced transaction costs or increased information
Price-based I (market reform)	Setting or modifying prices to incorporate the cost of ecosystem services	Eco-taxes	Measurable point source activities such as carbon emissions, water extraction, etc.
Price-based II (market design)	Utilizing market-mechanisms to allocate payments for ecosystem services	Auctions, tenders	Diffuse source environmental outcomes such as biodiversity, salinity mitigation, etc.
Quantity-based	Setting targets to achieve or maintain ecosystem services	Cap and trade mechanisms, tradable offsets	Measurable point source activities such as carbon emissions, water extraction, etc.

Source: Stewart Lockie, 'Market Instruments, Ecosystem Services, and Property Rights: Assumptions and Conditions for Sustained Social and Ecological Benefits', *Land Use Policy*, 31, March 2013, pp. 90–98. Reproduced by permission of Elsevier.

thus an outcome of market failure that ideally ought to be addressed through market means. MBIs that encourage market reform and full internalization of environmental costs are, by this logic, most desirable in the problem of overutilization. However, public expenditure may still be warranted if resource-users are to provide conservation that is clearly of a public good nature (e.g. preserving ecosystems for their cultural or scenic value rather than their contribution of sustainable production) or if more far-reaching market reform is impractical in the short to medium term. MBIs, under such circumstances, remain the most appropriate mechanisms through which to deliver public expenditure, since they allow policy makers to target instruments towards very specific environmental values; allow flexibility for individuals to choose the optimum amount and means of conservation they can provide depending on their own circumstances; provide continuing incentives to find innovative ways to further reduce environmental impacts; and thus provide the least-cost path to overall environmental outcomes. Regulation is dismissed by MBI proponents as cumbersome, blunt and ineffective, while direct subsidies for environmental protection are seen as potential de facto barriers to trade. Proponents argue that:

> With insights from newer ideas in economics, systematic use of field pilots and experimental economics, and new computational capabilities, it is now possible to design and create markets for previously intractable policy problems. Such approaches have led to the creation of 'designer' markets in areas of the economy where this was previously impossible ... These ideas and processes raise the prospect of developing completely new approaches to natural resource and environmental management.
>
> *(NMBIWG 2005: 5)*

Such claims position MBIs at the cutting edge of public policy. Alternative policy options are positioned as clumsy and inefficient political compromises to be minimized wherever possible.

Through a recursive process of experimentation, calculation and design, the discipline of economics promises to depoliticize hitherto conflict-ridden policy problems and to deliver public good outcomes at minimum cost. The following section will examine the often unstated assumptions behind these claims.

Market instruments: the constitutive power of economics

There are at least three reasons as to why it is not, in fact, obvious that environmental degradation should be seen as an indication of market failure or, to put this differently, that a properly functioning market would necessarily internalize the full cost of environmental conservation (Lockie 2013). First, it is only economically rational for the producers of goods and services to internalize those costs of conservation that support sustainable production. Conservation values unrelated to the productive activities of any particular business are more rationally, from an economic point of view, externalized. Second, cost–benefit assessment of conservation efforts that provide only indirect or long-term benefits may suggest that the economically rational strategy remains their externalization (Bromley 2007). Stern's (2007) conclusion, for example, that the long-term costs imposed by anthropogenic climate change would be higher than the short-term cost of mitigating GHG emissions was based on the use of discount rates lower than those conventionally used by economists to calculate the net present value of future benefits (in this case, avoided costs). Stern justified this decision by invoking the precautionary principle. Critics counter that uncertainty ought actually be cause to utilize higher discount rates and that Stern has consequently exaggerated the economic benefits of GHG mitigation. The point here is not to adjudicate this debate. The point is, rather, that valuing future economic benefits from conservation activity is a difficult process that does not necessarily provide a clear-cut case on the basis of economic rationality alone to absorb the costs of conservation. Third, the costs and benefits of conservation frequently fall on different actors (Bromley 2007). As the benefits of activities such as GHG mitigation are non-excludable across time and space, individual businesses (and countries) may well see a compelling economic case to free-ride on the mitigation efforts of others. There are, in sum, many compelling arguments as to why environmental conservation is important to the short and long-term well-being of people and our economies. However, the suggestion that markets, properly informed and uncorrupted by politics, would provide either sufficient or effectively targeted conservation is far from self-evident.

Of course, even if we do reject this particular aspect of economic triumphalism we may still accept the argument that market-based instruments remain the most effective and efficient policy options for addressing environmental degradation. Institutional economists argue that by establishing new markets for environmental services the market failures that lead to their undersupply are thereby resolved (Muradian et al. 2010). Such a claim rests on, among other things, the ability of various experiments in market-based governance to achieve demonstrable environmental outcomes. Market friction and eco-tax instruments aside, most MBIs function by establishing some sort of property right, either in the provision of ecosystem services or (perhaps counter-intuitively) in damage to ecosystem services.

Payment for ecosystem services (PES) programmes utilize mechanisms such as tenders and auctions to identify resource-users who are willing to provide a desired ecosystem service (e.g. biodiversity conservation, watershed protection) as well as the level of financial incentive they require to do so. Engel et al. (2008) nominate a number of widely cited conditions for the establishment of PES schemes. First, there must be a clear relationship between the actions resource-users agree to undertake (e.g. excluding livestock from waterways) and the ecosystem services the scheme is designed to provide (e.g. improved downstream water quality).

Second, participation must be voluntary. Third, monitoring systems must be in place to ensure actions are being undertaken to provide the desired service (the conditionality criterion). Fourth, these actions should be over and above the activities that resource-users would have undertaken in the absence of the PES scheme (the 'additionality' criterion). In most cases, the buyers of ecosystem services in PES schemes are government agencies. In theory, sellers will submit bids that reflect a calculus of the costs they will incur in providing the target service; the private benefits, if any, they derive from this service; and any personal interest in, or sense of responsibility to protect, the ecosystem service in question. Buyers are then able to assess the cost of bids relative to the quality of services on offer and thus achieve maximum conservation within their own resource constraints.

In practice, few existing PES schemes have been able to demonstrate greater efficiency or effectiveness than alternative policy instruments (Muradian et al. 2010; Pascual et al. 2010). Common pitfalls in PES implementation have been identified as poor targeting, offering insufficient incentives to encourage participation, displacing undesirable practices to alternative locations, paying for activities that would have been undertaken anyway (or, alternatively, undermining voluntary action), failing to guarantee long-term environmental improvements and dilution through multiple competing co-objectives (Engel et al. 2008; Lockie and Tennent 2010; Muradian et al. 2010; Wunder et al. 2008). Few existing schemes, further, actually fulfil the criteria proposed by Engel et al. (2008) to delineate a 'genuine' PES scheme (Muradian et al. 2010), raising questions over whether it is the market-based features of these schemes that lead to particular outcomes or whether outcomes should be attributed to non-market measures embedded within these schemes, such as educational support for participating resource users. Lockie (2013) argues that, in addition to those criteria established by Engel et al. (2008), market-based PES are likely to be effective only in situations in which multiple resource users are able to undertake desired actions, individual actions scale up effectively to provide the desired ecosystem service, resource users have adequate information and capacity to assess this information, resource users do not corroborate in the setting of prices, property rights and the responsibilities of property right holders are clear and institutions operating PES are considered competent and trustworthy. The socio-ecological complexity of natural resource management suggests that the situations in which all these criteria can be met are likely to be the exception more often than the rule. Where effective provision of a desired service or bundle of services requires the involvement of all, or most, members of a specific resource user group (e.g. all farmers within a watershed participating in programmes to protect soils, water quality and biodiversity), markets are not an appropriate mechanism through which to negotiate involvement or the conditions of that involvement. The more ecosystem services are required, the more narrowly defined the group of resource-users capable of providing that bundle of services will become and the more difficult, subsequently, for market-based PES to satisfy claims of effectiveness and efficiency.

Quantity-based MBIs such as cap-and-trade and tradable permit schemes establish limited rights to undertake activities that potentially damage ecosystem services; cap-and-trade schemes generally being applied to activities that discharge pollutants into ecosystems (e.g. GHGs) and tradable permits to activities that extract resources from ecosystems (e.g. water). These rights may then be bought and sold among resource-users in order to ensure, at least in theory, that discharges and extractions are kept within predetermined limits and that market forces will, over time, reallocate rights to their most economically optimal uses. Further, tradable limits on discharges and extractions create incentives among rights holders to innovate in ways that boost eco-efficiency (that is, the amount of production enterprises are able to achieve for any given unit of discharge or extraction) and/or which reduce the cost of minimizing pollution or extracting resources (Dargusch and Griffiths 2008). Increased eco-efficiency will allow enterprises either to produce (and sell) more from their existing discharge/extraction rights or to sell surplus rights to others. The price of rights will then reflect the value placed on them by other businesses based on the relative scarcity of these rights and the profitability of the uses to which they will be put. Cap-and-trade schemes may also set ceilings on discharge rights that lower slowly over time, increasing the scarcity (and cost) of rights in order to accelerate eco-efficiency innovation and encourage those enterprises incapable of sufficient innovation to redeploy their capital in other sectors. The integrity and effectiveness of markets in discharge/extraction rights will depend on a variety of factors, including sufficient scarcity and cost of rights to meet environmental objectives, and create incentives for innovation, monitoring and measurement of discharges or extractions and the enforcement of penalties for exceeding discharge or extraction rights (Henderson and Norris 2008). Henderson and Norris (2008) note that the cost-effectiveness of quantity-based MBIs is enhanced by the inclusion of as many potential participants as possible, although such inclusion may lead to environmental compromises by establishing the potential for localized hotspots of pollution or overextraction to develop.

There is generally better evidence for the effectiveness of quantity-based MBIs when applied to point source pollution and/or well-defined and readily exchangeable natural resources. In mitigating global climate change, for example, it matters little where GHG emissions are abated, or whether abatement is reasonably constant through time, provided total GHG concentrations in the atmosphere are reduced (Drechsler and Hartig 2011). As noted above, the two major implementation mechanisms associated with the Kyoto Protocol thus utilize tradable emission permits and offsets to help countries and businesses meet their GHG abatement targets. Emissions trading schemes were also operable, at the time of writing, in the European Union, New Zealand and several US states, and schemes were under development in Australia, China, Japan and Korea (Commonwealth of Australia 2011). As economic reforms, such schemes are comparable in scope and significance with rollback measures such as trade and labour market deregulation. Ross Garnaut (2008: 313) thus positions emissions trading as the 'next great reform agenda':

> The pervasive consequences of an emissions trading scheme make it a major reform of the Australian economy ... Previous reforms – such as trade liberalization, financial regulation and competition policy – were designed to raise incomes by allowing the allocation of resources to their most productive uses. By contrast, the climate change reform agenda must be focused on minimizing the potential for loss of income after the introduction of measures to limit the release of greenhouse gases.

Trading schemes have similarly been at the heart of major reform agendas in the allocation of water resources (although significant conflict has arisen over the consequences of such reform for equity and social welfare). Such schemes are limited, however, in their ability to provide ecosystem services such as biodiversity, which are likely to be compromised by habitat turnover associated with the movement of service providers in and out of a market (Drechsler and Hartig 2011). Biodiversity, and the multiple species and ecosystems that comprise it, require a high degree of spatial and temporal continuity of habitat provision in order to allow for species movement and evolution, ecosystem succession and restoration, recovery from extreme weather events, and so on. In this case, the temporal and spatial dynamics of efficiently operating markets are poorly matched with the temporal and spatial dynamics of resilient, biodiverse ecosystems.

Conclusion

Neoliberal rationality promises to depoliticize environmental policy, accommodating economic and environmental objectives within a single hybrid regime of governance. However, the compelling elegance of neoliberal rationality derives, in large part, from its circularity. Markets are construed as intrinsically superior mechanisms for the allocation of resources in a socially optimal manner and the failure of markets – including purpose-built markets in ecosystem services – is explained away as the consequence of political compromises and design flaws. After all, it is asserted, resource-users have an economic interest in resource sustainability. Further, trade liberalization and other rollback measures have already equipped resource-users with the entrepreneurial identities and capacities to calculate and regulate their own behaviour under the right circumstances; circumstances that can only be defined through the application of technical economic expertise. As neoliberalization has progressed through the rollback, rollout and designer phases, the reliance of environmental policy on economic expertise has deepened. Economics is no longer simply an analytical (or dismal, as it is so often characterized) science. It is architecture and engineering, with market-based incentives providing the practical expression of economists' work designing the institutions and bargaining mechanisms through which resource users are to deliver environmental outcomes as a logical consequence of their normal pursuit of economically rational self-interest. As this chapter has argued, many of the assumptions embedded within this logic are questionable.

Designer neoliberalization represents considerably more than incremental adjustment to the contradictions and legitimation challenges facing devolutionist policies. The embrace of market-based approaches to greenhouse gas mitigation brings environmental policy to the heart of neoliberal economic reform. Many would argue that this is precisely where environmental policy should be. But it is a very specific kind of 'environment' that is incorporated within the hybrid regimes of neoliberalism. As Pellizzoni (2011: 802) argues, the environment 'is no longer conceived as an objectively given, though cognitively mediated, reality, but as a constitutively fluid entity, a contingency purposefully produced and controlled for instrumental ends'. To put this another way, it is not just any environment, any nature or any natural resource that is incorporated within neoliberal regimes of governance. It is natures that can be defined through clearly delineated property rights; natures that can be reduced to simple and transferable common values; natures that comply with the requirement of markets for well understood and calculable risks; natures that are not, therefore, characterized by excessive uncertainty, complexity or the possibility of discontinuity and threshold effects. There is little doubt that anthropogenic climate change represents a substantial threat to the liveability of the Earth for us humans. The question is: How many dimensions of global environmental change and how many options to address it are marginalized or simply ignored as a consequence of prevailing ideas as to what constitutes effective governance?

References

Brenner, N., Peck, J. and Theodore, N. (2010) 'Variegated Neoliberalization: Geographies, Modalities, Pathways', *Global Networks*, 10, 2, pp. 182–222.

Brenner, N. and Theodore, N. (2002) 'Cities and the Geographies of Actually Existing Neoliberalisms', *Antipode*, 34, pp. 349–379.

Bromley, D. (2007) 'Environmental Regulations and the Problem of Sustainability: Moving beyond "Market Failure"', *Ecological Economics*, 63, pp. 676–683.

Commonwealth of Australia (2011) *Emissions Trading Schemes by Country*. Canberra, ACT: Australian Government Department of Climate Change. Available at www.climatechange.gov.au/government/international/global-action-facts-and-fiction/ets-by-country.aspx. Accessed 13 October 2011.

Dargusch, P. and Griffiths, A. (2008) 'Introduction to Special Issue: A Typology of Environmental Markets', *Australasian Journal of Environmental Management*, 15, pp. 70–75.
Dean, M. (1998) 'Administering Asceticism: Reworking the Ethical Life of the Unemployed Citizen', pp. 87–107 in M. Dean and B. Hindess (eds), *Governing Australia: Studies in Contemporary Rationalities of Government*. Cambridge, UK: Cambridge University Press.
Dean, M. (1999) *Governmentality: Power and Rule in Modern Society*. London: Sage.
Drechsler, M. and Hartig, F. (2011) 'Conserving Biodiversity with Tradable Permits Under Changing Conservation Costs and Habitat Restoration Time Lag', *Ecological Economics*, 70, pp. 533–541.
Engel, S., Pagiola, S. and Wunder, S. (2008) 'Designing Payments for Environmental Services in Theory and Practice: An Overview of the Issues', *Ecological Economics*, 65, pp. 663–674.
Foucault, M. (1991) 'Governmentality', pp. 87–104 in G. Burchell, C. Gordon and P. Miller (eds), *The Foucault Effect: Studies in Governmentality*, London: Harvester Wheatsheaf.
Garnaut, R. (2008) *The Garnaut Climate Change Review: Final Report*, Melbourne, VIC: Cambridge University Press.
Henderson, B. and Norris, K. (2008) 'Experiences with Market-Based Instruments for Environmental Policy', *Australasian Journal of Environmental Management*, 15, pp. 113–120.
Lemke, T. (2001) 'The Birth of Bio-Politics: Michel Foucault's Lecture at the Collège de France on Neo-liberal Governmentality', *Economy and Society*, 30, pp. 190–207.
Lockie, S. (2006) 'Networks of Agri-Environmental Action: Temporality, Spatiality and Identity within Agricultural Environments', *Sociologia Ruralis*, 46, 1, pp. 22–39.
Lockie, S. (2010) 'Neoliberal Regimes of Environmental Governance: Climate Change, Biodiversity and Agriculture in Australia', pp. 364–377 in M. Redclift and G. Woodgate (eds), *The International Handbook of Environmental Sociology*, 2nd ed. Cheltenham, UK: Edward Elgar.
Lockie, S. (2013) 'Market Incentives, Ecosystem Services, and Property Rights: Assumptions and Conditions for Sustained Social and Ecological Benefits', *Land Use Policy*, 31, pp. 90–98.
Lockie, S. and Higgins, V. (2007) 'Roll-out Neoliberalism and Hybrid Practices of Regulation in Australian Agri-Environmental Governance', *Journal of Rural Studies*, 23, pp. 1–11.
Lockie, S. and Tennent, R. (2010) 'Market Instruments and Collective Obligations for On-farm Biodiversity Conservation', pp. 287–301 in S. Lockie and D. Carpenter (eds), *Agriculture, Biodiversity and Markets: Livelihoods and Agroecology in Comparative Perspective*. London: Earthscan.
McCarthy, J. and Prudham, S. (2004) 'Neoliberal Nature and the Nature of Neoliberalism', *Geoforum*, 35, pp. 275–283.
Miller, P. and Rose, N. (1990) 'Governing Economic Life', *Economy and Society*, 19, 1, pp. 1–31.
Muradian, R., Corbera, E., Pascual, U, Kosoy, N. and May, P. (2010) 'Reconciling Theory and Practice: An Alternative Conceptual Framework for Understanding Payments for Environmental Services', *Ecological Economics*, 69, pp. 1202–1208.
National Market-Based Instrument Working Group (NMBIWG) (2005) *National Market-Based Instruments Pilot Programme: Round One. An Interim Report*, Canberra, ACT: National Action Plan for Salinity and Water Quality.
Pascual, U., Muradian, R., Rodríguez, L. and Duraiappah, A. (2010) 'Exploring the Links Between Equity and Efficiency in Payments for Environmental Services: A Conceptual Approach', *Ecological Economics*, 69, pp. 1237–1244.
Peck, J. and Tickell, A. (2002) 'Neoliberalizing Space', *Antipode*, 34, pp. 380–404.
Pellizzoni, L. (2011) 'Governing Through Disorder: Neoliberal Environmental Governance and Social Theory', *Global Environmental Change*, 21, pp. 795–803.
Prudham, S. (2004) 'Poisoning the Well: Neoliberalism and the Contamination of Municipal Water in Walkerton, Ontario', *Geoforum*, 35, pp. 343–359.
Stern, N. (2007) *The Economics of Climate Change: The Stern Review*, Cambridge, UK: Cambridge University Press.
Wunder, S., Engel, S. and Pagiola, S. (2008) 'Taking Stock: A Comparative Analysis of Payments for Environmental Services in Developed and Developing Countries', *Ecological Economics*, 65, pp. 834–852.

7
Dilemmas for standardizers of sustainable consumption

Magnus Boström and Mikael Klintman

As the debate over sustainable consumption among policy practitioners, scholars, non-government organizations (NGOs) and the media intensifies, the eighteenth-century notion (and ideal) of *Homo economicus* – a self-interested, individual making 'rational' choices on the market – has been ruthlessly mocked as both unrealistic and coldly amoral. Be that as it may, the notion of this square top hat figure remains relevant to the challenges facing sustainable consumption on one point: its emphasis on *knowledge* as the most fundamental component of the ability to act in any meaningful way on the market. But this poses yet another problem. Sufficient knowledge is in itself an immense challenge to the ideal of economically rational consumers making utility maximising choices. The addition of environmental and social concerns into consumer decision-making only serves to further increase knowledge complexity, which the environmentally and socially conscious consumer cannot manage individually. Thus, consumption has become increasingly socialised in terms of the actors and organizations involved in offering consumers information about the environmental and social implication of their purchases. Indeed, the jungle of environmental data needs to be translated into a manageable set of knowledge claims that can be used for daily decision making.

This chapter focuses on the role of labelling and certification. Globally, we see schemes for green and ethical issues in a wide range of sectors, including forestry, fishery, energy, mining, tourism, food, fashion, coffee, finance, chemicals and so on. Labelling and certification is a consumer-oriented and market-based strategy that relies on the standardization of principles and prescriptive criteria, as well as on the symbolic differentiation of sustainable and unsustainable products and services (Boström and Klintman 2008). Furthermore, this strategy relies on compromise among the key dimensions of sustainability: economic, social and environmental. Because labelling and certification is a consumer-oriented and market-based policy tool, it has to be adjusted to the logics of global capitalism. This is likely to result in a basic contradiction among the sustainability dimensions, as well as the fundamental fact that consumption, from an ecological point of view, needs to be reduced on a global scale. A number of scholars have discussed the inherent contradictions and limitations of using market-based governance and political action to foster sustainable development and tackle social and environmental problems (see Allen and Kovach 2000; Barham 2002; Dauvergne and Lister 2010; Guthman 2004, 2009;

Klooster 2010; Raynolds 2000; Shaw and Black 2010; Taylor 2004). In his investigation and comparison of Fairtrade and Forest Stewardship Council, Taylor (2004: 130) argues that:

> One of the most serious challenges of certification and labelling initiatives today is actually to be 'in the market but not of it', that is, to be able to pursue alternative values and objectives such as social justice and environmental sustainability without being captured by the market's conventional logic, practices and dominant actors.

In this chapter, a point of departure is the general observation that fundamental contradictions appear in labelling and certification strategy. The aim is to further identify what kind of key decision dilemmas emerge from this contradiction, and which ones the standardizers of sustainable consumption need to cope with. There are no universal solutions to these problems. Dilemmas cannot be eliminated, but we believe that revealing them rather than ignoring their existence will, in the long run, facilitate reflective and responsible consumerism.

This chapter is based on qualitative comparative, case-study-oriented research on eco-labelling systems at the global, regional and national levels. This includes organizational structures, decision-making processes and the debates around labelling processes (see Boström 2011; Boström and Klintman 2008; Klintman 2012, Klintman and Boström 2012; Tamm Hallström and Boström 2010). In this chapter, we will draw on several examples, including forest certification (Forest Stewardship Council; FSC), marine certification (Marine Stewardship Council; MSC), ecotourism, organic agriculture and other cases. The following discussion is presented according to the four general dilemmas identified, namely: (1) balancing inclusiveness and exclusiveness of social and environmental certification; (2) the scope of sustainability; (3) the need for local adaptation or global universality; and (4) simplifying the standard while preserving complexity.

For all or for the best ones?

In all sectors with social and environmental certification, from agriculture to tourism, the following dilemma emerges: should the criteria be formulated in an inclusive or an exclusive manner? At one extreme sits management standards associated with the International Organization for Standardization (ISO). This organization sets standards for both environmental (ISO 14000 series) and social (ISO 26000 guidance standard) performance. These standards emphasize harm reduction and continuous improvement. All kinds of organizations are supposed to be potential users of these standards, as long as they comply with certain management procedures. At the other extreme, environmental NGOs push for a strict differentiation between 'sustainable' and 'unsustainable' practices, to reward only the 'best' examples of sustainability within an industry, and to provide templates for others to follow. In this latter view, rhetorical adherence to generic sustainability goals and improved management procedures are not sufficient to distinguish between real forerunners and laggards. The use of prescriptive, substantive social and environmental criteria, including measurable thresholds, has to be part of the standard. This view also contends that, once a certain proportion of companies manage to reach the criteria, these standards should be raised even further so that the green progress and dynamics are kept alive.

There are potential pros and cons with either route. The exclusive 'best-in-class' principle has several advantages: first, it is perhaps (morally) intuitive that only the most environmentally and socially progressive products, services and companies should be awarded a label or certificate. Second, the exclusive route might make the sustainability certificate or label a particularly powerful tool of market distinction. This may trigger a particularly intense green competition among companies and service providers, as exclusive certificates and labels make it possible to introduce substantial price premiums, enhance the company's public relations, as well as promote visibility of certified companies, products and services. In that way, a 'race to the top' may be encouraged. Third, research indicates the importance of a clear role of green pioneers and frontrunners as role models acting as templates for business practices, as well as other policies and regulations (First and Khetriwal 2008; Herremans et al. 2009). Taken together, this may enhance the intensity of sustainability competition and innovations on the market, as well as foster 'rule-setting' in the market, benefiting society's struggle towards sustainable development.

On the other hand, a more inclusive approach has advantages in stimulating a wider range of actors to reduce their negative environmental and social impact. First, in terms of aggregated sustainability benefits, modest reductions of environmental and social harms of a large number of companies and organizations could be greater than dramatic reductions of harm by a small number of actors. Second, the inclusive approach avoids a dilemma that has both ideological and epistemological roots: What should be counted as the best? and What credible yardstick and procedures should we use to single out the real, objective forerunners? A related concern is that there may be tradeoffs involved. For instance, what might be best from an environmental sustainability viewpoint might not be perceived as the best strategy from a social sustainability perspective. A third dilemma is that very stringent criteria in terms of social and environmental conduct may create imbalances between demand and supply, and thus hamper market growth. This may result in the introduction of competing industry-dominated schemes into the market – something that has happened in several sectors.

Interestingly, the call for inclusive certificates through reductions of thresholds and substantive criteria have come from two diametrically opposed types of companies: the big, sometimes multinational companies (for example, large car producers and energy and food companies), as well as small and medium firms (SMEs), particularly in the Global South. To bring the big players on board is, in several sustainability schemes, presented as key to comprehensive reductions of environmental and social harm. The increased visibility that comes from the participation of multinational firms is an obvious rationale for inclusive strategies. For example, it was considered a milestone in the MSC's history when retail giant Walmart announced in early 2006 its intention to use MSC-labelled seafood in sourcing all of its wild-caught fresh and frozen fish for the North American market – a market in which the MSC previously had limited impact (Tamm Hallström and Boström 2010). Nevertheless, there are risks involved in such connections. Granting certificates or labels to organizations with controversial economic practices and advertising them as sustainable could undermine the legitimacy of the standard-setting organization. As a Greenpeace member ruminated about the MSC's so-called fast-growth strategy, 'I think there is a very dangerous trend coming up, that Wal-Mart and others are making big commitments to selling only certified fish within three years' (quoted in Tamm Hallström and Boström 2010: 158).

As for the smaller, less affluent players such as local SMEs, certification may require reductions of certain levels of criteria, such as those that are particularly costly. It is a common experience within labelling and certification schemes that 'economies of scale' create barriers to certification for SMEs. This is apparent in, for example, ecotourism (Klintman 2012), FSC-certification (Boström 2011), and organic agriculture (Guthman 2004, 2009). Even the Fairtrade system, which is explicitly aimed at redistributing economic values and altering trade relations along the mainstream commodity chain for the benefit of small-scale and family-based agriculture in developing countries, faces difficulties when it comes to avoiding barriers to entry and benefiting the farmers at the beginning of the product chain (Guthman 2009; Taylor 2004). In general, the main argument is that SMEs face too steep an uphill slope in terms of access to information, knowledge, skills, money, technical capacity and management structures to pass very stringent sustainability criteria. Small enterprises (often family-owned or community-based) have less financial flexibility and may be more averse to the economic risk involved with changing business practices. Various market issues may also encumber certification for SMEs. Butterfield and colleagues (2005) notice that small forest enterprises face difficulties in delivering large, consistent supplies of certified timber demanded by retailers or processing industries. Meidinger (1999) also concluded, in reference to a forestry case study, that the stricter the environmental standards, the greater the disparity between large, sophisticated corporations and small, often Third World businesses in their ability to meet them.

Consequently, there has to be room for learning processes, not least mutual ones, within the schemes where SMEs can be included and improve their sustainability record over time. Such measures as group certification, capacity building, financial donation, sliding scale of fees for members and the flexible interpretation of standards criteria are used in various systems (Boström 2011; Klintman 2012). Furthermore, lowering certain criteria in order to bring local SMEs in the Global South on-board as co-developers of the standards, and as users, brings legitimacy benefits for the bigger companies, often from the Global North, that are certified as well. Such inclusion increases external trust in the schemes by the market, and reduces the risk of them (and the companies involved) being criticized for having a Northern bias or for being a tool of neocolonialism (Klintman 2012).

A broad or narrow scope of sustainability?

A second dilemma is whether a sustainability certificate and labelling scheme should be broad or narrow in scope. Since sustainable development denotes ecological, social and economic dimensions – although they are often claimed to be inseparable – what ambitions should guide these schemes? Should (and could) a scheme cover all of these dimensions in any meaningful way? Could it even cover one of the dimensions in full?

How many dimensions?

The first issue to consider is whether schemes should focus on social, economic or environmental sustainability dimensions, or on all three. The FSC and ecotourism are cases in which there have been systematic ambitions to cover a broad spectrum of both social and environmental aspects. The MSC, organic agriculture and a number of common eco-labelling schemes (such as the Green seal, the EU flower, the German Blue Angel and the Nordic eco-label) emerged from a narrower environmental sustainability viewpoint, whereas the Fairtrade label came from a social one. The dilemma here is that the latter cases face bigger challenges in their legitimizing aspirations – to claim sustainability – although they are likely to handle internal processes more efficiently. There might be problems in terms of trust in a product or service using a label that on the one hand is presented as ethically superior to other products and services but that has only taken care of some aspects of sustainability associated with them. Also, narrow schemes that, for instance, only deal with chemical additives in products run the risk of competing with labels with an entirely different profile such as Fairtrade. Such competition between pieces of the sustainability puzzle seems incongruent in light of the foundational idea that sustainability integrates economic, social and environmental dimensions.

Perhaps it is no accident that we see increasing interaction between the environmentally oriented organic agriculture and the socially oriented Fairtrade, because both systems are subject to increasing pressure to better incorporate all sustainability dimensions. The MSC is no exception. The FSC (established in 1993) was seen as a positive model for the MSC (established in 1997). However, the founders of MSC found the FSC's strikingly democratic governance structure and processes too cumbersome and time-consuming, so they chose to set up a different kind of organization – a foundation rather than a membership organization – with a more narrow environmental focus (Tamm Hallström and Boström 2010). The internal management is much less complex and more straightforward within the MSC. Over the years, however, the MSC has been criticized for not including social stakeholders, such as organizations of fishing workers, and for the lack of social criteria in the MSC standards framework, as well as for its failure to reach the developing world (Tamm Hallström and Boström 2010). While they have not yet fully reorganized and reformulated their scope, various efforts have been made to respond to the demands and to make it easier for fisheries in developing countries to obtain certification.

What scope for social sustainability?

Social sustainability is generally seen as a fuzzy concept, and it is far from self-evident how it should be defined. What it means and what it entails is unclear and debated (see, for example, Davidson 2009; Dempsey et al. 2009; Lehtonen 2004; Littig and Grießler 2005). The list of what potentially can be included tends to be long (Boström 2012). Notions of social sustainability often refer to social welfare, quality of life, social justice, social cohesion, cultural diversity, democratic rights, gender issues, workers' rights, broad participation, development of social capital and individual capabilities and the like. It refers to both substantive and procedural issues, such as inclusive, democratic and transparent decision making. Addressing all of these aspects within a single labelling scheme would be an enormous task. Nevertheless, the FSC has ambitious targets, and can be used as an ideal model. It has defined principles and criteria for a broad array of aspects, including community and workers' rights, as well as the legal and customary rights of indigenous peoples to own, use and manage their lands, territories and resources. According to the FSC, the social costs and benefits of forest operations should be assessed in certification, including empowerment issues and the role of a diversified local economy. Organizing rights should be respected among workers and local civil societies and provision should be made for broad stakeholder input including local communities, indigenous groups and workers at every step of the planning process before certification. Nevertheless, the FSC has had mixed results (see Boström 2011 for an extended analysis and discussion).

The fuzzy and broad framing of social sustainability can easily result in high, but unrealizable, expectations (Boström 2012). Achieving social sustainability is an ambitious objective given the exceptionally problematic circumstances worldwide that the FSC is attempting to alter. Furthermore, it is very difficult, indeed nearly impossible, for such a market-based approach as labelling and certification to tackle the whole host of social issues: poverty reduction, literacy improvement, capacity building, wealth redistribution, lack of economic capital, local infrastructure, education and weak local civil societies and workers unions (Boström 2011; Klooster 2010). Indeed, various types of resources and institutions need to be already in place before certification can be initiated and accomplished. Thus, it is difficult to standardize and certify social sustainability without some degree of existing social and economic sustainability. In an FSC report, which assesses the environmental and social benefits of FSC certification, it is mentioned that stakeholders do not always recognize that FSC standards and certification necessarily result

85

from a compromise between social, environmental, and economic goals (FSC 2009: 216–220). Moreover, it is stressed that:

> FSC impact on the complex social realities is indeed often very critically measured against these *high expectations*. At the same time, internal FSC Working Groups and external observers are demanding that FSC 'raise the social bar'. These expectations are usually not addressed to other forest certification schemes with less prominent criteria for social impact.
>
> (FSC 2009: 84)

Evidently, within the social sustainability framing itself, a standards setter faces many dilemmas when trying to cover a broad spectrum of issues while making them achievable against a background of hopes and expectations expressed by various stakeholders.

What scope for environmental sustainability?

Eco-labelling schemes also face ambivalence as to whether they should aspire to cover all relevant environmental problems or only some. Eco-labels on food, including organic labels, are subject to intense debates on this matter (Klintman and Boström 2012). Here, it is useful to understand the history of organic farming and food. Organic food production originally developed as a reaction to intensified use of synthetic pesticides, herbicides and food additives in industrialized agriculture. Organic agriculture, therefore, started as an alternative that was free from such processes and additives, based on small-scale, local food production. This focus on pesticides and additives is still a major component of organic criteria. For instance, according to the European Union Regulation No. 834/2007, Article 4, on *Organic Production and Labelling of Organic Products*, the EU label for organic farming includes, among others, the following principles: the appropriate design and management of biological processes based on ecological systems using natural resources which are internal to the system, restriction of the use of external inputs and strict limitation of the use of chemically synthesized inputs to exceptional cases (European Community 2007).

However, in the last two to three decades, the dominant framing of environmental problems tied to food production has changed. The increasing dominance of the climate change discourse in the first decade of the millennium has led to intensified debates on pesticides and additives. Even the original local production ideal, although often simplistically equated to reduced carbon emissions due to shorter transportation distances, has been questioned: sometimes long-distance transportation of food produced in sunny areas with no need for heated greenhouses is said to be more climate-sound than local production in less conducive areas. Interestingly, this debate is congruent with the development where the ideology of 'small is beautiful' has given way to a more eco-modern view of sustainable development reflected in the upscaling of organic production. Organic producers and policymakers acknowledge the increasing demand for organic food, largely among the upper-middle class around the world, and have helped expand the organic criteria to converge across different parts of the world.

In contrast with this mainstreaming of large-scale organic agriculture and food, certain substantial consumer groups are questioning the very types of products awarded eco-labels, not least eco-labelled meat (Fox and Ward 2008). Should, for instance, beef be awarded an eco-label just because the cow has had pesticide-free feed and has been able to spend time outside of the barn? This question is part of a more radical criticism of meat production and the exploitation of animals, using a wide range of arguments from carbon emissions, animal welfare and land use to food security. All these put increasing pressure on organic food labels, and other eco-labels, to consider the wider ecological problems associated with the products (Klintman and Boström, 2012).

In sum, determining the scope of sustainability is a matter of finding a balance between including everything that is relevant from a sustainability point of view, and what is *achievable* as well as *feasible* to standardize. A more narrow scope might be much more feasible to handle in terms of governance, the gathering of relevant expertise, and technical solutions. On the other hand, the narrow scope may result in external legitimacy attacks by consumer groups and other stakeholders who feel that their own concerns are misrepresented. The broad-scope strategy may face legitimacy struggles as well, because the more one wants to include in the framework, the more likely it is that perceived gaps between achievements and expectations will appear eventually.

Locally adapted or universal?

The third dilemma has to do with a main tension surrounding sustainability-related policy development in general. Should the criteria be adapted to the particular local-regional ecology and culture? Or should there be a universal set of criteria (of sustainable forestry, fishery or tourism, for example) that can be applied to any culture or bio-region? The various schemes develop various mechanisms to handle this challenge, both *organizational*, such as setting up regional and national offices, and *regulatory*, such as allowing for schemes that are nationally or regionally adjusted from a global framework. However, no scheme can solve this tension permanently, as our illustrations here will show.

It is fair to say that most certification schemes, being tools on the market, are strongly affected by pressures towards international convergence. The liberal market principles point towards global translation, convergence and harmonization. Thus, few developments within a specific, nationally based eco-labelling scheme (for instance, the USDA-controlled green seal for organic food) can be fully understood without comparing these developments with processes in other regions (in this case, European and other international schemes for organic food). The common argument for the harmonization of standards in public debate is typically that there is a need to reduce consumer confusion. But international standards also facilitate trade and help to make companies as well as the NGOs involved more widely trusted among green political consumers globally (Crane 2005).

Nevertheless, the local–global dilemma remains. On the one hand, there is the general sustainable development ideal of marrying, integrating and cross-fertilizing the local and the global. On the other, there are worries, not least among several social and environmental NGOs, about the tension between the 'Local' South and the Global North. In the context of eco-standards and international eco-accreditation, this relationship has been portrayed as problematic in two contradictory ways: (1) the local, victimized South vs the global, eco-imperialist and exploitive North; and (2) the local, biased and uncontrolled South vs the Global North striving to develop controlled and neutral eco-standards to solve worldwide sustainability problems (Klintman 2012).

International meta-organizations (Ahrne and Brunsson 2008), such as the International Organization for Standardization (ISO), the International Federation of Organic Agriculture Movements (IFOAM), the Global Ecolabelling Network (GEN) and the International Social and Environmental Accreditation and Labelling Alliance (ISEAL Alliance) work hard to establish various kinds of international convergence. Individual labelling schemes then take part in these meta-organizations, which facilitate exchange and harmonization among members. For example, ISEAL has developed a *Code of Good Practice for Setting Social and Environmental Standards*, which its members, such as FSC, MSC, Fairtrade International and International Organic Accreditation

Service, use (Boström and Klintman 2008). But ISEAL Alliance faces the challenge of rapid growth in certification and labelling in general. With this growing number of standards comes the risk of any flaw or scandal among any individual labelling organization harming the entire movement (Sasha Courville, Executive Director of ISEAL Alliance, quoted in Boström and Klintman 2008: 102). This is a challenge all serious certification and labelling organizations face. Credibility issues thus drive harmonization.

One way of creating international convergence is to develop global accreditation of certification schemes. A large number (sometimes hundreds) of eco-labels and certificates can be scrutinized on the basis of certain very general criteria, and thus earn an international accreditation. This is far from unique to sustainable development projects. Some scholars argue that global accreditation schemes belong to a global trend in many sectors, including health care, higher education and technology. It has become so widespread that society as a whole has been called 'the audit society' (Power 1997).

When it comes to sustainable development issues, however, there are particular challenges involved in developing global accreditation and certification schemes. First, ecological and environmental problems differ substantially across local regions. For instance, water shortages are urgent problems in certain regions and almost nonexistent in others. Should requirements for economizing water be as strict in both regions as a basis for the certification and accreditation, then? Another obvious difference has to do with threats to biodiversity, a set of issues that vary substantially, too. Furthermore, in addition to the differences in ecology and environmental problems across local regions, some argue that cultural, social and economic differences also need to be considered. Therefore, should standards harmonize criteria for working hours, living wages and the minimum age of workers in certified companies across the globe? The typical response to the question of social uniformity is that certification and accreditation schemes should be based on minimum standards described in the United Nations or International Labour Organization resolutions in combination with the national regulations of the specific country in question. While some scholars and practitioners emphasize that global schemes, as well as eco-standards in general, could be a progressive force towards raising the social and economic criteria above the national regulations, others maintain that it would be more productive to use international eco-standards and accreditation to put pressure on local firms and certification bodies to live up to national regulations that are already in place. This, as the latter contends, would hopefully foster transparency and create market pressures on non-certified firms, as well as meet the current national regulations.

Simplifying the standard (the message) or preserving complexity?

The fourth dilemma is the following: To what extent should the sustainability standards, including the message sent to consumers, be simplified? To be sure, a label or seal is by definition, simple and categorical. A main reason why producers, retailers and consumers accept additional costs associated with the production or purchase of eco-labelled products and services is that someone has already done the job of analysing and assessing the sustainability impacts. The rest of the actors along the value chain should not have to make technical assessments and need only trust in the labelling scheme and its experts. Yet the issue cannot be entirely resolved within this answer. The standardizers themselves face the intricate challenge of translating an extraordinarily complex social and environmental situation into a simple, plain, categorical label. Furthermore, to what extent should consumers be informed about this process of translation?

The ethically labelled commodity could to some degree be seen as a fetish, Guthman (2009) argues, because of the many unveiled aspects of the labelling process. The typical solution presented to the simplicity of eco-standards is that they should be supplemented with more detailed, technical information about the actual reductions of environmental and social harms promised with this alternative product or service. Such substantive transparency is often called for in light of market worries about green inflation and greenwashing. Yet, as we have argued elsewhere (Klintman and Boström 2008), such transparency is rarely sufficient. In fact, technical information that is too detailed may obscure the important information and its implications. What is needed instead is a transparency that unveils the decision-making procedures as well as some of the technical details. By revealing which organizations have been part of the criteria setting, ideally with an inclusive approach to stakeholder participation in the schemes, and by indicating the political, strategic and scientifically influenced process that lead to the inclusion of some criteria (for example, chemical additives), and exclusions of others (such as carbon emissions), a reflective trust can be developed. This, in turn, can promote more responsible consumerism and avoid the kind of ethical fetishism that Guthman addresses.

As many social thinkers have stressed, market actors today, including citizen-consumers, are critical and reflective concerning their own and other actors' knowledge claims. Developers of tools for sustainable consumption need to, and should, consider the potential for reflective trust in standardization schemes among a broad group of the population (Boström and Klintman 2008, 2009). This is a trust that is incessantly provisional as well as conscious, based on insights into both substantive and procedural matters. Indeed, it is among the well-educated and reflective citizens that we can observe high levels of political consumerism (Boström and Klintman 2008, 2009). It is precisely among green consumers that one is most likely to find people with particularly reluctant attitudes towards green advertising and overly simplified messages (Zinkhan and Carlson 1995; see also Crane 2000) and who, therefore, have broad reflective potential.

Conclusions and discussion

In the introduction, the contradictions involved in labelling as a strategy to promote sustainable consumption were highlighted and related to its fundamental reliance on conventional market logics. The contradictions result in decision dilemmas that confront standardizers. The first dilemma – whether to formulate criteria that allow for many or few actors' compliance – has no simple answer. A pathway towards sustainability could theoretically be paved through each route, and standardizers need to reflect upon a multitude of issues including dynamic effects on the market, consequences for SMEs, and epistemological hurdles to be able to make claims about what is really most desirable. Neither does the second dilemma – to determine the scope among and within sustainability dimensions – have a straightforward answer. Comprehensiveness has to be weighed against what is achievable and feasible, and standardizers need to avoid enormous perceived gaps between visions, expectations and actual outcomes. The third dilemma is concerning the balance between adapting criteria very closely to the particular local-regional ecology and culture while keeping it universal. Although standardizers set up organizational and regulatory means to balance these double needs, it is easy to imagine never-ending calls for greater global harmonization and more fine-tuning of local adaptation at the same time. Fourth, whereas labelling fundamentally requires a translation process from complexity to simplicity, a dilemma arises around how much of this translation process should be revealed to consumers. After all, users of labels use labels because they simplify the purchase. The dilemmas, as such, cannot be avoided and need to be addressed. The solution would be to acknowledge them, as well as to foster a deliberate dialogue about them. Importantly, this dialogue should include the users. The last dilemma presented is perhaps one that is only superficial (if we go beyond the individual standard-setter's perspective). A broad public discussion about the very translation process (from complexity to simplicity) does not have to undermine the labelling process as such, but could on the contrary empower this tool in the broad repertoire of instruments for sustainable consumption. It could then demonstrate alternative green business practices and provoke broader debates on the topic (Guthman 2009; Klooster 2010). A layered transparency (Klintman and Boström 2008) and acknowledgement of a multifaced consumer role (Boström and Klintman 2008) could foster a more responsible consumerism. In this, we found great potential in green labelling for sustainable consumption. Finally, we (including standardizers) should also keep in mind that labelling is just one tool in the repertoire, which Dauvergne and Lister (2010: 146) emphasize:

> eco-labeling and green shopping *do* have value. Eco-consumerism can be an effective voluntary policy instrument to spur environmental improvements. Yet overestimating its potential on its own to produce *global change* can leave consumers overconfident in the power of their eco-purchases, thereby releasing pressure on governments and corporations for fundamental changes in the industrial use, marketing, and valuation of the world's natural resources.

References

Ahrne, G. and Brunsson, N. (2008) *Meta-organizations*. Cheltenham, UK: Edward Elgar.
Allen, P. and Covack, M. (2000) 'The Capitalist Composition of Organic: The Potential of Markets in Fulfilling the Promise of Organic Agriculture', *Agriculture and Human Values*, 17, pp. 221–232.
Barham, E. (2002) 'Towards a Theory of Values Based Labelling', *Agriculture and Human Values*, 19, pp. 349–360.
Boström, M. (2011) 'The Problematic Social Dimension of Sustainable Development: The Case of the Forest Stewardship Council', *International Journal of Sustainable Development and World Ecology*, 19, 1, pp. 3–15.
Boström, M. (2012) 'A Missing Pillar? Challenges in Theorizing and Practicing Social Sustainability: Introduction to the Special Issue', *Sustainability: Science, Practice, and Policy*, 8, 1, pp. 3–14.
Boström, M. and Klintman, M. (2008) *Eco-standards, Product Labelling and Green Consumerism*. Basingstoke, UK: Palgrave.
Boström, M. and Klintman, M. (2009) 'The Green Political Consumer of Food: A Critical Analysis of the Research and Policies', *Anthropology of Food*. Available at http://aof.revues.org/index6394.html. September 2009.
Butterfield, R., Hansen, E., Fletcher, R. and Nikinmaa, H. (2005) *Forest Certification and Small Forest Enterprises: Key Trends and Impacts – Benefits and Barriers*. Forest Trends and Rainforest Alliance. Available at www.rainforest-alliance.org/forestry/documents/forestcertpaper.pdf. Accessed 1 October 2010.
Crane, A. (2000) 'Facing the Backlash: Green Marketing and Strategic Reorientation in the 1990s', *Journal of Strategic Marketing*, 8, pp. 277–296.
Crane, A. (2005) 'Meeting the Ethical Gaze: Challenges for Orienting to the Ethical Market', pp. 219–232 in R. Harrison, T. Newholm and D. Shaw (eds), *The Ethical Consumer*. London: Sage.
Dauvergne, P. and Lister, J. (2010) 'The Prospects and Limits of Eco-Consumerism: Shopping Our Way to Less Deforestation? *Organization and Environment*, 23, 2, pp. 132–154.
Davidson, M. (2009) 'Social Sustainability: A Potential for Politics?' *Local Environment*, 14, 7, pp. 607–619.
Dempsey, N., Bramley, G., Power, S. and Brown, C. (2010) 'The Social Dimension of Sustainable Development: Defining Urban Social Sustainability', *Sustainable Development*, 19, 5, pp. 289–300.
European Community (2007) *European Union Regulation No 834/2007 on Organic Production and Labelling of Organic Products*, Official Journal of the European Union, Luxembourg. Available at http://eur-lex.europa.eu/LexUriServ/LexUriServ.do?uri=OJ:L:2007:189:0001:0023:EN:PDF. Accessed 2 September 2010.
First, I., and Khetriwal, D.S. (2008) 'Exploring the Relationship Between Environmental Orientation and Brand Value: Is the Fire or Only Smoke?' *Business Strategy and the Environment*, 19, pp. 90–103.
Fox, N., and Ward, K. (2008) 'Health, Ethics and Environment: A Qualitative Study of Vegetarian Motivations', *Appetite*, 50, 2–3, pp. 422–429.

Forest Stewardship Council (FSC) (2009) *FSC Reflected in Scientific and Professional Literature: Literature Study on the Outcomes and Impact of FSC certification*. FSC Policy Series No. 2009. P001. Forest Stewardship Council. Available at http://www.fsc.org/fileadmin/web-data/public/document_center/publications/FSC_Policy_Series/Impacts_report_-_Karmann_2009.pdf. Accessed 1 September 2010.

Guthman, J. (2004) *Agrarian Dreams: The Paradox of Organic Farming in California*. Los Angeles, CA: California University Press.

Guthman, J. (2009) 'Unveiling the Unveiling: Commodity Chains, Commodity Fetishism, and the "Value" of Voluntary, Ethical Food Labels', pp. 190–206 in J. Bair (ed.), *Frontiers of Commodity Chain Research*. Palo Alto, CA: Stanford University Press.

Herremans, I., Herschovis, M. and Bertels, S. (2009) 'Leaders and Laggards: The Influence of Competing Logics on Corporate Environmental Action', *Journal of Business Ethics*, 89, 3, pp. 449–472.

Klintman, M. (2012) 'Issues of Scale in the Global Accreditation of Sustainable Tourism Schemes: Toward Harmonized Re-embeddedness?' *Sustainability: Science, Practice, and Policy*, 8, 1, pp. 1–11.

Klintman, M. and Boström, M. (2008) 'Transparency Through Labelling? Layers of Visibility in Environmental Risk Management', pp. 178–197 in C. Garsten and M. Lindh de Montoya (eds), *Transparency in a New Global Order: Unveiling Organizational Visions*. Cheltenham, UK: Edward Elgar.

Klintman, M. and Boström, M. (2012) 'Political Consumerism and the Transition Towards a More Sustainable Food Regime: Looking Behind and Beyond the Organic Shelf', pp. 107–128 in G. Spaargaren, A. Loeber and P. Oosterveer (eds), *Food Practices in Transition*. London: Routledge.

Klooster, D. (2010) 'Standardizing Sustainable Development? The Forest Stewardship Council's Plantation Policy Review Process as Neoliberal Environmental Governance', *Geoforum*, 41, pp. 117–129.

Lehtonen, M. (2004) 'The Environmental–Social Interface of Sustainable Development: Capabilities, Social Capital, Institutions', *Ecological Economics*, 49, pp. 199–214.

Littig, B. and Grießler, E. (2005) 'Social Sustainability: A Catchword Between Political Pragmatism and Social Theory', *International Journal of Sustainable Development*, 8, pp. 65–79.

Meidinger, E. (1999) '"Private" Environmental Regulation, Human Rights and Community', *Buffalo Environmental Law Journal*, 7, pp. 123–237.

Power, M. (1997) *The Audit Society. Rituals of Verification*. Oxford, UK: Oxford University Press.

Raynolds, L. (2000) 'Re-embedding Global Agriculture: The International Organic and Fair Trade Movements', *Agriculture and Human Values*, 17, pp. 297–309.

Shaw, D. and Black, I. (2010) 'Market Based Political Action: A Path to Sustainable Development?' *Sustainable Development*, 18, 6, pp. 385–397.

Tamm Hallström, K. and Boström, M. (2010) *Transnational Multi-Stakeholder Standardization: Organizing Fragile Non-State Authority*. Cheltenham, UK: Edward Elgar.

Taylor, P.L. (2004) 'In the Market But Not of It: Fair Trade Coffee and Forest Stewardship Council Certification as Market-Based Social Change', *World Development*, 33, 1, pp. 129–147.

Zinkhan, G.M. and Carlson, L. (1995) 'Green Advertising and the Reluctant Consumer', *Journal of Advertising*, 24, 2, pp. 1–6.

Part II
Climate change, energy and adaptation

8

Climate, scenario-building and governance

Comprehending the temporalities of social-ecological change

Stewart Lockie

About 200 years ago the Earth entered the age of the Anthropocene, a geological epoch in which humans have come to rival the great forces of nature in the regulation of critical Earth-system processes (Crutzen 2002). With the advent of the industrial revolution, humans began to modify the Earth's atmosphere, biota, hydrology, nutrient cycles and surface-level energy balance to such an extent as to leave behind the relatively steady and favourable environmental state of the Holocene interglacial (see also Steffen et al. 2007). Following the Second World War, economic activity, energy use and, by extension, human influence on the environment grew exponentially in what Steffen et al. (2007: 617) refer to as the Great Acceleration, a stage they argue is reaching criticality. Of nine Earth-system processes identified as essential to the maintenance of a 'safe operating space' for humanity typical of the Holocene (Rockström et al. 2009), three appear to have already passed plausible boundaries for the avoidance of potentially dangerous environmental transformation; those processes being climate change, biodiversity loss and changes to the nitrogen cycle. At the same time, Steffen et al. (2007: 618) argue there is evidence that 'the intellectual, cultural and legal context that permitted the Great Acceleration after 1945 has shifted in ways that could curtail it' should the challenge to act be taken seriously.

We are used to thinking of modernity and associated processes of industrialization, differentiation and bureaucratization as a mere instant in time when compared with evolutionary, let alone geological, time. By contrast, the idea of the Anthropocene recasts modern humans as geological actors (Lövbrand et al. 2009). Traces of novel biotic, sedimentary and geochemical transformations are sufficiently distinct in the Earth's geology, according to the Geological Society of London, as to render this proposition scientifically credible (Zalasiewicz et al. 2010). In the longer term, the conceptual utility of 'the Anthropocene' will depend on its ability to explain observations that seem to signal the end of the Holocene. In the shorter term, its utility lies in its capacity as a 'vivid yet informal metaphor of global environmental change' to stimulate research, debate and policy that seeks to unpack and respond to the multiple temporal and spatial dimensions of human influence on Earth-systems (Zalasiewicz 2008: 7). The idea of the Anthropocene simultaneously stretches our thinking beyond intergenerational time horizons and linear models of global demographic and ecological change. It forces us to confront the possibility that continued modification of key Earth-system processes may push those systems towards thresholds, or tipping points, of rapid transformation into fundamentally altered system states. To put this more sociologically, the idea of the Anthropocene gives us a basic conceptual framework with which to begin rendering the Earth as a temporally and spatially co-evolving human-ecological system, both thinkable and governable (Lövbrand et al. 2009).

Superficially, at least, the proposed stages of the Anthropocene map reasonably neatly onto temporal categories used by social theorists to characterize major periods of social change: in particular, the ideas of modernity, late modernity, risk society, reflexive modernization and ecological modernization. This is not surprising given their common interest in explaining the consequences of the industrial revolution. What is more surprising is how little sustained attention social theorists have devoted to time, in general, and to the temporalities of social-ecological change, more specifically. The notion of temporality is used throughout this chapter to refer: first, to the unique and interdependent paces and rhythms of socio-ecological processes through time and space; and, second, to the ways in which we attempt to understand and control these rhythms through an assortment of conceptual frameworks, technologies and projects. The naming of epochs as one way to apprehend the temporality of major social-ecological transformations has a certain utility. However, it is not the only way to conceptualize temporality and it is not without consequence for our response to perceived threats. Anthropocene imagery evokes a pressing moral case for action to avoid undesirable Earth-system change but it may also obscure or avoid difficult questions over responsibility, democracy, citizenship, conflict and power (Lövbrand et al. 2009). The seemingly obvious proposition – that the geological force of humanity must be mitigated to reduce pressure on the Earth-system – is not obvious in its implications. The naming of time is a political and social act that warrants close sociological analysis. Climate models and scenarios synthesized through the Intergovernmental Panel on Climate Change (IPCC) assessment process present an alternative conceptualization of the temporality of environmental change, which evokes an equally compelling case for action. Climate scenarios, it is argued in this chapter, shape our response to the threat of Earth-system change but do so, again, in a manner that is not obvious in its implications.

The point of this chapter is not to problematize the notion of the Anthropocene, of climate scenarios, or any other means of conceptualizing the temporality of human-ecological co-evolution, for its own sake. The point, rather, is to review conceptualizations of time and temporality in social-ecological theory, and to begin to examine implications of these conceptualizations for our understanding of key environmental threats and for more effective environmental governance.

Social theory and the ecology of time

Time is a recurring theme in social theory (see Bourdieu 1990; Harvey 1996: Latour 1993; Thrift 1996). Marx, for example, argued that central to class struggle is the struggle to define and control labour time (Harvey 1996). For globalization and network theorists, the increasing velocity of material and symbolic flows compresses time (Thrift 1996). Drawing on Foucault (1991), governmentality theorists argue that the neoliberal subject is expected to internalize an entrepreneurial disposition towards time (anticipating, calculating and managing threats that may, or may not, materialize). Time is not, or is not only, a straightforward measure of the 'distance' between events. Like space, time is socially constructed through the recursive interplay of material circumstances, human actions and our conception of the two (Harvey 1996). Yet the temporalities of social practice – its tempos, rhythms and directions – are seldom subjected to the detailed investigation, Bourdieu (1990) argues, they deserve.

One of the most well-known and comprehensive attempts to develop a specifically socio-ecological theorization of time is provided by Adam (1998) (see also Altvater 1994; Harvey 1996). Adam's starting point is Giddens' and Beck's respective theories of risk and reflexive modernization. According to Beck (1996), modernization has generated a plethora of socio-ecological dangers that are progressively less amenable to comprehension and containment by the institutions of industrial society. The risk society is one in which conflict over the distribution of wealth and other social goods is increasingly accompanied and/or displaced by conflict over the distribution of exposure to hazards (Beck 1992). The spatial and social elements of this distribution are reflected in the environmental justice and popular epidemiology movements, community toxics campaigns, etc. Such social movements challenge institutional monopolies on science and knowledge. Science is problematized and democratized as the threats characterizing the risk society exceed the capacity of scientific institutions to determine and manage on our behalf. And yet, science is more indispensable than ever, as those same threats escape our senses and imaginations (Beck 1992). The risks associated with global environmental change, nuclear accidents, chemical contamination and so on cannot be limited in time or space (Beck 1996). Nor can they be understood according to 'established rules of causality, blame and liability' (Beck 1996: 31).

Conventional probabilistic risk management may be seen as an attempt to bring both time (particularly future time) and space back under control, as fundamental ecosystem processes are systematically transformed by human activity (Giddens 1994). But knowledge no longer brings control. Techno-industrial interventions in the global environment generate a plethora of intended and unintended consequences of unbounded reach and duration that, through feedback loops and interaction with other interventions, create new and unpredictable hazards. As Giddens (1994: 58) argues, the more we try to 'colonize the future', the more it is likely to 'spring surprises on us' due to these non-linear and unpredictable trajectories of change. A straightforward technical calculus of risk is often therefore impossible and/or misleading. Compounding this, in the shadows of many seemingly well-known risk events, lies a residual domain of uncertainty (Lockie and Measham 2012); that is, ambiguity over the likelihood, magnitude or consequences of a hazard at some temporal or spatial scales, despite a comparatively high level of certainty at others. In the case of anthropogenic climate change, for example, the high level of confidence expressed by the IPCC in the likelihood of global warming, and the causal processes behind it, is not matched by the same level of confidence in either the timing of change or its spatial distribution. The residual domain of uncertainty is great enough that in place of precisely calculated and manageable risks we generate an array of more-or-less-plausible scenarios. These scenarios again help to render the Earth, as a temporally and spatially co-evolving human-ecological system, thinkable and governable. These scenarios enable us to consider consequences, plan exigencies and debate responsibilities. Importantly, though, according to Giddens (1994), the plausibility of these scenarios is dependent not only on their empirical accuracy but also on the extent to which people become convinced of their plausibility and act in response to them. Scenarios such as those generated by the IPCC presuppose a reflexive influence on the phenomena they attempt to describe – a reassertion of control over the future.

Adam (1998) seeks to reinforce and build on the temporal dimensions of risk in the work of Beck and Giddens. She argues that the compression of space associated with techno-economic globalization is inseparable from both the compression of time and from conjoined processes of global environmental change. This argument is echoed by Thrift (1996), who notes that metaphors such as time–space entrainment, time–space convergence and time–space compression seek to do more than highlight the relationships *between* time and space. Such metaphors suggest, rather, that time and space are two sides of the same scalar coin; that 'the locale', 'the region' or, indeed, 'the epoch' or 'the times', are spatio-temporal configurations that make little sense stripped of either aspect of scale (see also Harvey 1996). Arguably, to fully make sense, these metaphors also need to take into account social and ecological scale – the former speaking to practices through which relationships with space and time are organized and rendered meaningful, the latter to biological, chemical and physical ecosystem processes and interactions which create their own spatio-temporal patterns and disruptions.

In the current epoch, industrialization homogenizes, accelerates and compresses time (Adam 1998). The near future is brought into the present and the distant future is colonized with the uncertain and indeterminate consequences of technological innovation. Through the deployment of technologies such as genetic engineering, nuclear power, synthetic chemicals, etc., industrial society manufactures its own risks and uncertainties. The scale of change precipitated by these technologies combines with accelerated cycles of innovation and obsolescence to create temporal mismatches between the goals and benefits of a particular technology and its long-term costs and consequences. Industrial society maintains faith that what is not known now will be known in time; that consequences will be averted, and exhausted resources substituted. The effect of innovation can never, according to Adam (1996), be discerned from the past and yet this is exactly what the mechanistic and linear conception of time embedded in industrial society encourages us to do. Industrialization and the Newtonian assumptions on which it is based impose a linear, decontextualized and abstract temporality that is blind to unpredictability, latency, threshold effects and irreversibility; all well-established features of ecosystem and Earth-system processes (Adam 1998).

Industrial/clock time is contrasted by Adam (1998) with the rhythmicity and seasonality of ecosystem processes and traditional societies. The point here is not to suggest that time circulates without changing (like money). Nor is it to romanticize traditional societies as living somehow closer to nature. The obverse of industrial/clock time is not an endless repetition of the same cycles, processes and events. Quite the contrary, it is industrial/clock time which itself assumes endless repetition – each minute and second followed in orderly and predictable fashion by functionally indistinguishable and interchangeable minutes and seconds. Rhythmicity, by contrast, speaks to the interplay of order and pattern, on the one hand, with unpredictability and transformation, on the other. Life on Earth, according to Adam (1998: 76) is 'orchestrated into a symphony of rhythms of varying speeds, durations and intensities'. Each movement contains the possibility of creativity, of not producing exactly that which preceded it, but constrained, nevertheless, by the potentialities embedded within what came before (see also Bourdieu 1990). Acorns do not grow into rabbits. Minutes and seconds march on, but the complexities, latencies, threshold effects and multiplicity of scales implicated in ecosystem processes always contain simultaneously the possibilities of reproduction, evolution and/or rapid systemic change (see also Harvey 1996). These possibilities are as interdependent as they are divergent. Systemic (indeed, catastrophic) transformations at one spatio-temporal scale may simply reproduce long-established patterns at another. Species evolve, migrate, (dis)appear. Waterways dry out, flood, shift course. Adam's conceptualization of a temporal perspective thus suggests that, as we confront the spatio-temporal effects of the industrial way of life, we must resist the temptation to conceive of sustainability as the preservation of past and present conditions. Industrial activity undermines the Earth's capacity for self-renewal and it is the bases for this creative capacity to which we must turn our collective attention (Adam 1998).

As noted above, Beck (1992) argues that the invisibility, irreversibility and definitional ambiguity of many contemporary risks increase our reliance on science while simultaneously calling the legitimacy of scientific institutions into question. Imperceptible and indeterminate threats create institutional dependencies at the same time that they erode trust, promote psychological distress and corrode relationships (see Freudenberg 1997). Science cannot resolve this paradox, according to Adam (1996), because it is captured, by and large, by the same materialist/empiricist epistemology that underpins the machine time of industrialization. Social scientists have failed to resolve this paradox because the industrial conception and commodification of time has been subject to insufficient critical analysis. While 'space is associated with visible matter and sense data, time is the invisible "other", that which works outside and beyond the reach of our senses' (Adam 1998: 10). Risk theorists name time, but they do little to theorize or interrogate it.

Opening the black box of time requires that existing assumptions be put aside in order to expose the 'implicit temporalities' embedded within knowledge practices: the knowledge practices behind technological innovation itself, and the knowledge practices behind our analyses of that innovation (Adam 1996: 85). Putting aside existing assumptions, according to Adam, allows us to see a number of theoretical and empirical problems in a new light. Most importantly, for our purposes here, opening the black box of time sensitizes us to temporal variability in exposures to and experiences of risk. It frees us from the tyranny of averages and the dangerous assumption that, for example, because mean exposures to radioactive isotopes for a given population are within safe ranges, all people, in all places and at all times, are therefore safe. Comprehending the complexity of time requires us to look beyond changing values at arbitrarily identified 'points in time' (implicitly assuming a linear temporal relationship between those points). Instead, Adam (1998: 205) suggests, we must find ways of incorporating the tempo, timing and rhythmicity of change along with 'the interpenetration of past, present and future' in the 'one moment of analysis'. Identifying this need is arguably Adam's key contribution to a social-ecological theory of time. What Adam does not do convincingly is demonstrate how we are to address it. As we have already seen, Adam argues that the process and epistemology of industrialization imposes a homogenized and accelerated temporality on techno-scientific innovation that is fundamentally at odds with the variable and indeterminate temporalities of its own consequences. This may well be the case, in a very broad sense, but this argument is limited, and limiting, in at least three important ways.

First, while technological innovation is critical to the maintenance of what Adam refers to as the industrial way of life, so too is institutional innovation (in particular, bureaucratic rationalization). Neglect of the latter is related, paradoxically, to an oft-made distinction in the work of Adam and others between risks that are 'natural' and those that are 'manufactured'. The point of this distinction is to demonstrate a qualitative difference between the anthropogenically induced consequences of industrialization and the plethora of environmental dangers humans have always faced (dangers which, it is implied, industrialization/modernization has mitigated). As Murphy (2001) points out, however, this distinction often obscures more than it reveals. In particular, it leads to neglect of the ways in which so-called 'natural hazards' are produced through specific institutional arrangements and socio-technical systems and the spatio-temporalities built into them. Murphy (2001) illustrates this argument through reference to the collapse of large parts of the Canadian electricity grid in 1998 because of 'abnormal' ice loading during an unusually long ice storm. Without power, homes went unheated and unlit for over three weeks with both direct and indirect impacts on public health. Previous ice storms of unusually long duration had caused comparatively minor disruption because 'pre-modern' energy systems were less integrated and had more built-in redundancy. Had he been writing a decade later, Murphy could just as well have illustrated this argument through reference to flooding of the Fukushima nuclear power plant in 2011 by a tsunami wave which exceeded the boundaries of official risk assessments despite historic evidence of such events. The risks peculiar to so-called modernity, therefore, stem not only from our disruption of ecosystem processes at higher scales (the sort of risk evident in anthropogenically induced global warming) but from increasingly tight coupling between the socio-technical systems on which we depend and the environmental conditions we consider 'normal' (Murphy 2001). Viewed over the longue-durée of intergenerational time, it is evident that whatever we consider 'normal' or 'predictable' now may be an unreliable guide to the conditions we face in either the near or distant future. In the end, improbable or simply unforeseen events are no less likely to happen today than they are decades from now. The rhythms of earth-system reproduction and transformation do not respect the linear extrapolations of probability-based models. As socio-technical systems such as electricity grids and power stations extend their spatial and temporal reaches, so too they extend their vulnerability to predictably unpredictable patterns of environmental change.

Second, in the absence of detailed analysis of the multiple and potentially competing temporalities embedded in specific processes of both technological and institutional innovation, industrial/clock time becomes a blunt and undifferentiated focus for well-worn critiques of science and technology. Industrial agriculture, for example, is criticized for the social and ecological consequences of practices which aim to erase spatial and temporal variability in food availability through an unholy alliance of chemical-intensive production methods, geno-technology, post-harvest treatments, international trade, over-processing, futures exchanges and economistic/industrial 'habits of mind' (Adam 1998). In arguing that the commoditization (and acceleration) of time is implicated as deeply here as the commoditization of space and social relations, Adam is not alone. Numerous authors have made exactly the same point. However, they have also shown that, despite the very best efforts of agribusinesses and techno-scientific institutions, resistance to industrialization remains evident among agro-ecologies, producers, consumers and others (Goodman et al. 1987). The imposition of machine/clock time is significant, but it is also spatially and sectorally uneven in its reach and its consequences. The point here is not that Adam's critique is wrong. The point is, rather, that a 'temporal gaze' is not, by itself, sufficient to highlight hitherto unrecognized hazards of technological (or institutional) development nor the multiple temporalities that may persist or evolve in the face of industrialization and modernization.

Third, it follows from the above that the very ideas of industrialization and modernization deserve more sustained critical attention. Modernization evokes an implicit temporality of de-traditionalization and progress, of constant revolution and innovation. But what if, as Latour (1993) argues, we have never been modern? What if, we might add, the complex rhythmicity of ecosystem processes and so-called traditional societies is not the antithesis of industrial society but constitutive of it? Latour (1993: 76) argues that '[t]he idea of an identical repetition of the past and that of a radical rupture with any past are two symmetrical results of a single conception of time'. For Adam, this single conception is the linear/clock time of industrialization; its uniform and interchangeable units as disposable as they are durable. For Latour, this single conception of time certainly is central to the 'modern constitution' and its Janus-faced separation of the natural and social worlds. However, as much as modernism may assert the transcendence of nature and history, its paradigmatic moments of transformation (the Enlightenment, the Scientific and Industrial Revolutions) have failed to cleave us from our embeddedness in nature/culture. So too have they failed to cleave us from what came before. Latour thus shifts focus from the deployment of temporality as an explanatory category to the exploration of particular temporalities as contingent and potentially unstable achievements. He argues that:

> We can never move either forward or backward. We have always actively sorted out elements belonging to different times. We can still sort. *It is the sorting that makes the times, not the times that make the sorting.*
>
> *(Latour 1993: 76, emphasis in original)*

Linear/clock time does not, therefore, explain the idiotically short-sighted transformation of Earth-system processes by us humans. As one potential temporality among many, it is linear-clock time that itself begs explanation.

How then are we to unpack the black box of time in a manner that contributes to the understanding and resolution of contemporary social-ecological challenges? How, following Adam (1998), are we to expose the temporalities embedded within knowledge practices in a manner that incorporates the tempo, timing and rhythmicity of change in the one moment of analysis? The first steps, it is argued here, are to set aside pretensions of macro-social explanation and to concentrate on specific attempts to inform and enact environmental governance. With this in mind, the following section explores the temporalities embedded in climate governance through the use of climate modelling and scenario-building, techniques that are designed to render climate change knowable and governable.

Climate change, scenario-building and temporality

The temporal aspects of anthropogenic climate change are among its most controversial. While few would argue that climate change ought to be ignored on the basis that some parts of the world may be affected more than others, many argue that we do not *yet* know enough about the magnitude and timing of climate impacts; that predictions based on climate models are unreliable; that the 'jury is still out' on the causes of climate change and the potential for measures to mitigate it. Many argue that 'others' ought to take more responsibility because they contribute more greenhouse gases (GHG) to the atmosphere, because they have *historically* contributed more, or because they *will* contribute more. Some argue that future impacts should be discounted; that a rational cost–benefit analysis suggests investment to mitigate climate change in the future offers poor value when compared with investment to alleviate poverty and other problems now. Temporal and spatial uncertainties alike are compounded by mismatches between the costs and benefits of mitigation measures (for discussion of the consequences of cost–benefit analysis, see Lockie 2013; Norgaard 2010).

Scientific understanding of the temporality of climate change is dominated by climate models and scenarios synthesized by the IPCC. Scenarios, as already noted, help to render the Earth as a co-evolving human-ecological system both thinkable and governable. As straightforward predictions of future climatic conditions, scenarios are almost certain, eventually, to be proven wrong. To date, most climate modelling has underestimated subsequently observed climate changes (Schneider 2009). Even as understanding of the ecosystem processes implicated in climate change deepens, climate scenarios will continue both to depend on projected levels of human 'socio-economic development' and to pre-empt our mitigation and adaptation responses (Giddens 1994). Scenarios provide us with a tangible reference point. They bring the future into the present in a manner that enables us not only to estimate the impacts of expected climate change on people and ecosystems but also to debate responsibilities and mitigation measures. In doing so, however, scenarios encourage us to engage with and potentially undermine the assumptions on which they are based. None of this will be news to climate modellers. They will respond that models will continue to improve and that the scenarios generated by those models will constantly be updated in light of changing real-world observations of GHG emissions, changing agreements on mitigation measures, and so on. They will acknowledge uncertainty over the timing and distribution of climate change impacts, but they will also express frustration at the reluctance of so-called denialists to respond to what we *do know* about the relationships between atmospheric GHG concentrations and the Earth's surface temperature (Schneider 2009).

Critically, scenarios not only enable us to think about and respond to climate change; they also shape our thinking and responses to climate change in specific ways. This begs the question: Do scenarios encourage us to ignore potentially useful ways of conceptualizing and responding to climate change? While the increasing politicization of public views towards climate change (McCright and Dunlap 2011) may serve to discourage critical scientific engagement with the social and political processes through which climate knowledge is constructed – for fear of fuelling the denialist cause – it is nevertheless incumbent on us to undertake such engagement if we are to contribute to a deeper understanding of key environmental threats and to more effective environmental governance. To open the black box of climate knowledge is not to engage in a senseless act of deconstruction or to assert the absurd proposition that climate change has no reality outside the semantic frame of various institutionalized knowledge claims. It is simply to recognize that knowledge is necessarily mediated through the interplay of scientific practices and conventions, theories, instruments and institutions (Latour 1993). Knowledge is always partial and provisional (Demeritt 2006). The implications of this partiality and provisionality for understanding climate change are discussed in detail by Demeritt (2001, 2002, 2006). Here, we are concerned more specifically with the implications and limitations of construing the temporality of climate change through climate scenarios.

Climate scenarios synthesized through IPCC assessments have been useful for framing climate change as a problem amenable to action through the negotiation of international agreements. Models that extrapolate the implications of various atmospheric GHG concentrations for climate change (including temperature, precipitation, sea level and extreme events), and thence for ecosystem and human well-being, have provided a basis from which to establish targets for emissions reduction that in turn have fed into the design of policy responses. However complex the algorithms embedded within various climate models, scenarios present governments with a simple policy equation: X per cent reduction in GHG emissions equals Y parts per million atmospheric concentration of CO_2 equivalents, holding the world to Z degrees of anthropogenic climate change. It follows from this that all units of carbon dioxide and other greenhouse gases are functionally equivalent and that transferable liabilities (i.e. markets) for emissions reduction are therefore an appropriate and efficient means through which to mitigate climate change. The only questions remaining, it would seem, are how much climate change decision makers are willing to tolerate, and how responsibility to cut emissions to a commensurate level should be distributed among nation states. However, the temporal linearity of climate change inferred through this equation is not without consequence. Clark (2010), for example, argues that the assumption embedded in international climate negotiations and regulation of a gradual transition to future climate states systematically ignores evidence regarding the possibility of abrupt and unexpected climate transitions. Linear models reduce sensitivity to threshold effects in the climate system and ignore altogether the potential influence of non-anthropogenic climate drivers. Nature may yet spring surprises on us. And, while many would argue that taking the risk of abrupt change seriously only strengthens the case to invoke the precautionary principle and to dramatically reduce GHG emissions, non-linear climate dynamics complicate our accounting of responsibilities and thus the established basis of transnational climate governance (Clark 2010).

The translation of all GHG emissions into transferable carbon dioxide equivalents (CDEs) is reflected in treatment of their impact as a global and regional problem measured in changing global and regional long-term climate averages. While this recognizes that the effects of climate change are spatially differentiated, it provides comparatively little insight into the dynamics of climate at finer temporal scales. The residual domain of uncertainty referred to earlier in this chapter concerning the timing and magnitude of climate change is largest precisely at those temporal and spatial scales at which people's well-being is most intimately connected with prevailing environmental conditions. Food security, public safety, amenity and a host of other values are impacted as much by short-term variability and capacity to deal with it as they are by long-term climate trends. Such variability is, again, a predictably unpredictable feature of the Earth's climate system. Floods, heatwaves, droughts, ice storms and so on are as ordinary as they are erratic, indiscriminate and destructive. How are we to know if any recent experience of climatic variability is 'normal' or 'unprecedented', 'natural' or 'anthropogenic'? Only long-term movement in averages can tell us anything about climate change but, in important respects, the world of climate averages is not the world we actually inhabit. Climate modelling and scenario-building cannot be expected to address this incongruity. However, the notion that no individual weather event can be linked to anthropogenic climate change makes less and less sense as we move deeper into the Anthropocene. The temporality of climate modelling and scenario-building does not speak to the temporality of living in an already variable and capricious climate.

As a consequence of living in an already variable climate, adaptation to change and recovery from extreme events is an ever-present demand on socio-ecological communities. While the need for adaptation to unavoidable climate change is recognized in IPCC assessments, the predominant focus of international climate policy remains mitigation of GGH emissions (Urwin and Jordan 2008; see also Adger et al. 2007). This is not simply an oversight. The temporality implicit within climate scenarios suggests that abatement ought always take priority over adaptation. Whatever the challenges of adaptation in the short-term, failure to mitigate immediately will lead almost certainly to more pronounced climate change and thus to even greater adaptation challenges in the future. Scenarios focus our thinking on the importance of reducing GHG emissions in order to keep these challenges at bay. However, scenarios simultaneously promote the deferral or deprioritization of policy and research to guide adaptation and to a corresponding lack of integration in policy and research of mitigation and adaptation strategies. To the extent that researchers and policy makers have engaged with climate change adaptation they have done so, for the most part, through the concepts of adaptive capacity, resilience and vulnerability. It is a sign of how little theoretical development has been achieved in this domain that all three of these concepts mean essentially the same thing. Methodologically, each depends on the investigation of historic climate variability and the characteristics and responses of households or communities which appeared to cope with or recover from that variability (Adger et al. 2003). This serves as a useful reminder that the motive for implementing climate adaptation measures is usually, in practice, to cope with current extreme events (Adger et al. 2007). However, identifying the resources and responses associated with past adaptation provides no guarantee that those resources and responses will be relevant to future climate challenges; that they will be enhanced by changes in the regulatory, institutional or economic environments; nor that they will be sufficient to overcome thresholds at which the magnitude or duration of climatic variability exceed historic levels, tipping ecosystems into fundamentally altered states (Adger et al. 2007).

Conclusion

Humans have altered key Earth-system processes – in particular those associated with climate, biodiversity and the nitrogen cycle – to a degree that dangerous environmental transformation is considered increasingly likely; that is, if such transformation has not already occurred. Comprehending the temporality of potentially dangerous environmental changes in terms of their pace and rhythm is as fundamental to the IPCC and other experiments in Earth-system governance as is comprehension of the spatial distribution of environmental change. However, as this chapter has argued, the goals of avoiding and/or adapting to dangerous socio-environmental transformation also require us to engage more reflexively with the conceptual frameworks, technologies and projects through which we attempt to understand and control Earth-system processes. This suggests a shift in focus for sociological analyses of time: from the critique of modernity to the temporalities implicit in specific attempts at enacting environmental governance and the knowledge practices that inform them.

Climate modelling and scenario-building have proven to be powerful means through which to comprehend the temporality of climate change; to bring the future into the present in order to plan responses and calculate responsibilities. The purpose of interrogating such techniques and responses in this chapter has been to illustrate the partiality of any knowledge that we bring to bear on environmental problems and the associated need to maintain a constructively critical stance towards our knowledge, the conceptual frameworks with which we organize it, and the policy responses it informs. It has not been to dismiss either the techniques and responses themselves or the underlying notion of anthropogenic climate change as 'mere' social constructs; this would be neither insightful nor useful.

In the case of climate governance, the chapter has argued that the focus of the IPCC on scenario-building has generated a sense of urgency around mitigating GHG emissions that has tended to crowd out serious attention to adaptation, other disruptions to Earth-system processes such as biodiversity loss, and the possibility that climate will change in ways that undermine predominantly market-based policy responses. Moreover, the reliance on modelling and scenarios contributes to a disjuncture between discourses of *future* climate change and people's experience of living in and needing to adapt to an *already* variable climate. The challenge is to bring such temporalities together in meaningful ways, helping us avoid the false choice between adaptation and mitigation, which would likely only further fuel political division over anthropogenic climate change.

References

Adam, B. (1996) 'Re-vision: The Centrality of Time for an Ecological Social Science Perspective', pp. 84–103 in S. Lash, B. Szerszynski and B. Wynne (eds), *Risk, Environment and Modernity: Towards a New Ecology*. London: Sage.

Adam, B. (1998) *Timescapes of Modernity: The Environment and Invisible Hazards*. London: Routledge.

Adger, W.N., Huq, S., Brown, K., Conway, D. and Hulme, M. (2003) 'Adaptation to Climate Change in the Developing World', *Progress in Development Studies*, 3, pp. 179–195.

Adger, W.N., Agrawala, S., Mirza, M.M.Q., Conde, C., O'Brien, K., Pulhin, J., Pulwarty, R. Smit, B. and Takahashi, K. (2007) 'Assessment of Adaptation Practices, Options, Constraints and Capacity', pp. 717–743 in M.L. Parry, O.F. Canziani, J.P. Palutikof, P.J. van der Linden and C.E. Hanson (eds), *Climate Change 2007: Impacts, Adaptation and Vulnerability. Contribution of Working Group II to the Fourth Assessment Report of the Intergovernmental Panel on Climate Change*. Cambridge, UK: Cambridge University Press.

Altvater, E. (1994) 'Ecological and Economic Modalities of Time and Space', pp. 76–90 in M., O'Connor (ed.), *Is Capitalism Sustainable? Political Economy and the Politics of Ecology*. New York: Guilford Press.

Beck, U. (1992) *Risk Society*. London: Sage.

Beck, U. (1996) 'Risk Society and the Provident State', pp. 27–43 in S. Lash, B. Szerszynski and B. Wynne, B. (eds), *Risk, Environment and Modernity: Towards a New Ecology*. London: Sage.

Bourdieu, P. (1990) *The Logic of Practice*. Cambridge, UK: Polity Press.

Clark, N. (2010) 'Volatile Worlds, Vulnerable Bodies: Confronting Abrupt Climate Change', *Theory, Culture and Society*, 27, pp. 31–53.

Crutzen, P. (2002) 'Geology of Mankind', *Nature*, 415, p. 23.

Demeritt, D. (2001) 'The Construction of Global Warming and the Politics of Science', *Annals of the Association of American Geographers*, 91, pp. 307–337.

Demeritt, D. (2002) What is the 'Social Construction of Nature'? A Typology and Sympathetic Critique', *Progress in Human Geography*, 26, pp. 767–790.

Demeritt, D. (2006) 'Science Studies, Climate Change and the Prospects for Constructivist Critique', *Economy and Society*, 35, pp. 453–479.

Foucault, M. (1991) 'Governmentality', pp. 87–104 in G. Burchell, C. Gordon and P. Miller (eds), *The Foucault Effect: Studies in Governmentality*. London: Harvester Wheatsheaf.

Freudenberg, W. (1997) 'Contamination, Corrosion and the Social Order: An Overview', *Current Sociology*, 45, 3, pp. 19–39.

Giddens, A. (1994) 'Living in a Post-Traditional Society', pp. 56–109 in U. Beck, A. Giddens and S. Lash (eds), *Reflexive Modernization: Politics, Tradition and Aesthetics in the Modern Social Order*. Stanford, CA: Stanford University Press.

Goodman, D., Sorj, B. and Wilkinson, I. (1987) *From Farming to Biotechnology: A Theory of Agro-Industrial Development*. Oxford, UK: Basil Blackwell.

Harvey, D. (1996) *Justice, Nature and the Geography of Difference*. Malden, MS: Blackwell.

Latour, B. (1993) *We Have Never Been Modern*. Cambridge, MA: Harvard University Press.

Lockie, S. (2013) 'Market Incentives, Ecosystem Services, and Property Rights: Assumptions and Conditions for Sustained Social and Ecological Benefits', *Land Use Policy*, 31, 90–98.

Lockie, S. and Measham, T. (2012) 'Social Perspectives on Risk and Uncertainty: Reconciling the Spectacular and the Mundane', pp. 1–13 in T. Measham and S. Lockie (eds), *Risk and Social Theory in Environmental Management*. Canberra, ACT: CSIRO Publishing.

Lövbrand, E., Stripple, J. and Wiman, B. (2009) 'Earth System Governmentality: Reflections on Science in the Anthropocene', *Global Environmental Change*, 19, pp. 7–13.

McCright, A. and Dunlap, R. (2011) 'The Politicization of Climate Change and Polarization in the American Public's Views of Global Warming, 2011–2010', *Sociological Quarterly*, 52, pp. 155–194.

Murphy, R. (2001) 'Nature's Temporalities and the Manufacture of Vulnerability: A Study of a Sudden Disaster with Implications for Creeping Ones', *Time and Society*, 10, 2/3, pp. 329–348.

Norgaard, R. (2010) 'Ecosystem Services: From Eye-Opening Metaphor to Complexity Blinder', *Ecological Economics*, 69, pp. 1219–1227.

Rockström, J., Steffen, W., Noone, K., Persson, A., Chapin, S., Lambin, E., Lenton, T., Scheffer, M., Folke, C., Joachim Schellnhuber, H., Nykvist, B., de Wit. C., Hughes, T., van der Leeuw, S., Rodhe, H., Sörlin, S., Snyder, P., Costanza, R., Svedin, U., Falkenmark, M., Karlberg, L., Corell, R., Fabry, V., Hansen, J., Walker, B., Liverman, D., Richardson, K., Crutzen, P. and Foley, J. (2009) 'A Safe Operating Space for Humanity', *Nature*, 461, pp. 472–475.

Schneider, S. (2009) *Science as a Contact Sport: Inside the Battle to Save the Earth's Climate*. Washington, DC: National Geographic Society.

Steffen, W., Crutzen, P. and McNeill, J. (2007) 'The Anthropocene: Are Humans Now Overwhelming the Great Forces of Nature?' *AMBIO*, 36, 8, pp. 614–621.

Thrift, N. (1996) *Spatial Formations*. London: Sage.

Urwin, K. and Jordan, A. (2008) 'Does Public Policy Support or Undermine Climate Change Adaptation? Exploring Policy Interplay Across Different Scales of Governance', *Global Environmental Change*, 18, pp. 180–191.

Zalasiewicz, J., Williams, M., Smith, A., Barry, T., Coe, A., Brown, P., Brenchley, P., Cantrill, D., Gale, A., Gibbard, P., Gregory, F., Hounslow, M., Kerr, A., Pearson, P., Knox, R., Powell, J., Waters, C., Marshall, J., Oates, M., Rawson, P. and Stone, P. (2008) 'Are We Now Living in the Anthropocene?', *GSA Today*, 18, 2, pp. 4–8.

Zalasiewicz, J., Williams, M., Steffen. W. and Crutzen, P. (2010) 'The New World of the Anthropocene', *Environmental Science and Technology Viewpoint*, 44, pp. 2228–2231.

9

From Rio to Copenhagen

Multilateral agreements, disagreements and situated actions

Chukwumerije Okereke and Sally Tyldesley

Governing climate change is arguably one of the most complex problems, environmental or otherwise, that the global community has had to contend with. There are at least five key factors that together produce this complexity. First, despite dramatic advances since the first IPCC report in 1990, scientific uncertainty remains over the precise magnitude of future climate changes and the consequences of these changes (Metz et al. 2007). Second, the causes and effects of climate change are global in nature; thus the entire global community needs to be engaged in the search for credible solutions. Third, the relative contributions of different countries and regions vary widely, as do the impacts of climate change. Critically, the worst effects of climate change will be felt largely in those countries least responsible (at least historically) and least able to cope with them. Fourth, the production of greenhouse gases is inherently linked to a wide variety of human activities and is embedded particularly in activities considered vital to economic growth and national security. Last, the impacts of climate change are long-term. This raises the problem of attribution but also means that the consequences of the decisions made in climate policy today will mostly be borne by future generations who have no way of taking part in current decision-making processes. Given these scientific, political and moral complexities, it is little wonder that there are large variations in the views of different countries on the optimum way to deal with the challenge.

On top of these issues is the fact that there have been dramatic shifts in the global political and economic dynamics since the United Nations Framework Convention on Climate Change (UNFCCC) was adopted in 1992. Perhaps the most significant of these are: the onset of the global economic crisis in 2007; and the emergence of the new group of economic powers, namely China, India, Brazil, Mexico, South Africa and Indonesia, which now challenge the dominance of the US and Western Europe in crafting regimes of governance. One obvious implication of these shifts is that some of the architecture and nomenclature around which the original UNFCCC was constructed are no longer applicable to the reality of today's world. The unprecedented challenges posed by climate change within the context of the dynamic and anarchical nature of international politics require that a successful regime of global climate governance must be characterized by innovations that address and temper competing interests and political imperatives.

This chapter highlights the innovations in governance that have characterized the global climate change regime as it sought to respond to and manage these complexities, political imperatives and competing interests. We suggest that the key contestations and innovations within climate governance can be understood in terms of four themes/questions:

- Who should take responsibility or how to allocate responsibility for climate change?
- Who has the power/authority to act or participate in decision making?
- What are the best approaches in terms of policy tools and institutions for tackling the challenge?
- How should effort between mitigation and adaptation be divided?

The rest of the chapter considers these questions, respectively. This is followed by a brief discussion and conclusion.

The quest for justice and equity within the regime

Contestations over responsibility and justice have been a central feature of climate change negotiations from the UNFCCC's inception in Rio, through Copenhagen, to Cancun. Perhaps no issue in the regime has been as divisive as how to allocate responsibility for the causation of, and solutions to, climate change. There remain two different stances on this issue, split broadly between developed and developing countries. In preparing for the Rio Convention in 1992, developing countries saw climate change negotiations as an opportunity to address issues of injustice and inequity in the global economic system (Dasgupta 1994). They argued that, since developed nations are both historically responsible for, and have benefited hugely from, the various processes that have caused climate change, it is only fair that the burden of action should lie with the industrialized countries. What's more, given that climate change will impose new constraints on the development trajectory of the Third World countries, they have continued to maintain strongly that an equitable climate regime warrants North–South compensation through financial and technology transfers and contributions to capacity building.

Developed countries, on the other hand, pointed out that whilst they are indeed historically responsible for a large amount of the greenhouse gas emissions these were produced at a time when the consequences were not well known. They argued further that it is unjust to hold today's generation responsible for the actions of the past generations. In addition, given that the high population growth and industrialization of the large developing countries means that the balance of emissions is changing (den Elzen et al. 2005), the West now strongly insists that it would be worthless from the point of view of stabilizing global emissions to allow the large developing countries to be exempt from emissions reductions.

The most notable innovation constructed to mediate between these competing narratives within the regime was the concept of 'Common but Differentiated Responsibilities' (CDR). Though woolly and imprecise, the concept appeared to incorporate the sentiments of both the developed and developing nations, thus helping to keep the two blocs of countries at the negotiating table long enough to agree on the UNFCCC in 1992. CDR has remained a central component of the climate change negotiations, at least through the Sixteenth Conference of Parties (COP) at Cancun. In addition to CDR, there have also been many other equity concepts, ideas and policy innovations within the regime that have served to define the new political and social-ecological relations implicated in the global governance of climate change. These include the notions of per capita emission, historic responsibility, technology transfer, new and additional funding, capacity building, the Clean Development Mechanism (CDM), the division of countries into Annex I and non-Annex I countries, and the subsequent exemption of developing countries from quantified emission reduction targets in the Kyoto Protocol.

While these terms remain regular features of climate negotiations and official texts, their exact meaning and the policies they should engender have been subject to intense debate, multiple interpretations and controversy. Take CDR, for example. Developed countries view this concept in terms of aid or assistance; as recognition of the fact that they have greater financial and technological *capabilities*, and, as such, have a greater *ability* to deal with climate change. Critically, CDR is interpreted in terms of greater ability rather than obligation (Harris 1999; Okereke 2008). Developing countries, on the other hand, see CDR as endorsement of the idea of global equity and the notion that distributive justice should be the cornerstone for international climate policy (Anand 2004). The view is held that developed countries are obligated not only to reduce emissions but also to make substantial transfers of resources to assist poor countries (Okereke 2010). This difference in interpretation has manifested in deep wrangling over nearly all policy and institutions crafted to deal with climate change since Rio 1992.

As this is being written, negotiations are focusing on what will happen when the first commitment period of the Kyoto Protocol ends in 2012. It is unlikely that developed nations will agree to a second commitment period of emissions reductions without the involvement of developing countries (Bodansky 2010). Russia and Japan have categorically stated that they will not be signing up to a second commitment phase of the Kyoto Protocol and developing countries have responded with equal force arguing that they will resist any attempt by the developed countries to 'kill' Kyoto and the CDR concept. However, if developed countries do not renew their commitment to emissions reductions, even while developing countries are being asked to take more responsibility for theirs, it will seem like a move away from the CDR principle (Rajamani 2010).

At COP15 in Copenhagen in 2009, an Accord bearing that city's name was produced. On the basis of this Accord, a total of 76 emission targets were submitted by both developed and developing countries (Rajamani 2010). The Accord and submissions, given a formal status under the UNFCCC process at COP16 in Cancun, mark the first occasion when the rapidly emerging economies and the USA have put forward mitigation actions and have accepted any type of internationalization of their climate change policies. In addition, it is the first time that the emerging economy countries, such as the BASIC countries (Brazil, South Africa, India and China), agreed to any international consultation or analysis concerning their emissions reduction actions (Bodansky 2010). Interestingly, while these Agreements included targets from both developed and developing nations, they still maintain the language of CDR. The new understanding appears to be that the emissions reductions agreed to by developing countries are voluntary. Regardless, the fact that the larger developing countries in particular have taken on emission reduction targets reflects changes in global climate political dynamics, including acceptance of a radical redefinition of the concept of CDR and movement away from the principle of global climate justice.

It is not only the larger developing countries that are undertaking mitigation actions. Outside of the UNFCCC, a group of highly vulnerable developing nations has formed the V11 – the Climate Vulnerable Forum – consisting of Bangladesh, Barbados, Bhutan, Ghana, Kenya, Kiribati, the Maldives, Nepal, Rwanda, Tanzania and Vietnam. The group was formed in 2009, to show 'moral leadership' in beginning to 'green' their economies in order to achieve carbon neutrality (V11 2009). These actions may reflect many nations' increasing understanding and acceptance of the growing scientific knowledge on the effects on climate change. In addition, it indicates a frustration with the 'deadlock' that has resulted from the debilitating focus to date on CDR and the North–South divide in climate negotiations. These changes may also reflect a shift in political attitudes towards climate change: acting on climate change is often seen now as being in a country's interest, rather than hindering economic growth (Townshend et al. 2011).

In sum, it is probably fair to suggest that the concept of global climate justice is losing its original appeal and that the notion of responsibility has shifted in ways that appear to privilege the developed countries. The move in recent years towards actions at a national level, or what some call the 'bottom-up' approach (Dubash 2009), means that, while they continue to be bandied about, it is not clear now if concepts like historical debt and per capita emissions will ever lead to significant North–South transfers. In addition, with developing countries now submitting national mitigation action plans to the UNFCCC, it may only be a matter of time before these plans are converted into soft obligations and conditions for accessing climate finance. More broadly, the failure of equity to gain serious traction within the global climate regime, despite the glaring issues of responsibility implicated by climate change, raises serious questions about the normative dimensions of international cooperation and in particular the prospects of global regimes of governance to serve as medium for international distributive justice (Barrett 2003; Helm 2008).

The problem of delineating agency, scale and authority

Next in importance to responsibility in international climate negotiations is the issue of agency and action on climate change. Here, we refer to such questions as: Who has the authority to negotiate and act on climate change? How can participation be made more inclusive? And at what scale should the search for climate action be concentrated?

Given the global nature of the problem, the international regime has been the main forum for negotiating action on climate change. From the beginning, the regime approach has been decidedly state-centric, with only states having the legitimacy and the legal status to participate in negotiations and take decisions. The negotiation of the climate agreement under the umbrella of the United Nations and the adaptation of the 'one nation, one vote' rule of the UN were supposed to be procedural innovations designed to address the concern of some weak states for equal voice. However, this quest for inclusivity has never quite been realized, despite the innovations. Rather, despite the 'one nation, one vote' policy, international negotiations over binding emissions have been dominated by the most powerful players, notably the USA and the European Community (EC). In addition, the focus on mitigation in the initial establishment of the international regime resulted in a lower level of involvement from countries with low emission levels. The majority of developing countries had little to do with the formation of the Kyoto Protocol, in the course of which OECD countries bargained with each other over their agreed allocations (Roberts 2007).

With increasing awareness of climate impact, developing countries have sought to increase their voice and impact rulemaking processes. But, severely hamstrung by lack of technical capacity and financial resources, those countries have remained unable to send enough delegates to conventions. For example, it has been reported that at COP6 the USA brought 99 formal delegates and the EC brought 76, while many small island and African states were represented by one-, two- or three-person delegations at the most (Roberts 2007).

Regardless of the number of delegates a state sends, a serious issue for discussion has been whether the state can or should represent all sub-national actors. Many have alluded to what is often called 'participation deficit' in the climate regime, in that actors involved in implementing the regime often were not involved in negotiating the rules (Bäckstrand 2008; Bulkeley and Betsill 2005). The clearest example of this is the United States. In the USA, though actions and commitments at a national level have been minimal, policies have been implemented at a regional level to reduce carbon emissions. The prime example is the Regional Greenhouse Gas Initiative (RGGI), a cap-and-trade programme to reduce carbon dioxide emissions from electricity-generating power plants, in which ten states participate (Hahn 2009). In addition, twenty-three states are in the process of developing and implementing mandatory regional carbon trading schemes. Many already have energy efficiency standards, renewable portfolio standards and climate plans, an example being California's 2006 Global Warming Solutions Act, which aims to reduce state emissions to 1990 levels by 2020 (Hahn 2009). Despite the fact that the thrust of climate action in the USA is at state and regional levels, these actors and entities have no official negotiating status within the climate regime.

Another major innovation that has characterized the regime for climate change is the proliferation of transnational agency comprising sub-national and community-level groups. The involvement of sub-national actors, mainly cities and state/provincial governments, as noted, has been largely instigated, on the one hand, by the lack of progress in international and national action on climate change, notably the refusal of the USA to sign up to the Kyoto Protocol and, on the other hand, by the determination of a number of state governors and mayors to take action. The World Mayors Council on Climate Change, formed in 2005, is an alliance of committed local government leaders advocating an enhanced recognition and involvement of mayors in multilateral efforts addressing climate change and related issues of global sustainability. As of this writing, there were over seventy members of the Council, representing a vast network of local governments on every continent working to reduce global greenhouse gas emissions.[1]

However, the relationship between these and many similar platforms with the international climate regime remains unclear. But, while the involvement of cities and regional governments has raised serious issues of agency, scale and authority (Okereke et al. 2009) it has been the activities of the indigenous community that have caused the greatest political controversy. Indigenous communities have been incensed that, while they are prone to the direct impact of climate change policies – especially with regard to forestry and land use – the rules of this regime have been negotiated by national government officials who do not understand or espouse views that reflect their perspectives and concerns. Their agitation eventually led to the indigenous community being allowed to make a representation for the first time at the floor of the COP14 plenary in Poznań, Poland, in 2008. This move appears to confer state status on groups in this sector and raises all sorts of problems regarding legitimacy and authority to negotiate. It is not clear whether or how this precedence will be followed in the future, as countries such as Australia that have aboriginal and indigenous communities have fought hard to ensure that the event in Poland is not repeated.

The difficulty faced by the indigenous communities in making their voices heard contrasts sharply with the experience of the business actors who have direct access to national governments and many times are grafted as official members of national delegations. It is evident that business actors have in some cases exploited this 'special relationship' with governments to their advantage, even sometimes gaming the regime. For example, in the initial phase of the EU's Emissions Trading System (ETS), emissions allocations were handed out for free, allowing many companies to make windfall profits (Hepburn and Stern 2008). However, it is difficult to see how the state authority can avoid the overbearing influence of business actors given the role the latter play as 'on the ground' implementers of the climate rules. In 1997, nations signed up to the Kyoto Protocol and the detailed rules of implementation were decided upon at COP7 in 2001, at Marrakesh. These rules provided for 'flexible mechanisms' such as emissions trading, joint implementation and the Clean Development Mechanism (CDM). The EU ETS allows member nations to reduce emissions through limiting the amount of carbon individual companies can emit. The CDM allows companies to invest in emission-reduction projects in developing countries in order to earn Certified Emission Reduction (CER) credits, each equivalent to one tonne of CO_2. These CERs can be traded and sold, and used by industrialized countries to meet a part of their emission reduction targets under the Kyoto Protocol. In essence, such flexible mechanisms mean that the private sector occupies an important platform in the bid to achieve global and state emissions reductions targets. In fact, given the private sector's central role in implementing the climate regime, it may be argued that the World Business Council for Sustainable Development and other global business bodies also need elevated status in the international climate negotiating arena.

But, while states like to explain the huge involvement the private sector on the basis of the efficiency and innovation of enterprise, most industrial sectors have both historically and currently depended on high levels of emissions of largely unregulated greenhouse gas to thrive. The concern this creates is that the success of key instruments for emission reduction appears to depend on the very actors who ordinarily should have the incentive to undermine these instruments. In the light of the huge resources committed to EU ETS – by far the single largest carbon trading scheme – with little emission reduction and given the windfall profits which accrued to companies in the first phase of trading (Helm 2010; Okereke and McDaniels 2012), these concerns appear well founded. Furthermore, many have noted that the CDM appears to have provided opportunity for some companies to make money without leading to emission reduction (Hepburn and Stern 2008; Wittneben 2009). However, companies insist that the major reason why they are hamstrung in taking action is that governments have failed to provide the kind of regulatory certainty required to make massive investments (Blyth et al. 2007). For their part, governments have tended to accuse companies of hypocrisy and ambivalence. The charge is that companies often purport to be supporting strong action for climate while at the same time undermining efforts at stringent climate regulations (Blyth et al. 2007), the most obvious examples being the lobby seeking to undermine the science of climate change in the USA or the coal lobby in Australia. It remains to be seen what innovations might be engineered to navigate the obvious inherent tensions this approach poses.

Important negotiations at recent COPs have addressed how corporate actors may be best mobilized to take action on climate change and how this would relate to the setting of targets and taking of actions by nation states. These discussions are most critical with respect to the provision of finance for climate mitigation and adaptation, design of rules for enhancing private investment in clean technology, potential use of trade rules, and dealing with the issue of intellectual property rights to facilitate technology diffusion.

Lastly, growing public awareness of climate change has increased the amount of action taken at an individual level. In the UK, charity sector campaigns such as the 10:10 campaign, which aimed to reduce UK emissions by 10 per cent by 2010, have involved individuals, companies and public sector organizations. In addition, whilst the actual role of civil society in influencing the negotiations is unclear, it has been suggested that the unprecedented public attention focused on COP15 at Copenhagen – led by civil society organizations and environmental NGOs – resulted in an increase in the number of climate-related policies made by governments (Jackson and McGoldrick 2010). Arguably, the process of negotiations on climate change by nation states in the UNFCCC forum has produced very few results in terms of mitigation. The failure of the Copenhagen conference to produce a legally binding agreement in particular led some to question the UNFCCC as a forum for decision making. Despite this, though actions by individuals and businesses can produce results, the scale of emissions reductions required still necessitates the involvement of the nation state. A lot of debate has therefore focused on who should lead action on climate change: the green state or the corporate sector.

Selecting the best tools and institutions for tackling climate change

A third major challenge for international climate negotiations and around which much experimentation has taken place concerns selection of the best policy tools and institutions for managing this challenge at both global and national levels. From Rio onwards, many policy and institutional mechanisms have been canvassed and a variety of ideas have been discussed. One of the most drastic to be proposed was that of a global governing body resembling a world government. This idea was put forward at The Hague conference on climate change in 1989 (Bodansky 1994). This governing body or authority would focus on environmental protection. However, the idea of a global government was considered too far-reaching and unworkable. Eventually the discussions came to be based around whether to establish a target-based protocol or a framework convention without targets and timetables. The discussions were largely between the USA and the EC – the latter led mostly by Germany, the Netherlands and Denmark (Depledge 2005). The stance of the USA was based upon the argument that the scientific uncertainties surrounding climate change meant that a target-based approach was unwarranted (Bodansky 1994). The suggestion, instead, was that a more cautious approach should be taken, allowing for a greater understanding of the economic and social consequences of various options before solid targets were agreed. The EC, however, pushed hard for targets and timetables for emissions reductions to be set within the Convention for all developed countries. Their key argument was that scientific uncertainty was not enough reason for inaction and that the precautionary principle should apply. A few EC members even agreed to unilateral targets for emissions reductions before the negotiations (Depledge 2005).

The developing countries were divided on this issue, with various blocs expressing different viewpoints. On the one hand, the Small Island States and others particularly vulnerable to climate change impacts were keen to see a binding agreement with emissions reduction targets and timetables in place. On the other, the oil-producing and -exporting countries (e.g. many OPEC members) did not wish to have an agreement that could lead to a tax or price being put on carbon that could have potential effects upon their economies. The latter countries therefore preferred a weak climate change regime or none at all (Okereke 2010).

In the end, the UNFCCC, with mention only of a desire to cut emissions, resembled the kind of instrument that the USA had been pushing for. This was largely because the EC countries did not want to proceed with an agreement without the USA. Eventually, though, the EC did succeed in getting the UNFCCC parties to agree the Kyoto Protocol, with targets and timetables, despite strong opposition from the USA (Okereke 2010). The merits and limitations of Kyoto have been subject to great discussion. Some criticize the regime for being too top-down, wasteful and infective as a means of tacking climate change (Prins and Rayner 2007; Victor 2001). But many others defend the regime and arguing that it played a major role in mobilizing international effort and providing a framework for serious action in combating climate change (Müller et al. 2009).

With the Kyoto Protocol expiring in 2012, the type of agreement that will be used after the end of its first commitment period has been a huge talking point. Negotiations have proceeded along two tracks. The first – the Ad hoc Working Group on Further Commitments for Annex I Parties under the Kyoto Protocol (AWG-KP) – began in 2005 and has focused on negotiating improvements to the Kyoto Protocol and emissions targets for developed countries, post-2012. This would not affect the US since it did not commit to reducing emissions for the first commitment period. The second track began in 2007, launched by the Bali Action Plan to work on an 'agreed outcome' under the UNFCCC. The second track is known as the Ad Hoc Working Group on Long-Term Cooperative Action (AWG-LCA). The AWG-LCA has aimed to develop a comprehensive outcome covering a shared, long-term vision; mitigation commitments from developed countries; nationally appropriate mitigation actions (NAMAs) from developing countries; financial measures; adaptation and technology transfer measures; and a system for measurement, reporting and verification (MRV). In the negotiations, the sticking point has been on whether and how these two tracks could be brought together, and with what result. In particular, it is not clear whether the outcome for replacing the Kyoto Protocol will be a single agreement, or two separate agreements under a common umbrella regime: one agreement for developing countries and the USA, and the second for other developed countries. Developing countries favour two separate agreements, while developed countries are pushing for a single agreement.

While discussions are continuing as this is being written, indications from the Copenhagen Accord and the Cancun Agreements are that states are moving away from internationally prescribed targets towards domestically decided emissions reductions. The political stumbling blocks at the international level remain too much to be overcome in terms of negotiating emissions targets and timetables. It seems that the domestically defined emissions targets submitted to the Copenhagen Accord and integrated into the UNFCCC process in the Cancun Agreements may be the primary way forward in terms of emission reductions, as a stalemate has been reached in the negotiations over what the next commitment phase of the Kyoto Protocol will look like (Grubb 2010). The main problem is that the 'pledge and review' approach does not come close to anything near what climate scientists say needs to be done to avert catastrophic climate change. Besides, but related to the question of the nature of the overarching regime or framework within which the international community should address the problem, are questions about policy instruments and institutions for driving down greenhouse gas emissions, raising and managing climate finance, and facilitating technology transfer and diffusion.

In terms of policy instruments, the key debate had been between national and (possibly) global carbon taxes on the one hand, and cap and trade of carbon emissions on the other. In the run-up to the negotiation of UNFCCC in Rio, Europe had strong preference for national- and continental-based carbon taxes. Many developing countries (excluding the OPEC countries) supported this view and wanted, in addition, a global tax possibly on fossil fuel and/or on international aviation. The USA was strongly opposed to carbon tax under any guise, however, and fiercely advocated for 'market-based solutions', in the form of cap and trade. The argument was that a carbon tax would be far too rigid and inefficient as a means of tackling climate change. Cap and trade, in contrast, was projected as flexible and allowing action on climate change to occur without hindering economic growth. By not controlling specifically where and how emissions are reduced, the cap and trade system was offered as the cheapest method of reducing emissions and the necessary condition for bringing the USA on board to an extended Kyoto agreement. In addition, use of emissions trading schemes was seen as the best way of encouraging private sector investment in low-carbon technologies, which in turn would lower the cost of emissions reductions in the long run.

Persuaded by the USA, the EU did support (albeit reluctantly) cap and trade as a key instrument for addressing global climate change. Following the initial adoption of Kyoto, the EU invested massively in establishing a trading scheme – the largest and most ambitious ever continental emissions trading system (Helm 2010). Other trading schemes include the New Zealand emissions trading scheme (launched July 2010); the northeastern US states' Regional Greenhouse Gas Initiative (RGGI) cap-and-trade scheme (launched January 2009); and the Tokyo metropolitan trading scheme (launched April 2010). Together these schemes were worth about US$135 billion at the time of this writing. In 2009, the US House of Representatives narrowly passed a climate bill that sought to establish a national trading scheme, but that bill stalled in the US Senate. The huge investment in (and accrued financial value of) these schemes notwithstanding, it is not clear that such market mechanisms do in fact lead to significant emissions reductions. Moreover, many developing countries have argued that the use of market mechanisms to tackle climate change as preferred vehicle for North–South transfer creates a loophole for industrialized countries to avoid their obligation to provide required assistance to developing countries.

Another serious issue related to the carbon tax vs cap-and-trade argument is the design of institutions that could be used for financing climate adaptation measures in, and facilitating transfer of resources to, developing countries. Here, again, a number of ideas have jostled for recognition and supremacy from Rio to Cancun. A series of proposals have revolved around the idea of a global fund that would provide resources for action on climate change and would facilitate this resource transfer from developed countries to developing. Potential sources of funding included a tax or levy on fossil fuel consumption in industrialized countries, a global carbon tax, or a percentage of a country's gross national product. These suggestions were largely mooted by the developing nations and were linked to the concept of historical responsibility for climate change with the suggestion that nations' contributions to the fund could be linked to their historical responsibility (Okereke and Schroeder 2009). Developing countries did in fact win a few concessions when the Special Climate Fund (SCF) was established to help fund adaptation in the least developed countries, but this was clearly branded an experiment by the industrialized countries. As the climate change negotiations developed, however, and especially with the onset of global economic recession, these suggestions were increasingly opposed by the developed world. Developed countries have long favoured an approach that relies more on the market and private-sector-led investment to generate climate finance.

The Adaptation Fund was one of the innovations that served to bridge the commitment of the industrialized countries to market-generated finance and the developing countries' need for predictable funding. The new instrument was established to finance specific adaptation projects and programmes in developing country parties to the Kyoto Protocol that are particularly vulnerable to the adverse effects of climate change. The Adaptation Fund was designed to get its resources from a 2 per cent levy on Certified Emission Reduction credits generated from projects carried out under the CDM. The CDM projects, as noted, are carried out largely by the private sector; as such, the Adaptation Fund comes close to a private sector tax. This innovative design for sourcing climate adaptation financing is similar to that proposed by many developing countries at the beginning of international climate negotiations.

Developing countries' sustained push for large-scale and predictable funding resulted in the establishment of the Green Climate Fund at Cancun, first proposed at COP15 in Copenhagen. Developed nations were to provide US$30 billion during the period, 2010–12, increasing to a total of US$100 billion by 2020. Individual developed countries pledged sums to this total amount. This money is to be 'new and additional' to that already given to the developing nations in aid. This mode of financing leaves concerns about the extent to which developed countries in fact will meet the amounts that they have pledged, a concern that perhaps would be less likely to occur if the finance were to come from a fossil fuel levy or tax. Already there have been claims that the money put forward by some developing countries cannot be classified as 'new and additional', but instead is redirected aid from other parts of the national budget (Liverman and Billett 2010). The recipients of this fund are to be the 'poor developing countries'. This definition represents the changing global dynamics since the initial discussions of a fund of this type and makes a distinction between the poorest developing countries and emerging economies such as the BASIC countries.

Discussions over the type of finance mechanisms used to support systems to reduce emissions from deforestation and forest degradation (REDD) took place at COP15 and COP16.

Debates again largely split between developed countries supporting market mechanisms and developing countries pushing for a fund. As of the time of this writing, negotiations were leaning towards market-based mechanisms, as the private sector's involvement is likely required due to the sizeable funding needed. Decisions on this matter were to have been made at COP17 in Durban.

It is likely that the two funding mechanisms will continue side by side. One threat to the 'market mechanisms' approach was the lack of an agreement on what will happen after the first commitment phase of the Kyoto Protocol. Continuation of the CDM requires an extended agreement. On the other hand, the EU ETS likely will survive, as a large amount of political capital and time has been put into its formation and its negotiators will not let it go lightly. In addition ETS has been highly profitable for those who have participated in emissions trading, even while debates continue about its actual contributions to reductions in emissions.

The false divide between mitigation and adaptation

It is now widely accepted that both mitigation and adaptation are necessary to deal with climate change. The relative emphasis on each has varied over time. Initially, negotiations focused almost solely on mitigation. As discussed above, early negotiations centred on debates between the USA and EU over quantified emission reduction commitments. Mitigation targets are a key component of both the UNFCCC and the Kyoto Protocol. The focus on mitigation occurred to such an extent that some observers commented that adaptation seemed a 'taboo subject' due to an apparent fear that greater focus on this issue would lessen efforts to reduce emissions (Pielke et al. 2007). Developing countries, however, made a big push to recognize adaptation, in COP7 in Marrakesh.

The heavy focus on mitigation has now shifted, representing one of the most notable changes in the discourse of global climate negotiation. There has been a realization that even if deep cuts in emissions were achieved, there would still likely be a rise in temperature that would require adaptation measures to be taken. In fact, many developing countries claim that climate change is already having devastating impacts on their population and severely threatening their development. As with much of climate change negotiations, there has been a North–South divide on the subject and tensions have been high. One issue that has complicated the debate in recent years is that developed countries now wish to differentiate between the poorest developing countries and the larger industrializing countries which are quickly becoming their economic competitors. Developed countries not only want these countries to make their own emissions reduction targets and commitments, but also do not want to support their competitors by financing adaptation projects or transferring new technologies to them (Okereke 2010).

Another factor that has hindered action on adaptation is scientific uncertainty on the precise nature of the climate changes in specific locations, particularly on a small scale. It is difficult to adapt when you don't know what you are adapting to (Yamin et al. 2001). In addition, there are difficulties in determining what problems are caused by climate change and what would have occurred anyway. Developing countries have been instrumental in pushing adaptation back to the forefront of the debate (Depledge 2005; Jackson and McGoldrick 2010). A higher focus on adaptation can be traced back to the COP at Marrakech in 2001. Since then, adaptation has become a much greater focus of the negotiations, to the extent that the Nairobi COP in 2006 was unofficially named the adaptation conference (Depledge 2005). In Montreal, at the COP11 in 2005, the developing countries negotiated for the establishment of a Work Programme on Adaptation and an Adaptation Fund. These were designed to help a variety of adaptation measures in developing countries. The Green Climate Fund – proposed at COP15 in Copenhagen – was to be split between mitigation and adaptation.

Conclusion: looking ahead

There have been a variety of innovations within the global climate change regime as it sought to deal with the unprecedented and complex challenges posed by anthropogenic global warming. The concept of Common but Differentiated Responsibilities (CDR) has been present at every stage of the negotiations; how it is conceived and interpreted now is undergoing a sea change. The notions of who has agency in negotiations and who has authority to act on climate change also have varied over time. While nation states have largely dominated the international negotiations, this dominance has been increasingly challenged by the private sector, civil society and indigenous communities. Negotiations over climate change have indeed highlighted the fragmented and contested nature of power, authority and agency at both national and international levels. With the fragmentation of agency and authority has come a variety of mechanisms for dealing with climate change – including global targets, carbon taxes, CDM cap and trade, and voluntary carbon offset schemes.

Given that the sum of these efforts still falls short of what scientists calculate is needed to avert dangerous climate change, the greatest need facing the climate regime is development of a series of innovations that together will achieve scale-up and establish regime effectiveness. The challenge is now even more daunting because global effort no longer can concentrate solely on mitigation but now must focus also on how to adapt to unavoidable climate change, by far the more urgent concern for developing nations. All of this is highly complicated by the heightened suspicion and distrust that characterize international climate negotiations following, on the one hand, many decades of failed promises by the North and, on the other hand, the increasing political and economic power of BASIC countries, especially China and India. The outcome of the Conference of Parties in Copenhagen, which saw an Accord negotiated by the BASIC countries and the USA, with the EC and the rest of the world playing little role, may indeed indicate that the geopolitics of climate change and international environmental cooperation is changing for good.

Note

1 World Mayors Council on Climate Change (2010) 'Members' List', www.worldmayorscouncil.org/members/members-list.html.

References

Anand, R. (2004) *International Environmental Justice: A North–South Dimension*. Aldershot, UK: Ashgate Publishing.
Bäckstrand, K. (2008) 'Accountability of Networked Climate Governance: The Rise of Transnational Climate Partnerships', *Global Environmental Politics*, 8, 3, pp. 74–102.
Barrett, S. (2003) *Environment and Statecraft: The Strategy of Environmental Treaty Making*. Oxford, UK: Oxford University Press.
Blyth, W., Bradley, R., Bunn, D., Clarke, C., Wilson, T. and Yang, M. (2007) 'Investment Risks Under Uncertain Climate Change Policy', *Energy Policy*, 35, 11, pp. 5766–5773.
Bodansky, D. (1994) 'Prologue to Climate Change Convention', pp. 45–75 in I.M. Mintzer and J.A. Leonard (eds), *Negotiating Climate Change: The Inside Story of the Rio Convention*. Cambridge, UK: Cambridge University Press.
Bodansky, D. (2010) 'The Copenhagen Climate Change Conference: A Post-Mortem', *American Journal of International Law*, 104, 2, pp. 230–241.
Bulkeley, H. and Betsill, B. (2005) 'Rethinking Sustainable Cities: Multilevel Governance and the "Urban" Politics of Climate Change', *Environmental Politics*, 14, 1, pp. 42–63.
Dasgupta, C. (1994) 'The Climate Change Negotiations', pp. 129–148 in I.M. Mintzer and J.A. Leonard (eds), *Negotiating Climate Change: The Inside Story of the Rio Convention*. Cambridge, UK: Cambridge University Press.

Depledge, J. (2005) *The Organization of Global Negotiations: Constructing the Climate Change Regime*. London: Earthscan.

Dubash, N.K. (2009) 'Climate Change and Development: A Bottom-Up Approach to Mitigation for Developing Countries?' pp. 172–178 in R.B. Stewart, B. Kingsbury and B. Rudyk (eds), *Climate Finance: Regulatory and Funding Strategies for Climate Change and Global Development*. New York, NY: New York University Press.

Elzen, M. den, Fuglestvedt, J., Höhne, N., Trudinger, C., Lowe, J., Matthews, B., Romstad, B., Pires de Campos, C. and Andronova, N. (2005) 'Analysing Countries' Contribution to Climate Change: Scientific and Policy-Related Choices', *Environmental Science and Policy*, 8, 6, pp. 614–636.

Grubb, M. (2010) 'Copenhagen: Back to the Future?' *Climate Policy*, 10, pp. 127–130.

Hahn, R.W. (2009) 'Climate Policy: Separating Fact from Fantasy', *Harvard Environmental Law Review*, 33, 2, pp. 557–593.

Harris, P.G. (1999) 'Common but Differentiated Responsibility: The Kyoto Protocol and United States Policy', *NYU Environmental Law Review*, 17, 1, pp. 27–48.

Helm, D. (2008) 'Climate-Change Policy: Why Has so Little Been Achieved?' *Oxford Review of Economic Policy*, 24, 2, pp. 211–238.

Helm, D. (2010) 'Government Failure, Rent-Seeking, and Capture: The Design of Climate Change Policy', *Oxford Review of Economic Policy*, 26, 2, pp. 182–196.

Hepburn, C. and Stern, N. (2008) 'A New Global Deal on Climate Change', *Oxford Review of Economic Policy*, 24, 2, pp. 259–279.

Jackson, E. and McGoldrick, W. (2010) 'Global Climate Policy Post-Copenhagen: Progress and Prospects', Discussion Paper. Sydney: Climate Institute.

Liverman, D. and Billett, S. (2010) 'Copenhagen and the Governance of Adaptation', *Environment Magazine*, 52, pp. 28–36.

Metz, B., Davidson, O.R. Bosch, P.R., Dave, R. and Meyer, L.A. (eds) (2007) *Contribution of Working Group III to the Fourth Assessment Report of the Intergovernmental Panel on Climate Change*. Cambridge/New York: Cambridge University Press.

Müller, B., Höhne, N. and Ellermann, C. (2009) 'Differentiating (Historic) Responsibilities for Climate Change', *Climate Policy*, 9, 6, pp. 593–611.

Okereke, C. (2008) 'Equity Norms in Global Environmental Governance', *Global Environmental Politics*, 8, 3, pp. 25–50.

Okereke, C. (2010) 'The Politics of Interstate Climate Negotiations', pp. 42–61 in M.T. Boykoff (ed.), *The Politics of Climate Change: A Survey*. London: Routledge.

Okereke, C., Bulkeley, H. and Schroeder, H. (2009) 'Conceptualizing Climate Change Governance Beyond the International Regime', *Global Environmental Politics*, 9, 1, pp. 58–78.

Okereke, C. and McDaniels, D. (2012) 'To What Extent are EU Steel Companies Susceptible to Competitive Loss Due to Climate Policy?' *Energy Policy*, 46, pp. 203–215.

Okereke, C. and Schroeder, H. (2009) 'How Can the Objectives of Justice, Development and Climate Change Mitigation be Reconciled in the Treatment of Developing Countries in a Post-Kyoto Settlement?' *Climate and Development*, 1, pp. 10–15.

Pielke, R., Prins, G., Rayner, S. and Sarewitz, D. (2007) 'Climate Change 2007: Lifting the Taboo on Adaptation', *Nature*, 445, pp. 597–598.

Prins, G. and Rayner, S. (2007). 'Time to Ditch Kyoto', *Nature*, 449, pp. 973–975.

Rajamani, L. (2010) 'III. The Making and Unmaking of the Copenhagen Accord', *International and Comparative Law Quarterly*, 59, 3, pp. 824–843.

Roberts, J.T. (2007) 'Globalization: The Environment and Development Debate', pp. 3–18 in C. Okereke (ed.), *The Politics of the Environment: A Survey*. London: Routledge.

Townshend, T., Fankhauser, S., Matthews, A., Feger, C., Liu, J. and Narciso, T. (2011) *GLOBE Climate Legislation Study*. London: Global Legislators Organisation, Grantham Research Institute.

V11 (2009) *Declaration of the Climate Vulnerable Forum*. Available at http://daraint.org/wp-content/uploads/2010/12/Declaration-of-the-CVF-FINAL2.pdf. Accessed 7 January 2012.

Victor, D. (2001) *The Collapse of the Kyoto Protocol and the Struggle to Slow Global Warming*. Princeton, NJ: Princeton University Press.

Wittneben, B.B.F. (2009) 'Exxon Is Right: Let Us Re-examine Our Choice for a Cap-and-Trade System Over a Carbon Tax', *Energy Policy*, 37, 6, pp. 2462–2464.

Yamin, F., Burniaux, J. and Nentjes, A. (2001) 'Kyoto Mechanisms: Key Issues for Policy Makers for COP-6', *International Environmental Agreements*, 1, 2, pp. 187–218.

10
Marriage on the rocks
Sociology's counsel for our struggling energy–society relationships

Debra J. Davidson

In January 2012, amid much fanfare, US President Barack Obama rejected an application to build a new pipeline, the Keystone XL, for transporting synthetic crude oil from Alberta, Canada to Nebraska. Just two months later, however, he granted approval to extend the existing Keystone Pipeline from Alberta all the way to Texas. In December 2012, meanwhile, China commenced transport of gas through the world's longest natural gas pipeline, constructed at a cost US$22 billion and traversing 8,700 kilometres through Central Asia to the Yangtze River delta. Further North, Canada had its largest contingent of troops ever in the Arctic, where Norway moved its military command centre in 2010. Canada and Norway are just two of several countries that see the melting Arctic in terms of its resource exploitation possibilities. On the other side of the world, Nigeria exports some two million barrels a day of oil and experienced GDP growth in 2010 of 8.4 per cent, but nonetheless bore a poverty rate of 70 per cent, a life expectancy of 47.5 years and an infant mortality rate of 92 deaths per 1,000 live births (CIA 2012). Back in North America, over three million Mexicans confronted food insecurity in the face of one of the worst droughts in generations in 2012, a harbinger of events likely to increase in frequency and intensity due to our changing climate. Despite an extraordinary expenditure of political resources in the past two decades, global greenhouse gas (GHG) emissions in 2012 were the highest ever, at about 50 gigatonnes.

There are a few good news stories – Apple is working on a solar-powered iPad, and the mayor of New York donated US$50 million to the Sierra Club's anti-coal campaign in 2011. But there are good reasons for social scientists to direct their gaze toward society's crumbling relationship with energy. This relationship is most certainly complex. However, underlying today's headlines is a stark fact we are yet to accommodate. The disruptive effects of fossil fuel dependence are approaching an order of magnitude beyond previous experience. We have consumed approximately half of our global oil reserves, and our supplies of natural gas are not much brighter (Leggett 2005). We still have plenty of coal reserves, but only adamant climate change deniers take heart in this fact. Peak oil refers to the point at which global oil production begins a one-way downward path; a point that, according to many scientists, we have already reached. Despite recent years of high prices that *should* motivate increased productivity, global oil production has largely levelled off since 2005 (Kerr 2011). This does not suggest an abrupt end to oil consumption, but it does mean that new discoveries are smaller and harder to access. Remaining oil reserves require more inputs to recover and process, generating more negative externalities in the process. One key input is energy itself. A large proportion of our remaining oil reserve has a much lower energy return on investment (EROI) than the conventional sources we have relied upon to date. The most foreboding of externalities is in very large part, but not entirely, associated with this most foreboding of inputs. The CO_2 embodied in the numerous non-conventional oil reserves we have begun to exploit promises an *increase* in GHG intensity. But there are numerous other externalities, both social and ecological.

Meanwhile, as with previous energy crisis eras, planners across the globe tend to focus almost exclusively on supply-side management (Nye 1998), taking demand (and its projected future increase) as a given to which the Earth must simply adjust. Without a full-scale change of political heart, the smattering of disruptions to be catalogued here promise to swell into a growing tide, starting with the economic and livelihood effects of escalating fuel prices.

This chapter traces social-scientific consideration of what I consider to be three major drivers in the energy–society relationship: the escalation of fossil fuel dependency's social disruption; the escalation of fossil fuel dependency's environmental disruption; and macro-social transition prospects. The chapter argues that sociology and the other social sciences are in a position to offer much-needed contributions to the mediation of society's faltering marriage with energy. However, there are more holes than fabric in the canvas of social scientific understanding, and thus the chapter ends with a call to arms that suggests directions for future research.

Escalation of fossil fuel dependency's social disruption

The escalating social footprint of fossil fuel reliance can be observed all along the commodity chain, and from the local to the global scale. The distorting effects of export dependency on state polities are stark, even in comparison to other raw material sectors, so much so that nations in the grip of such dependency have a special name – 'petro-states'. Petro-statism has not been with us since the emergence of petro-dependency; the first reserves to be exploited lying within countries whose economies and civil societies were already well established. At least since the Second World War, however, newcomers to the global oil marketplace have tended to be post-colonial nations with weak civil societies and governing apparatuses, and limited industrial development, such that oil dependency quickly became the defining feature of their development processes (Shelley 2005). Detailed accounts of the social impacts of petro-statism can be found elsewhere (Shelley 2005). Arguably, the most significant social impacts lie in the negative relationship between petro-statism and democracy (Silver 2011). Such conditions bode poorly for equity, health care, the status of women and sustainable economic planning, but even more worrisome are the relationships between petro-statism, violence and militarism. The presence of rent-based economies – economies in which profits are secured not through productivity and efficiency but through the exercise of political and economic power – fosters political and economic conflict (see Huber 2011; Tanuro 2010).

In Nigeria, for example, the heavy footprint of transnational capital has at one and the same time bolstered the Nigerian state in the global political economy while instigating domestic challenges to the legitimacy of that very state at home (Watts 2004). By the early 2000s, some US$30 billion in petro dollars had been extracted from the Niger Delta by multi-national energy corporations (Watts 2004). What minuscule portion of that total was retained and reinvested locally is hard to say, but such investments are certainly not reflected in current employment, infrastructure or services, particularly for the Ogoni people inhabiting the Niger Delta.

> Few Ogoni households have electricity, there is one doctor per 100 000 people, child mortality rates are the highest in the nation, unemployment is 85 per cent, 80 per cent of the population is illiterate and close to half of Ogoni youth have left the region in search of work.
>
> *(Watts 2004: 208)*

Between 1970 and 2000 (precisely overlapping the emergence of Nigeria as a global oil producer), the number of people subsisting on less than US$1 a day grew from 19 million to 90 million (or from 36 per cent to over 70 per cent of the population). Environmental protection, as might be presumed, is all but nonexistent. What the Ogonis have inherited is not a source of economic well-being but an ecological dead zone.

At the heart of the matter lies the Oil Empire or, more broadly, 'petro-capitalism' (Watts 2004: 195). The USA has its sights on controlling the entire producing region between China and India, exacerbating a growing confrontation with China (Amineh and Houweling 2007). Of course, maintaining control over the global oil economy is no easy task, particularly considering the US penchant for private enterprise. The US state can 'pave the way', but bringing reserves to market is left to private energy corporations – ExxonMobil, British Petroleum and Royal Dutch Shell – three of the four largest corporations in the world and bankroll for the climate denial machine (e.g. McCright and Dunlap 2010).

The Americans' pretences toward imperialism are elusive for another reason as well. The geopolitical terrain is becoming more and more complex and volatile. For one, it is becoming more populated, with newly industrializing buyers and formerly self-reliant countries that have depleted their domestic reserves. For another, declining reserves motivate commercialization of ever smaller pools in ever more spatially dispersed locations, and the resource requirements of maintaining control over such terrain are being spread thin. The recent entrance of new producing nations in politically volatile places like Venezuela and Nigeria, eleventh and fifteenth largest global oil producers respectively, are just two of several new fronts of regional political tension. And the buyers vying for control over these dispersed pools are no longer just the big three energy corporations, but a number of aggressive state-owned energy companies like China's CNOOC Ltd, attempting to secure the country's economic future without the criteria of shareholder profit limiting procurement strategies (Klare 2008). In addition to emerging petro-states, two multi-jurisdictional reserves are becoming hotbeds of political tension. One is the Caspian Sea, where Russia, Iran, Turkmenistan, Azerbaijan and Kazakhstan are all vying for jurisdiction; a land-locked pool, moreover, requiring pipeline transport across some of the most politically precarious terrains one could point to on a map. And then we have the Arctic Circle, where eight countries are urgently mapping the seabed to determine who gets which slice of that melting pot (literally). The Arctic Circle is quietly becoming among the most militarized places on the planet (Huebert 2010).

In short, for very good reason, Foster (2008:12) describes our current moment as 'a dangerous new era of energy imperialism'. Meanwhile, in stark contrast, the political careers of US elected officials in the heart of the oil consumption vacuum are being decided by the price of a gallon of gas, already the lowest in the Western world, among consumers with one of the highest per capita incomes in the world. Here too, though, the mark of social disruption can be measured in multiple costs associated with car dependency. Today, we appear to have surpassed a critical threshold of functionality on North America's roadways, with enormous costs to public coffers for maintaining the 5.5 million or so kilometres of North American asphalt

(Thün and Velikov 2009), costs which are barely covered by struggling industrial cities and rural communities. The 18 lanes of the busiest highway on the North American continent – the 401 in southern Ontario – carry some 420,000 vehicles on a typical day, much of the time in gridlock (Thün and Velikov 2009). The freedom and convenience of the private automobile costs the commuter in time, stress and highway fatalities, while the suburbanization enabled by cheap oil has left a legacy of decaying urban infrastructure and an abandonment of efforts to co-exist in a socially diverse landscape (Beauregard 2006).

Escalation of fossil fuel dependency's environmental disruption

As with social disruptions, the environmental disruptions caused by the extraction and consumption of fossil fuels dates from the beginning of the coal era. The resulting urban smog left its mark in the pages of history and the lungs of city dwellers throughout the Industrial Revolution. And, while the GHG emissions from such processes likewise date back to the first days of fossil fuel burning, it is the growing intensity of this linkage between fossil fuel dependence and climate change that defines the ultimate contradiction of this form of economic development, as volumes of fossil fuel consumed follow an exponential growth trajectory. A number of social scientists have remarked upon the centrality of climate change in contemporary societies (e.g. Beck 2008; Giddens 2009; Urry 2011). In addition to remarking upon the obvious (that the primary sources of GHG emissions are fossil fuels), several sociologists have harkened back to energy institutions in their analyses of the climate crisis and our constrained efforts in climate mitigation, including, for example, the power of various energy lobby groups to hamstring ratification of the Kyoto Protocol (e.g. Fisher 2006), and the role of energy corporation-sponsored think tanks in supporting climate scepticism (McCright and Dunlap 2010).

Attention to other forms of energy-based environmental disruption by social science has largely followed the direction of attention by civil society, with a non-fossil fuel – nuclear power – attracting the lion's share (e.g. Freudenburg and Rosa 1984; Gamson and Modigliani 1989), although the social responses to perceived environmental disruption have been the focus of research rather than the environmental disruption itself. Nuclear power has certainly inspired many of our most organized and successful environmental new social movements (Jacob 1990). Persistent concerns among a large proportion of the populace regarding the risks associated with nuclear power and waste has likewise rendered nuclear power a regular feature of social-psychological analysis (e.g. Slovic et al. 1991). The catastrophe at Japan's Fukushima Daiichi nuclear power plant in March 2011 offered resounding validation of public concerns. But one of the more compelling findings is nuclear power's aptitude for inciting social disruption, in the form of stress and anxiety, even in the absence of real exposure (Freudenburg and Jones 1991).

The social responses to nuclear power have been unquestionably warranting of scholarly attention. However, this has left several unfortunate gaps in inquiry. The coal industry, for one, imposes quite possibly the most significant ecological footprint of the modern era in terms of landscape disruption, toxic emissions and health implications (see, e.g., Younger 2004). Sociologists have been largely silent both on the ecological devastation incurred by modern coal-mining methods – including mountaintop removal – and on the (growing tide of) social responses to them. Coal mining continues to be among the most dangerous of occupations, as well, with tonnage extracted marked by health and fatality costs in developed and developing countries alike (e.g. Homer 2009). Further, while coal burning is now recognized as one of the leading sources of GHG emissions, the air quality effects have largely been ignored by social scientists (cf. Pobodnik 2006). Unlike oil, we still have plenty of coal in the ground, and China for one is capitalizing on their plentiful coal beds to electrify the nation. Technological enthusiasts are banking on carbon capture and storage to ameliorate coal's impacts on the climate, although, even if this does become a reliable means of capturing carbon, the costs of doing so would likely trump coal's primary attraction as a relatively inexpensive source of energy.

Two other processes characterizing the modern fossil fuel era bode particularly ill for environmental well-being, both of which have received scant attention from social scientists. First, the reckless environmental behaviours of energy corporations, enabled in no small part by weak environmental governance apparatuses in emerging petro-states, have been nothing short of alarming. Even in strong states like China, the regulatory apparatus has not kept pace with the explosion of new coal mines across the country (Andrews-Speed et al. 2003). The resulting ecological devastation, moreover, has often occurred in highly populated, impoverished social settings in which the primary sources of local livelihood have been rendered toxic (Watts 2004).

The second process of interest has to do with the growth in proportion of remote and non-conventional fossil fuel sources in our global energy portfolio. Remote sources include, for example, the deep-water oil reserves requiring drills some eight kilometres long to access. In their analysis of the 2010 blowout of BP's Macondo well, Freudenburg and Gramling (2010) point out the convenient tendency for the media to attribute such disasters to human error, diverting attention from the inherent risks of operating such complex and precarious infrastructures, particularly when escalating costs motivate money-saving but risk-enhancing practices among major corporations, and complicity from regulators. Transpose the Macondo well into the Caspian or the Arctic, both characterized by particularly deep deposits and notoriously rough weather, and the economic benefits of bringing these pools to market quickly wane.

The main non-conventional fuel sources – of which, as cornucopian enthusiasts will rightly point out, there are plenty – include heavy oil, bitumen and shale gas. In this case, however, plenitude is no blessing. As Pobodnik (1999: 163) states:

> Simply put, there may be sufficient petroleum reserves in the earth to cause serious environmental damage if they are consumed. From an environmental perspective, in short, the real problem facing the world economy appears to be the overabundance of petroleum and other hydrocarbon energy resources rather than their scarcity.

Their name does little to describe what is unique about this suite of fuel sources. Their uniqueness has to do with the fact that they are of markedly lower quality than conventional sources. It takes far more work to get them out of the ground, a greater volume of material per unit output needs to be extracted from the earth, and it takes more energy-intensive and pollution-producing processing to turn these low quality fuels into materials suitable for economic use.

Shale gas, the latest 'energy independence' calling card of the Obama administration, is just beginning to attract attention for its water- and land-consuming features (Charman 2010). But the heel of non-conventional fuel's ecological footprint lies in northeastern Alberta, Canada. Each day, 1.5 million barrels of *synthetic* crude oil – so called because it resembles so faintly the conventional stuff – is produced, 75 per cent of which is exported to the USA. The footprint that has accumulated to date in order to produce this oil includes open pit mines covering 602 square kilometres, water stored in tailings ponds covering 50 square kilometres that is expected to remain unusable for roughly a century, consumption of approximately 176 million cubic metres of freshwater per year, the emission of 5 per cent of Canada's total GHG emissions, and a cocktail of toxic effluents spewed into the water and air (Davidson and Gismondi 2011).

State and industry proponents have set their sights on increasing production to anywhere from 5 to 10 million barrels per day over the next 15 years, which, without the benefit of miraculous advances in technology, would increase the size of this footprint accordingly. Even the most rudimentary of cost accounting exercises suggest shale gas is of questionable social, environmental and perhaps even economic benefit when the costs of remediation are considered (Davidson and Gismondi 2011).

Transition prospects

The quintessential social question is, of course, what happens now? What is perhaps most remarkable about our fossil fuel dependence is the extent to which it has become normalized; the ready availability of heat, light, mobility and large industrial machinery – which have only really been available in the West for about a century, and remain unavailable in many other parts of the world today – are taken for granted (Crosby 2006). Yet the scale of our dependence on energy reveals itself in its absence, as it did across the northeastern United States and southern Canada in the winter of 2003. This single event which, in some locations, left households, businesses, even entire communities, without power for days on end but, in most locales, amounted to intermittent blackouts of a few hours each, brought the entire regional social system to a standstill (Murphy 2009). Peak oil, combined with climate change, holds promise not only for an escalation of the social and environmental disruptions described above, but also regular occurrences of quite literal social disruptions in the form of energy shortages and failures.

But is the 2003 blackout a harbinger of the future of modernity, or do we have the capacity to end our marriage with fossil fuels in a way that does not invoke bedlam? Various strands of social inquiry contribute to our understanding of prospects for transition, including extensive research on attitudes and behaviours toward conservation and alternative fuels dating back to the 1970s oil embargo (Heberlein 2011), accounts of social mobilization in support or opposition to various forms of energy development (e.g. Toke 2011), and a small but growing number of systemic forecasting analyses of crisis impacts and transition (e.g. Elzen et al. 2004; Shove and Walker 2010). Contemplation of our energy and society relationship, however, would be incomplete without consideration of the macro-scale social change that is on the horizon. There is very good reason, therefore, to invest more of our conceptual energy in grand theory, or 'big picture' gazing.

Big picture gazing

Questions regarding the extent to which certain of our disciplinary forefathers had any energy intelligence continue to engage contemporary scholars, with Spencer and Marx in particular receiving attention from theorists today, both of whom offered energetic perspectives that remain relevant. Marx's focus was essentially on the metabolism of the individual labourer, and the aggregate systemic contradictions that evolve when the energy demands imposed upon labourers by capital extend beyond labour power's metabolic-energetic limits (Burkett and Foster 2006). Labour's replacement by machines exacerbates this contradiction, generating an unprecedented demand for raw materials, including, first and foremost, a new, non-human energy source. Herbert Spencer was relatively less preoccupied with the potential for crisis, but nonetheless spent a great deal of conceptual energy tracing the role of energy as a defining feature of the character and evolution of social systems. For Spencer, the inorganic, organic and social (or 'supra-organic') share the same evolutionary processes, all governed by flows of energy. Basically, humans (and society) persist through countering the forces that constrain them. As summarized by McKinnon (2010: 452):

> Herbert Spencer brings the question of energy to the forefront of sociological theory by conceiving society as an energetic system: trade and production ... are shaped by the necessary energy expenditures and the flow of goods and people along paths of least resistance. As such, energy is not something secondary to social organization, but its vital Force in its struggle to survive and thrive.

McKinnon (2010) suggests strengthening Spencer's theory of social evolution by acknowledging that increasing social complexity has been enabled by increasing dependence on non-human forms of energy. Consequently, while the continued trajectory of human energy use need not be taken as inevitable (see also McKinnon 2007; Pobodnik 2006), the historical course of increasing energy-dependent complexity 'does suggest that considerable effort and ingenuity will be necessary in order to alter this path of least resistance' (McKinnon 2010).

Energy has also had a central place in a number of comprehensive, quasi-empirical treatments of civilization's history, including some foundational pieces in environmental sociology, such as Catton's *Overshoot* (1982), and Schnaiberg's *The Environment* (1980). These studies highlight through history the repeated realization of what was considered unthinkable: major social crises induced by society's fumbling interactions with its ecological foundations, including energy supply. Cottrell's (1955) work continues to stand out in this field as possibly the most well-developed theoretical treatment of the energy–society relationship before or since. He traced energy's central contribution to social change from 'low-energy' societies to modern industrial societies that are highly dependent on multiple, energy-intensive institutions. Notably, Cottrell highlighted the multiple social contingencies that allowed for highly variable patterns of development, as well as the likelihood for energy surpluses to support not only material development, but also concentrated power structures and symbolic pursuits.

Most sociological studies have focussed on the disruptive effects of fossil fuels. At the other end of the spectrum are optimistic accounts in which institutional change (via human ingenuity read technology) is held to occur more or less autonomously in times of need. Pobodnik (1999, 2006) is well represented here. Pobodnik's first major energy-related work (1999) offers a detailed historical analysis of three long-wave dynamics – those of geopolitical rivalry, commercial competition and social conflict – that have interacted to produce global energy regime shifts during specific historical periods. The historic value of this work stands on its own, but in one of his more recent treatments, Pobodnik (2006) begins with the premise of impending energy crisis, and then confidently forecasts a transition to renewables on the basis of these dramatic historic energy shifts; shifts that were at the time considered just as inconceivable as a renewable energy shift appears today. Considering the historic basis of the work, such an account is surprisingly ahistorical, with no consideration of the much larger, energy-hungry population today (bordering on 7 billion people, projected to reach at least 9 billion before stabilizing); the non-interchangeability of energy sources (coal cannot fuel internal combustion engines, for example); the low energy density characterizing renewable energy sources; and the increasingly complex network of geopolitical relations in the energy domain with the entrance of new producing and importing countries.

As with other macro-system scholars on both sides of the debate, critical evaluation of by what *means* the much-heralded technological innovation will be invoked and diffused is left unexamined. Of course, Pobodnik is not alone. To offer just one other example, Nuttall and Manz (2008) predict a forthcoming renewable energy transition borne by carbon capture and storage, hybrid vehicles, and so on, their confidence premised on the willingness and ability of Western countries to adjust in response to rapid and serious climate change. If only!

Fortunately, the literature does include closer looks at agents of change and resistance, starting with the sheer bureaucratic complexities associated with policy development and how they vary by political and historical context. Such complexities are bound to escalate, particularly with the inevitable conjunction between energy and climate change as policy issues (Harriss-White and Harriss 2006). Of course one mustn't forget brute political might, either. Several historical accounts of energy politics have highlighted the central roles of states and corporations in directing energy development onto particular self-serving pathways to the detriment of social well-being (e.g. Klare 2008; Leggett 2005). Tracing the coincident development of solar power and nuclear power in the post-Second World War West, Etzkowitz (1984) clearly shows that, despite solar's technical viability, state investments favoured nuclear power development for political reasons.

Consumption: attitudes, behaviours and consumer culture

The largest body of work focusing on agency looks at individual attitudes and behaviours pertaining to energy consumption. In the words of one commentator:

> When we speak of pressures on the natural environment, we should speak more about home loans, old-age income, and women drivers, more about shrinking households and all-night shopping, and perhaps less about coal mines and pulp mills.
>
> *(Schipper 1996: 113)*

With the exception of a suite of studies conducted at the height of the 1970s oil embargo that indicated high levels of public support for conservation (e.g. Olsen 1981), much of this research more or less puts the kibosh on the optimism for easy transition expressed by Pobodnik and others, while at the same time providing necessary correctives to the determinism embedded in Limits to Growth and energetic perspectives.

Attitudes and behaviours pertaining to energy consumption can be situated within a broader field of empirical quantitative research on consumption, which includes, in addition to social factors driving consumption behaviours, attention to inequities in energy access. One consistent observation highlighted by Rosa et al. (1988) continues to motivate research in this vein: beyond the threshold energy requirements of modernity, the amount of energy consumed in different cultural and socio-political contexts varies considerably. Wide discrepancies are noted from the household to the national level, attributed to everything from car ownership (Mogridge 1985), age and family structure (Schipper 1996; York 2007) to per capita GNP (Dunlap et al. 1994) and national land area (Green 2004). Most crucially, the development options for lower income countries are limited by access to global energy supplies, while those same countries, ironically, will pay the most severe climatic prices of energy consumed elsewhere (e.g. Pobodnik 2002). Synthesizing several decades of energy research, Thomas Heberlein (2011) concludes that price has a far greater impact on consumption than even the most persuasive education campaign. Despite a large body of research suggesting that environmental concern is an expression of deeply held values and beliefs (e.g. Stern et al. 1999), this finding emphasizes that attitudes are frequently poor predictors of behaviours (Lutzenhiser 1994).

One response to the attitude–behaviour gap is to blame the consumer herself (and it usually is the female household members held up for scrutiny). However, blaming the individual 'ignores the existence of a pre-existing cultural system that exerts a profound impact on the supposedly free marketplace choices of citizens' (Dunlap et al. 1994: 37). Much attitude–behaviour work has been criticized for its tendency, as in economic formulations, to treat the individual as the basic unit of analysis (Lutzenhiser 1994), as well as its atheoretical focus and general inability to offer a systematic understanding of energy consumption (Rosa et al. 1988). More recent work that focuses not on consumers as such but rather the institutionalized consumer culture within which consumption practices take place has made some noteworthy advances. It may well be the case that what is unique about the high consumption levels of Americans (of energy and other material goods) is not Americans per se, but the uniquely American suite of social structures that nurture consumptive behaviours.

In short, the exuberant energy consumption levels of western societies are driven by the normalization of consumption to meet the expectations of the distinctly suburban way of life (Beauregard 2006), the exceptional privilege of which becomes re-configured as entitlement, so much so that rising gas prices become an affront to freedom. This strong identity association with energy-consumptive practices serves the interests of Big Oil and their political pundits remarkably well, of course, and car buyers are bombarded with advertising that has nothing whatsoever to do with automobiles and everything to do with fulfilment of one's expressive (and mobility) destinies. In fact, nothing represents this lifestyle more than the car. As noted by Huber (2011), outrage over high gasoline prices in the USA is far out of proportion to their actual financial impact on all but the lowest income households (who tend not to be the outrage expressers at any rate), and has far more to do with the fact that:

> the use-value of gasoline is not simply about the instrumental need to move from point A to point B, but has become entangled within wider imaginaries of work, home, mobility, freedom, and a specifically 'American way of life'; in short, ideas that go to the heart of capitalist ideology.
>
> *(Huber 2009: 466)*

None of this is unique to the USA, of course. Most importantly, culture is more process than structure, and it can change. Relatively unremarked upon by sociologists are the tremendous lengths Western leaders have gone to construct through discourse a relationship between investments in foreign (military) policy and everything from terrorism to drug use, and concealing the mundane reality of the matter; namely, the desire to maintain the tremendous economic payoff offered by the American oil-dependent consumer culture (Campbell 2005). Recent research on the emergence of green consumption – the exception that proves the rule – has identified multiple hurdles faced by that minority which has attempted to engage in a lifestyle that abides by an ecological worldview (Moisander 2007).

Conversely, this body of research on consumption and consumers offers much-needed attention to the demand side of the energy commodity chain, identifying possibilities for intervention in energy–society relations which do not solely involve technological efforts to increase supply (Shove and Walker 2010). Understanding the prospects for realization of such possibilities involves looking beyond values and attitudes and paying greater attention to, for one thing, trust – or, more aptly, what Freudenburg (1993) terms 'recreancy'. In the author's words, it is 'the failure of institutional actors to carry out their responsibilities with the degree of vigour necessary to merit the societal trust they enjoy' (Freudenburg 1993: 909). Recreancy comes into play in public support for various policies in the energy realm, and may well serve as a linchpin for contemporary transition efforts emanating from civil society. A recent review of the literature regarding UK public attitudes to a wide array of alternative energy technologies and consumer behaviours concluded that members of the public are largely resistant to making substantive behavioural changes in the absence of leadership from government and industry that is perceived to be sound and trustworthy (Whitmarsh et al. 2011). Citizens' support for carbon reduction and energy efficiency policies in particular has been found to have as much to do with how people conceptualize those policies, and the governance approaches underpinning them, as with their values, attitudes and perceptions of the environment (Fischer et al. 2011).

Social mobilization

Organized opposition to energy development (motivated, at least in part, by a decline in recreancy in many developed countries) marks a prospective change agent that ostensibly has much greater potential than individual-level behavioural change, on the basis of its ability to influence the cultural milieu in which such behaviours take place. Social movement responses to nuclear power provide one ready example of the potential influence of civil society over the trajectory of energy development. Offshore oil development also tends to inspire organized opposition (Freudenburg and Gramling 2002; Gramling 1996). More recently, Carruthers (2007) highlights the emerging environmental justice framing espoused by local opponents in energy production zones. Growing resistance in such unlikely places as Appalachia are noteworthy, although have not yet been taken up by the academy to a large extent. Social movements have also been shown to play an instrumental role in the expansion of renewable energy projects, such as wind energy development in Denmark (Toke 2011). On the other hand, local residents have been among the first to point out that renewables are no panacea, and wind energy projects in particular have their own negative impacts, inspiring their own opposition (Devine-Wright 2005).

Other forms of social mobilization represent proactive efforts to engender alternative, low-footprint lifestyles, including emerging trends like Transition Towns, a movement barely six years old originating in the Irish countryside and yet already embraced by several hundred communities across the world (Chatterton and Cutler 2008). Transition Towns have less to do with political mobilization and more to do with community-level praxis, premised on a collective commitment to low carbon living and represents a compelling trend away from mainstream energy–society relations warranting greater sociological attention.

Forecasting analysis

The daunting challenge facing such motivated citizen organizations is outlined by a number of scholars engaged recently in forecasting the likely future transition trajectories available to us given current geopolitics and some basic demographic maths. Coming to the fore in such accounts are the aggressive development plans invoked by India and China. China has, since early this century, the notorious status of being the world's largest GHG source and, if current development trends continue, the upward pathways for both fossil fuel consumption and GHG emissions are in their early days in this Asian behemoth. Kuby et al. (2011) show that, despite China's concerted efforts at reducing the *intensity* of GHG emissions (and energy consumption), which some have argued are among the most progressive observed today (e.g. Zhang et al. 2007), the massive increase in overall electrification and industry more than offsets these efforts. Parallel research in the area of energy analysis offers limited enthusiasm for renewables to supply the energy needs of societies over the next century, and forecasts the continued reliance on fossil fuels, particularly oil, due to a grim lack of alternatives (Baghat 2008).

Conclusion: promising trends in energy–society research

Sociology's traditional theoretical unease with nature and technology – an unease shared to a greater or lesser extent with other social sciences – has been well documented (e.g. Lutzenhiser 1994: 58). Even environmental sociology has had comparatively little to say about peak oil and the dangers of non-conventional fuels (for exceptions, see Crosby 2006; Dennis and Urry 2009; Pobodnik 1999, 2002). As noted by Lutzenhiser (1994), energy analysis has been dominated by fields of inquiry that tend to diminish the importance of all things social.

So where to from here? To begin, I suggest that social scientists need to move beyond debates over *whether* energy scarcity is 'real' or 'constructed' and consider instead the ways in which scarcity is both materially and symbolically *produced*. As Mulligan (2010) observes, talk of energy security has shifted in political discourse from a material perspective to one in which economic and geopolitical narratives prevail, reducing the likelihood for concerted policy response. While scarcity only has meaning in the context of social demand and is not an ecological condition per se, the relationship between supply and demand is also an objective one. Geographer Michael Watts, who has conducted some exceptional research on social conflict in resource zones, argues that:

> Oil is, of course, biophysical; it is also a commodity that enters the market with its price tag, and as such is the bearer of particular relations of production. And, not least, oil harbours fetishistic qualities: it is the bearer of meanings, hopes, expectations of unimaginable powers.
> *(Watts 2004: 203)*

Innovative new conceptual work that explicitly or implicitly embraces a critical realist perspective offers a promising trend including a small but growing body of systems-based work. A number of recent critical realist studies acknowledge the objective conditions underpinning society's relationship with energy, while also situating consideration of societal responses to those conditions within a complexity framework that encompasses the multiple political, economic, cultural and organizational filters through which such responses must travel. One key distinction of this work, as with other critical realist efforts, is a move away from the tendency to favour structure or agency and to focus instead on the mutually constitutive relation between them. I will highlight just a handful of works in this vein.

Powells (2009) adopts a complexity systems perspective to evaluate the effectiveness of the UK's energy efficiency commitment. He highlights the unanticipated outcomes of tinkering with complex systems, including multiple interactive, emergent processes that unfold – particularly the marginalization of certain groups – when governments attempt to control the energy/carbon dynamic. Dennis and Urry (2009) are more ambitious in attempting to move beyond analysis of current complexities and to engage in forecasting the likely futures of our car-dominant transportation system. The system is described as having reached a point of self-organized criticality: hovering in a state of precarious stability, when non-average actions can instigate a dramatic shift to a post-car transportation structure. The authors posit three possible future scenarios, and then, by 'back-casting', or tracing back the potential transition pathways to the major driving forces of society today (global climate change, peak oil, urbanization, and digitization/virtualization), explore the plausibility of each. Davidson and Gismondi situate contemporary conflicts over the development of the Alberta tar sands into a global systems network of productive and consumptive spaces and the flows of materials and information connecting them, to inform our assessment of the likelihood for curtailment of tar sands production and societal transition away from fossil fuel dependence more broadly (Davidson and Gismondi 2011). Finally,

Sonnenfeld and Mol (2011) have compiled a number of innovations in sociology in a recent special edition of *Global Environmental Change* that, while not exclusively focused on energy, provide much-needed groundwork for reconsideration of environment–society theories 'in the new world (dis)order'. Sonnenfeld and Mol (2011: 773) argue that:

> [prevailing theoretical] traditions are of limited value in fully understanding current environmental crises and related institutional challenges. Contemporary social theory needs to be adapted to reflect the new social, economic and political architecture underlying both causes of and solutions for today's environmental challenges. The growing understanding of (a sense of) a new, planetary world (dis)order has stimulated social theorists to begin to fundamentally adapt conventional schemes as well as develop new interpretative frameworks.

If anything, recent research only serves to highlight the magnitude of the work that remains. Future work is needed that encompasses critical consideration of crisis drivers destabilizing the current energy–society relationship. These drivers include, to name just a few: (1) increasingly tense geopolitical dynamics as newly industrializing states seek to secure their share of the declining pool of remaining fossil fuel reserves while, at the same time, advanced industrial states seek to maintain their own share; (2) side effects associated with non-conventional fuel production, particularly the shale gas that has received such enthusiastic attention of late; (3) the likely critical thresholds defining our car-dependent transportation system, marked by urban air pollution, gridlock, escalating vehicular deaths and increased infrastructure costs; and (4) the multiple possible sources of contradiction and complementarity between energy security mandates and climate change mitigation mandates.

Finally, it would behove us to engage in further critical but non-normative inquiry into nuclear power. Numerous climate scientists and energy analysts view nuclear power as indispensable to our transition away from greenhouse gas-producing fossil fuels (e.g. Weaver 2008). The plausible argument that nuclear energy is one of few viable energy options does nothing to eradicate its many ecological and socio-political challenges. The foreboding risk portfolio that creates such challenges demands (re)consideration from the social sciences, particularly in light of the recent accident in Japan and Germany's decision to phase out nuclear power shortly afterwards.

References

Amineh, M.P. and Houweling, H. (2007) 'Global Energy Security and its Geopolitical Impediments: The Case of the Caspian Region', *Perspectives of Global Development and Technology*, 6, 1–3, pp. 365–388.

Andrews-Speed, P., Yang, M., Shen, L. and Cao, S. (2003) 'The Regulation of China's Township and Village Coal Mines: A Study of Complexity and Ineffectiveness', *Journal of Cleaner Production*, 11, 2, pp. 185–196.

Baghat, G. (1999) 'Oil Security at the Dawn of the New Millennium', *Journal of Social, Political and Economic Studies*, 24, 3, pp. 275–290.

Baghat, G. (2008) 'Energy Security: What Does It Mean? And How Can We Achieve It?' *Journal of Social, Political, and Economic Studies*, 33, 1, pp. 85–98.

Beauregard, R.A. (2006) *When America Became Suburban*. Minneapolis, MI: University of Minnesota Press.

Beck, U. (2008) *World at Risk*. London: Polity Press.

Burkett, P. and Foster, J.B. (2006) 'Metabolism, Energy and Entropy in Marx's Critique of Political Economy: Beyond the Podolinski Myth', *Theory and Society*, 35, 1, pp. 109–156.

Campbell, D. (2005) 'The Biopolitics of Security: Oil, Empire and the Sports Utility Vehicle', *American Quarterly*, 57, 3, pp. 943–972.

Carruthers, D.V. (2007) 'Environmental Justice and the Politics of Energy on the US–Mexico Border', *Environmental Politics*, 16, 3, pp. 394–413.

Catton, W.R., Jr. (1982) *Overshoot: The Ecological Basis of Revolutionary Change*. Urbana, IL: University of Illinois Press.

Central Intelligence Agency (CIA) (2012) *The World Factbook*. Available at www.cia.gov/library/publications/the-world-factbook/index.html. Accessed 8 June 2012.

Charman, K. (2010) 'Trashing the Planet for Natural Gas: Shale Gas Development Threatens Freshwater Sources, Likely Escalates Climate Destabilization', *Capitalism, Nature, Socialism*, 21, 4, pp. 72–82.

Chatterton, P. and Cutler, A. (2008) *The Rocky Road to a Real Transition: The Transition Towns Movement and What it Means for Social Change*. Available at http://trapese.clearerchannel.org/resources/rocky-road-a5-web.pdf. Accessed 7 January 2012.

Clarke, L. (1985) 'The Origins of Nuclear Power: A Case of Institutional Conflict', *Social Problems*, 32, 5, pp. 474–487.

Cottrell, W. (1955) *Energy and Society: The Relation Between Energy, Social Changes, and Economic Development*. New York, NY: McGraw-Hill.

Crosby, A.W. (2006) *Children of the Sun: A History of Humanity's Unappeasable Appetite for Energy*. New York, NY: W.W. Norton and Co.

Davidson, D.J. and Gismondi, M. (2011) *Challenging Legitimacy at the Precipice of Energy Calamity*. New York, NY: Springer.

Dennis, K. and Urry, J. (2009) *After the Car*. Cambridge, UK: Polity Press.

Devine-Wright, P. (2005) 'Beyond NIMBYism: Towards an Integrated Framework for Understanding Public Perceptions of Wind Energy', *Wind Energy*, 8, 2, pp. 125–139.

Dunlap, R., Lutzenhiser, L. and Rosa, E.A. (1994) 'Understanding Environmental Problems: A Sociological Perspective', pp. 27–50 in B. Bürgenmeier (ed.), *Economy, Environment, and Technology*. Armonk, NY: M.E. Sharpe.

Elzen, B., Geels, F.W. and Stirling, A. (2004) *System Innovation and the Transition to Sustainability*. Cheltenham, UK: Edward Elgar.

Etzkowitz, H. (1984) 'Solar Versus Nuclear Energy: Autonomous or Dependent Technology?' *Social Problems*, 31, 4, pp. 417–434.

Fischer, A., Peters, V., Vávra, J., Neebe, M. and Megyesi, B. (2011) 'Energy Use, Climate Change and Folk Psychology: Does Sustainability Have a Chance? Results from a Qualitative Study in Five European Countries', *Global Environmental Change*, 21, 3, pp. 1025–1034.

Fisher, D.R. (2006) 'Bringing the Material Back in: Understanding the United States Position on Climate Change', *Sociological Forum*, 21, 3, pp. 467–494.

Foster, J.B. (2008) 'Peak Oil and Energy Imperialism', *Monthly Review*, July/August, pp. 12–33.

Freudenburg, W.R. (1993) 'Risk and Recreancy: Weber, the Division of Labor, and the Rationality of Risk Perceptions', *Social Forces*, 71, 4, pp. 909–932.

Freudenburg, W.R. and Gramling, R. (2002) 'How Crude: Advocacy Coalitions, Offshore Oil, and the Self-Negating Belief', *Policy Sciences*, 35, 1, pp. 17–41.

Freudenburg, W.R. and Gramling, R. (2010) *Blowout in the Gulf*. Cambridge, MA: MIT Press.

Freudenburg, W.R. and Jones, T.R. (1991) 'Attitudes and Stress in the Presence of Technological Risk: A Test of the Supreme Court Hypothesis', *Social Forces*, 69, 4, pp. 1143–1168.

Freudenburg, W.R. and Rosa, E.A. (1984) *Public Reactions to Nuclear Power: Are There Critical Masses?* Boulder, CO: Westview/AAAS.

Gamson, W.A. and Modigliani, A. (1989) 'Media Discourse and Public Opinion on Nuclear Power: A Constructionist Approach', *American Journal of Sociology*, 95, 1, pp. 1–37.

Giddens, A. (2009) *The Politics of Climate Change*. London: Polity Press.

Gramling, R. (1996) *Oil on the Edge: Offshore Development, Conflict, Gridlock*. Albany, NY: State University of New York Press.

Green, B.E. (2004) 'Explaining Cross-National Variation in Energy Consumption: The Effects of Development, Ecology, Politics, Technology, and Region', *International Journal of Sociology*, 34, 1, pp. 9–32.

Harriss-White, B. and Harriss, E. (2006) 'Unsustainable Capitalism: The Politics of Renewable Energy in the UK', pp. 72–101 in L. Panitch and C. Leys (eds), *Coming to Terms with Nature*. London: Merlin Press.

Heberlein, T.A. (2011) *Navigating Environmental Attitudes*. New York, NY: Oxford University Press.

Homer, A.W. (2009) 'Coal Mine Safety Regulation in China and the USA', *Journal of Contemporary Asia*, 39, 3, pp. 424–439.

Huber, M.T. (2011) 'Enforcing Scarcity: Oil, Violence and the Making of the Market', *Annals of the Association of American Geographers*, 101, 4, pp. 816–826.

Huebert, R. (2010) *The Newly Emerging Arctic Security Environment*. Ottawa: Canadian Defence and Foreign Affairs Institute. Available at www.cdfai.org/PDF/The%20Newly%20Emerging%20Arctic%20Security%20Environment.pdf. Accessed 7 January 2012.

Jacob, G. (1990) *Site Unseen: The Politics of Sitting a Nuclear Waste Repository*. Pittsburgh, PN: University of Pittsburgh Press.

Kerr, R.A. (2011) 'Peak Oil Production May Already Be Here', *Science*, 331, pp. 1510–1511.

Klare, M.T. (2008) *Rising Powers, Shrinking Planet: The New Geopolitics of Energy*. New York, NY: Metropolitan Books.

Kuby, M., He, C., Trapido-Lurie, B. and Moore, N. (2011) 'The Changing Structure of Energy Supply, Demand and CO_2 Emissions in China', *Annals of the Association of American Geographers*, 101, 4, pp. 795–805.

Leggett, J. (2005) *Half Gone: Oil, Gas, Hot Air and Global Energy Crisis*. London: Portobello Books.

Lutzenhiser, L. (1994) 'Sociology, Energy and Interdisciplinary Environmental Science', *American Sociologist*, 25, 1, pp. 58–79.

McCright, A.M. and Dunlap, R.E. (2010) 'Anti-Reflexivity: The American Conservative Movement's Success in Undermining Climate Science and Policy', *Theory, Culture and Society*, 26, 2–3, pp. 100–133.

McKinnon, A.M. (2010) 'Energy and Society: Herbert Spencer's "Energetic Sociology" of Social Evolution and Beyond', *Journal of Classical Sociology*, 10, 4, pp. 439–455.

McKinnon, A.M. (2007) 'For an "Energetic" Sociology, or, Why Coal, Gas and Electricity Should Matter for Sociological Theory', *Critical Sociology*, 33, pp. 345–356.

Mogridge, M.J.H. (1985) 'Transport, Land Use, and Energy Interaction', *Urban Studies*, 22, 6, pp. 481–492.

Moisander, J. (2007) 'Motivational Complexity of Green Consumerism', *International Journal of Consumer Studies*, 31, 4, pp. 404–409.

Mulligan, S. (2010) 'Energy, Environment and Security: Critical Links in a Post-Peak World', *Global Environmental Politics*, 10, 4, pp. 79–100.

Murphy, R. (2009) *Leadership in Disaster: Learning for a Future with Global Climate Change*. Montreal: McGill-Queen's University Press.

Nuttall, W.J. and Manz, D.L. (2008) 'A New Energy Security Paradigm for the Twenty-First Century', *Technological Forecasting and Social Change*, 75, 8, pp. 1247–1259.

Nye, D.E. (1998) *Consuming Power: A Social History of American Energies*. Cambridge, MA: MIT Press.

Olsen, M.E. (1981) 'Consumer's Attitudes Toward Energy Conservation', *Journal of Social Issues*, 37, 2, pp. 108–131.

Pobodnik, B. (1999) 'Toward a Sustainable Energy Regime: A Long-Wave Interpretation of Global Energy Shifts', *Technological Forecasting and Social Change*, 62, 3, pp. 155–172.

Pobodnik, B. (2002) 'Global Energy Inequalities: Exploring the Long-Term Implications', *Journal of World Systems Research*, 8, 2, pp. 252–274.

Pobodnik, B. (2006) *Global Energy Shifts: Fostering Sustainability in a Turbulent World*. Philadelphia, PN: Temple University Press.

Powells, G.D. (2009) 'Complexity, Entanglement and Overflow in the New Carbon Economy: The Case of the UK's Energy Efficiency Commitment', *Environment and Planning A*, 41, pp. 2342–2356.

Rosa, E.A., Machlis, G.E. and Keating, K.M. (1988) 'Energy and Society', *Annual Review of Sociology*, 14, pp. 149–172.

Schipper, L. (1996) 'Life-Styles and the Environment: The Case of Energy', *Daedalus*, 125, 3, pp. 113–138.

Schnaiberg, A. (1980) *The Environment: From Surplus to Scarcity*. Oxford, UK: Oxford University Press.

Shelley, T. (2005) *Oil: Politics, Poverty and the Planet*. London: Zed Books.

Shove, E. and Walker, G. (2010) 'Governing Transitions in the Sustainability of Everyday Life', *Research Policy*, 39, 4, pp. 471–476.

Silver, N. (2011) 'Egypt, Oil and Democracy', *New York Times*, 31 January. Available at http://fivethirtyeight.blogs.nytimes.com/2011/01/31/egypt-oil-and-democracy/. Accessed 7 January 2012.

Slovic, P., Flynn, J.H. and Layman, M. (1991) 'Perceived Risk, Trust, and the Politics of Nuclear Waste', *Science*, 254, pp. 1603–1607.

Sonnenfeld, D. and Mol, A. (2011) 'Social Theory and the Environment in the New World (Dis)Order', *Global Environmental Change*, 21, 3, pp. 771–775.

Stern, P.C., Dietz, T., Abel, T., Guagnano, G.A. and Kalof, L. (1999) 'A Value–Belief–Norm Theory of Support for Social Movements: The Case of Environmentalism', *Human Ecology Review*, 6, 2, pp. 81–97.

Tanuro, D. (2010) 'Marxism, Energy and Ecology: The Moment of Truth', *Capitalism, Nature, Socialism*, 21, 4, pp. 89–101.

Thün, J. and Velikov, K. (2009) 'The Post-Carbon Highway', pp. 164–211 in J. Knechtel (ed.), *Fuel*. Cambridge, MA: MIT Press.

Toke, D. (2011) 'Ecological Modernisation, Social Movements and Renewable Energy', *Environmental Politics*, 20, 1, pp. 60–77.

Urry, J. (2011) *Climate Change and Society*. London: Polity.

Watts, M.J. (2005) 'Righteous Oil? Human Rights, the Oil Complex and Corporate Social Responsibility', *Annual Review of Environment and Resources*, 30, pp. 373–407.

Watts, M.J. (2004) 'Antinomies of Community: Some Thoughts on Geography, Resources and Empire', *Transactions of the Institute of British Geographers*, 29, pp. 195–216.

Weaver, A. (2008) *Keeping Our Cool: Canada in a Warming World*. Toronto: Viking Canada.

Whitmarsh, L., Upham, P. Poortinga, W., McLachlan, C., Darnton, A., Devine-Wright, P., Demski, C. and Sherry-Brennan, F. (2011) *Public Attitudes, Understanding and Engagement in Relation to Low-Carbon Energy: A Selective Review of Academic and Non-Academic Literatures*. Report for Research Councils UK Energy Programme. Available at www.rcuk.ac.uk/documents/energy/EnergySynthesisFINAL20110124.pdf. Accessed 7 January 2012.

Whitney, V.H. (1956) 'Some Interrelations of Population and Atomic Power', *American Sociological Review*, 21, 3, pp. 273–279.

York, R. (2007) 'Demographic Trends and Energy Consumption in European Union Nations, 1960–2025', *Social Science Research*, 36, 3, pp. 855–872.

Younger, P.L. (2004) 'Environmental Impacts of Coal Mining and Associated Wastes: A Geochemical Perspective', *Geological Society, London, Special Publications*, 236, pp. 169–209.

Zhang, L., Mol, A.P.J. and Sonnenfeld, D.A. (2007) 'The Interpretation of Ecological Modernization in China', *Environmental Politics*, 16, 4, pp. 659–668.

11

Sustainability as social practice
New perspectives on the theory and policies of reducing energy consumption

Harold Wilhite

The spectres of climate change and resource depletion create an urgent need for deep reductions in energy consumption in the rich countries of the world. Rapidly increasing energy use in developing countries for human and economic development (poverty reduction and the provision of basic energy services such as health care and schools) increase the urgency for deep reductions in the energy consumption of OECD countries. The need for rapid and radical change is disheartening given the deeply anchored associations in both research and policy between more consumption, economic progress and better lives. These associations have contributed to a stripping down of energy savings to questions of efficiency, both economic and technical. Greater efficiency promises reductions in energy use without threatening the prospects for economic expansion. The research domain that has focused on sustainable energy has been dominated by technologists and economists, and by an agenda based on assumptions about economically rational actors and the power of markets to reorder the social contexts around consumption (Shove and Wilhite 1999; Wilhite and Norgaard 2004; Wilhite et al. 2000).

The debates about the strength of this techno-economic paradigm to deliver results are no longer hypothetical. After forty years of research and policies based on this paradigm, energy consumption in OECD countries has been only marginally reduced. This empirically demonstrable result, together with the increasing urgency of climate change, has grudgingly created an opening for new thinking on energy consumption, and fledgling engagement with broader theories of socio-technical change.

This chapter presents a theoretical framework that draws together perspectives on social, cultural and material contributions to consumption: social practice theory. I explore its potential for inspiring new thinking on how we conceptualize home energy consumption. The important bedrock concepts in practice theory are discussed, including agency, routine, behaviour, reflexivity and habit. Particular attention is given to how practices form, stabilize and change. A distinction is drawn between strong and weak habits, important for policies directed at catalysing change. The chapter concludes with a brief discussion of the policy implications of the social practice approach.

Harold Wilhite

Social practice theory

Over the past decade, social scientists from various academic disciplines have contributed to the development and application of practice theory as related to everyday energy consumption (Røpke 2009; Shove 2003; Warde 2005; Wilhite 2008). This scholarship has moved the theory of energy consumption from its focus on economic rationality and technical efficiency to encompass the ways that people and things interact and how that interaction is mediated by social contexts. These efforts draw on newer refinements in the theory of practice, such as those of Reckwitz (2002). He defines a practice as 'a routinized type of behaviour which consists of several elements, interconnected to one another: forms of bodily activities, forms of mental activities, "things" and their use, a background knowledge in the form of understanding, know-how, states of emotion and motivational knowledge' (2002: 249, cited in Warde 2005). This perspective provides a way for addressing the contributions of non-reflexive knowledge, informal learning and cultural specificity. As Seyfang et al. (2010: 8) put it, from a practice perspective:

> Individuals ... are no longer either passive dupes beholden to broader social structures, or free and sovereign agents revealing their preferences through market decisions, but instead become knowledgeable and skilled 'carriers' of practice who at once follow the rules, norms and regulations that hold practice together, but also, through their active and always localised performance of practices, improvise and creatively reproduce and transform them.

An important shift in practice theory is from a focus on individual agents to a deployment of the concept of agency, defined by Ortner as the potential to influence acts (Ortner 1999, 2006a, 2006b).[1] In the agent-based approach to consumption, individual consumers are free agents whose intentions and actions make consumption happen. The attention is on reflexivity, cognition and conscious decision making. In a practice theoretical approach, agency can be said to be distributed between individuals; the things with which they interact; and the routines and habits that develop in that interaction. Consumption is conceptualized as the result of interaction between the consuming agent (with her preferences and predispositions) and the material environment, mediated by the socio-cultural context in which the consumption practices are performed.

The form for agency possessed by things, routines and contexts has been referred to, variously, as practical knowledge, practical consciousness, tacit knowledge and embedded knowledge. In Pierre Bourdieu's writings, he conceptualized practical knowledge as knowledge that accumulates in and through social relations and interactions. This knowledge is tacit in the sense that it is in the form of a potential or predisposition for action (1977, 1998). He used the term *habitus* to capture this field of structured predispositions. Bourdieu's emphasis was on how predispositions are embedded through 'practice, action, interaction, activity, experience, and performance' (1998: 3) The ways we dress, eat, clean, organize space and time are all saturated by practical knowledge. Bourdieu has been criticized for 'downplaying the agency of the subjective meaningfulness of action' and for viewing practical knowledge as being overly deterministic (Warde 2011: 11). However, I agree with Warde's assessment that *habitus* ought to be rehabilitated in the emerging effort to understand the relationship between lived experience, practical knowledge and action. I return below to a different form for practical knowledge as theorized in Science of Technology Studies (STS), but first want to draw attention to the body, an important site of practical knowledge that is strongly agentive in practices.

Embodiment

Both Bourdieu and anthropologist Marcel Mauss were interested in the body as a site of practical knowledge and in understanding how bodies become knowledgeable. Their work goes against the grain of the bulk of sociological research on behaviour, because, as Crossley (2007: 81) writes, the assumption has been that 'Action, behaviour, interaction, practice and praxis have both embodied and mindful aspects, without any implication that these aspects derive from separate sources or "substances"'. Practical knowledge has been subordinated, because 'culturally appropriate bodily action and coordination "just happens" and falls below the threshold of perception and reflective knowledge' (Crossley 2007: 83). Mauss was interested in fleshing out what 'just happens' when people pick up and hold a baby, pick up and use a fork, apply cosmetics, all of which involve predisposed, 'embodied' agency. Each of these are particular to cultural settings and are capable of being performed without the application of reflexive knowledge. Take eating, for example: when North Americans sit down to eat a meal, they take the fork in their left hand and the knife in their right. They cut their food, transfer the fork to their right hand, and then convey it their mouth. Europeans typically move the food to their mouth without making the transfer of fork from left to right hand. In both of these culturally specific 'eating techniques', the action happens below the threshold of conscious thought. Consumers need not think through their movements in order to accomplish the action. In fact, while performing these movements, people often have their cognitive attention directed elsewhere, for example, to a conversation with a dinner partner or a television programme.

Embodied knowledge can also form through purposive training for competitive sports and other skill-based activities. Training consists of repetition and the honing of tacit skills. While swimming laps in a swimming pool with protected lanes, the experienced breaststroker should be able to accomplish lap after lap without giving any conscious thought as to how to negotiate movements. The successful performance is dependent on a stable socio-material context. For the experienced swimmer, when swimming in a river, lake or ocean, an encounter with rough seas or high waves might require a cognitive override and adjustments in movements to account for changing conditions. The fingers of an experienced typist produce text without reflection, given no disruptions. A talkative colleague, a blaring radio or a malfunctioning word processor might elicit engagement of the cognitive self. Thus the strength of a body technique is related to the strength and uniformity of the cultural predispositions (hiatus) as well as to the nature of the socio-material context in which the action is performed. In the next section I give attention to the role of materially embedded predispositions in practices.

The knowledge embedded in things

Madeleine Akrich (2000: 208) wrote that technologies bring to practices 'scripts' or 'scenarios', which she defined as 'a framework of action together with the actors and the space in which they are supposed to act'. This insight from STS is important for theorizing the agency of embedded knowledge in practices. Curiously, while things and material contexts are named as important contributors to practice, little effort has been made to merge or connect the insights of STS with theories of socially embedded agency. As Marcia Dobres (1999: 8) writes, 'theories of agency are practically mute on the active role of material culture and technological endeavour'. Many social scientists ignore technology because of concerns about technological determinism, anthropomorphizing of objects or perhaps simply because objects do not talk. As anthropologist Ted Ingold (1999: ix) wrote, 'technology tends to be associated with the mechanical replication of the given rather than the creative production of novelty, and hence with what is objective and determined rather than what is subjective and spontaneous'. As a result, social science has been most interested in how individuals exert agency on technologies, for example, how household technologies are 'domesticated' or 'appropriated' by their users (see Lie and Sørensen 1996). This may be illustrated by how people misunderstand room thermostats and use them like on–off switches; how they override movement-sensitive or natural-light sensitive lighting systems by manually manipulating lighting; or by how people open windows in thermostatically controlled buildings to regulate heat. However, little attention has been given to the capacity of technologies to reshape practices once they are taken into use.

Archaeologists (anthropologists of the past) are an exception. They have a more robust theory of things for very good reasons: things are most often all archaeologists have to work with in their efforts to construct past practices. Archaeologists attempt to dig out (literally and figuratively) the ways in which objects reflect the practices of their users and have influenced social life. Marcia Dobres expresses it this way:

> Because technology is an ever unfolding *process*, a 'becoming,' as it were, it necessarily interweaves the experiential making and use of material culture with the making and remaking of social agents.
>
> *(1999: 3, emphasis in original)*

My research in South India illustrates how knowledge embedded in things can affect practices over time (Wilhite 2008). There is a longstanding food ideology in South India, with roots in India's Ayurvedic health tradition, which associates the storing of prepared foods with the accumulation of substances which cause laziness and stupidity. This ideology contributed to a lack of enthusiasm for the refrigerator when it became widely available in India in the 1960s. Those who purchased the first generation of refrigerators were more interested in their space-saving properties (eliminating the need for storage rooms and cabinets for raw foods like eggs and vegetables) than in their capacity to store cooked foods and reheat them for consumption at later meals. However, a social change is taking place in India in which women increasingly are taking work outside the home, yet retain full responsibility for food preparation and other household chores. The resulting squeeze on women's time has spurred interest in the refrigerator.[2] Once installed in home and in social practices, the refrigerator offers many possibilities for saving time, but some of these conflict with ideas about healthy food. Many families with refrigerators are still conflicted about exploiting the refrigerator's full time-saving potential, for example, to store cooked foods for later reheating and for storing ready-made foods. Many women still insist on cooking food from scratch for each meal and using the refrigerator mainly to store raw foods and dairy products. However, generational differences are beginning to emerge. Many young women are now routinely making food in bulk, storing uneaten portions and reheating them for later meals. This new practice has paved the way for the microwave oven, which in 2004 was the fastest-selling household appliance in India. The embedded potential in the refrigerator has not only contributed to the change in several food-related practices; it has also created an opening for a regime of refrigeration-related technologies in household consumption (microwave, freezer) as well as in food provision (expanded refrigeration in convenience stores) and distribution (refrigerated transport between wholesalers and retailers).

Habit formation

How do practices form and change, and how do practices become habituated? Over time, the ways people cool or heat their homes, prepare foods and wash their bodies (to name only a few examples) have changed dramatically. Shove (2003) uses bathing as an example of a practice that has undergone dramatic changes over the past century. Nineteenth-century Europeans bathed rarely due partly to concerns that baths were sources of impurities. In the twentieth century, bathing became more frequent as health sciences proposed that water washes away impurities, and then, at mid-century, soap manufacturers invented 'body odour' and 'germs' as targets of the bathing experience. Today most Europeans bath at least once a day and the practice is accomplished with perfumed soaps, shampoos and deodorants.

Reducing the timeframe to that of a couple of generations within a household, many practices, including bathing, are stable. This stability is related to the frequency of performance, as well as to the complexity in terms of space, material contributors and time. Practices become habituated through repetition and in the process a form for practical knowledge develops. The bath or shower; the ways we light and heat the house; how we commute and shop; and the ways we wash our clothes and our houses. Body techniques are strong habits. One reason for this is the uniformity of the socio-material backdrop. This is also true of some of the consumption practices that take place inside the four walls of the home, such as bathing and dishwashing. The complexity of the material, temporal and spatial variables associated with heating, cooking and clothes cleaning, weakens the agency of embedded knowledge and weakens the habit. Routines that extend beyond the borders of the home, such as commuting and shopping, involve bigger spaces, longer time intervals and more opportunities for reflexive decisions. These are weak habits.

We should not forget Bourdieu's point that the socio-cultural context (*habitus*) has an effect on the strength of habits. In Wilhite et al. (2001), my co-authors and I examined the cultural importance of lighting and bathing in Japan and Norway. We determined that highly energy-intensive lighting practices in Norway were deeply anchored in home culture, in which the preferred aesthetic uses small lamps or point lighting to create light and shadow. Light is not used to produce lumens, but rather a particular aesthetic in living areas. This contrasts with Japanese households who used lighting in a more functional way to light up spaces using ceiling lights. Concerning bathing, Japanese associate the bath as much with relaxation as with washing, and spend long intervals bathing, soaping, rinsing and relaxing in the bath. Norwegians are more interested in getting (and smelling) clean and are more likely to take relatively brief showers. These cultural considerations contribute to the strength of lighting habits in Norway and bathing habits in Japan. In these examples, even though the temporal and material components are complex, the grounding in the cultural *habitus* makes them strong habits.

The performance of habits create an inertia, or resistance to change, but nonetheless can change through a change in the socio-cultural context or a change in the material constituents. A move to a new dwelling, the purchase of a new technology or a change in *habitus* can lead to a change in habits. An example from my research on household consumption in India revealed how long-term exposure to a new cultural context can lead to changes in both strong and weak habits. In South India, a large proportion of the workforce spends years abroad working in places like Saudi Arabia and other countries around the Gulf of Amman. These are most often men. Their wives and families remain in South India, where they stay with members of the extended family and, as income accrues, establish their own family homes. In their places of work, migrants are exposed to new household technologies and new ways of accomplishing cooking, heating, cooling and mobility. Over time, migrants live themselves into new habits. A form for 'double *habitus*' is created which connects the work residence to the household in India. Both new household appliances and new routines are transferred through this double *habitus* to the South India home. My research has shown that the ways homes are cooled, the way food is prepared and the ways people transport themselves are examples of habits which have reformed and established themselves as normal practice in India (Wilhite 2008).

Conclusion: encouraging sustainable practices

Such insights on habit are important to the development of policy levers which can be used to move practices in a sustainable direction. The implications are, first, that policy should give consideration to the sources of the embedded knowledge that hold practices together (cultural context, technology agency, repeated performance). Second, there should be an attempt to assess the strength of habits. Once this has been accomplished, appropriate forms for information, technology policies, economic incentives and regulations (laws) can be developed. There is a vast potential for new forms for information based on social learning; new technology designs with 'scripts' that foster less energy intensive habits (not just greater energy efficiency); and new forms for economic interventions which acknowledge the inertia in habits and their inherent resistance to change. The precise nature of these kinds of interventions constitutes a new agenda for sustainable energy policy and need to be developed. Here I give a few suggestions to exemplify new approaches.

Concerning the cultural contribution to household habits, not all practices are equally culturally important. Above, I gave the examples of the importance of lighting in Norway and bathing in Japan. In North America and in many European countries, automobile-centred mobility is important, whereas in Norway car use is related more to functional issues than to social signalling or personal identity. There is a potential in Norway to accelerate change from use of the car to other forms of mobility by making the latter cheaper and more convenient. The technology design element is important, providing the material infrastructure and alternative transport systems which are fast and reasonable.[3] Other types of alternative mobility practices, such as car sharing or appliance leasing, also deserve greater attention. Such practices demystify non-ownership and have demonstrated that sharing does not necessitate radical changes in convenience or the quality of life (Attali and Wilhite 2001; Jelsma and Knot 2002).

Another policy lever based on Ackrich's concept of the technology script is to increase support to the design and implementation of technologies with low energy scripts. The low-energy house and smart-house technologies are examples (Goodchild and Walshaw 2011). Experience shows that it is important to take cultural considerations into account and not give the impression that life in one of these houses would resemble life in a space station. It will be important to emphasize that not only do energy use and charges decline, but that new practices related to these technologies will still be comfortable, convenient and not radically different from life in a conventional home. Smart cities offer further examples of using technology design to affect transport practices (Bulkeley et al. 2011). The provision of bicycle infrastructures and subsequent increase in bicycling is an example of smart mobility. In Copenhagen, after decades of work on making bicycling safe and convenient, over 50 per cent of commuters today commute by bicycle; reduced automobile use hangs together with provision of convenient alternatives in the form of safe walking and biking and fast, convenient public transportation.

Changes in practice can be facilitated with new forms for information that play on what could be called social learning: when people make decisions they rely on the experiences of people who have made similar decisions. Two types of information that draw on the social learning idea hold promise. One involves the conveyance of examples of transitions to successful low-energy practices. For example, many cities around the world are enforcing a new regulation that closes off all or parts of a city to automobiles at certain times of the day or week. These have been almost uniformly well received (Topp and Pharoah 1994). Demonstration projects also play on the social learning idea. They were widely used in the 1980s but are now largely forgotten. In Davis, California, in the 1970s and 1980s, great strides were made in home weatherization after demonstration homes were set up in neighbourhoods around the city. People were able to observe and experience first-hand how life in a thermally tight house was cooler (in the hot climate) and that the retrofits (windows, insulation, weather stripping) did not degrade notions of what was a cosy home aesthetic. Yet another form for information using the social learning principle provides households with a benchmark by which they can compare and assess their levels of energy use with other households living in similar dwellings. Observing that one's own household energy consumption is higher than that of others living in a similar house can be a stimulus to digging into household habits, assessing the energy consequences, and making a change, whether it be the way energy is managed (i.e. thermostats) or a new purchase (energy efficient fridge or wall insulation) (see Wilhite and Ling 1995; Fischer 2007).

Another insight from a social practice perspective is that tacit knowledge in household practices gets challenged when a family moves from one house and neighbourhood to another. A move often initiates a flurry of projects involving organization of the home's spatial layout, the purchase of new appliances and changes in routines (Wilk and Wilhite 1985; Wilhite and Ling 1992). Further reflection begins when people are preparing to have a child, or later in the family cycle when children move out of the home. Policy for sustainability should give more attention to households in these transitions.

In conclusion, social practice theory offers new insights on stability and change in consumption by accounting for the dynamic relationship between material artefacts, social contexts and individual consumers. Social practice theory acknowledges the co-presence of subjects and objects in the world and gives attention to the field of opportunities and obstacles formed in their interrelationship. It offers a new theoretical foundation for policy that is enabling yet, at the same time, challenging. It is one thing to acknowledge the power of practical knowledge and yet another to find ways to influence and move associated practices. There is a dire need for further development of practice perspectives in future research on sustainable consumption.

Notes

1 In examining the relationship between learning and practices, linguistic anthropologist Ahern has defined agency as 'the socio-culturally mediated capacity to act' (2001: 18). In a recent paper, Gordon Walker (2010) draws parallels to Amartya Sen's (1999) capability theory, in which human development is theorized as providing people with the capacity to improve their lives (education, health, economic opportunity and so on).
2 This is reminiscent of the increasing popularity of the refrigerator and other household appliances in the 1950s and 1960s in Europe and North America, also related to time pressures on housewives (Cowen 1989).
3 This potential has unfortunately not been grasped in Norway, where the rail system is old, underdimensioned, expensive and inefficient.

References

Ackrich, M. (2000) 'The De-scription of Technical Objects', pp. 205–224 in W. Bijker and J. Law (eds), *Shaping Technology/Building Society*. Cambridge, MA: MIT Press.
Ahern, L.M. (2001) 'Language and Agency', *Annual Review of Anthropology*, 30, pp. 109–137.
Attali, S. and Wilhite, H. (2001) 'Assessing Variables Supporting and Impeding the Development of Car Sharing', *Proceedings of the ECEEE 2001 Summer Study*. Paris: European Council for an Energy Efficient Economy.
Bourdieu, P. (1977) *Outline of a Theory of Practice*. Cambridge, UK: Cambridge University Press.
Bourdieu, P. (1998) *Practical Reason*. Cambridge, UK: Polity Press.
Bulkeley, H., Castán Broto, V., Hodson, M. and Marvin, S. (eds) (2011) *Cities and Low Carbon Transitions*. London: Routledge.
Cowen, R. (1989) *The Ironies of Household Technology from the Open Hearth to the Microwave*. London: Free Association Books.

Crossley, N. (2007) 'Researching Embodiment by Way of "Body Techniques"', *Sociological Review*, 55, pp. 80–94.
Dobres, M. (1999) 'Introduction' in M. Dobres and C.R. Hoffman (eds), *The Social Dynamics of Technology: Practice, Politics, and World Views*. Washington, DC: Smithsonian Institute Press.
Dobres, M. (2000) *Technology and Social Agency*. Oxford, UK: Blackwell.
Fischer, C. (2007) 'Influencing Electricity Consumption via Consumer Feedback: A Review of Experience', *Proceedings of the 2007 ECEEE Summer Study*. Stockholm: European Council for an Energy Efficient Economy.
Giddens, A. (1979) *Central Problems in Social Theory: Action, Structure and Contradiction in Social Analysis*. Berkley, CA: University of California Press.
Goodchild, B. and Walshaw, A. (2011) 'Towards Zero Carbon Homes in England? From Inception to Partial Implementation', *Housing Studies*, 26, 6, pp. 933–949.
Ingold, T. (1999) 'Foreword', in M. Dobres and C. R. Hoffman (eds), *The Social Dynamics of Technology: Practice, Politics, and World Views*. Washington, DC: Smithsonian Institution Press.
Jelsma, J. and Knot, M. (2002) 'Designing Environmentally Efficient Services; A "Script" Approach', *Journal of Sustainable Product Design*, 2, pp. 119–130.
Lie, M. and Sørensen, K.H. (eds) (1996) *Making Technology Our Own? Domesticating Technology into Everyday Life*. Oslo: Scandinavian University Press.
Mauss, M. (1973) 'Techniques of the Body', *Economy and Society*, 2, 1, pp. 70–89.
Ortner, S. B. (1999) '"Thick Resistance": Death and the Cultural Construction of Agency in Himalaya Mountaineering', pp. 136–165 in S. Ortner (ed.), *The Fate of 'Culture': Geertz and Beyond*. Berkeley, CA: University of California Press.
Ortner, S. (2006a) 'Updating Practice Theory', pp. 1–18 in S. Ortner (ed.), *Anthropology and Social Theory: Culture, Power and the Acting Subject*. Durham, NC: Duke University Press.
Ortner, S. (2006b) 'Power and Projects: Reflections on Agency', pp. 129–154 in S. Ortner (ed.), *Anthropology and Social Theory: Culture, Power and the Acting Subject*. Durham, NC: Duke University Press.
Reckwitz, A. (2002) 'Toward a Theory of Social Practices: A Development of Culturist Theorizing', *European Journal of Social Theory*, 5, 2, pp. 243–263.
Røpke, I. (2009) 'Theories of Practice: New Inspiration for Ecological Economic Studies on Consumption', *Ecological Economics*, 68, 10, pp. 2490–2497.
Sen, A. (1999) *Development as Freedom*. New York, NY: Anchor Books.
Seyfang, G., Haxeltine, A., Hargreaves, T., Longhurst, N. and Baldwin, R. (2010) 'Understanding the Politics and Practice of Civil Society and Citizenship in the UK's Energy Transition', Paper presented at the SPRU Conference (unpublished).
Shove, E. (2003) *Comfort, Cleanliness + Convenience: The Social Organization of Normality*. Oxford, UK: Berg.
Shove, E. and Wilhite, H. (1999) 'Energy Policy: What It Forgot and What It Might Yet Recognize', *Proceedings from the ECEEE 1999 Summer Study on Energy Efficiency in Buildings*, Paris: European Council for an Energy Efficient Economy.
Topp, H. and Pharoah, T. (1994) 'Car-Free City Centres', *Transportation*, 21, 3, pp. 231–247.
Walker, G. (2010) 'Inequalities, Capabilities and Sustainable Practices: Doing Everyday Life – Doing (In) Justice', Paper presented at the workshop *Practice Theory and Climate Change*, Lancaster University, July.
Warde, A. (2005) 'Consumption and Theories of Practice', *Journal of Consumer Culture*, 5, 2, pp. 131–153.
Warde, A. (2011) 'Social Science and Sustainable Consumption', Symposium Prospectus, Paper presented at the symposium *Social Science and Sustainable Consumption*, Helsinki Collegium for Advanced Studies, Helsinki, Finland, January.
Wilhite, H. (2008) *Consumption and the Transformation of Everyday Life: A View from South India*. Basingstoke, UK: Palgrave Macmillan.
Wilhite, H. (2009) 'The Conditioning of Comfort', *Building Research & Information*, 37, 1, pp. 84–88.
Wilhite, H. and Ling, R. (1992) 'The Person Behind the Meter: An Ethnographic Analysis of Residential Energy Consumption in Oslo, Norway', *Proceedings of the ACEEE 1992 Summer Study on Energy Efficiency in Buildings*. Washington, DC: American Council for an Energy Efficient Economy.
Wilhite, H. and Ling, R. (1995) 'Measured Energy Savings from a More Informative Energy Bill', *Energy and Buildings*, 22, 2, pp. 145–155.
Wilhite, H., Nakagami, H., Masuda, T., Yamaga, Y. and Haneda, H. (2001) 'A Cross-Cultural Analysis of Household Energy-Use Behavior in Japan and Norway', pp. 159–177 in D. Miller (ed.), *Consumption: Critical Concepts in the Social Sciences*. London: Routledge.

Wilhite, H., Nakagami, H. and Murakoshi, C. (1997) 'Changing Patterns of Air Conditioning Consumption in Japan', pp. 149–158 in P. Bertholdi, A. Ricci and B. Wajer (eds), *Energy Efficiency in Household Appliances*. Berlin: Springer.

Wilhite, H. and Norgaard, J. (2004) 'Equating Efficiency with Reduction: A Self-Deception in Energy Policy', *Energy and Environment*, 15, 6, pp. 991–1009.

Wilhite, H., Shove, E., Lutzenhiser, L. and Kempton, W. (2000) 'The Legacy of Twenty Years of Demand Side Management: We Know More about Individual Behavior But Next to Nothing about Demand', pp. 109–126 in E. Jochem, J. Sathaye and D. Bouille (eds), *Society, Behaviour and Climate Change Mitigation*. Dordrect: Luwer Academic Press.

Wilk, R. and Wilhite, H. (1985) 'Why Don't People Weatherize Their Homes? An Ethnographic Solution', *Energy: The International Journal*, 10, 5, pp. 621–630.

12
Environmental migration
Nature, society and population movement

Anthony Oliver-Smith

Migration, whether permanent or temporary, has long been a response or survival strategy for people experiencing environmental change (Hugo 1996). Throughout human existence, adapting to environmental fluctuations, sometimes expressed in 'natural' disasters, has been a consistent necessity for societies around the world. In some cases, migration has been seen as an adaptive option. Environmental changes have opened up new, more inviting prospects in other climes. Indeed, anatomically modern humans migrated from Africa in the middle Palaeolithic as the northern regions of the world warmed in the late Pleistocene (Templeton 2002). In other cases, environmental changes propelled the abandonment of environments. The last ice age, between 22,000 and 10,000 years ago, saw the depopulation of much of Northern Europe, Asia and North America (Stringer 1992). In addition, seasonal environmental fluctuations guided the movements of Neolithic hunters and gatherers as they harvested both plants and animals. Among the complex societies of prehistory, changes in rainfall regime over a 200-year span contributed to the massive depopulation of the Yucatan peninsula (Medina-Elizalde and Rohling 2012).

Nonetheless, the role of environment in migration, particularly in the modern era, cannot be reduced to a simple cause and effect relationship. In most circumstances, environmental, social, economic and political forces combined to increase the risk of uprooting for many vulnerable populations in exposed regions. In recent history, for example, the Great Flood of 1927 in the lower Mississippi Valley displaced nearly 700,000 people, approximately 330,000 of whom were African Americans who were subsequently interned in 154 relief 'concentration camps', where they were forced to work. However, the flood and its aftermath were only some of many reasons for African Americans to leave the South (Barry 1997: 417). Dust bowl migrations to California were due as much to the economic depression of the 1930s as they were to drought (Egan 2006). In Hurricane Katrina, the displacement of hundreds of thousands of people, many of them permanently, was due as much to human destruction of the environment's natural protections and inadequate local, state and federal policy and practice as they were to the hurricane itself (Elliott and Pais 2010).

Today the impacts of societal development are driving environmental changes that are potentially more extreme than at any time in recorded history, bringing with them a serious potential for uprooting large numbers of people. The complex interplay of social and economic factors in the environment has resulted in greater environmental change and vulnerability of people to those changes. The linkages between environment and society have grown ever more complex, making it difficult to speak of direct environmental causality in human migration. The relationship between environmental change and migration is embedded in the complexity of both and in the nature of causality between such complex phenomena. And, as with all things human, culture and society play crucial mediating roles between a population and the environment it inhabits (or leaves) (Oliver-Smith 2009). Indeed, local culture derived from lived experience with their surroundings is fundamental to understanding how environmental change is perceived, responded and adapted to (Enfield and Morris 2012).

The issue of environment and migration is among the most discussed and debated dimensions of the impact of global environmental change on human beings. The contingent nature of prediction of environmental impacts, the complex question of causality, the elusive nature of definitional issues, the vast disparities in predictions of numbers of people to be affected, and the overall complexity of human–environment relations, all present serious challenges to researchers attempting to analyse the relationship between environment and migration.

Environmental change and the potential for displacement and migration

Although environmental change does take place through natural disturbances and cycles (Holling 1994), the enormous impact of modern industrial societies on the environment over the last 200 years is especially well documented. The structure and organization of Western economic institutions, in particular, ideologically buttressed by concepts of the human domination and rational control of nature, essentially entrained a process of global environmental transformation. The result of embracing the rationality of pursuing self-interest in the use of natural resources and the mobilization of human labour has been unprecedented extremes of wealth and poverty, unprecedented levels of environmental destruction and the rapid amplification of socially constructed vulnerability (Oliver-Smith 2002).

Through misguided or various forms of direct and indirect coercion, human exploitation of the environment has severely impacted many environments. Most recently, the Millennium Ecosystem Assessment (MA 2005a) concluded that 15 of 24 assessed ecosystem services were being degraded or used unsustainably, with serious effects for poor, resource-dependent communities. Among the issues the MA calls attention to is the fact that 10 to 20 per cent of drylands are already degraded, affecting as many as two billion people. Increasing pressure on dryland ecosystems will affect the provision of ecosystem services such as food, and water for humans, livestock, irrigation and sanitation. There will likely be increases in water scarcity, as well, due to climate change in highly populated regions that are already under water stress. Droughts are also increasing in frequency and their continuous reoccurrence can overwhelm community coping capacities. When coping capacities and adaptation strategies of communities are overcome by the loss of ecosystem services, droughts and loss of land productivity can act as triggers for the movement of people from drylands to other areas (MA 2005b; Renaud and Bogardi 2007; Warner et al. 2010).

Furthermore, the Intergovernmental Panel on Climate Change (IPCC 2007a, 2007b) asserts that human-induced factors are generating significant increases in temperatures around the world, producing increases in the rate of sea level rise, increases in glacial, permafrost, Arctic and Antarctic ice melt, more rainfall in specific regions of the world and worldwide, more severe droughts in tropical and subtropical zones, increases in heatwaves, changing ranges and incidences of diseases and more intense hurricane and cyclone activity. Data from the Emergency Events Database (EM-DAT) at the Centre for Research on the Epidemiology of Disasters

(CRED 2008) suggest that floods, droughts, storm surges and other natural and technological agents are impacting greater numbers of people and increasing damages globally, although fatalities are reportedly on the decline (CRED 2008). The real and potential impacts of these changes are also predicted to generate natural and social processes that represent a significant potential to displace large numbers of people, obliging them to migrate as individuals and families or permanently displacing them and/or relocating them as communities.

The 'nature' of socio-ecological systems

Partially as a result of these interlinked changes, it is an increasingly accepted scientific tenet that nature and society are no longer seen as separate interacting entities, but rather as mutually constituting components of a single system, often referred to as a socio-ecological system. Understanding environmental change and its effects, such as population displacement and migration, therefore requires recognition that nature and society are inseparable, each implicated in the life of the other, each contributing to the resilience and vulnerability of the other (Oliver-Smith 2002). Some have suggested that human action now dominates and that we are living in a new geological epoch referred to as the Anthropocene (Crutzen and Stoermer 2000; Vitousek et al. 1997). Global climate change, driven by excess production of greenhouse gases, would appear to support that assertion. In effect, for all intents and purposes, natural processes are now in interaction with social processes in the production of global and specific vulnerability, environments and problems. The recognition of the human influence on global climate patterns now confirms that human action both purposefully and inadvertently shapes natural systems into human constructed, although not entirely controlled, environments. Human action notwithstanding, however, we still have to contend with forces within nature, albeit inflected profoundly by human processes, that clearly transcend any social efforts to transform or control. The agency of a nature that has been profoundly socialized complicates an adequate theorization of the relationship between environment and migration. Developing an appropriate theoretical framework for this task is difficult because the behaviour of socio-ecological systems cannot be understood unless both sides are treated as endogenous (Kotchen and Young 2007: 150).

The endogeneity of both sides is the challenge. Both society and nature are highly interactive, incorporating dimensions of the other in their own processes. Therefore, environmental features and ecological processes – such as earthquakes, hurricanes, floods, soil erosion or climate change – must be recognized as features of social life; and social and cultural elements – such as racism, religion and politics or commodities, land markets and currency circulation – must be seen as functioning ecologically (Harvey 1996: 392).

The endogeneity of socio-ecological systems further complicates seeking single-agent, direct causality in the environment, since it tends to elide the fact that environmental resources, as well as hazards, are always channelled for people through social, economic and political institutions and practices. Thus, it is difficult to point to the environment, even in natural agent disasters, as the single cause of anything. Seeking single-agent causality to such complex phenomena would seem a doomed effort in any context. By the same token, eliminating environment factors as the single cause of forced migration hardly warrants discounting them as part of a multiplicity of interacting drivers, or in some cases the triggering event, that combine to generate forced migration.

Debating environmental migration

Despite the fact that the reality of environmental change, and specifically climate change, is generally accepted, the impacts of actual and projected effects are still much debated in both scientific and political forums. There is considerable uncertainty about local manifestations of global environmental change and what necessary adjustments will be induced in natural and human systems (Dessai et al. 2007: 1). The uncertainty, in fact, characterizes the problem both at the level of physical impacts and at the level of responses and adaptations in human communities. Indeed, the projected effects of environmental change, particularly as they pertain to specific human communities, have entered as much into political controversy as they have into academic and scientific debate.

The research and scholarship focusing on the relationship between environment and migration is shot through with controversy due to this uncertainty, centring largely around the issues of predicted numbers, appropriate terminology for people uprooted by environment and the political implications of both research and policy pertaining to environmentally displaced people. There is considerable debate about what exactly constitutes an environmentally induced move and how to measure and explain it. The actual processes through which major population dislocations might occur are still only partially understood (Adamo 2008: 2). The United Nations High Commissioner for Refugees (UNHCR 2009: 4) sees five displacement scenarios emerging in the near future: hydrometeorological disasters, population removal from high risk areas, environmental degradation, the submergence of small island states and violent conflict.

However, some scholars assert that it is erroneous to attribute causality to the environment, since migration is always the result of multiple factors, including social, economic and political as well as environmental forces, underscoring the fact that human demographic movement is both a social and an ecological phenomenon, both impacted by and impacting the environment (Black 2001; Castles 2002; Kibraeb 1997). There are legal objections to the term 'environmental refugee'. The 1951 United Nations Convention Relating to the Status of Refugees legally defined a 'refugee 'as a person who flees their country of nationality for fear of persecution based on race, religion, nationality, ethnic or social group or political opinion. People displaced by environmental causes do not qualify under the UN convention definition of refugee. Moreover, critics also fear that applying the term 'refugee' to environmentally displaced people will mask the political causes of displacement and allow states to evade their obligation to provide asylum. Other scholars object to the term politically, because of instances when the term 'refugee' has nourished xenophobic and racist perspectives, pointing to the fear of climate-induced migration that has recently entered European and North American political discourse (Hartmann 2009; Wisner 2009).

Complexity and causality in environment and migration

While the substance of all these assertions, both pro and con, on environmental migration, may be questioned, the concerns they express are valid and reflect the difficulties of developing appropriate political, policy and practical responses for environmentally displaced peoples in the near future. The relationship between environment and migration is far from linear or straightforward and understanding it presents a number of conceptual challenges. These challenges are embedded in the complexity of the relationship between social and ecological systems and in the nature of causality between such complex phenomena. It is also clear that the 'environmental refugee' controversy is both highly charged and deeply embedded in the way complex human–environment relations are understood by scholars, politicians and the general public.

The complexity of socio-ecological systems and environmental migration necessarily obliges us to deal with the issue of causality. Causality is a much-discussed and -disputed concept whose mathematical and philosophical parameters have been debated since Aristotle. Direct relationships of causality are hard to come by. In the strictest sense of the word, if A causes B, then

A must always be followed by B. In common parlance, when we say A causes B, as in smoking causes cancer, what we should really say is that smoking causes an increase in the probability of B (Spirtes et al. 2000). In other words, A increases the risk of B. In this case radical environmental change increases the risk of displacement and the incidence of migration. Migration research has shown that the reasons for migration are highly complex, often combining a variety of social, economic, demographic and political factors, acting either to push or pull, and sometimes both, people from their original location towards another. In effect, drivers may be multiple and often intertwined. The complexity of our social and psychological makeup makes reducing human behaviour to single causes always risky.

A finer-grained understanding of both environmental forces and their effects must be based on an approach that recognizes the environment as a socially mediated context experienced by people both positively and negatively just as society expresses itself environmentally both positively and negatively. Particularly in climate change, people will be displaced by a set of processes created and driven by human agency, specifically massive production of greenhouse gases that have entrained a series of processes that are transforming global climate and therefore nature. The fact that these processes manifest themselves in and as events that transpire in the environments that we live in or in ways that take the form of natural processes (wind, rain, drought, erosion, etc.) obscures their partial human origins. Under no circumstances should they be interpreted as natural. They are most certainly environmental processes that combine human and natural forces and features.

Environment and the multiple drivers of migrations

In effect, rather than trying to identify and isolate environmental factors as single drivers, the more relevant task has become analysing the role that environmental factors play along with other political, economic, demographic and social drivers of migration (Black et al. 2011). Economic drivers, perhaps most often given the greatest importance, can act as both push factors, where local livelihoods or employment become reduced, or pull factors, where economic activities are more vital in other regions. Clearly environmental factors, whether in the form of local hazard impact or better environment services availability elsewhere can interact with economic drivers to provoke migration. Similarly, political factors that run the gamut from policies to reduce social services to the breakout of war can degrade environments inducing people to migrate. Demographic factors also may contribute to migration through increasing densities on available land, reducing surplus carrying capacity and stressing environmental services. Social and cultural features may encourage migration, as well. The so-called 'bright lights' theory refers to the attractions that urban environments represent for rural people who desire to engage in more contemporary lifestyles (Byerlee 1974). And finally, environmental events and processes may also drive migration. The impact of natural hazard occurrence, such as Hurricane Mitch in Honduras, which uprooted thousands, many of whom migrated to the United States in 1998. In some cases emergency evacuation either before or after a hazard event can result in permanent migration, such as occurred in Hurricane Katrina. Slower onset processes such as droughts have also stimulated various sorts of migration in Africa, in some cases leading to the sedentarization of nomadic pastoralist populations, and in others turning seasonal adaptive migrations into permanent resettlement (Kenny 2002; Merryman 1982). Moreover, many environmental impacts that uproot people are often shown to be far from naturally generated, but rather have their origins in human policies and practices, as the environmental destruction in New Orleans tragically demonstrated in Hurricane Katrina. Indeed, in human communities environmental factors are always mediated through social frames.

Therefore, Black's (2001) critique that emphasizing environmental factors diminishes the role played by political and economic factors in migration is well taken, and coincides with the viewpoint of most disaster researchers today that highlights the political or economic forces that together with natural agents produce disasters or, for that matter, any forced migration that might ensue. In the face of such complexity then, the question thus becomes how causality is to be reckoned. Hilhorst (2004) contends that the fact that disasters involve the interaction of multiple adaptive subsystems within social and natural systems renders them acutely unpredictable in their development and outcome, if not entirely so in their occurrence. We now understand that most environmental changes, particularly those generated by climate change, are similar.

Elusive definitions

Compounding the complexity of socio-ecological systems and the often intertwined causal factors of migration, the failure to reach a consensus definition of environmental migration has further impaired efforts to diminish the uncertainty that surrounds the issue. Since the 1980s, researchers have linked the issue of environmental change with human migration, designating as 'environmental migrants', 'environmental refugees', 'climate migrants', or 'environmentally displaced peoples', people who are forced to leave their homes, temporarily or permanently, due to the threat, impact or effects of a hazard or environmental change. There have been many attempts at definition, but none has fully succeeded in being generally accepted. Without the parameters of an established definition, it is difficult to state whether migrating populations are actually environmental migrants or, for example, economic migrants. Most of the definitions offered in the literature address the issue of an environmental disruption, whether a sudden disaster from the occurrence of a natural hazard or a slower onset process of resource degradation of either natural or anthropogenic origin. There has also been debate over the appropriate terminology to use. Are environmentally uprooted people refugees (see below) or migrants, or disaster victims?

The disparate causes of environmental migration, including disasters, environmental degradation, contamination and climate change, as well as the different forms and trajectories the migratory process may take, have proved challenging to bracket within one overarching definition. Some researchers are reluctant to include migrants who were temporarily displaced and those who permanently relocated in the same category. Further distinctions have been drawn between those who leave voluntarily and those who are forced to migrate. Others have objected to the inclusion in one category of disaster victims and people displaced by environmental degradation. While the authors who have advanced definitions for environmentally displaced people number more than a dozen, the following are representative. El-Hinnawi (1985), who first coined the term 'environmental refugee', used it to describe people who had been temporarily displaced, those who had been permanently displaced and those who migrated because their home environment no longer could sustain basic needs. Myers (2002) defined environmental refugees as people who can no longer gain a secure livelihood in their homelands because of drought, soil erosion, desertification, deforestation and other environmental problems, together with the associated problems of population pressures and profound poverty. Renaud et al. (2007) constructed a typology of environmental migrants that distinguishes between: (1) an environmentally motivated migrant who chooses to leave a steadily deteriorating environment to pre-empt the worst outcome; (2) an environmentally forced migrant who must leave to avoid the worst outcome; and (3) those who must flee the worst outcome.

The dynamics of environment and migration

Moreover, efforts to define environmental migration have difficulty accounting for the dynamics of the process. Environmental changes expected to pressure people to migrate may be adapted to through social and cultural means, thereby avoiding the uprooting process. Cultures around the world have adapted to the seasonal environmental fluctuations by expanding their ecological niches, adopting 'famine foods', risk-sharing institutions and seasonal temporary migrations (Torry 1979). Both adaptation and its related concept, mitigation, entail changes in social, technological and environmental relations. Mitigation is proactively concerned with strategies to minimize impact and loss, and to facilitate recovery and thus increasing the resilience of a society. Adaptation, on the other hand, is a process that offers possible adjustments that may enable people to safeguard livelihoods and welfare.

Adaptation, however, because it is deployed in numerous institutional and environmental contexts, is a complex issue. Adaptation is the fundamental conceptual nexus in human–environment relations. It is through the process of adaptation that humans and natural systems conjointly construct socio-ecological systems, or environments. Humans interact with and adapt to both a socio-cultural (institutional) environment as well as a natural environment. That is, our institutions are at once part of our overall adaptation, but must be adapted to as well. These local circumstances of both natural and social nature are the basis of a community's ability to mitigate or adapt to environmental change to avoid displacement or migration.

When used in the social scientific sense, adaptation refers mainly to changes in belief and/or behaviour in response to altered circumstances to improve the conditions of life (or survival). In that sense, adaptation in general is reactive, adjusting primarily to novel conditions. Human adaptations to environmental change are largely social organizational and technological. Faced with environmental change human beings assess options, make decisions and implement strategies, based on existing knowledge and technology for exploiting an environment's energy potentials, or where they are lacking, for abandoning it (Bennett 1996; Holling 1994).

Estimating the hard-to-count

One of the complications of the lack of a consensus definition is the enormous disparity in estimates of people who have been, or will be, displaced by the effects of environmental change. Estimates are at least in part contingent on how environmental migration is defined and who will fall under a given definition. The range of estimates is considerable, as illustrated in Table 12.1.

Moreover, the failure of most of those who have offered estimates to specify the methods by which they arrived at their numbers has generated significant scientific debate and presented

Table 12.1 Estimates of people displaced by the effects of environmental change

Source	Estimated displacement
El-Hinnawi (1985)	50 million
Almeria Statement (1994)	135 million
Myers (2005)	200 million
The Stern Review (Stern 2006)	200 million by 2050
Friends of the Earth (2007)	200 million by 2050
Christian Aid (2007)	250 million
Global Humanitarian Forum (2009)	20 million in 2009

policymakers with confusing data. There is also no commonly agreed upon methodology (Gemenne 2011). Some estimates are based on field reports from relief agencies. Some are based on population figures in areas that are experiencing environmental change. Few have solid bases for the numbers that are estimated. Probably the most reliable figures at present are produced by the International Displacement Monitoring Centre and the Norwegian Refugee Council, which place the number of people displaced by natural disasters at over 42 million people, using a baseline of events from the EM-DAT database to produce a core data-set for events where over 50,000 people were affected. Data on the displaced from each event is then sought from organizations involved in relief for those events (Yenotani 2011).

Most estimates of environmentally displaced people gloss over distinctions that many of the definitions point to, such as the differences between temporary or permanent migration, or between voluntary and involuntary displacement. Nor do most of the estimates establish whether the numbers are for a given year or cumulative for a given time period. Still another problematic dimension is that neither the definitions nor the estimates have captured the fact that most people have been uprooted by a mix of environmental with other economic, political, social and/or demographic factors. The debate over this issue, with claims of millions of environmental refugees being produced versus counter-claims that the evidence is uneven, unconvincing and counterproductive, has been active since the 1980s. However, it is clear that to develop adequate responses to these issues and uncertainties regarding the social impacts of environmental change we must begin by addressing them at the multiple levels at which they exist, and particularly in the complex interrelationships between nature and society, both conceptually and specifically as expressed in local contexts.

Social vulnerability, environmental change and forced migration

Research on social vulnerability since the 1990s has made clear that exposure to hazards alone does not determine where the serious effects of any hazard will most likely be experienced. Social vulnerability refers the characteristics of a person or group in terms of their capacity to anticipate, cope with, resist, and recover from the impact of a natural hazard. It involves a combination of factors that determines the degree to which someone's life and livelihood is put at risk by a discrete and identifiable event in nature or society (Wisner et al. 2004).

What can we expect from future effects of environmental change for specific regions and communities? To answer that question is difficult because of the numerous variables and the non-linearity of their interactions. The challenge lies in determining not just absolute exposure and absolute exposed population but specific lands and populations in different socially configured conditions of resilience or vulnerability. In fact, in many areas, conditions of vulnerability are accentuating rapidly due to increasing human induced pressures on ecosystems. Moreover, the vulnerability of a nation to environmental change effects is partially a function of its level of development and per capita income (Nicholls et al. 2007: 331). Less-developed countries have a significantly higher level of vulnerability to climate change effects.

However, the problem with assessing the exposure of both lands and people to climate change is that, not only are we dealing with projected environmental change effects, but also with projections about various physical, societal and infrastructural trajectories including greenhouse gas emissions, in the case of climate change, demographic change, migration trends, infrastructural development, mitigation strategies, adaptive capacities, vulnerabilities and patterns of economic change, all of which will play out in different ways, according to the political, economic and socio-cultural dispositions of national governments, international organizations and general populations (Nakicenovic and Swart 2000).

These difficulties in establishing both exposure and vulnerability to specific localities and populations notwithstanding, environmental change and its impacts have serious human rights implications. However, these implications tend to be subsequent to the human rights violations that pre-date climate change (Adger et al. 2006). The problems of Andean agro-pastoralist peasant farmers or the slum dwellers of Mumbai do not start with environmental or climate change, but climate change will make their problems worse by any measure, resulting in many cases in likely displacement and migration.

The politics of environmental displacement

Given the increasing urgency in global climate change predictions and the expansion of hazards and disasters that threaten to generate population displacement, the debate on environmental migration has not only sharpened, but has acquired political overtones. Despite the attention that the issue of environmental displacement has garnered in recent years, there are no legally binding internationally recognized instruments that pertain to the needs of people displaced by environmental causes. Recognition of this lack has prompted a number of proposals for appropriate forms of governance pertaining to environmentally displaced peoples (Biermann and Boas 2010; Koivurova 2007). Recognizing the problem of complicating the status of legally defined refugees, these proposals argue against including environmentally displaced peoples under the 1951 Geneva Convention Relating to the Status of Refugees. Instead, they propose new legal instruments designed specifically to address the needs of environmentally displaced peoples.

There are other political objections to the linkages being made between environment and migration. Some researchers have concerns that the term 'environmental refugee' is depoliticizing, dehistoricizing and Malthusian (Hartmann 2009). It is felt that its use, particularly by political interests and the media, naturalizes these crises, allowing social factors responsible to escape responsibility. The dangers in the potential misuse of issues of environment and migration are unquestionable. Representations in the media of scientific findings are frequently problematic. Indeed, today in the United States, science reporting is considered to be in crisis (Mooney and Kirshenbaum 2009). In today's journalism, the more dramatically the implications of scientific findings can be framed, the better.

The issue of causality has also been manipulated by politicians for a variety of motives. This would not be the first time politics has misused science, particularly where findings are exploratory and contingent. Environmental migration has been used to alarm the developed nations of the north, particularly Europe and the United States, that they will be inundated by millions of environmentally displaced peoples from the south. Some politicians make these claims to generate support for anti-immigrant policies, with the triage or lifeboat ethic that is covertly associated with that perspective. Others use the spectre of millions of unfortunate refugees rushing over the US borders to generate support for stabilization of greenhouse gases and other forms of climate change mitigation. Clearly related, the distortions that politics and the media engage in when discussing environmental migration constitute a serious concern and it is incumbent on environment and migration researchers to clarify issues of causality when discussing the complexity and interrelationships of drivers in the displacement of populations.

Conclusion

There is little question that some environmental processes force people to migrate, but they do not do so in a socio-ecological vacuum. Both socio-natural and technological disasters, sometimes in combination as recently occurred in Japan, may uproot communities by sudden destruction (Button 2011). The South Asian tsunami displaced millions and Hurricane Katrina uprooted more than 1 million people and left many hundreds of thousands permanently displaced. Their displacement, however, is not due to environmental causes alone, but also to the political economy of reconstruction.

In climate change, nature will not be displacing people, but rather an array of human-generated forces driven by massive production of greenhouse gases that are transforming global climate and therefore nature. The wind, rain, drought, erosion, etc. that displace people resemble natural forces, but their origins are becoming as much human as they are natural. They are processes transpiring in environments that combine human and natural forces and features. Therefore, although it may seem obvious, environmental change and, particularly, climate change are not things 'out there' but are fundamentally tied to both social and ecological processes driven by human action. Nevertheless, the language often used to discuss environmental migration continues to reflect an interacting but still dualistic separation, eliding the endogeneity of nature and society, particularly when discussing causality.

Although the recent report from the Global Humanitarian Forum (2009) estimated that as many as 20 million people would be displaced by climate change in 2009, at the moment in most cases environmental change effects such as migration are hard to quantify, but there is little doubt that they will make the daily challenges of survival worse for the world's most vulnerable people. Where displacement is occurring, it is generally the outcome of multiple factors, including environmental, political and economic causes. In fact, at present the problems afflicting, for example, people as disparate as the slum dwellers of Mumbai or the pastoralists of the high Andes, are not primarily climate change, but rather the conditions of poverty and exclusion that they are consigned to by the larger political economy encompassing their region, nation and the world.

While the numbers associated with environmentally displaced people are unsubstantiated, it would be incautious to say the least to dismiss environmental migration because of the difficulties in defining it or quantifying its effects. Environmental change and migration is a complex issue, but it is happening. Research on environment and migration is politically volatile and vulnerable to misuse and misrepresentation, but despite that it must be taken absolutely seriously because the potential outcomes are serious. If predictions from the IPCC (IPCC 2007a, 2007b) and other research organizations are even half right, and confidence in estimates for the degradation of fragile environments, sea-level rise, coastal erosion, desertification and other forces that may displace people is considerably higher than that, then we must be prepared for significant increases in the role environmental factors will play in displacement and migration in the relatively near future.

References

Adamo, S.B. (2008) 'Addressing Environmentally Induced Population Displacements: A Delicate Task', A Background Paper for the *Population–Environment Research Network Cyberseminar on Environmentally Induced Population Displacements*, 18–29 August 2008. Available at www.populationenvironmentresearch.org/seminars082008.jsp. Accessed 10 January 2013.

Adger, W.N., Paavola, J., Huq, S. and Mace, M.J. (eds) (2006) *Fairness in Adaptation to Climate Change*. Cambridge, MA: MIT Press.

Almeria Statement (1994) 2006 II International Symposium Desertification and Migrations. Available at www.sidym2006.com/eng/eng_doc_interes.asp. Accessed 10 January 2013.

Barry, J.M. (1997) *Rising Tide: The Great Mississippi Flood of 1927 and How It Changed America*. New York, NY: Simon and Schuster.

Bennett, J. (1996) *Human Ecology as Human Behavior*. New Brunswick, NJ: Transaction Publishers.

Biermann, F. and Boas, I. (2010) 'Preparing for a Warmer World: Towards a Global Governance System to Protect Climate Refugees', *Global Environmental Politics*, 10, 1, pp. 60–68.

Black, R. (2001) 'Environmental Refugees: Myth or Reality?' *UNHCR Working Papers*, 34, pp. 1–19.
Black, R., Adger, W.N., Arnell, N.W., Dercon, S., Geddes, A. and Thomas, D.S.G. (2011) 'The Effect of Environmental Change on Human Migration', *Global Environmental Change*, 21s, pp. S3–S11.
Button, G. (2011) *Disaster Culture: Knowledge and Uncertainty in the Wake of Human and Environmental Catastrophe*. Walnut Creek, CA: Left Coast Press.
Byerlee, D. (1974) 'Rural–Urban Migration in Africa: Theory, Policy and Research Implications', *International Migration Review*, 8, 4, pp. 543–566.
Castles, S. (2002) 'Environmental Change and Forced Migration: Making Sense of the Debate', *UNHCR Working Papers*, 70, pp. 1–14.
Centre for Research on the Epidemiology of Disasters (CRED) (2008) *EM-DAT: The International Disaster Database*. Brussels, Belgium: CRED, Catholic University of Louvain. Available at www.emdat.be/. Accessed 10 January 2013.
Christian Aid (2007) *Human Tide: The Real Migration Crisis*. Available at www.christianaid.org.uk/Images/human-tide.pdf. Accessed 10 January 2013.
Crutzen, P.J. and Stoermer, E.F. (2000) 'The Anthropocene', *Global Change Newsletter*, 41, pp. 2–7.
Dessai, S., O'Brien, K. and Hulme, M. (2007) Editorial: On Uncertainty and Climate Change', *Global Environmental Change*, 17, pp. 1–3.
Egan, T. (2006) *The Worst Hard Time*. New York, NY: Houghton-Mifflin.
El-Hinnawi, E. (1985) *Environmental Refugees*. Nairobi: United Nations Environmental Programme.
Elliot, J.R. and Pais, J. (2010) 'When Nature Pushes Back: Environmental Impact and the Spatial Redistribution of Socially Vulnerable Populations', *Social Science Quarterly*, 91, 5, pp. 1187–1202.
Friends of the Earth (2007) *A Citizen's Guide to Climate Refugees*. Melbourne, VIC: Friends of the Earth, Australia. Available at www.safecom.org.au/pdfs/FOE_climate_citizens-guide.pdf. Accessed 10 January 2012.
Gemenne, F. (2011) 'Why the Numbers Don't Add Up: A Review of Estimates and Predictions of People Displaced by Environmental Changes', *Global Environmental Change*, 21s, pp. S41–S49.
Global Humanitarian Forum (2009) *The Anatomy of a Silent Crisis*. Geneva: Global Humanitarian Forum.
Hartmann, B. (2009) 'Climate Refugees and Climate Conflict: Who's Taking the Heat for Global Warming?' pp. 142–155 in S. Mohamed (ed.), *Climate Change and Sustainable Development: New Challenges for Poverty Reduction*. Cheltenham, UK: Edward Elgar.
Harvey, D. (1996) *Nature, Justice and the Geography of Difference*. Oxford, UK: Blackwell.
Holling, C.S. (1994) 'An Ecologist View of the Malthusian Conflict', pp. 79–103 in K. Lindahl-Kiessling and H. Landberg (eds), *Population, Economic Development, and the Environment*. New York, NY: Oxford University Press.
Hugo, G. (1996) 'Environmental Concerns and International Migration', *International Migration Review*, 30, 1, pp. 105–131.
Intergovernmental Panel on Climate Change (IPCC) (2007a) *Climate Change 2007: The Physical Science Basis, Summary for Policy Makers, Contribution of Working Group I to the Fourth Assessment Report of the Intergovernmental Panel on Climate Change*. Paris: IPCC.
Intergovernmental Panel on Climate Change (IPCC) (2007b) *Climate Change 2007: Climate Change Impacts, Adaptation and Vulnerability, Summary for Policy Makers, Contribution of Working Group II to the Fourth Assessment Report of the Intergovernmental Panel on Climate Change*. Brussels: IPCC.
Kenny, M.L. (2002) 'Drought, Clientelism, Fatalism and Fear in Northeast Brazil', *Ethics, Place and Environment*, 5, 2, pp. 123–134.
Kibraeb, G. (1997) 'Environmental Causes and Impact of Refugee Movements: A Critique of the Current Debate', *Disasters*, 21, 1, pp. 20–38.
Koivurova, T. (2007) 'International Legal Avenues to Address the Plight of Victims of Climate Change: Problems and Prospects', *Journal of Environmental Law and Litigation*, 22, pp. 267–299.
Kotchen, M. and Young, O.R. (2007) 'Meeting the Challenges of the Anthropocene: Toward a Science of Coupled Human–Biophysical Systems', *Global Environmental Change*, 17, pp. 149–151.
Medina-Elizalde, M. and Rohling, E.J. (2012) 'Collapse of Classic Maya Civilization Related to Modest Reduction in Precipitation', *Science*, 335, pp. 956–959.
Merryman, J.L. (1982) 'Pastoral Nomad Resettlement in Response to Drought: The Case of the Kenya Somali', pp. 105–120 in A. Hansen and A. Oliver Smith (eds), *Involuntary Migration and Resettlement: The Problems of Dislocated Peoples*. Boulder CO: Westview Press.
Millennium Ecosystem Assessment (MA) (2005a) *Ecosystems and Human Well-Being: Synthesis*. Washington, DC: Island Press.

Millennium Ecosystem Assessment (MA) (2005b) *Ecosystems and Human Well-Being: Desertification Synthesis*. Washington, DC: World Resources Institute.

Mooney, C. and Kirshenbaum, S. (2009) 'Unpopular Science', *The Nation*, 289, 5, pp. 20–24.

Myers, N. (2002) 'Environmental Refugees: A Growing Phenomenon of the 21st Century', *Philosophical Transactions of the Royal Society B*, 357, 1420, pp. 167–182.

Myers, N. (2005) 'Environmental Refugees: An Emergent Security Issue', Presented to *Thirteenth Economic Forum, Prague (23–27 May) Session III Environment and Migration*. Available at www.osce.org/eea/14851. Accessed 10 January 2013.

Nakicenovic, N. and Swart, R. (2000) *Emissions Scenarios. Special Report of the Intergovernmental Panel on Climate Change*. Cambridge, UK: Cambridge University Press.

Nicholls, R.J., Wong, P.P., Burkett, V.R., Codignotto, J.O., Hay, J.E., McLean, R.F., Ragoonaden, S. and Woodroffe, C.D. (2007) 'Coastal Systems and Low-Lying Areas', pp. 315–356 in M.L. Parry, O.F. Canziani, J.P. Palutikof, P.J. van der Linden and C.E. Hanson, (eds), *Climate Change 2007: Impacts, Adaptation and Vulnerability. Contribution of Working Group II to the Fourth Assessment Report of the Intergovernmental Panel on Climate Change*. Cambridge, UK: Cambridge University Press.

Oliver-Smith, A. (2002) 'Theorizing Disasters: Nature, Culture, Power', pp. 23–48 in S.M. Hoffman and A. Oliver-Smith (eds), *Culture and Catastrophe: The Anthropology of Disaster*. Santa Fe, NM: School of American Research Press.

Oliver-Smith, A. (2009) 'Nature, Society and Population Displacement: Toward an Understanding of Environmental Migration and Social Vulnerability', *InterSecTions*, no. 8/2009. Bonn: United Nations University Institute for Environment and Human Security. Available at www.ehs.unu.edu/file/get/5130. Accessed 10 January 2013.

Renaud, F. and Bogardi, J. (2007) 'Forced Migrations due to Degradation of Arid Lands: Concepts, Debate and Policy Requirements', pp. 24–34 in C. King, H. Bigas and Z. Adeel (eds), *Desertification and International Policy Imperative. Proceedings of a Joint International Conference*, Algiers, Algeria, 17–19 December 2006. UNU Desertification Series no. 7. Tokyo: United Nations University.

Renaud, F., Bogardi, J., Dun, O. and Warner, K. (2007) 'Control, Adapt or Flee: How to Face Environmental Migration?', *InterSecTions*, no. 5. Bonn: United Nations University Institute for Environment and Human Security.

Spirtes, P., Glymour, C. and Scheines, R. (2000) *Causation, Prediction and Search*. Cambridge, MA: MIT Press.

Stern, N. (2006) *The Stern Review: On the Economics of Climate Change*. Cambridge, UK: Cambridge University Press.

Stringer, C.B. (1992) 'Evolution of Early Modern Humans', pp. 241–251 in S. Jones, R. Martin and D.R. Pilbeam (eds), *The Cambridge Encyclopaedia of Human Evolution*. Cambridge, UK: Cambridge University Press.

Templeton, A. (2002) Out of Africa Again and Again', *Nature*, 416, p. 45.

Torry, W.I. (1979) Anthropological Studies in Hazardous Environments: Past Trends and New Horizons', *Current Anthropology*, 20, 3, pp. 517–541.

United Nations High Commissioner for Refugees (UNHCR) (2009) *Climate Change, Natural Disasters and Human Displacement: A UNHCR Perspective*. Geneva: UNHCR. Available at www.unhcr.org/cgi-bin/texis/vtx/home/opendocPDFViewer. html?docid=4901e81a4&query=displacement%20scenarios. Accessed 10 January 2013.

Vitousek, P.M., Mooney, H.A., Lubchenco, J. and Melillo, J.M. (1997) 'Human Domination of the Earth's Ecosystems', *Science*, 277, pp. 494–499.

Warner, K., Hamza, M., Oliver-Smith, A., Renaud, F. and Julca, A. (2010) 'Climate Change, Environmental Degradation and Migration', *Natural Hazards*, 55, pp. 689–713.

Wisner, B. (2009) *Climate Change and Migration: Scientific Fact or Leap of (Bad) Faith? Invitation to a Debate and Radix Collection of Materials Elucidating Debate and the Assumptions and Politics in the Back Ground*. Radix. Available at http://radixonline.org/ccm.html. Accessed 10 January 2013.

Wisner, B., Blaikie, P., Cannon, T. and Davis, I. (2004) *At Risk: Natural Hazards, People's Vulnerability and Disasters, Second Edition*. New York, NY: Routledge.

Yenotani, M. (2011) *Displacement Due to Natural Hazard-Induced Disasters: Global Estimates for 2009 and 2010*. Oslo: International Displacement Monitoring Centre and Norwegian Refugee Council.

Part III
Urban environmental change, governance and adaptation

13
Climate change and urban governance
A new politics?

Harriet Bulkeley[1]

Over the past two decades climate change, perhaps as no other contemporary environmental concern, has attracted sustained political attention. From the initial ambitions of the international community to create a legally binding agreement between nation states to the complex picture of climate governance occurring at multiple levels and between state and non-state actors that we currently witness, the issue has been one in which the intimate connections between processes of social and environmental change have been starkly illustrated. Perhaps owing to this realization, it is no longer possible to speak of climate change as if it were (merely) a global problem. Rather, recognizing the connection between social dynamics – of our modes of transportation, heating our homes, disposing of waste and so on – it has increasingly been realized that addressing climate change requires interventions at multiple scales of social action. This chapter focuses on one such site – the city. With estimates suggesting that between 70 and 80 per cent of anthropogenic emissions of global carbon dioxide emissions related to energy use may be attributed to urban areas (IEA 2008; Stern 2006), and significant concerns about the potential vulnerability of cities to the effects of climate change, it is perhaps not surprising that cities should have become the focus for significant climate governance efforts. In the first half of this chapter, the shifting regimes of urban climate governance are examined, and the argument is made that we have witnessed a shift from a period of *municipal voluntarism* to one that could be described as *strategic urbanism*. The second half of this chapter examines the current era of strategic urban climate governance in greater depth. Drawing on recent work, which suggests that globally connected cities are seeking to secure their resources and critical infrastructures in the face of climate change (Hodson and Marvin 2010; While et al. 2010), the chapter considers the role of governance innovations or experiments in cities as one means through which this new politics is emerging (see also Hoffman 2011). This analysis suggests that, while such discourses have increasing traction, other forms of political mobilization around the nexus of climate and resilience are emerging in cities, which contain the potential for alternative urban futures.

Governing climate change in cities: shifting regimes?

The emergence and development of urban climate change governance has taken place in parallel to the international policy process and growing concern about the issue across many national governments over the past twenty years. While initiated and still dominated by municipal authorities and their representative bodies, more recently a range of private and non-governmental actors have become involved in urban climate governance. This change reflects both a shift in the ways in which municipalities are addressing the issue, but also a growing interest in the urban as an arena in which climate governance can be accomplished by non-state actors. In the loose sense of the term, it is possible to identify a shift in the regime of urban climate governance from one dominated by the largely voluntary actions of municipalities primarily focused on their own operations or driven by the imperative of addressing prominent local environmental, economic and social concerns at the same time as climate change, to one where municipalities and other actors are engaging in more strategic efforts to address the implications of climate change for their cities while also innovating with urban responses (Bulkeley 2010; Hodson and Marvin 2009, 2010; While et al. 2010).

Municipal voluntarism

The earliest urban responses to the issue of climate change emerged in Europe and North America in the late 1980s and early 1990s, with municipal governments, individually and collectively through EU funded projects and incipient transnational municipal networks, seeking to take action to mitigate, or reduce, greenhouse gas (GHG) emissions (Bulkeley and Betsill 2003, 2005; Kern and Bulkeley 2009). Between 1991 and 1993 the International Council for Local Environmental Initiatives, now known as ICLEI Local Governments for Sustainability, organized the *Urban CO_2 Reduction Project*,[2] funded by the US Environmental Protection Agency, the City of Toronto, and several private foundations, with the aim of developing city-level plans and tools for the reduction of GHG emissions (Bulkeley and Betsill 2003). By the mid-1990s, three transnational municipal networks had been formed – ICLEI Cities for Climate Protection, Climate Alliance and Enérgie-Cités. Behind this growing movement was the rationale that urban areas represented concentrations of fossil-fuel-intensive activities – transportation and energy use – and that municipal governments had, at least in most cases, some oversight of planning, transport, energy and waste policies through which such sources of GHG emissions might be addressed. The urban scale was also regarded as offering the potential for testing new technologies and policies. Furthermore, a growing interest in issues of (local) sustainability, together with the democratic mandate of local government, and the growing phenomenon of partnership working and public participation, meant that municipalities could be regarded as a means through which the wider involvement of local communities in responding to the issue might be mobilized (e.g. Allman et al. 2004; Bulkeley and Betsill 2003; Collier 1997; DeAngelo and Harvey 1998).

Across North America, Europe and Australia, where these activities were then concentrated, local governments were increasingly being encouraged to embrace the logic of new public management, which celebrated 'evidence-based' policy and the development of targets and indicators as a means of ensuring good governance (Bulkeley and Kern 2006). In this context, and given the growing emphasis internationally on targets and timetables for the reduction of GHG emissions, during the 1990s municipalities stressed the importance of defining the baseline of urban GHG emissions, setting targets, creating action plans and monitoring progress towards them. The ICLEI Cities for Climate Protection (CCP) programme is perhaps the most prominent example of this approach (see Table 13.1, which depicts the most recent iteration of this model), but variations were also developed by individual municipal governments and also by the other two transnational municipal networks in operation at this time: energie-cites and the Climate Alliance. The international attention given to the need to mitigate climate change,

Table 13.1 The ICLEI Cities for Climate Protection milestone methodology

Milestone 1	Establish an inventory and forecast for key sources of GHG emissions in the corporate (municipal) and community areas, and conduct a resilience assessment to determine the vulnerable areas based on expected changes in the climate.
Milestone 2	Set targets for emissions reduction and identify relevant adaptation strategies.
Milestone 3	Develop and adopt a short- to long-term Local Action Plan to reduce emissions and improve community resilience, addressing strategies and actions for both mitigation and adaptation.
Milestone 4	Implement the Local Action Plan and all the measures presented therein.
Milestone 5	Monitor and report on GHG emissions and the implementation of actions and measures.

Source: ICLEI Local Governments for Sustainability (2011).

together with the potential benefits to be realized for cities in the Global North from such strategies, in terms of financial savings through energy efficiency, the reduction in air pollution and congestion, together with opportunities for urban redevelopment in line with principles of sustainability, meant that it was this issue, rather than climate adaptation, that dominated urban climate governance at this time.

In Europe, membership of climate change networks grew steadily during the 1990s and had reached a plateau by the end of the decade (Kern and Bulkeley 2009). Elsewhere, new regional campaigns by ICLEI CCP saw membership increase significantly in Australia and the USA, as well as in Asia and Latin America. Despite the growth in membership, however, reported actions were primarily focused on the reduction of GHG emissions from within the municipality – a 'self-governing' approach (Bulkeley and Kern 2006), albeit one that has led to new mechanisms for financing projects, accounting for carbon, the deployment of novel technologies, and a growing political awareness about the issue of climate change. In seeking to roll out comprehensive approaches to addressing climate change across urban communities, municipal governments were faced with challenges of ascertaining the baselines against which performance should be measured, limited competencies in critical areas relating to GHG emissions, a lack of technical and financial capacity to implement measures, and political struggles concerning how, if at all, climate change should have a bearing on increasing demands for transport, economic growth and energy use (Bai 2007; Betsill and Bulkeley 2007; Bulkeley et al. 2009; Romero Lankao 2007; Schreurs 2008). In confronting these challenges, municipalities sought to 'reframe' climate change as an issue through which other significant local agendas – air pollution, health, congestion, energy security and so on – might be addressed. In this context, energy efficiency is a particularly potent issue as it can 'advance diverse (and often divergent) goals in tandem' (Rutland and Aylett 2008: 636), serving to translate various interests into those concerning climate change and effectively forging new partnerships. Where action was forthcoming, municipal governments primarily sought to develop an enabling mode of governing through which business and communities were encouraged to act in, and on behalf of, the city (Bulkeley and Kern 2006). This again reflected broader approaches to urban governance, where efforts to 'roll back the state' had led to an emphasis on the need for local governments to engage a wider range of partners in delivering their core services, within the broad ideology that the state should be 'steering' society rather than acting directly for individuals and communities. In this context, it is perhaps not surprising to learn that the number of municipalities that have pursued a comprehensive, planned, approach to climate governance are few and far between while 'numerous cities, which have adopted GHG reduction targets, have failed to pursue such a systematic and structured approach and, instead, prefer to implement no-regret measures on a case-by-case basis' (Kern and Alber 2008: 173; see also Jollands 2008). The very nature of enabling as a mode of governing means that it is more likely to operate through discrete projects that gather interests together, while a focus on co-benefits of addressing climate change may lead to the relabelling of discrete initiatives under a climate change umbrella. While the ideal of the municipal voluntarism that characterized urban climate governance during this decade was based on an integrated, evidence-based approach to climate planning and policy, the challenges of institutional capacity and of political economy that were encountered as authorities sought to engage in responding to climate change beyond their own operations led to a more piecemeal and opportunistic approach. While some cities were able to develop sufficient capacity and political will to overcome such barriers and to draw others together to sustain a programmatic approach to climate change in the city, many witnessed a growing gap between the rhetoric of a need for an urgent response and the realities of governing climate change on the ground.

Strategic urbanism?

By 2000, membership of the three transnational climate governance networks – ICLEI CCP, Climate Alliance and energie-cites – had begun to level off, and in the face of international uncertainty over the fate of the 1997 Kyoto Protocol and the extent to which agreement would be forthcoming, movement at the urban level remained rather constrained. Momentum came from several, rather unlikely, sources. In the USA, the growing intransigence of the George Bush administration with regard to climate change led some progressive municipal governments to form the US Mayors Agreement (Gore and Robinson 2009). While it was in 2000 that the US Conference of Mayors first noted the significant role that mayors could take in addressing climate change, it was in 2005 that the Mayor of Seattle, Greg Nickels, challenged mayors across the USA to take action on the issue (Gore and Robinson 2009: 142). Following an initial agreement amongst ten of the leading US cities on climate change, a further call to action attracted over 180 mayors, and by 2009 over 900 mayors had signed up to the Climate Protection Agreement (Gore and Robinson 2009: 143). Likewise, in Australia, the Howard government's position that Australia should not sign the Kyoto Protocol served to spur many local authorities to join the CCP programme in Australia and to engage their political representatives in recognizing the importance of the issue. This approach of engaging locally elected politicians with the climate change agenda has been replicated globally, most recently with the launch in 2009 of the European Covenant of Mayors, which requires signatories to pledge to go beyond the EU target of reducing CO_2 emissions by 20 per cent by 2020 through the formation and implementation of a Sustainable Energy Action Plan (CoM 2011a). In 2011, the plan had more than two thousand members (CoM 2011b).

This more overtly political stance is also evident in the engagement of global cities with the climate change agenda, primarily in the form of the C40 Cities Climate Leadership Group. This network was instigated by the then mayor of London, Ken Livingstone and his Deputy, Nicky Gavron, together with the Climate Group, a not-for-profit organization based in London. It was formed by eighteen cities in 2005 as a parallel initiative to the Group of Eight (G8) Gleneagles summit on climate change. In 2007, this network entered into a partnership with the Clinton Climate Initiative and expanded its membership to include forty of the largest cities in the world. Through such networks, and also on their own initiative, there is evidence that a broader range of private sector interests are becoming involved in urban climate governance – for example, in the form of collaboration of the C40 network with

Microsoft to produce software for greenhouse gas emissions accounting at the city scale; or in HSBC's Climate Change Partnership, which involves activities in five of its global centres: New York, London, Hong Kong, Mumbai and Shanghai. Together with the renewed expansion of the existing transnational climate networks, these new developments have led to a growing engagement with issues of climate change in cities in the Global South. For example, some fifty local authorities in India have participated in ICLEI South Asia's Roadmap project to conduct emissions inventories and develop climate change action plans. In addition, twenty of the forty cities included in the C40 Cities Climate Leadership group are located in countries in the Global South. While these networks have continued to focus on climate mitigation, adaptation is increasingly on the urban agenda. Existing networks, most notably ICLEI, have begun to focus on climate adaptation and are seeking to engage cities through the concept of 'resilience', epitomized in the annual conference on Resilient Cities first held in 2010, and through new toolkits, strategies and programmes focusing on adaptation across their regional offices. In addition, the Rockefeller Foundation has established the *Asian Cities Climate Change Resilience Network*, a network of ten cities explicitly focused on climate adaptation, and UN-Habitat is working with cities in Asia, Africa and Latin America through its *SUD-Net: Cities and Climate Change Initiative* to support urban adaptation responses. This is unusual, not only in the urban context, given the predominant focus on mitigation, but also in terms of climate governance more broadly, where most transnational governance initiatives have also been focused on mitigation.

Over the past two decades, then, a subtle, but nonetheless important, shift may be discernible in the ways in which transnational networks, municipal governments and other urban actors are seeking to govern climate change in cities. The limits of municipal capacity, the political economy of urban climate governance, the growing involvement of forms of private authority, the participation of cities in the Global South, and the broadening of the agenda to include adaptation, have all served to promote an alternative regime within which climate change is at once a more strategic concern and one that is more closely aligned to the concerns of urban growth and resource security that dominate urban agendas (Hodson and Marvin 2009, 2010). In this vein, While et al. (2010: 82) suggest that a processes of eco-state restructuring is taking place, focused on 'carbon control' and creating a:

> distinctive political economy associated with climate mitigation in which discourses of climate change both open up, and necessitate an extension of, state intervention in the spheres of production and consumption.
>
> *(While et al. 2010: 87)*

In a similar vein, Hodson and Marvin (2009: 195–196) suggest that issues of climate change mitigation and adaptation are becoming a key strategic concern for urban authorities, provoked by discourses of the urban causes and consequences of environmental problems and facilitated through the restructuring of the state and the creation of 'new state spaces' within which to secure the provision of resources and infrastructure. This issue, they contend, is leading 'the world's largest cities' to begin 'to translate their strategic concern about their ability to guarantee resources into strategies designed to reshape the city and their relations with resources and other spaces' (Hodson and Marvin 2009: 200). While the types of municipal voluntarism that characterized early responses to climate change in cities continue, this analysis suggests that a new regime of urban climate governance is emerging that may have profound implications for the organization of urban economies and everyday life in particular cities.

Governance experiments and the new urban politics of climate change

The recognition of the emergence of new forms of urban climate governance, orchestrated around strategic concerns for urban development and resource security, alongside continued concern for the environmental and social dimensions of climate change, raises important questions about how and with what implications such forms of governance are being pursued. Rather than seeking to address their own GHG emissions, municipal authorities and other urban actors are increasingly seeking to engage in strategies, programmes and projects that address broader questions of urban adaptation and mitigation. This process has entailed the emergence of new discourses that seek to align urban resource and security issues with the priorities of other actors, as well as the development of specific initiatives and interventions aimed at providing alternative low carbon or resilient infrastructures and to enable other actors to undertake appropriate actions. While strategic in their orientation, and often framed or funded through municipal plans, such initiatives are often explicitly experimental in character, seeking to offer an alternative to governance or development as usual within the city and as the starting point for a broader transition to a low carbon, resilient economy (Bulkeley et al. 2011; Hoffman 2011). Such interventions can be described as 'experiments' in that they explicitly attempt to test technologies, provide some form of learning or create novel experience concerning what it might mean to address climate change in the city. At the same time, other forms of experiment in the city seek to frame the climate change issue in radically different ways, offering an alternative to its alignment with the priorities of dominant business and government interests. This section explores these different types of climate change experiment and their implications for the politics of governing climate change in the city.

Securing low carbon urbanism

The development of a more strategic approach to low carbon urbanism has been particularly marked in some of the world's largest cities (Hodson and Marvin 2010). In London, the opportunity afforded by the development of the C40 network also served to catalyse responses within the city. While individual borough councils across the Greater London area had been involved in climate change policy and initiatives during the preceding decade, the formation of the Greater London Authority (GLA) and the office of the Mayor of London in 2000 provided the political arena within which climate policy could develop in a more strategic manner. In 2004, the Mayor released a Climate Change Adaptation Strategy and an Energy Strategy for the city, noting the need to decentralize and decarbonize the energy supply system. Following the successful development of the C40 network, as well as increasing interest in climate change issues in the international arena, by national government, within the financial sector of the City of London and the wider public, policy ambitions were increased. In 2005, the London Climate Change Agency (LCCA) was established as 'a municipal company wholly owned and controlled by the London Development Agency (LDA) and chaired by the Mayor' (LCCA 2007: 3–4). The Agency received private sector support from 'BP, Lafarge, Legal and General, Sir Robert McAlpine, Johnson Matthey, and the City of London Corporation' as well as from 'Rockefeller Brothers' Trust, KPMG, Greenpeace, the Climate Group, the Carbon Trust and the Energy Savings Trust' (LCCA 2007: 3–4). The resulting London Climate Change Action Plan contained the ambitious goal of reducing London's GHG emissions by 60 per cent by 2025. This goal was to be partly achieved through a radical decentralization of energy supply within the city:

> The Mayor's top priority for reducing carbon emissions is to move as much of London as possible away from reliance on the national grid and on to local, lower-carbon energy

supply (decentralised energy, including combined cooling heat and power networks, energy from waste, and onsite renewable energy – such as solar panels) … The Mayor's goal is to enable a quarter of London's energy supply to be moved off the grid and on to local, decentralised systems by 2025, with more than half of London's energy being supplied in this way by 2050.

(GLA 2007: 105)

The LCCA was in itself innovative in its design as a municipally owned company, in a UK context where such entities are uncommon, and in its explicit engagement of private interests in the governing of climate change in the UK, reflecting in turn the growing strategic importance assigned by these actors, not only to the issue of climate change but to an urban position on the issue. The GLA and the LCCA sought to pursue these targets in traditional terms, through new planning requirements and a statutory duty for the Mayor to produce an annual climate change strategy, but also through experimenting with different forms of public–private partnership, including for the delivery of green homes, behavioural change programmes, building refurbishment and, notably, new economic models for achieving energy efficiency and decentralized energy generation in the city. Here, the formation of the London Energy Services Company (ESCO) is particularly notable. It was:

established to design, finance, build and operate local decentralized energy systems for both new and existing developments. It has been established as a private limited company with shareholdings jointly owned by the London Climate Change Agency Ltd (with a 19% shareholding) and EDF Energy (Projects) Ltd (with an 81% shareholding).

(LCCA 2007: 5–6)

By offering an alternative model through which energy systems and building development could take place, the London ESCO served as a means through which the LCCA sought to pursue its wider targets for decentralizing energy generation in the city and achieving its low carbon ambitions. While the LCCA was disbanded in 2009 and the London ESCO is now fully owned by Environmental Defense Fund (EDF), following the change in London's political leadership, the policy goals for energy decentralization and self-sufficiency remain. In this case, experimentation in institutional terms has led to new ways of considering what urban development in the city should involve, even though some of the institutional structures have been disbanded. In this manner, such forms of innovation serve as a means through which strategic concerns for low carbon urbanism can be made tangible, further securing this discourse about certain types of urban futures.

Low carbon experiments are also being undertaken by a range of non-state actors. In Bangalore, T-Zed (Towards Zero Carbon Development) is a project that seeks to provide an alternative form of urban development, targeted at the higher income residents of the city who primarily work in the IT and related sectors. The development was initiated by a private company, Biodiversity Conservation India (BCIL). It occupies a location at the outskirts of the city and, like many other developments in this area, is a gated community or compound. The development includes sixteen single-family houses and seventy-five apartments and incorporates numerous social and technical innovations with the intention of reducing GHG emissions and increasing its self-sufficiency, in order in part to reduce its dependence on the city's (often unreliable and limited) resources. For example, the design includes measures to reduce the embodied energy of the building (including the use of local materials), renewable energy generation, energy-efficient technologies (including air conditioning, refrigeration and lighting), rainwater harvesting, as well as landscaping and conservation. In addition to these technical forms of innovation, the development seeks both to offer 'no compromise' in lifestyle while also engaging residents in a suite of low carbon practices, such as behavioural changes to reduce energy and water use, organic vegetable gardening and community-based activities (Castán Broto and Bulkeley 2011).

T-Zed reflects a complex process of assembling actors, land and materials through the promise of securing resources and minimizing the carbon footprint of urban development, and ongoing processes of work to maintain its innovative character and relations to broader urban networks and with the community living in the compound. In as much as it seeks to design an alternative form of urbanism for Bangalore, it can also be regarded as a strategic intervention in the city. Rather than being accomplished from the 'top down', this form of innovation works through the socio-technical networks that comprise the urban. First, T-Zed has created a space for climate change innovation, through the replication of T-Zed-like development projects, seeding new companies (e.g. Flexitron, a light innovation company), and through the consideration of principles and practices designed at T-Zed in policy at local and national levels. Second, the presence of T-Zed within the urban landscape of Bangalore is serving to transform and reconfigure urban infrastructure systems more broadly. Here, the project relies on simultaneous processes of disconnection – of the archetypal 'gated' community and autonomous energy production – and of reconnection – through urban land disputes, to the water table through the boreholes which the development eventually required, and through fostering relations with neighbouring communities regarding the potential of low carbon living. Third, and related to this process, T-Zed has provided an arena in which new discourses of responsibility and carbon control for middle-class residents has flourished. By adopting an approach that suggests that low carbon living is compatible with modern urbanism in Bangalore, the development provides a 'showcase' for other middle-class communities. However, it is a contested process as, on the one hand, residents accept forms of oversight from their peers but challenge the direct intervention of BCIL, for example in terms of their use of baths or domestic appliances, in their everyday lives. Innovative approaches for governing climate change in the city come, therefore, not only through the usual channels of municipal authority, but are also being forged through the private sphere. In this case, as elsewhere, the outcomes are deeply ambivalent. On the one hand, such strategic interventions to secure low carbon futures serve to provide an alternative model of urbanism, while on the other the very processes of disconnection/reconnection upon which they rely equally maintains development as usual.

Experimenting with alternatives

There are, however, alternative means through which innovations for addressing climate change are emerging in the urban arena. One prominent example in the UK is the Transition Towns movement, which was established in the small town of Totnes, Devon, in 2006, and has since spread to over 300 communities in the UK as well as North America, Europe, Australia, Japan and Chile (Bailey et al. 2010; North 2010; Transition Towns 2011). Like T-Zed, Transition Towns advocates a mode of self-sufficiency as a means of achieving both resilience and low carbon urbanism. It also seeks to approach this challenge through assembling technical and social innovations to demonstrate the potential of alternatives. However, rather than being orchestrated by a single actor or municipal authority, the process of being formally constituted as a Transition Town relies on the engagement of a community (variously and ambiguously identified) in adopting sixteen criteria, devised by its original founders in Totnes, that set out the principles of Transition. As Transition Town Brixton, London, explains:

> Transition Town Brixton is a community-led initiative that seeks to raise awareness locally of Climate Change and Peak Oil. TTB proposes that it is better to design that change, reduce impacts and make it beneficial than be surprised by it. We will vision a better low energy/carbon future for Brixton. We will design a Brixton Energy Descent Plan. And then we will make it happen.
>
> *(Transition Town Brixton 2011a)*

Once adopted, a wide range of actions and activities are undertaken under the banner of Transition Towns, primarily in relation to issues of local food production, energy and the development of alternative forms of local economy. In Brixton, sixteen different groups have been formed, addressing issues including education, arts and culture, recycling and reusing materials, energy conservation, local food production, and the development of a local currency, the Brixton Pound (Bailey et al. 2010; North 2010; Seyfang 2009). Rather than being concerned with the strategic dimensions of climate security and low carbon development for the city, in Brixton the Transition Towns group focuses on issues of individual and community resilience. For example, the energy and buildings group has a 'draught-busting' initiative whereby members of the local community engage with householders in draught-proofing their homes in order to save energy, carbon and money, and also provide loans of smart meters so that householders can assess the effectiveness of their own efforts to reduce energy use (Transition Town Brixton 2011b). Like other Transition Town initiatives, within Brixton there is a strong focus on the development of alternative sources of food within the community, including the development of community gardens, beekeeping, seed sharing and planting edible trees.

While initiatives such as these based on ideals of self-sufficiency and community offer an alternative discourse of the challenges of, and appropriate responses to, climate change, they remain largely orchestrated by the concerns of middle-class urban residents concerned about issues of overconsumption and the carbon content of their everyday lives. A further set of urban innovations in response to climate change explicitly seek to address those communities and households who experience some level of social and economic hardship and for whom both climate change and society's response to it pose potentially significant challenges. In a developed country context, such initiatives focus on residents who experience a degree of energy 'poverty' or vulnerability, in terms of their ability to afford the basic services which energy provides (e.g. heating, cooling, power, cooking, cleanliness). Although programmes to address so-called 'fuel poverty' are relatively common in the UK and in some other parts of North America, Europe and Australia, it has only been recently that such initiatives have sought either to take an explicit place-based focus or to consider how energy vulnerability and responses to climate change go hand in hand. One case in point is the work being undertaken in Melbourne, as a partnership between the Moreland Energy Foundation and the Brotherhood of St Lawrence. Here, addressing climate change is seen both in terms of enhancing the resilience of the poorer communities in society and in terms of building an alternative economy:

> The Brotherhood's climate change proposals focus on:
>
> - giving low-income households better access to energy efficiency measures that will reduce their vulnerability to price increases resulting from an emissions trading scheme, increase comfort at home, and lessen the effect of damp, heatwaves and severe winters on their health

- promoting the growth of green collar jobs and related training, to ensure that disadvantaged job seekers and long-term unemployed people can join the green workforce.
    (*Brotherhood of St Lawrence* 2011)

As part of their Moreland Solar City project, funded by the Australian Federal Government's Solar City programme, these organizations have established a 'Warm Homes, Cool Homes' programme to provide residents with free advice and installation service to increase the energy efficiency of homes across the region and hence the vulnerability to both cold and hot weather that the city experiences (Moreland Solar City 2011). There is also evidence that experiments that seek to address the challenges of poverty and economic development alongside climate change are emerging in cities in countries in the Global South. One of the most well-known examples is the Kuyasa project in the Khayelitsha area of Cape Town. Led by the NGO SouthSouthNorth, the project involved providing an energy upgrade to low income housing, including retrofitting ceilings, energy-efficient light bulbs and solar-hot water heating, which together reduced energy use in households (hence yielding carbon savings) and energy poverty, providing direct financial benefits. Unlike other projects of this nature which are taking place in cities in the Global South, the Kuyasa initiative is particularly innovative because of its use of the Clean Development Mechanism, a financial instrument agreed as part of the climate treaty – the Kyoto Protocol – as a means through which countries in the North can finance projects in the Global South that reduce greenhouse gas emissions. As a result, the project has created local employment opportunities as well as providing direct financial and carbon savings (see SouthSouthNorth 2011).

While the sorts of strategic experiments discussed above posit low carbon urbanism not only as compatible with, but as essential to existing patterns of economic growth (Hodson and Marvin 2010), these alternative forms of innovation challenge this dominant regime in two important ways. First, they seek to provide an alternative model of low carbon living, where forms of social and technical innovation are put to work to create new forms of economic and community relations. Second, they explicitly recognize that resource security is an essentially contested and unequal concept, with the result that vulnerability and resilience are highly differentiated within the city. Rather than witnessing the straightforward emergence of a homogeneous and dominant regime for governing climate change in cities, the presence of these alternative forms of innovation points to a more fractured landscape, where strange bedfellows (e.g. international carbon finance and low income households in South Africa) are conjoined in developing new discourses of security and resilience, and where the potential for contestation and conflict is ever-present.

Conclusion

Climate change raises significant governance dilemmas at multiple sites of social organization. Focusing specifically on the urban governance challenge posed by this interconnected social and ecological challenge, this chapter has documented the shifting governance regimes that have emerged over the past two decades as cities have sought to address these dilemmas. It has argued that a discernible shift has taken place from a more municipal, voluntary approach to one in which climate change is a strategic issue for many different actors seeking to develop specific forms of urbanism in cities globally. As the second half of the chapter explored in more detail, this shift has been orchestrated in part through the development of specific governance innovations or climate change experiments, many of which seek to promulgate discourses and practices in which urban elites seek to secure resources and infrastructure systems for distinct forms of economic development. However, such discourses have also opened up the space for alternative framings of the social and environmental implications of climate change and how these should be addressed which pay more specific attention to issues of equity and justice and which proffer alternative visions of economic development.

With this shift to a more strategically important place on certain urban agendas for the climate change issue comes significant challenge. For most cities, and arguably especially those in the Global South, where the impacts of climate change may be most significant, it remains a marginal concern. This status in part reflects the continuing dominance of the mitigation agenda in relation to issues of adaptation, though there are some signs that the ground here is also shifting. As more cities in the Global South engage with the issue, and the challenges of addressing with the climate change that we have already committed to emerge, adaptation is growing on urban agendas. Calls on the one hand to mainstream climate change within urban agendas and on the other to integrate mitigation and adaptation may appear sensible and straightforward in their design, but mask many potential conflicts within and between different aspects of the climate change agenda while also running the risk that the issue becomes simply another 'wash' over economic development as usual. At the other extreme, perhaps, is the danger that the increasing priority being given to climate change could serve to undermine other important social, economic and environmental agendas locally, as it forms a master narrative through which all questions of urban development must be viewed. Clearly, neither approach is desirable. As some of the alternative discourses emerging around climate change in cities have shown, there are ways and means of integrating a more radical sense of urban economic development with the climate change agenda, and it is perhaps in these few instances that the seeds for a more just and appropriate form of urban climate governance may be emerging.

Notes

1 The ideas and material upon which this chapter is based have been developed during my ESRC Climate Change Fellowship, *Urban Transitions: climate change, global cities and the transformation of socio-technical networks* (Award Number: RES-066-27-0002). I am grateful to Vanesa Castán Broto for her work on this project, and also to Mike Hodson and Simon Marvin, whose ideas have also informed my thinking in this area. The usual disclaimers apply. For further information about this project, please see: www.geography.dur.ac.uk/projects/urbantransitions.
2 Fourteen municipalities from North America and Europe participated in the *Urban CO$_2$ Reduction Project*: Ankara, Turkey; Bologna, Italy; Chula Vista, California, USA; Copenhagen, Denmark; Dade County, Florida, USA; Denver, Colorado, USA; Hannover, Germany; Helsinki, Finland; Minneapolis, Minnesota, USA; Portland, Oregon, USA; Saarbrucken, Germany; Saint Paul, Minnesota, USA; and Toronto, Canada.

References

Allman, L., Fleming, P. and Wallace, A. (2004) 'The Progress of English and Welsh Local Authorities in Addressing Climate Change', *Local Environment*, 9, 3, pp. 271–283.

Bai, X. (2007) 'Integrating Global Environmental Concerns into Urban Management: The Scale and Readiness Arguments', *Journal of Industrial Ecology*, 11, 2, pp. 15–29.

Bailey, I., Hopkins, R. and Wilson, G. (2010) 'Some Things Old, Some Things New: The Spatial Representations and Politics of Change of the Peak Oil Relocalisation Movement', *Geoforum*, 41, 4, pp. 595–605.

Betsill, M. and Bulkeley, H. (2007) 'Looking Back and Thinking Ahead: A Decade of Cities and Climate Change Research', *Local Environment: The International Journal of Justice and Sustainability*, 12, 5, pp. 447–456.

Brotherhood of St Lawrence (2011) *Equity and Climate Change*. Available at www.bsl.org.au/equity-and-climate-change.aspx. Accessed 11 January 2013.

Bulkeley, H. (2010) 'Cities and the Governing of Climate Change', *Annual Review of Environment and Resources*, 35, pp. 229–253.

Bulkeley, H. and Betsill, M. (2003) *Cities and Climate Change: Urban Sustainability and Global Environmental Governance*. Oxon, NY: Routledge.

Bulkeley, H. and Betsill, M. (2005) 'Rethinking Sustainable Cities: Multilevel Governance and the "Urban" Politics of Climate Change', *Environmental Politics*, 14, 1, pp. 42–63.

Bulkeley, H., Castán Broto, V., Hodson, M. and Marvin, S. (eds) (2011) *Cities and Low Carbon Transitions*. London: Routledge.

Bulkeley, H. and Kern, K. (2006) 'Local Government and the Governing of Climate Change in Germany and the UK', *Urban Studies*, 43, 12, pp. 2237–2259.

Bulkeley, H., Schroeder, H., Janda, K., Zhao, J., Armstrong, A., Chu, S.Y. and Ghosh, S. (2009) 'Cities and Climate Change: The Role of Institutions, Governance and Urban Planning', Presented to *World Bank Urban Research Symposium Cities and Climate Change*. Marseille, 28–30 June. Available at http://siteresources.worldbank.org/INTURBANDEVELOPMENT/Resources/336387-1256566800920/6505269-1268260567624/Bulkeley.pdf. Accessed 11 January 2013.

Castán Broto, V. and Bulkeley. H. (2011) 'Case Study Report: Towards Zero Carbon Development in Bangalore', *Urban Transitions Project*. Durham, UK: University of Durham/ESRC.

Collier, U. (1997) 'Local Authorities and Climate Protection in the European Union: Putting Subsidiarity into Practice?' *Local Environment*, 2, 1, pp. 39–57.

Covenant of Mayors (CoM) (2011a) *About the Covenant of Mayors*. Available at www.covenantofmayors.eu/about/covenant-of-mayors_en.html. Accessed 11 January 2013.

Covenant of Mayors (CoM) (2011b) *Welcome*. Available at www.eumayors.eu/home_en.htm. Accessed 1 March 2011.

DeAngelo, B.J. and Harvey, L.D.D. (1998) 'The Jurisdictional Framework for Municipal Action to Reduce Greenhouse Gas Emissions: Case Studies from Canada, the USA and Germany', *Local Environment*, 3, 2, pp. 111–136.

Gore, C. and Robinson, P. (2009) 'Local Government Response to Climate Change: Our Last, Best Hope?' pp. 138–158 in H. Selin and S.D. VanDeveer (eds), *Changing Climates in North American Politics: Institutions, Policymaking and Multilevel Governance*. Cambridge, MA: MIT Press.

Greater London Authority (GLA) (2007) *Action Today to Protect Tomorrow: The Mayor's Climate Change Action Plan*. London: GLA.

Hodson, M. and Marvin, S. (2009) '"Urban Ecological Security": A New Urban Paradigm?' *International Journal of Urban and Regional Research*, 33, 1, pp. 193–215.

Hodson, M. and Marvin, S. (2010) *World Cities and Climate Change: Producing Urban Ecological Security*. Milton Keynes, UK: Open University Press.

Hoffman, M.J. (2011) *Climate Governance at the Crossroads: Experimenting with a Global Response*. New York, NY: Oxford University Press.

ICLEI Local Governments for Sustainability (2011) *The Five Milestone Processes*. www.iclei.org/index.php?id=810. Accessed 11 January 2013.

International Energy Agency (IEA) (2008) *World Energy Outlook 2008*. Paris: IEA.

Kern, K. and Alber, G. (2008) 'Governing Climate Change in Cities: Modes of Urban Climate Governance in Multi-level Systems', pp. 171–196 in *Proceedings of the OECD Conference on Competitive Cities and Climate Change*, Milan, 9–10 October.

Kern, K. and Bulkeley, H. (2009) 'Cities, Europeanization and Multi-level Governance: Governing Climate Change Through Transnational Municipal Networks', *Journal of Common Market Studies*, 47, 2, pp. 309–332.

Jollands, N. (2008) 'Cities and Energy: A Discussion Paper', pp. 135–146 in *Proceedings of the OECD Conference on Competitive Cities and Climate Change*, Milan, 9–10 October.

London Climate Change Agency (2007) *Moving London Towards a Sustainable Low-Carbon City: An Implementation Strategy*. London: London Climate Change Agency.

Moreland Solar City (2011) *Energy Hub*. Available at www.morelandsolarcity.org.au/index.php?nodeId=23. Accessed 11 January 2013.

North, P. (2010) 'Eco-Localisation as a Progressive Response to Peak Oil and Climate Change: A Sympathetic Critique', *Geoforum*, 41, 4, pp. 585–594.

Romero Lankao, P. (2007) 'How Do Local Governments in Mexico City Manage Global Warming?' *Local Environment*, 12, 5, pp. 519–535.

Rutland, T. and Aylett, A. (2008) 'The Work of Policy: Actor Networks, Governmentality, and Local Action on Climate Change in Portland, Oregon', *Environment and Planning D: Society and Space*, 26, 4, pp. 627–646.

Schreurs, M.A. (2008) 'From the Bottom Up: Local and Subnational Climate Change Politics', *Journal of Environment and Development*, 17, 4, pp. 343–355.

Seyfang, G. (2009) *The New Economics of Sustainable Consumption: Seeds of Change*. New York, NY: Palgrave Macmillan.

SouthSouthNorth (2011) *Project Portfolio and Reports*. Available at www.southsouthnorth.org. Accessed 1 March 2011.

Stern, N.S. (2006) *Stern Review: The Economics of Climate Change*. London: HM Treasury.

Transition Town Brixton (2011a) *About 'Transition'?*. Available at www.transitiontownbrixton.org/whats-is-transition-about/. Accessed 1 March 2011.

Transition Town Brixton (2011b) *Buildings and Energy*. Available at www.transitiontownbrixton.org/category/groups/buildingsandenergy/. Accessed 1 March 2011.

Transition Towns (2011) *Official Initiatives by Number*. Available at www.transitionnetwork.org/initiatives/by-number. Accessed 11 January 2013.

While, A., Jonas, A.E.G. and Gibbs, D. (2010) 'From Sustainable Development to Carbon Control: Eco-State Restructuring and the Politics of Urban and Regional Development', *Transactions of the Institute of British Geographers*, 35, pp. 76–93.

14
Recovering the city level in the global environmental struggle
Going beyond carbon trading

Saskia Sassen

Opening up to the city level in existing formal governance framings carries significant implications. It can help overcome the narrow nationalisms of inter-state negotiated agreements. Large complex cities share far more with other cities across the world in terms of challenges and the resources they need than they share with the rest of the areas within their national states. Cities share a specific position in a multi-scalar global governance system. The ongoing elaboration of the European Union has brought this issue to the fore, notably in the need for subsidiarity regimes that go from the European to the local level. Given this transnational affinity among cities when it comes to challenges and needed resources, bringing the city level into global regimes might be one of the most effective ways of achieving protection without national protectionism.

If we are to open up macro-level regimes to this sub-national scale, it becomes critical to recognize the specificity and the specialized difference of the local level. Three features stand out in this regard. First, the city level makes possible the implementation and application of forms of scientific knowledge and technological capacities that are not practical at a national level. The city's multiple ecologies enable the mixing of diverse forms of knowledge and diverse technologies in ways that the more abstract 'national space' does not. This difference also means that the city introduces a type of environmental governance option that takes a radically different approach from the common and preferred choice of an international carbon-trading regime. The aim becomes addressing the carbon and nitrogen cycles in situ by implementing measures that reduce damage in a radical way (e.g. Sassen and Dotan 2011).

Second, introducing the city level into global governance regimes enables what Kern and Moll (2013) refer to as horizontal governance, encompassing regimes that can work alongside traditional vertical forms of governance. Numerous and diverse urban initiatives have been a critical component of climate governance even though they have not formally been part of the global regime. City mayors and leaders have *had* to confront the direct and indirect impacts of global conflict and environmental crises on the movement of people; in contrast, national governments and international organizations rarely do. This is one challenge that has led city mayors to become part of international networks and to find urban solutions (Toly 2010).

Third, opening up the global climate regime to the city level brings into the frame a range of troublesome developments that can be bypassed, as if they did not exist, at the very general global level. These include challenges such as anti-immigration sentiment, racism, extreme forms of inequality and a proliferation of new types of urban violence (i.e. militarized gang warfare, extreme forms of inequality with their associated displacements of people, the urbanizing of war associated with asymmetric war, that is, war between a conventional army and irregular combatants).

But the global regime has made little formal room for these types of issues. I begin with an examination of this absence and then focus on the potentials of a focus on cities. One critical issue here is that the city level, especially in the case of large complex cities, incorporates the global scale. Much of what is seen as a global trend in international treaties actually takes place and becomes concrete and urgent in cities.

The hole in the current climate change governance framework

Neither the Kyoto Protocol nor the United Nations Framework Convention on Climate Change (UNFCCC) contain specific references to local government or city-level actions to meet Protocol commitments. There are just a few references to local-level involvement, with Article 10 of the Kyoto Protocol recognizing that regional programmes may be relevant to improve the quality of local emission factors. The Copenhagen UN Climate Conference (COP15) did not advance matters significantly; one positive step was to add a Local Government Climate Change initiative that did introduce local issues in some of the debates and briefings. Similarly, the one significant contribution of the recently held Rio+20 conference (June 2012) was the recognition of cities as key players in any effort towards environmental sustainability. However, in their formal features, these international conferences remain the domain of national governments, and thus, perhaps not surprisingly, get caught up in nationalisms and protectionisms that drown out the major issues that we need to address.

Despite the fact that neither the Kyoto Protocol nor the UNFCCC consider any role for cities or local governments, the latter have in fact established and built up financial and fiscal incentives, local knowledge and education, and other municipal frameworks for action through the actual practical obligations and opportunities that municipal-level governments encounter. Based on their legal responsibilities and jurisdictions, local governments have developed targets and regulations and, in this work, they have tended to go beyond national and state jurisdictional obligations.[1] Since 2007, the Local Government Climate Roadmap (a consortium of global municipal partnerships) has focused on the failure of international climate negotiations to recognize cities. One basic premise in this effort is that including the local government level would ensure that the full chain of governance, from national to local, should be involved in the implementation of a climate agreement.

Further, and very illuminating as to a specific urban structural condition, some of these local initiatives go back to the 1980s and 1990s, when major cities, notably Los Angeles and Tokyo, implemented clean-air ordinances. This was not because their leaderships were particularly enlightened but because they had to for public health reasons. The global initiative Cities for Climate Protection, developed by the ICLEI–Local Governments for Sustainability Network, was active as far back as 1993, supporting mostly result-based, quantified and concrete local climate actions long before the Convention and Kyoto Protocol for national governments came into force.[2] Local governments held Municipal Leadership Summits in 1993, 1995, 1997 2005 and 2012, parallel to the official negotiations for the climate regime (which are called Conferences of Parties (COP) or Meetings of the Parties (MOP)). Thereby the Local Government and Municipal Authority Constituency have built upon their role as one of the first non-governmental organizations (NGOs) to act as observer to the official UNFCCC international climate negotiations process.[3]

These interactions have led to an increasing recognition of a role for local governments and authorities, particularly regarding discussions on reducing emissions from deforestation and forest degradation in developing countries (REDD)[4] and the Nairobi work programme[5] on adaptation within the new and emerging concepts of the international climate negotiations. There is by now a rather extensive set of studies showing that cities and metro regions can make a large difference in reducing global environmental damage, focused mostly on greenhouse gas emissions (GHG). But the international level, whether the Kyoto Protocol or the post-2012 UNFCCC negotiations, fails formally when it comes to recognizing this potential. Neither is this potential built into draft agreements (Newman 2006; Sovacool and Brown 2010). The effort focused on mitigation and adaptation needs to be localized, with cities one key locality. This would involve both a bottom-up (information from the local level) and a top-down understanding of how existing protocols and post-2012 agreements integrate cities.

But ultimately, I argue, there is a need to go well beyond these governance frameworks. Cities make this need visible and urgent. We cannot simply redistribute carbon emissions. Nor are mitigation and adaptation directives enough. We need to bring in the knowledge that diverse natural sciences have accumulated, including practical applications, to address the major environmental challenges.

At the level of the city, using this knowledge is a far more specific and domain-interactive effort than at the level of national policy. Further, it will entail an internationalism derived from the many different countries that are leaders in these scientific discoveries and innovations. But this will be an internationalism that runs through thick local spaces, each with their own political and social cultures for implementing change. Finally, as I argue, capturing the complexity of cities in their multi-scalar and multi-ecological composition will allow for many more vectors for implementation than just about any other level, whether national, international or sub-urban, such as the neighbourhood. This should, in turn, allow us to go well beyond adaptation and mitigation as currently understood.

The urbanizing of global governance challenges

Many of today's major global governance challenges become tangible, urgent and practical in cities worldwide. Urban leaders and activists have had to deal with many issues long before national governments and inter-state treaties addressed them. Cities are sites where these challenges can be studied empirically and where policy design and implementation often is more feasible than at the national level. Among these global governance challenges are those concerning the environment; human insecurity, including the spread of violence against people of all ages and a proliferation of racisms; and the sharp rise in economic forms of violence. Cities also constitute a frontier space for new types of environmentally sustainable energy sources, construction processes and infrastructures. Finally, cities are critical for emerging inter-city networks that involve a broad range of actors (NGOs, formal urban governments, informal activists, global firms and immigrants) that potentially could function as a political infrastructure with which to address some of these global governance challenges.

Cities also enter the global governance picture as sites for the enactment of new forms of violence resulting from various crises. In the dense and conflictive spaces of cities, we foresee a variety of forms of violence that are likely to escape the macro-level norms of good governance. For instance, drug-gang violence in São Paulo and Rio de Janeiro points to a much larger challenge than inadequate local policing. So do the failures of the powerful US army in Baghdad to institute order. To explain this away as simply acute anarchy is inadequate and facile. It will take much effort to maintain somewhat civilized social environments in cities. In discussing global governance questions, one challenge is to push the macro-level frames of international treaties such as the Kyoto Protocol to account for, and factor in, the types of stress that arise from violence and insecurity in dense spaces in everyday life – the type of issue that global governance discourse and its norms do not quite capture. Yet, it is critical that such everyday conditions be incorporated in the global governance framing, since some of these conditions may eventually feed into micro- and macro-style armed conflicts, and this will only make matters worse.

More than nation states, cities will be forced to the frontlines by global warming, energy and water insecurity, and other environmental challenges (see Reuveny 2008; Warner et al. 2009). The new kinds of crises and, possibly, ensuing violence will be felt particularly in cities because of the often extreme dependence of cities on complex systems. City life depends on massive infrastructures (electricity for elevators and abundant public transport) and institutional support (e.g. hospitals, water-purifying plants). Infrastructure ranging from apartment buildings to hospitals, vast sewage systems, vast underground transport systems and entire electric grids depend on computerized management that is vulnerable to breakdown. NASA's study *Severe Space Weather Events: Understanding Societal and Economic Impacts* shows the vulnerabilities of the computerized systems that manage the US electricity grid and the potential for negative effects on cities and households.[6] We already know that a rise in water levels will flood some of the densest areas in the world. When these realities hit cities, they will hit hard and preparedness will be critical. These realities are overtaking the abstract norm-oriented arguments of global governance debates that consist largely of future-oriented 'oughts' – what we ought to do.

These challenges are emergent at the national level, but acutely present in cities. In this sense, cities are in the frontline and will have to react to global warming, whether or not national states sign on to international treaties. The leadership of cities is quite aware of this issue.

Can we bridge the ecologies of cities and the biosphere?

The enormously distinctive presence that is urbanization is changing a growing range of nature's ecologies, from the climate to species diversity and ocean purity. It is creating new environmental conditions – heat islands, ozone holes, desertification and water pollution. We have entered a new phase. For the first time, humankind is the major consumer in all the significant ecosystems and urbanization has been a major instrument (e.g. Girardet 2008). There is now a set of global ecological conditions that have never been seen before. Major cities have become distinct socio-ecological systems with a planetary reach. Cities have a pronounced effect on traditional rural economies and their longstanding cultural adaptation to biological diversity. Rural populations have become consumers of products produced in the industrial economy, which is much less sensitive to biological diversity. The rural condition in much of the world has evolved into a new system of social relationships, one that does not work with biodiversity (Girardet 2008). These developments all signal that the urban condition is a major factor in any environmental future. It all amounts to a radical transformation in the relationship between humankind and the rest of the planet.

But is it urbanization per se or the particular types of urban systems and industrial processes that we have instituted? That is to say, is it the urban format marked by agglomeration and density dynamics or what we have historically and collectively produced partly through processes of path-dependence that kept eliminating options as we proceeded? Are these global ecological conditions the results of urban agglomeration and density, or are they the results of the specific types of urban systems that we have developed to handle transport, waste disposal, building, heating and cooling, food provision, and the industrial processes by which we extract, grow, make, package, distribute and dispose of the foods, services and materials that we use?

It is, doubtless, the latter – the specific urban systems that we have made. One of the outstanding features, which is evident when we examine a range of today's major cities, is the pronounced differences among cities in terms of environmental sustainability. These differences result from diverse government policies, economic bases, patterns of daily life, and so on. In addition to these differences are a few foundational elements that now increasingly dominate our way of doing things. One is the fact that the entire energy and material flux coursing through the human economy returns in altered form as pollution and waste to the ecosphere. The rupture at the heart of this set of flows is *made* and can, thus, be unmade – and some cities are working on it. This rupture is present in just about all economic spaces, from urban to non-urban. However, it is in cities where it has its most complex interactions and cumulative effects, because cities contain a vast number of diverse economic sectors. This fact makes cities a source of most of the environmental damage, and some of the most intractable conditions that feed the damage. Nevertheless, it is also the complexity and internal diversity of cities that is part of the solution.[7]

It is now imperative to make cities and urbanization part of the solution. We need to use and build upon those features of cities that can reorient the material and organizational ecologies of cities to positive interactions with nature's ecologies. These interactions, and the diversity of domains that they cover, are themselves an emergent socio-ecological system. Elsewhere (Sassen 2006; Sassen and Dotan 2011), I have conceptualized the interaction between cities and the biosphere as a series of bridges that connect the diverse ecologies of each. (For longstanding work on biospheric capacities that cities could mobilize, see for example, Danso 2007; Garcia et al. 2000; Jonkers 2007; Van Veenhuizen and Danso, 2007.) Part of the effort is needed to maximize the probability of positive environmental outcomes. (for the breadth of longstanding issues this entails, see e.g. Dietz et al. 2009; Mgbeoji 2006; Gupta 2004). Specific features of cities that help are economies of scale, density and the associated potential for greater efficiency in resource use and important, but often neglected, dense communication networks that can serve as facilitators to institute environmentally sound practices in cities. More theoretically, one can say that, in so far as cities are constituted through various processes that produce space, time, place and nature, they also contain the transformative possibilities embedded in these same processes. For example, the temporal dimension becomes critical in environmentally sound initiatives: ecological economics helps us recognize that what is inefficient or value-losing according to market criteria with short temporal frames is actually positive and value-adding if we use environment-driven criteria.[8]

Cities as complex systems

As has been well documented, cities have long been sites for innovation and for developing and instituting complex physical and organizational systems. It is within the complexity of the city that we must find the solutions to much environmental damage and the formulas for reconfiguring the socio-ecological systems that constitute urbanization. Cities contain the networks and information loops that may facilitate communicating, informing and persuading households, governments and firms to support and participate in environmentally sensitive programmes and radically transformative institution building.

Urban systems also entail systems of social relationships that support the current configuration.[9] Aside from adoption of practices, such as waste recycling, it will take a change in these systems of social relationships themselves to achieve greater environmental sensitivity and efficiency. For instance, a crucial issue is the massive investment around the world promoting large projects that damage the environment. Deforestation and construction of large dams are perhaps among the best-known problems. The scale and the increasingly global and private character of these investments suggest that citizens, governments and NGOs lack the power to alter these investment patterns. However, there are structural platforms for acting and contesting these powerful corporate actors (Sassen 2005). The geography of economic globalization is strategic rather than all-encompassing and this is especially true in the managing, coordinating, servicing and financing of global economic operations. The fact that it is strategic is significant for a discussion of the possibilities of regulating and governing the global economy. There are sites in this strategic geography, such as the network of global cities, where the density of economic transactions and top-level management functions come together to form a strategic geography of decision making. Such a geography of decision making is also a strategic geography for demanding accountability for environmental damage from major corporate actors (Sassen 2005). It is precisely because the global economic system is characterized by an enormous concentration of power in a finite number of large multinational corporations and global financial markets that makes for concentrated (rather than widely dispersed) sites for accountability and changing investment criteria. Engaging the headquarters is a very different type of action than engaging the thousands of mines and factories and the millions of service outlets of such global firms. This engagement is facilitated today by the recognition of an environmental crisis by consumers, politicians and the media. Certainly, it leaves out millions of small local firms that are responsible for much of the environmental damage. However, they are more likely to be controllable by means of national regulations and local activism.

A crucial issue raised by the aforementioned discussion is the question of the scale at which damage is produced and intervention or change should occur. This scale may, in turn, differ from the levels and sites for responsibility and accountability. The city is, in this regard, an enormously complex entity. Cities are multi-scalar systems where many of the environmental dynamics that concern us are constituted and which, in turn, constitute what we call the city. It is in the cities where different policy levels, from the supra- to the sub-national, are implemented. Further, specific networks of mostly global cities also constitute a key component of the global scale and, hence, can be thought of as a network of sites for accountability of global economic actors.

Urban complexity and diversity are further augmented by the fact that urban sustainability requires engaging the legal systems and profit logics that underlie and enable many of the environmentally damaging aspects of our societies (Sassen 2008: Chapters 4 and 5). The question of urban sustainability cannot be reduced to modest interventions that leave these major systems untouched. The actual features of these systems vary across countries and across the North–South divide. While it is possible to deal just with scientific knowledge when the aim is, for instance, cleaning a body of polluted water, this option is not possible when dealing with cities. Non-scientific elements are a crucial part of the picture, as are questions of power, poverty and inequality, ideology and cultural preferences. This becomes even more of a challenge in the current era, with global firms and the spread of markets to more and more institutional realms. Questions of policy and proactive engagement possibilities have become a critical dimension of treatments of urban sustainability, whether they involve asking people to support garbage recycling or demanding accountability from major global corporations that are known to have environmentally damaging production processes.

A debate that gathered heat, beginning in the 1990s and remaining unresolved, pits the global against the local or vice versa as the most strategic scale for action. Redclift (2009) argued that we cannot manage the environment at the global level. Global problems are caused by the aggregation of production and consumption, much of which is concentrated within the world's urban centres. For Redclift, we first need to achieve sustainability at the local level. He argues that the flurry of international agreements and agencies are international structures for managing the environment that bear little or no relationship to the processes through which the environment is being transformed. Not everyone agrees. Satterthwaite (Satterthwaite et al. 2007) has long argued that we need global responsibilities, but cannot have such without international agreements. Gleeson and Low (2011) add that we have a global system of corporate relationships of which city administrations are increasingly part. This complex cross-border system is increasingly responsible for the health and destruction of the planet. Today's processes of development bring into focus the question of environmental justice at the global level, a question that would have referred to the national level in the early industrial era.

Conclusion

Bringing the city level into larger governance regimes could enable a better use of the properties of cities in the work of securing a more environmental sustainability. Full use of the complexity of cities – their multi-scalar and ecological features – becomes critical. I do not think we are close to such a full use, but there is the beginning of a mobilization in this direction.

A second strategic element concerns the city as a social and power system – with laws, extreme inequalities and vast concentrations of power. Implementing environmental measures that go beyond current modest mitigation and adaptation efforts will require engaging legal systems and the profit logics that underlie and enable many of the environmentally damaging aspects of our societies. Any advance towards environmental sustainability is necessarily implicated in these systems and logics. The actual features of these systems vary across countries and across the North–South divide. While in some of the other environmental domains it is possible to confine the discussion of the subject to scientific knowledge, this is not the case when dealing with cities.

Cities are complex systems in their geographies of consumption and waste production. This complexity makes them essential for the creation of solutions. Some of the geographies for sound environmental action in cities can operate worldwide. The network of global cities is a space on a global scale for the management of investments, but also potentially for the re-engineering of environmentally destructive global capital investments into more responsible investments. It contains the sites of power of some of the most destructive actors, but also potentially the sites at which to demand accountability of these actors. The scale of the network differs from the scale of the individual cities that comprise this network.

Finally, a focus on cities enables more complex applications of mixes of policy and scientific knowledge than international regimes. Making the local application of scientific knowledge more central to the governance discussion counteracts the excessive weight of markets (i.e. carbon trading) as a means to address the environmental crisis. Carbon trading leads to unhelpful nationalisms and misplaced protectionism of a country's 'right' to pollute – including to pollute more than the carbon-trading regime allows for and than genuine environmental sustainability can tolerate. Effective action needs to bring in actual change on the ground. The articulation between 'urban ground' and international regimes can generate a third type of environmental governance vector: a multi-sited global regime centred in the development of new kinds of urban capacities regardless of (sovereign) country.

Notes

1 See, for instance, the *Global Status Report on Local Renewable Energy Policies*, Institute for Sustainable Energy Policies, Tokyo.
2 See, for instance, ICLEI Climate Program, www.iclei.org/index.

3 The UNFCCC is focused on a successor to the climate protection agreement following 2012, also known as the post-Kyoto or post-2012 agreement, http://unfccc.int/adaptation/nairobi_work_programme/items/3633.php. Accessed on 8 July 2012.
4 UNFCCC, REDD Web Platform, http://unfccc.int/methods_science/redd/items/4531.php. Accessed on 8 July 2012.
5 UNFCCC http://unfccc.int/adaptation/nairobi_work_programme/items/3633.php. Accessed on 8 July 2012.
6 http://science.nasa.gov/science-news/science-at-nasa/2009/21jan_severespaceweather/
7 That it is not urbanization per se that is damaging, but the mode of urbanization also is signalled by the adoption of environmentally harmful production processes by pre-modern rural societies. Until recently, these had environmentally sustainable economic practices, such as crop rotation and choosing to forgo the use of chemical fertilizers and insecticides. Further, our extreme capitalism has made the rural poor, especially in the Global South – so poor that, for the first time, many now are also engaging in environmentally destructive practices, notably practices that lead to desertification.
8 For some of the foundational concepts and logics of ecological economics, see Daly (1977), Daly and Farley (2003), Porter et al. (2009), Rees (2006), Schulze (1994) and Porter et al. (2009).
9 This is a broad subject. For studies that engage a range of aspects, see Beddoe et al. (2009), Girardet (2008), Morello-Frosch et al. (2009), Sassen (2001, 2005), Satterthwaite et al. (2007).

References

Beddoe, R., Costanza, R., Farley, J., Garze, E., Kent, J. et al. (2009) 'Overcoming Systemic Roadblocks to Sustainability: The Evolutionary Redesign of Worldviews, Institutions, and Technologies', *PNAS*, 106, 8, pp. 2483–2489.
Daly, H.E. (1977) *Steady-State Economics: The Economics of Biophysical Equilibrium and Moral Growth*. San Francisco, CA: W.H. Freeman and Company.
Daly, H.E. and Farley, J. (2003) *Ecological Economics: Principles and Applications*. Washington, DC: Island Press.
Dietz, T., Rosa, E.A. and York, R. (2009) 'Environmentally Efficient Well-Being: Rethinking Sustainability as the Relationship Between Human Well-Being and Environmental Impacts', *Human Ecology Review*, 16, 1, pp. 114–123.
Etsy, D.C. and Ivanova, M. (2005) 'Globalisation and Environmental Stewardship: A Global Governance Perspective', in F. Wijen, K. Zoeteman, J. Pieters, and P. van Seters, (eds), *A Handbook of Globalisation and Environmental Policy: National Government Interventions in a Global Arena*. Cheltenham, UK: Edward Elgar.
Garcia, J., Mujeriego, R. and Hernández-Mariné, M. (2000) 'High Rate Algal Pond Operating Strategies for Urban Wastewater Nitrogen Removal', *Journal of Applied Phycology*, 12, pp. 331–339.
Girardet, H. (2008) *Cities People Planet: Urban Development and Climate Change*, 2nd ed. Amsterdam: John Wiley and Sons.
Gleeson, B. and Low, N.P. (eds) (2001) *Governing for the Environment: Global Problems, Ethics and Democracy*. Basingstoke, UK: Palgrave.
Gupta, A.K. (2004) 'WIPO-UNEP Study on the Role of Intellectual Property Rights in the Sharing of Benefits Arising from the Use of Biological Resources and Associated Traditional Knowledge', Geneva: World Intellectual Property Organization and United Nations Environmental Programme.
Jonkers, H.M. (2007) 'Self Healing Concrete: A Biological Approach', pp. 195–204 in S. van der Zwaag (ed.), *Self Healing Materials: An Alternative Approach to 20 Centuries of Materials Science*. Delft: Springer.
Kern, K. and Mol, A.P.J. 2013. 'Cities and Global Climate Governance: From Passive Implementers to Active Co-Decision Makers.' pp. 288–307 in Joseph E. Stiglitz and Mary Kaldor (eds), *The Quest for Security: Protection without Protectionism and the Challenge of Global Governance*. New York: Columbia University Press.
Mgbeoji, I. (2006) *Biopiracy: Patents, Plants, and Indigenous Knowledge*. Vancouver: University of British Columbia Press.
Morello-Frosch, R., Pastor, M., Sadd, J. and Shonkoff, S.B. (2009) *The Climate Gap: Inequalities in How Climate Change Hurts Americans and How to Close the Gap*. Los Angeles, CA: USC Program for Environmental and Regional Equity. Available at http://college.usc.edu/geography/ESPE/documents/The_Climate_Gap_Full_Report_FINAL.pdf. Accessed 12 January 2013.
Newman, P. (2006). 'The Environmental Impact of Cities', *Environment and Urbanization*, 18, 2, pp. 275–295.
Porter, J., Costanza, R., Sandhu, H., Sigsgaard, L. and Wratten, S. (2009) 'The Value of Producing Food, Energy, and Ecosystem Services within an Agro-Ecosystem', *Ambio*, 38, 4, pp. 186–193.

Redclift, M. (2009) 'The Environment and Carbon Dependence: Landscapes of Sustainability and Materiality', *Current Sociology*, 57, 3, pp. 369–387.

Rees, W.E. (2006) 'Ecological Footprints and Bio-Capacity: Essential Elements in Sustainability Assessment', pp. 143–158 in J. Dewulf and H. Van Langenhove (eds), *Renewables-Based Technology: Sustainability Assessment*. Chichester, UK: John Wiley and Sons.

Reuveny, R. (2008) 'Ecomigration and Violent Conflict: Case Studies and Public Policy Implications', *Human Ecology*, 36, pp. 1–13.

Sassen, S. (2001) *The Global City, Second Edition*. Princeton, NJ: Princeton University Press.

Sassen, S. (2005) 'The Ecology of Global Economic Power: Changing Investment Practices to Promote Environmental Sustainability', *Journal of International Affairs*, 58, 2, pp. 11–33.

Sassen, S. (ed.) (2006) 'Human Settlement and the Environment' in *EOLSS Encyclopedia of the Environment*, vol. 14. Oxford, UK: EOLSS/UNESCO.

Sassen, S. (2008) *Territory, Authority, Rights: From Medieval to Global Assemblages*. Princeton, NJ: Princeton University Press.

Sassen, S. and Dotan, N. (2011) 'Delegating, not Returning, to the Biosphere: How to Use the Multi-Scalar and Ecological Properties of Cities', *Global Environmental Change*, 21, 3, pp. 823–834.

Satterthwaite, D., Huq, S., Pelling, M., Reid, H. and Lankao, P. R. (2007) 'Adapting to Climate Change in Urban Areas: The Possibilities and Constraints in Low- and Middle-Income Nations', *Human Settlements Discussion Paper Series*, London: International Institute for Environment and Development. Available at www.iied.org/pubs/pdfs/10549IIED.pdf. Accessed 12 January 2013.

Schulze, P.C. (1994) 'Cost–Benefit Analyses and Environmental Policy', *Ecological Economics*, 9, 3, pp. 197–199.

Sovacool, B.K. and Brown, M.A. (2010) 'Twelve Metropolitan Carbon Footprints: A Preliminary Comparative Global Assessment', *Energy Policy*, 38, 9, pp. 4856–4869.

Toly, N.J. (2008) 'Transnational Municipal Networks in Climate Politics: From Global Governance to Global Politics', *Globalizations*, 5, 3, pp. 341–356.

Van Veenhuizen, R. and Danso, G. (2007) *Profitability and Sustainability of Urban and Peri-Urban Agriculture*. Rome: Food and Agriculture Organization of the United Nations.

Warner K., Ehrhart, C., Sherbinin, A., Adamo, S. and Chai-Onn, T. (2009) *In Search of Shelter: Mapping the Effects of Climate Change on Human Migration and Displacement*. Geneva: CARE International. Available at www.careclimatechange.org/files/reports/CARE_In_Search_of_Shelter.pdf. Accessed 12 January 2013.

15

Hybrid arrangements within the environmental state[1]

Dana R. Fisher and Erika S. Svendsen

This chapter explores ways that collaboration and hybrid arrangements are emerging within the environmental state today. Drawing from relevant theories of the environmental state, we explore ways that scholars have analysed the roles in environmental governance played by different social actors. Our analysis builds on extant research on collaborative and networked governance, parsing out ways that hybrid arrangements have emerged to explain the manifold forms of partnerships taken among social actors working to protect the natural environment. Examples of specific types of arrangements from the stewardship regime in New York City are provided to illustrate how hybrid environmental governance is taking place in the real world.

Governance and the environmental state

A good deal of work within political sociology, largely European in origin, calls for the emergence of an 'environmental state' (Mol and Buttel 2002; see also Buttel 2000; Frank et al. 2000a, 2000b; Goldman 2001). As discussed in previous studies (e.g. Fisher and Freudenburg 2004), this work includes three main branches – reflexive modernization, ecological modernization and postmaterialism – that are largely independent of each other. Even with their lack of integration, these theories tend to share two commonalities. First, much work in all three branches reflects the view that, in the words of Anthony Giddens, environmental protection is becoming 'a source of economic growth rather than its opposite' (1998: 19). Second, all three branches tend to include an expectation that advanced or industrialized nation states will treat environmental protection 'as a basic state responsibility' (Frank et al. 2000b: 96), with the state seen, at least implicitly, as having enough autonomy and capacity to carry out that responsibility.

Of these analyses of the environmental state, perhaps the theory of ecological modernization focuses most specifically on issues related to governance. Ecological modernization scholars have noted that new forms of environmental governance involve the participation of emergent coalitions in multiple tiers of policymaking. Although ecological modernization theory initially looked to 'modern institutions such as science and technology and state intervention' to lead the way to environmental governance (Mol and Spaargaren 1993: 454–455), more recent work explores the role of civil society actors, including social movements and non-governmental organizations (NGOs) (see particularly Mol 2000; Mol et al. 2009; Sonnenfeld 2000, 2002;

Sonnenfeld and Mol 2002; see also van Tatenhove and Leroy 2003 for a full discussion). Mol and Spaargaren (1998: 3) (see also Spaargaren and Mol 2008), for example, suggest that ecological modernization involves changes in the traditional roles of actors; in particular,

> various transformations regarding the traditional central role of the nation-state in environmental reform ... [with] more opportunities for non-state actors to take over traditional tasks of the nation-state.

In short, the theory has expanded to identify coalitions forming across the state, market and civil society sectors that make environmental policymaking possible.

These studies overlap to some degree with research on forms of new civic engagement, public reasoning and collective learning that have led to shifting roles and positions of the state, market and civil society. In light of new environmental decision-making networks, van Tatenhove and Leroy suggest there has been an institutionalization of 'interference zones', particular spaces within the political domain where the meaning and nature of public participation, state rule-making and market influences are reconfigured. In their words:

> Society no longer seems separate from the state and that can be governed by it. Instead, the subsystems of civil society and market and their respective agencies are now conceptualized in terms of 'networks', 'associations', 'public–private partnership' and the like, in which the state negotiates with non-state agencies, either from the market or society, in order to formulate and implement an effective and legitimate policy.
>
> (Tatenhove and Leroy 2009: 199)

Ecological modernization theory has expanded to address how the state, market and civil society sectors work together to create governing frameworks at the sub-national, national and international levels (e.g. Fisher et al. 2009; Sonnenfeld and Mol 2002). Such collaboration includes what Mol and Spaargaren (2006), in their work on environmental flows, call 'hybrid arrangements' among different social actors, including from civil society.

To date, numerous empirical studies have noted the expansion of scales at which environmental policymaking is taking place. Scholars coming from an ecological modernization perspective have looked at diverse issues and international environmental regimes. Some comparative research has focused on the international level, addressing issues related to climate change regime formation (Fisher and Freudenburg 2004) and the effect of Europeanization on countries in Central and Eastern Europe (Andersen 1999). Other research has focused on the national level, looking at countries as diverse as Japan (Barrett 2005), Finland (Jokinen 2000; Sairinen 2002), Hungary (Gille 2000), Lithuania (Rinkevicius 2000a, 2000b) and Vietnam (Frijns et al. 2000). At the same time, a limited number of studies have explored ecological modernization processes taking place at the sub-national level. Gonzalez (2001), for example, focuses on automobile emission standards in California. Also looking at various states in the United States, Scheinberg analyses recycling (2003; but see Pellow et al. 2000, for a contrasting view).

Collaboration within the environmental state

Combining research on the environmental state with the broader literature on collaborative governance is helpful in developing a deeper understanding of hybrid arrangements and how they work. Governance is used as an overarching term to describe the governing arrangements between the state, non-governmental actors and individual citizens as they shape the vested interests of other organizations (DiMaggio and Powell 1991). It encompasses a range of social actors engaging in formal and informal decision making, as well as policy setting, around a variety of social issues. Shifts in the hierarchical position of government have been described as a 'hollowing out of the state', where non-governmental organizations and business entities are becoming more centralized or operate as part of a network in urban regimes (Rhodes 1994). Specifically, these actors are shaping new governing structures that have a direct impact on resource allocation and land use (Jessop 1995). Others have referred to non-state coalitions as the 'shadow state', where non-governmental sectors act outside traditional democratic processes and assume state responsibilities (e.g. Wolch 1990). In this case, the state retains a significant role in rule-making authority but yields some of its power in exchange for capital and labour.

In societies with advanced notions of self-governance and coordination, there is an expectation that different types of social arrangements will take shape (Hirst 2000). Here, collaborative governance is defined as a governing arrangement whereby public agencies and civic actors engage in public deliberation, share expertise, leverage assets, develop strategic partnerships and assure accountability in the co-production of public decision making (see particularly Ansell and Gash 2008; Sirianni 2009). In the words of Ansell and Gash (2008: 543), collaborative governance 'brings public and private stakeholders together in collective forums with public agencies to engage in consensus-oriented decision making'. Collaborative governance has been studied from the perspective of grassroots groups (Weber 2003), participatory natural resource management (Vira and Jeffery 2001), partnership (Leach et al. 2002), co-management (Singleton 2000) and third party or 'new governance' (Salamon 2002). In several of these cases, the form that collaboration takes is dependent on past histories of conflict or compromise, stakeholder incentives, resource imbalances, leadership styles and organizational structures (Ansell 2003; Ansell and Gash 2008). Successful outcomes of collaborative governance often are attributed to whether or not an issue can be defined as transboundary and/or whether the structure and resources required for group decision making are sufficient (Hirst 2000; Koontz et al. 2004). Others have found important differences in policy outcomes depending upon the positioning of civil society in relationship to the state (Dryzek et al. 2003; see also Kearns 1991).

It has been suggested that new forms of collaborative governance arise naturally from the complexity of problems, diverse publics and the persistent need for voice and inclusion (Sirianni 2009). As Ansell and Gash (2008) argue, collaborative governance differs from traditional public–private partnerships in that the former expands beyond the provision of public services to include official rulemaking and agenda-setting over public space. In this sense, collaboration becomes a form of agenda-setting. Perhaps Harry Boyte (2005: 536) best explains how the shift from government to governance holds real promise for reframing democracy:

> The shift involves a move from citizens as simply voters, volunteers, and consumers to citizens as problem solvers and co-creators of public goods; from public leaders, such as public affairs professionals and politicians to providers of services and solutions to partners, educators, and organizers of citizen action; and from democracy as elections to democratic society.

Recent research on collaborative governance focuses on issues related to the environment, which some scholars call 'environmental governance'. Within this work, there has been a noticeable evolution from formal, command-and-control government procedures and debates over deregulation in the United States to instances of collaborative governance of natural resources that reveal a complex web of state and non-state actors interacting at multiple scales

(Koontz et al. 2004). Not unlike the new governance structures that are forming at the global environmental scale, sub-national and local environmental organizations are working among a much broader set of partners and perspectives (Ansell 2003; Ansell and Gash 2008; Boyte 2005; Jänicke and Jörgens 2009; Pierre 2000)

Case studies of environmental governance reveal a range of institutional arrangements between civil society and government, including the state as both an actor and institution, state-led community collaborations, civic-led initiatives and those efforts facilitated by coalition groups and alliances, and other emerging forms of governance (Koontz and Thomas 2006; Koontz et al. 2004). The prevalence of collaborative governance is often attributed to instances where civil society has successfully pursued progressive and discursive approaches to planning, environmental management and economic development (e.g. Gibbs and Jonas 2000; Healey 1996; Innes and Booher 1999; Rhodes 1996, 2000; Wondolleck and Yaffee 2000).

Increasingly, this type of environmental governance is understood to operate within a networked structure comprised of individuals and organizations, or what Ostrom (2010) would call 'polycentric networks' (see also Connolly et al. 2013). In fact, many argue that we have entered the 'age of networked governance' as citizens, non-governmental organizations, market forces and the state work in coalitions rather than as single, organizational entities (Sirianni 2009). Advocates of network governance find that this polycentric structure enables actors to embrace complexity (Ansell 2003; Ernston et al. 2008; Innes and Booher 1999), become more adaptive (Barthel et al. 2005; Leach et al. 2002), foster democracy (Sorenson and Torfing 2005) and to manage public goods sustainably, especially those goods that could be classified as natural resources (Ostrom 2010). Although participants acting within networked governance hold varying degrees of political power, 'democratic polities can be networked; networks cannot be democratic polities' (Taylor 2009: 827). There are strong critiques of networked governance that include concerns over equity or, more specifically, the question of how the needs and concerns of those located at the periphery become included in democratic decision-making processes (Davies 2011; Ernston 2008; Roelofs 2009).

As outcomes vary based on geography, scale, social norms, discourse, local economies, natural resource properties and other factors, debates over environmental governance models continue (Hajer and Wagenaar 2003; Koontz and Thomas 2006; Koontz et al. 2004; Ostrom 1999; Pierre 2000). There is much evidence to suggest, however, that the social and spatial structure of networks is highly relevant to the emergence of local hybrid arrangements (Bulkeley 2005; see also Pincetl 2003).

Building on scholarship on collaborative and networked forms of environmental governance, we examine emerging hybrid arrangements within the environmental state. At any given moment, multi-sector actors are operating at varying spatial scales that will ultimately inform decision-making structures and policy. We will argue that there is a continuum of arrangements among social actors that emerge within environmental states. These hybrid arrangements involve multiple social actors working together in a diversity of ways. Such partnerships between state and non-state actors not only lead to innovative governing arrangements but, in some cases, they create new civic forms. The forms vary in terms of the degree of state, business and civic involvement. Moreover, the diversity of forms may be associated with different public policy outcomes. Thinking of hybrid arrangements in this way enables us to put a sharper point on the notion of collaborative governance, acknowledging that civil society and the state, once discrete entities, today are forever changed through association. In the roles and responsibilities of civic groups, business interests and state organizations, there is often a 'blurred line' between public and private. In some hybrid arrangements, the lines become so blurred that entirely new organizational forms emerge.

Hybrid arrangements within New York's stewardship regime

How do hybrid arrangements work in practice? This section presents examples from the stewardship regime in New York City, where civic groups collaborate with governmental agencies – and, at times, businesses – to conserve, manage, monitor, restore, advocate for and educate the public about a wide range of issues related to sustaining the local environment.[2] Hybrid arrangements take a diversity of forms within New York City. The discussion below focuses on the range of variation in hybrid arrangements that can be seen within this one city. The examples represent points along the continuum of collaborative, environmental governance, with non-governmental groups differing in their autonomy from the state. The range extends from *completely autonomous* civic groups working with the government, but outside of it; to civic groups working *in tandem* with government actors; to civic groups that are *borne from* governmental entities. New York City's governing arrangements are ever-dynamic. As such, many of the arrangements continue to shift in terms of their programmes and partnerships, in accordance with the availability of resources, changes in leadership and new demands for urban environmental change.

Data for these cases were collected as part of a multi-year project studying environmental stewardship in New York City. A survey of all stewardship groups was conducted in 2007 (n = 506). Based on these data, the groups that emerged as the most central 'organizational nodes' within the stewardship regime participated in open-ended, semi-structured interviews, in 2010.[3] A summary of findings from these interviews provides instructive examples of groups representing points on the continuum. Table 15.1 presents a distribution of types of hybrid arrangements among the organizational nodes within the New York City stewardship regime. Each type will be discussed in detail in the sections that follow.

Civic groups external to government

Civic groups are formed and operate external to governmental entities. Traditionally, many civic organizations were created in opposition to government rather than working alongside government initiatives (Taylor 2009). In New York City, we find a surprising number of independent civic organizations that operate programmes supporting local, state and federal government activities. In our sample, four of the organizational nodes, constituting one third of all groups, represented this perspective. These programmes involve developing infrastructure and support for programmes like greenways, bike path development, tree planting events and neighbourhood groups that support public parks. Some groups are focused on short-term issues such as advocating for saving a specific government programme or service.

Other programmes run by civic groups perform services or have developed expertise in areas that benefit government – such as public land acquisition, raising private funds and cultivating spheres of influence – but that could represent a conflict of interest for government agencies. Not surprisingly, an increasing number of these groups report that they have good relations with government officials and rarely see the need to engage in oppositional politics. Rather,

Table 15.1 Distribution of hybrid arrangements

Type of hybrid arrangement	Number in sample (%)
Civic groups external to government	4 (33%)
Civic groups working in tandem with government	7 (58%)
Civic groups nested within the state	1 (8%)

many civic groups have invited public officials into their decision-making processes and, in some cases, they have offered agency representatives ex officio or advisory status within their organizations.

Examples of this type of arrangement include civic groups such as the Green Guerillas, a city-wide community gardening support programme, and Just Food, a programme that connects local farms and city residents, providing sustainably grown food. These groups have developed their mission and purpose either in absence of government policy or in areas that are traditionally outside of local policymaking, drawing support directly from individuals and other civic groups. These groups often work with government and, at times, receive government funding, but retain a degree of independence that enables them to engage in external critiques of government entities. In one example in the late 1990s, representatives of the Green Guerillas were highly critical of the City of New York's sponsored sale of community gardens, suing the City to save the gardens. Another example is Just Food, which addresses deficiencies in food access and security in New York City. The organization has, in some instances, developed a platform that challenges current policy and procedures. Such examples notwithstanding, at any given time both the Green Guerillas and Just Food work cooperatively with the City and local partners on a range of activities intended to promote well-being and community development.

Civic groups in tandem with government

The most common hybrid arrangements in New York City are environmental organizations that have taken on work that historically was the responsibility of government actors and agencies (seven organizational nodes, or 58 per cent of the groups in our sample). Such work includes direct management of urban public land, dissemination of knowledge and information, administration of programmes designed to ensure the health of urban waterways and promotion of efforts to establish local food security. The degree of professionalization within this category varies significantly, including small groups of neighbours and friends managing a community garden, as well as organizations with full-time professional staff and revenue generating mechanisms (for a full discussion of the professionalization of civic stewardship groups in New York City, see Fisher et al. 2012). In the case of the larger, professional civic organizations, the private business sector tends to play a substantial role in their organizational structure. For example, private business leaders participate on executive boards, advisory groups or provide volunteers and materials for specific projects. It is often the case that all actors – civic, state and business – benefit from their interaction. Of the three types of hybrid arrangements highlighted here, this type of civic organization typically remains the most autonomous as the organization has attracted both business and government involvement through highly specific functions and benefits.

Two examples of this type of hybrid arrangement include the Trust for Public Land and the Park Slope Civic Council, both of which have created long-term bases of support from civil society while, at the same time, performing direct services for government. The Trust for Public Land cultivates private donors and specializes in the type of legal work necessary for acquiring conservation land for a public purpose. Government is unable to perform this function independently. Another example of civic groups working in tandem with government is the century-old Park Slope Civic Council. Having gained the trust of many public agencies through its many years of community advocacy, the Council represents a constituency of neighbours and local organizations while working with those agencies to understand better government policies and procedures. In such efforts, the Council acts as a broker between government and civil society. In this type of hybrid arrangement, civic groups have taken over fiscal and programmatic responsibilities from government and business actors. In most cases, government has not relinquished its authority as landowner or regulator, per se. Increasingly, however, it has transferred expertise to organizations in exchange for the latter performing a public service. As a result of this hybrid arrangement, all of the social actors involved in the relationship become highly dependent upon each other and decision making follows a very collaborative and collective model.

Civic groups nested within the state

The least common type of hybrid arrangement in our sample occurs when civic groups become nested within the state (only one of our organizational nodes, or 8 per cent of all groups in our sample). Although this type of arrangement occurred only once in our sample, other respondents gave examples of now-defunct or lesser-known civic programmes that were initiated by a government agency, housed within a government office, but retained some degree of independence. We suspect that there may be more examples of this type of arrangement in cities where the non-profit sector is less robust than in New York City. This type of nested arrangement is the result of representatives from within government agencies identifying the need for a civic group to supplement government funding or for greater flexibility in programming. In these cases, civic organizations are given office space and access to expertise, data and equipment within government agencies to fulfil duties that originally may have been expected to have been completed by the agencies themselves. In exchange, the agencies have greater control over civic groups' development and implementation of activities in support of agreed-upon needs. In such circumstances, civic groups have to navigate between the two worlds of government policy and civic participation.

GrowNYC, an organizational node in our sample, is one example of a civic group nested within the state. Founded decades ago within the Mayor's Office of the City of New York, the group reportedly functions today more like a non-profit organization than a government-sponsored programme. At times, GrowNYC has been at odds with government agencies in the city. Nonetheless, the group has been able to maintain access to information and resources because it is perceived to be 'quasi-governmental'.

Other government and civic organizations in New York City also have developed programmes and initiatives that have spun off into new civic forms. Although it did not emerge as an organizational node in our stewardship census of New York City, another good example is Partnerships for Parks, a civic group dedicated in part to supporting community groups' efforts to care for neighbourhood parks. This programme is supported with public and private funding through the City Parks Foundation, itself an independent, non-profit organization. A second critical mission for Partnership for Parks is to help communities better understand the plight of City Parks workers. The result is an arrangement that is neither civic nor governmental, but straddles the line between the two.

In contrast to the other, more popular hybrid arrangements, the sustainability and effectiveness of civic groups nested within the state is uncertain. Some organizations formed thirty years ago or longer remain in existence while others are now in rapid decline. Many of the older organizations that still are active have moved away from their governmental origins and are operating well within the civic realm. However, few of the government-initiated organizations tend to be critical of public agencies, preferring to continue to work in collaboration. Such uncertainties notwithstanding, we expect the number of nested governing arrangements to grow as civic organizations continue to engage in multi-level partnerships, leveraging resources and garnering a wider range of support to accomplish their goals.

Conclusion

As local, multi-scaled and multi-sector arrangements are formed to address complex and overlapping issues relevant to wider circles of constituents, the environmental state expands to include new forms governance. The growing diversity of arrangements within the environmental state appears to be an inevitable progression of our understanding of complexity. In fact, the environmental state itself has become an active proponent of new social forms by creating the space for hybrid arrangements to flourish. Findings from our study of hybrid environmental governance arrangements in New York City are consistent with theories of ecological modernization that suggest that actors within the state acknowledge their institutions' limitations in solving persistent environmental problems and allow and encourage market-based, technological solutions and hybrid arrangements (Jänicke and Jörgens 2009). As the types of hybrid arrangements appear along a dynamic continuum, it is unclear to what extent the rate of change in the formation of such arrangements in organizations is related to their effectiveness. The question of who ultimately leads or sustains governing coalitions is important. Also, such collaborations may reveal critical lessons on how power is negotiated across social, economic and physical geographies and scales.

From the perspective of the civic groups surveyed in our study, hybrid arrangements are pursued for their efficiency and to leverage limited resources. In all such types of arrangements in New York City, there is an undercurrent of cooperation that may be important to understand more deeply through future research. Which actors are, and are not, cooperating and why? Who is, and is not, included among these hybrid arrangements? In the absence of regulatory and traditional governing arrangements, and given the resource limitations facing states around the world, highly creative and innovative forms of environmental governance seem to be emerging from a growing social network of actors and organizations. As hybrid arrangements become de rigueur in certain areas, it will become increasingly important to examine the effectiveness and outcomes of such arrangements, as well as issues of equity, voice, inclusion and democracy with respect to the participation of various civic, governmental and private sector actors.

Notes

1 This project was supported by funding from the USDA Forest Service Northern Research Station and the US National Science Foundation (DEB-0948451).
2 For details on our definition of stewardship, see Fisher et al. (2007).
3 For more details on the data, methods and findings of this project, see Fisher et al. (2012); see also Connolly et al. (2013).

References

Andersen, M.S. (1999) 'Governance by Green Taxes: Implementing Clean Water Policies in Europe 1970–1990', *Environmental Economics and Policy Studies*, 2, 1, pp. 39–63.
Ansell, C.K. (2003) 'Community Embeddedness and Collaborative Governance in the San Francisco Bay Area Environmental Movement', pp. 123–144 in M. Diani and D. McAdams (eds), *Social Movements and Networks: Relational Approaches to Collective Action*. Oxford, UK: Oxford University Press.
Ansell, C.K. and Gash, A. (2008) 'Collaborative Governance in Theory and Practice', *Journal of Public Administration Research and Theory*, 18, 4, pp. 543–571.
Barrett, B.F.D. (2005) *Ecological Modernization and Japan*. London: Routledge.
Barthel, S., Colding, J., Elmqvist, T. and Folke, C. (2005) 'History and Local Management of a Biodiversity-Rich Urban, Cultural Landscape', *Ecology and Society*, 10, 2, art. 10.
Boyte, H.C. (2005) 'Reframing Democracy: Governance, Civic Agency, and Politics', *Public Administration Review*, 65, 5, pp. 536–546.
Bulkeley, H. (2005) 'Reconfiguring Environmental Governance: Towards a Politics of Scales and Networks', *Political Geography*, 24, 8, pp. 875–902.

Buttel, F.H. (2000) 'World Society, the Nation-State, and Environmental Protection', *American Sociological Review*, 65, 1, pp. 117–121.

Connolly, J., Svendsen, E., Fisher, D.R. and Campbell, L. (2013) 'Organizing Urban Ecosystem Services Through Environmental Stewardship Governance in New York City', *Landscape and Urban Planning*, 109, pp. 76–84.

Davies, J.S. (2011) *Challenging Governance Theory: From Networks to Hegemony*. Bristol, UK: Policy Press.

DiMaggio, P.J. and Powell, W.W. (eds) (1991) *The New Institutionalism in Organizational Analysis*. Chicago, IL: University of Chicago Press.

Dryzek, J.S., Downes, D., Hunold, C. and Schlosberg D. (2003) *Green States and Social Movements. Environmentalism in the United States, United Kingdom, Germany, and Norway*. Oxford, UK: Oxford University Press.

Ernstson, H., Sörlin, S. and Elmqvist, T. (2008) 'Social Movements and Ecosystem Services: The Role of Social Network Structure in Protecting and Managing Urban Green Areas in Stockholm', *Ecology and Society*, 13, 2, art. 39.

Fisher, D.R., Campbell, L.K., and Svendsen, E.S. (2007) 'Towards a Framework for Mapping Urban Environmental Stewardship', presented to the International Symposium on Society and Resource Management, Park City, UT, 17–21 June.

Fisher, D.R., Campbell, L.K. and Svendsen, E.S. (2012) 'The Organizational Structure of Urban Environmental Stewardship', *Environmental Politics*, 21, 1, pp. 26–48.

Fisher, D.R. and Freudenburg, W.R. (2004) 'Post-Industrialization and Environmental Quality: An Empirical Analysis of the Environmental State', *Social Forces*, 83, 1, pp. 157–188.

Fisher, D.R., Fritsch, O. and Andersen, M.S. (2009) 'Transformations in Environmental Governance and Participation', pp. 141–155 in A.P.J. Mol, D.A. Sonnenfeld and G. Spaargaren (eds), *The Ecological Modernisation Reader*. London: Routledge.

Frank, D.J., Hironaka, A. and Schofer. E. (2000a) 'Environmentalism as a Global Institution: Reply to Buttel', *American Sociological Review*, 65, 1, pp. 122–127.

Frank, D.J., Hironaka, A. and Schofer. E. (2000b) 'The Nation-State and the Natural Environment over the Twentieth Century', *American Sociological Review*, 65, 1, pp. 96–117.

Frijns, J., Phuong, P.T. and Mol, A.P.J. (2000) 'Ecological Modernization Theory and Industrializing Economies: The Case of Viet Nam', pp. 257–291 in A.P.J. Mol and D.A. Sonnenfeld (eds), *Ecological Modernisation Around the World: Perspectives and Critical Debates*. Essex, UK: Frank Cass.

Gibbs, D. and Jonas, A.E. (2000) 'Governance and Regulation in Local Environmental Policy: The Utility of a Regime Approach', *Geoforum*, 31, pp. 299–313.

Giddens, A. (1998) *The Third Way: The Renewal of Social Democracy*. Cambridge, UK: Polity Press.

Gille, Z. (2000) 'Legacy of Waste or Wasted Legacy? The End of Industrial Ecology in Post-Socialist Hungary', pp. 203–234 in A.P.J. Mol and D.A. Sonnenfeld (eds), *Ecological Modernisation Around the World: Perspectives and Critical Debates*. Essex, UK: Frank Cass.

Goldman, M. (2001) 'Constructing an Environmental State: Eco-Governmentality and Other Transnational Practices, of a "Green" World Bank', *Social Problems*, 48, 4, pp. 499–523.

Gonzales, G.A. (2001) 'Democratic Ethics and Ecological Modernization: The Formulation of California's Automobile Emission Standards', *Public Integrity*, 3, 4, pp. 325–344.

Hajer, M. A. and Wagenaar, H. (eds) (2003) *Deliberative Policy Analysis: Understanding Governance in the Network Society*. Cambridge, UK: Cambridge University Press.

Healey, P. (1996) 'Consensus-Building Across Difficulty Divisions: New Approaches to Collaborative Strategy Making', *Planning Practice and Research*, 11, 2, pp. 207–216.

Hirst, P. (2000) 'Democracy and Governance', pp. 13–35 in J. Pierre (ed.), *Debating Governance: Authority, Steering, and Democracy*. Oxford, UK: Oxford University Press.

Innes, J.E. and Booher, D. (1999) 'Consensus Building and Complex Adaptive Systems: A Framework for Evaluating Collaborative Planning', *Journal of the American Planning Association*, 65, 4, pp. 412–423.

Jänicke, M. and Jörgens, H. (2009) 'New Approaches to Environmental Governance', pp. 156–189 in A.P.J. Mol, D.A. Sonnenfeld and G. Spaargaren (eds), *The Ecological Modernisation Reader: Environmental Reform in Theory and Practice*. London: Routledge.

Jessop, B. (1995) 'The Regulation Approach, Governance and Post-Fordism: Alternative Perspectives on Economic and Political Change', *Economy and Society*, 24, 3, pp. 307–333.

Jokinen, P. (2000) 'Europeanisation and Ecological Modernisation: Agro-Environmental Policy and Practices in Finland', pp. 138–170 in A.P.J. Mol and D.A. Sonnenfeld (eds), *Ecological Modernisation Around the World: Perspectives and Critical Debates*. Essex, UK: Frank Cass.

Kearns, A.J. (1991) 'Active Citizenship and Urban Governance', *Transactions of the Institute of British Geographers*, 17, 1, pp. 20–34.

Koontz, T.M., Steelman, T.A., Carmin, J., Korfmacher, K.S., Moseley, C. and Thomas, C.W. (2004) *Collaborative Environmental Management: What Roles for Government?* Washington, DC: Resources for the Future.

Koontz, T.M. and Thomas, C.W. (2006) 'What Do We Know and Need to Know about the Environmental Outcomes of Collaborative Management', *Public Administration Review*, 66, 6, pp. 109–119.

Leach, W.D., Pelkey, N.W. and Sabatier, P.A. (2002) 'Stakeholder Partnerships as Collaborative Policymaking: Evaluation Criteria Applied to Watershed Management in California and Washington', *Journal of Policy Analysis and Management*, 21, 4, pp. 645–670.

Mol, A.P.J. (2000) 'The Environmental Movement in an Era of Ecological Modernisation', *Geoforum*, 31, 1, pp. 45–56.

Mol, A.P.J. and Buttel, F.H. (2002) *The Environmental State Under Pressure*. Greenwich, CT: JAI Press.

Mol, A.P.J. and Spaargaren, G. (1993) 'Environment, Modernity and the Risk Society: The Apocalyptic Horizon of Environmental Reform', *International Sociology*, 8, 4, pp. 431–459.

Mol, A.P.J. and Spaargaren, G. (1998) 'Ecological Modernization Theory in Debate: A Review', presented to the Fourteenth World Congress of Sociology, Montreal, 26 July–1 August.

Mol, A.P.J., Sonnenfeld, D.A. and Spaargaren, G. (eds) (2009) *The Ecological Modernisation Reader*. London: Routledge.

Mol, A.P.J. and Spaargaren, G. (2006) 'Towards a Sociology of Environmental Flows. A New Agenda for Twenty-First-Century Environmental Sociology', pp. 39–84 in G. Spaargaren, A.P.J. Mol and F.H. Buttel (eds), *Governing Environmental Flow: Global Challenges for Social Theory*. Cambridge, MA: MIT Press.

Ostrom, E. (1999) 'Polycentricity, Complexity, and the Commons', *Good Society*, 9, 2, pp. 36–40.

Ostrom, E. (2010) 'Beyond Markets and States: Polycentric Governance of Complex Economic Systems', *American Economic Review*, 100, 3, pp. 641–672.

Pellow, D.N., Schnaiberg. A. and Weinberg, A.S. (2000) 'Putting the Ecological Modernization Thesis to the Test: The Promises and Performances of Urban Recycling', pp. 109–137 in A.P.J. Mol and D.A. Sonnenfeld (eds), *Ecological Modernisation Around the World: Perspectives and Critical Debates*. Essex, UK: Frank Cass.

Pierre, J. (ed.) (2000) *Debating Governance: Authority, Steering, and Democracy*. New York, NY: Oxford University Press.

Pincetl, S. (2003) 'Nonprofits and Park Provision in Los Angeles: An Exploration of the Rise of Governance Approaches to the Provision of Local Services', *Social Science Quarterly*, 84, 4, pp. 979–1001.

Rhodes, R.A.W. (1994) 'The Hollowing out of the State: The Changing Nature of the Public Service in Britain', *Political Quarterly*, 65, 2, pp. 138–151.

Rhodes, R.A.W. (1996) 'The New Governance: Governing without Government', *Political Studies*, 44, 4, pp. 652–667.

Rhodes, R.A.W. (2000) 'Governance and Public Administration', pp. 54–90 in J. Pierre (ed.), *Debating Governance: Authority, Steering and Democracy*. Oxford, UK: Oxford University Press.

Rinkevicius, L. (2000a) 'Ecological Modernisation as Cultural Politics: Transformation of Civic Environmental Activism in Lithuania', pp. 171–202 in A.P.J. Mol and D.A. Sonnenfeld (eds), *Ecological Modernisation Around the World: Perspectives and Critical Debates*. Essex, UK: Frank Cass.

Rinkevicius, L. (2000b) 'The Ideology of Ecological Modernization in "Double-Risk" Societies: A Case Study of Lithuanian Environmental Policy', pp. 163–186 in G. Spaargaren, A.P.J. Mol and F.H. Buttel (eds), *Environment and Global Modernity*. London: Sage.

Roelofs, J. (2009) 'Networks and Democracy: It Ain't Necessarily So', *American Behavioral Scientist*, 52, 7, pp. 990–1005.

Sairinen, R. (2002) 'Environmental Governmentality as a Basis for Regulatory Reform: The Adaptation of New Policy Instruments in Finland', pp. 85–103 in A.P.J. Mol and F.H. Buttel (eds), *The Environmental State Under Pressure*. Greenwich, CT: JAI Press.

Salamon, L.M. (ed.) (2002) *The Tools of Government: A Guide to the New Governance*. New York, NY: Oxford University Press.

Scheinberg, A. (2003) 'The Proof of the Pudding: Urban Recycling in North American as a Process of Ecological Modernisation', *Environmental Politics*, 12, 4, pp. 49–75.

Singleton, S. (2000) 'Co-operation or Capture? The Paradox of Co-Management and Community Participation in Natural Resource Management and Environmental Policy-Making', *Environmental Politics*, 9, 2, pp. 1–21.

Sirianni, C. (2009) 'The Civic Mission of a Federal Agency in the Age of Networked Governance: US Environmental Protection Agency', *American Behavioral Scientist*, 52, 6, pp. 933–952.

Sonnenfeld, D.A. (2000) 'Contradictions of Ecological Modernisation: Pulp and Paper Manufacturing in South-East Asia', *Environmental Politics*, 9, 1, pp. 235–256.

Sonnenfeld, D.A. (2002) 'Social Movements and Ecological Modernization: The Transformation of Pulp and Paper Manufacturing', *Development and Change*, 33, 1, pp. 1–27.

Sonnenfeld, D.A. and Mol, A.P.J. (2002) 'Globalization and the Transformation of Environmental Governance', *American Behavioral Scientist*, 45, 9, pp. 1318–1339.

Sorensen, E. and Torfing, J. (2005) 'Network Governance and Post-Liberal Democracy', *Administrative Theory and Praxis*, 27, 2, pp. 197–237.

Spaargaren, G. and Mol, A.P.J. (2008) 'Greening Global Consumption: Redefining Politics and Authority', *Global Environmental Change*, 18, 3, pp. 350–359.

Taylor, B. (2009) '"Place" as Prepolitical Grounds of Democracy: An Appalachian Case Study in Class Conflict, Forest Politics and Civic Networks', *American Behavioral Scientist*, 52, 6, pp. 826–845.

van Tatenhove, J.P.M. and Leroy, P. (2003) 'Environment and Participation in a Context of Political Modernisation', *Environmental Values*, 12, 2, pp. 155–174.

van Tatenhove, J.P.M and Leroy, P. (2009) 'Environment and Participation in a Context of Political Modernisation', pp. 190–206 in A.P.J. Mol, D.A. Sonnenfeld and G. Spaargaren (eds), *The Ecological Modernisation Reader: Environmental Reform in Theory and Practice*. London: Routledge.

Vira, B. and Jeffery, R. (eds) (2001) *Analytical Issues in Participatory Natural Resource Management*. New York, NY: Palgrave Macmillan.

Weber, E.P. (2003) *Bringing Society Back in: Grassroots Ecosystem Management, Accountability, and Sustainable Communities*. Cambridge, MA: MIT Press.

Wolch, J. (1990) *The Shadow State: Government and the Voluntary Sector in Transition*. New York, NY: Foundation Center.

Wondolleck, J.M. and Yaffee, S.L. (2000) *Making Collaboration Work: Lessons from Innovation in Natural Resource Management*. Washington, DC: Island Press.

16

The new mobilities paradigm and sustainable transport

Finding synergies and creating new methods

Rachel Aldred

Increasingly, transport is perceived as having important environmental consequences, which are proving even harder to resolve than in other sectors of production and consumption. For example, within the European Union, CO_2 emissions from manufacturing are in decline, but emissions from transport stubbornly refuse to fall. On a global level, transportation energy use continues to increase dramatically (Woodcock et al. 2009). This is due to the increasing movement both of people and of goods. While hopes have been raised that this might be offset by the increasing movement of information, so far the evidence for this is limited. Transport is a major and growing contributor to global CO_2 emissions, but is also implicated in a range of other environmental problems, including (but not limited to) air pollution, noise pollution, light pollution, resource depletion, community severance, decreasing biodiversity and soil erosion.

Although the overall picture is of increasing motorization, diversity exists within and between regions, creating different challenges and opportunities for change. Among highly motorized countries, some cities have encouraged commuting by active transport (walking and cycling) while, in others, public transport is relatively dominant (Aldred in press a). However, all such countries are strongly dependent upon fossil fuel-powered transport. Middle-income countries have been motorizing quickly, with traditional ways of travelling swiftly marginalized and injury rates high. In poor countries, only the elite is motorized and much of the population lack access to key intermediate technologies such as bicycles. Foreign aid often centres around building highways, although this is not of benefit to majority populations whose only transport is their feet (Roberts 2007).

There are major social justice issues that continue to be poorly recognized and addressed. Clearly, it is not possible for all the Earth's citizens to consume fossil fuel-powered transport resources at the same rate as Europeans, let alone people in the USA or Australia. The spectre of environmental terrorism directed at air travellers has been raised as a possible desperate response in the future by those who have seen their homes destroyed partly by air travel, the use of which is massively unequal within and between countries. Within nation states, there are major inequalities in terms of the costs and benefits of transport. In many poor countries, it is largely the rich who drive and the poor who suffer the consequences. Even in highly motorized countries such as the UK, the poor – who drive least – have the highest risks of being injured or killed on the roads (Edwards et al. 2006) and who experience the most transport-related air pollution.

In response to this, it is now increasingly recognized that transportation systems urgently require transition. For example, within transport modelling a focus on small and incremental changes has been supplemented by the rise of 'visioning', with an explicit concern for the social. The EPSRC-funded *Visions of Walking and Cycling in 2030* project was an example of this in the UK context. One motivation behind the project was the desire to communicate the need for a qualitative shift in people's lives, and to demonstrate the different possible futures this might entail. While these sustainable futures would involve less transport, they would involve mobility being understood differently, and could involve more of other things that people value. Just as the transition to the motor car involves substantial shifts in social values and social practices, so will the transition away from the motor car. This is what makes it (a) so challenging, and (b) so interesting for social scientists.

This chapter discusses what the 'new mobilities paradigm' (NMP) can offer the growing body of work within the field of sustainable transport. The NMP and sustainable transport approaches can both be seen as critiques of traditional transport studies from different perspectives and for different reasons. They remain rather separate literatures, even though key figures in each have sought to bring them together. John Urry (2008, 2010), for example, has written extensively on transport and climate change from an NMP perspective, and David Banister (2005) provides an intriguing discussion about transport and culture in his book *Unsustainable Transport*.

This chapter highlights areas where the NMP and sustainable transport perspective share common themes, and identifies other areas where NMP contributes additional foci. There are points of disagreement, particularly around issues of normativity, and neglected areas where a combined approach might prove fruitful. These include a mobilities approach to 'freight', a sociology of mobilities movements, and a more in-depth treatment of the question of levels of analysis within mobility regimes or systems. Both NMP and sustainable transport approaches share similar roots but have different perspectives. Combining the two creates opportunities for critical dialogue which can yield new areas of work.

The new mobilities paradigm

History and key concerns

The NMP emerged in the early twenty-first century, largely led by sociologist John Urry (2007).[1] On one level it represents the continuation of an internal critique of key sociological concepts in response to Manuel Castells' (1996) work, which posits the end of 'society' and its replacement by networks. But one key innovation made by Urry and others writing within a mobilities approach focused on the role of transportation in this posited move beyond society. I will suggest that, at least in Urry's formulation of the NMP, there was a reorientation of structuralist sociology: while 'society' may have vanished, social systems remain in Urry's concept of the 'car-system'.

The NMP's critique of 'sedentarism' was not purely about asserting the importance of the physical movement of people. In Urry's (2007) *Mobilities*, he argues for an equal focus on other types of movement, including virtual and imaginary movement, and the movement of things. But research explicitly or implicitly located within the paradigm has tended to concentrate upon the movement of people, including detailed ethnographies of public and private transport journeys (see, for example, Aldred and Jungnickel 2012; Jungnickel 2005; Laurier et al. 2008; Watts 2008). In recent years, there has been a particular flourishing of work on 'alternative' mobilities, although much work continues to study motor transport.

Societies or networks?

Although Urry's *Mobilities* (2007) brought the new paradigm to the fore, it was his *Sociology Beyond Societies* (2000) that laid the basis for it. Here, (Urry 2000: 5) points out that 'if sociology does possess a central concept, it is surely that of society', yet sociologists disagree on what exactly society *is*. He argues that 'the material reconstitution of the social presumes a sociology of diverse mobilities' (Urry 2000: 20). For Urry, sociology has theorized globalization while ignoring what made globalization possible:

> strangely the car is rarely discussed in the 'globalization literature', although its specific character of domination is more systemic and awesome in its consequences than what are normally viewed as constitutive technologies of the global, such as the cinema, television and especially the computer.
>
> *(Urry 2006: 25)*

Drawing on the work of globalization theorists such as Castells (1996) and actor–network theory (ANT), Urry proposes that sociology should replace static with mobile metaphors of social life. For Kaufmann (2010), removing 'society' from sociology opens up questions of space, as spatial attributes of the 'social' are brought into question rather than assumed to be located within (national) 'society'. Places should then be seen as:

> a set of spaces where ranges of relational networks and flows coalesce, interconnect and fragment ... These propinquities and extensive networks come together to enable performances in, and of, particular places.
>
> *(Urry 2000: 140)*

This sense of the 'cultural production of space' connects with work by David Harvey and Henri Lefebvre in seeing places as socially produced (Richardson and Jensen 2003), although they place more emphasis on localized contestation and inequality.

Instead of structures, Urry claims that networks provide a better way of thinking about the organization of social life. These networks are not purely social but include, for example, the transmission of quantitative information by computers (Urry 2007: 34). Urry makes a distinction between scapes and flows, where the former are socio-technical infrastructures that shape and organize flows of people or things. One issue here relates to sociology's long-running dichotomy of structure and agency: does this approach risk black-boxing the agency of actants that flow (see Böhm et al. 2006)? Are they merely propelled along the circuits of scapes?

From transport to mobilities

Urry's (2000) critique of 'society' exists in the context of broader shifts within the social sciences from the 1970s onwards, such as the reaction against 'methodological nationalism'. However, it was his emphasis on mobilities (*Sociology Beyond Societies* is subtitled *Mobilities for the Twenty-First Century*) that defined the subsequent 'mobilities turn' (Urry 2007). For sociology, transport had long been seen as a mundane and marginal area of research, almost as if sociologists agreed with transport economists that time spent travelling was purely a cost to be minimized. Yet, in recent times, transport has become a key area of interest, albeit redefined as 'mobilities'.

The concept of 'mobilities' broadens the idea of 'transport', covering not just the movement of material goods and people but also virtual and imaginary movement (Urry 2007). This has helped to reorient academic fields, bringing policy transfer literature into a new sociological space and incorporating migration studies with work on transport-related inequalities (for example, in the journal *Mobilities*, co-edited by Urry). 'Mobilities' has broadened the scope of social science: physical movement becomes sociologically meaningful, rather than dead space that people traverse on the way to meaningful places. Ole Jensen (2006, 2010) has drawn upon classic sociological theory to argue for the application of interactionist sociology to street experiences between differently mobile citizens in different kinds of infrastructural settings. A very different example of this revitalized sociology of the streets is Peter Merriman's (2007) cultural history of the M1 that rewrites the motorway as a meaningful space.

The move from transport to mobilities also implies a methodological shift. Jon Shaw and Markus Hesse (2010) note that the mobilities literature tends to foreground experiential and non-rationalistic aspects of movement, while transport geography is often more quantitative, positivist and oriented towards existing policy paradigms. While bringing qualitative methods into transport has been extremely fruitful, there is a potential problem: the reproduction of social science's long-running methodology wars, which is less fruitful and potentially inhibits collaboration. Intriguingly, like the car-system itself, the mobilities turn has a tendency to spread out and mutate: 'mobile methods' or 'mobile methodologies' (Büscher et al. 2011; Fincham et al. 2010) are currently fashionable and can encompass both methods that are used 'on the move' (like video ethnography) and methods of studying movement (which might themselves be sedentary, such as the more traditional sit-down interviews, or, potentially, quantitative methods).

Finally, 'mobilities' is linked, as suggested above, to sociological and geographical narratives about social change and related changes in what might be termed 'structures of feeling' (Williams 1961). This 'mobilities' narrative itself is open to question. For example, Pooley et al. (2005) found that mobility has not changed dramatically in qualitative terms in Britain during the twentieth century: people make similar types of journeys, although they travel longer distances to get there. This raises the question of to what extent 'mobilities' are assumed to be inherently meaningful to participants, and to what extent this is a matter for empirical investigation in different contexts. Of course, the sociologically meaningful is not necessarily coterminous with what is meaningful for individuals. Yet sometimes narratives around changing mobilities may imply that everyday life has been transformed to a greater extent than is the case (Pooley et al. 2005; see also Cresswell 2010).

Identities

The NMP constructs mobilities as at least potentially meaningful for the (sub)cultural identities of individuals and groups. This can be seen as a two-way process – transport influences social identities, while existing social identities shape how people use and understand transport (Skinner and Rosen 2007). The former is explored in David Gartman's (2004) characterization of 'three ages of the automobile' in relation to the changing significance of the car as a consumption object. A different slant on the same question can be found in my article on 'cycling citizenship' characterizing relationships to the city that might be enabled through the practice of cycling (Aldred 2010).

The pluralization of 'mobilities' implies diversity rather than commonality. Much mobilities literature discusses ways in which different identities impact mobility experiences. Melissa Butcher's recent paper, for example, includes a discussion of how gender and class distinctions are expressed, transformed and erased through the use of the Delhi Metro (Butcher 2011). Similarly, Green et al.'s (2012) paper on cycling in London analyses how cycling in this context might be viewed differently depending on social identities such as class, gender and ethnicity. This literature is important in terms of making transport more sustainable, because it stresses that strategies and policies appealing to one social group may have contrary effects upon others. It highlights the importance of thinking about culture and identity in relation to travel choices. However, as will be discussed later, the pluralization of mobility experiences in turn comes into contention with the sustainable transport-inflected concept of 'car culture' as singular.

As the literature on mobility and identity is becoming extensive, here I outline some themes through examples focusing on the car to indicate its foci and its contribution. While cars have traditionally been gendered as a masculine possession, still expressed in differential car ownership and use, cars containing children are gendered as women's space (Barker 2009). There is a gendered division of travel space and labour when the family travels by car, while passengers are rarely as passive as the description would suggest (Laurier et al. 2008). Yet, family car travel imposes certain legal and physical limits. Children 'never have independent and autonomous access to cars, are unable to drive, are embedded within specific micro-political power relations within families and are also subject to broader restrictions regarding their age' (Barker 2009: 74; see also Bonham 2006). These processes and the norms they engender may then affect broader perceptions of childhood and the family, indicating connections between how specific identities shape transport use and broader social processes.

Daniel Miller's influential book *Car Cultures* (2001) sees cars as connected to social identities in diverse ways. It includes two articles that examine how cars are used as objects of rebellion against ascribed cultural identities of gender and race (Garvey 2001; Gilroy 2001). Both articles – especially the latter – represent an application of cultural criticism in that they suggest cars are a form of commodified rebellion standing in for true resistance. Gilroy's analysis of race and automobility is somewhat at odds with Miller's focus on diversity of meanings: it implies that there are common understandings of what cars mean, associated with deeply held values. The drivers he discusses are not attaching different meanings to cars, but claiming existing dominant associations (speed and power) for themselves. Similarly, the Norwegian women discussed in the book use dangerous driving to reject dominant ideals of femininity.

Other work has discussed cars as national objects. Rudy Koshar (2004) compares German and British discourses around cars in the inter-war period. One of relatively few analysts to take a nationally comparative approach, he links cultures and stereotypes to the changing fortunes of the German and British car industry and inter-war economic and political change more broadly. More work in this vein, both historical and contemporary, could prove useful in addressing the gap between broader social claims made in theoretical work and often relatively small-scale empirical case studies. This area of research has generated some of the most productive themes stemming from NMP-influenced studies, and debates over how to conceptualize the connections between (sub)cultural mobility identities and broader social processes look set to continue.

Systems

While Urry's 'new mobilities paradigm' abolished society, it did not abolish structures and systems. For all the focus on identity in the mobilities literature, Urry's founding framework can be seen as leaning towards a structuralist approach. Indeed, some mobilities writers have criticized Urry's emphasis on the systemic as producing a 'rather one-sided' view of car use, ignoring the multiple ways in which the car is embraced by individuals (Dant and Martin 2001: 148). Conversely, Urry has stressed the ways in which the system perpetuates itself:

> Automobility can be conceptualized as a self-organizing autopoietic, non-linear system that spreads world-wide, and includes cars, car-drivers, roads, petroleum supplies and many novel objects, technologies and signs. The system generates the preconditions for its own self-expansion.
>
> *(Urry 2006: 27)*

Urry has predicted a shift beyond the car, citing pressure from declining oil supplies to develop electric powered public transport and smart cars. But, by stressing the systemic nature of automobility, one risks downplaying the political dimensions of automobility, as discussed by Böhm et al. (2006), Paterson (2007) and Rajan (2006). While concurring with Urry in characterizing automobility as 'one of the principal socio-technical institutions through which modernity is organized', Böhm et al. (2006: 1) propose an alternative concept – 'regimes of automobility'. In contrast with the idea of a single car-system, they posit interconnected regimes that interact and conflict at multiple levels. Similarly, critiquing the perceived novelty and ubiquity of the 'system', Tim Cresswell (2010: 26) proposes the concept of 'constellations of mobility', which focuses on 'historical senses of movement [being] attentive to movement, represented meaning, and practice, and the ways in which these are interrelated'.

A related question is the link between automobility and other mobilities highlighted by Urry's work but potentially lost in its emphasis on the *car*-system. Within different countries, non-car forms of mobility play distinctive roles within an automobilized system. One might think of Denmark and the Netherlands, where cycling is integrated into city life without longer-distance personal and freight travel being displaced from the system. Yet, the mobilities paradigm does imply the need to pay comparative attention to trends within regions and countries, which may or may not point beyond the car-system. For example, what kinds of 'structures of feeling' might contribute to a transition towards a post-car-system? This is an important question given that in transport, as in many other areas, richer people express more environmentalist attitudes while continuing to behave in more environmentally damaging ways.

The challenge of sustainable transport

History and key concerns

The new mobilities paradigm is a newcomer to the social sciences, only establishing itself in the twenty-first century. By contrast, what Banister (2008) has called the 'Sustainable Mobilities Paradigm' draws upon a more explicit critique of the environmentally problematic implications of contemporary transportation dating back to the 1970s. The oil crisis that started in 1973 fuelled an ecological critique of transportation. This critique was environmental and moral in nature, criticizing the effects that, it argued, mass motorization had on cultures and communities. Influential work from this period was framed sociologically, philosophically and geographically by writers such as Andre Gorz (1979), Ivan Illich (1974), Henri Lefebvre (2008) and Michel de Certeau (2001).

Groups campaigning on issues of transport and the environment were created or revitalized. In the UK, Cyclebag (which later became a large charity called Sustrans) was formed and the Cycle Campaign Network was created. Across Europe, campaigns forced action to restrict the encroachment of the car into city space. The focus was frequently on issues of local liveability. In the Netherlands, for example, the campaign 'Stop the Child Murders' concentrated on child road deaths on residential streets. Change seemed to be in the air. The 1976 UK report *Cycling: A New Deal*, produced by the hardly radical Sports Council with the Cycling Council of Great Britain, spoke of 'growing numbers of people who recognize the bicycle as the great liberator' (Sports Council 1976: 2). While this period ultimately came to represent a false start in the case of the UK (Golbuff and Aldred 2011), organizations and ideas active in the 1970s were to return in the 1990s in a modified form.

1990s: transport policy discourse becomes 'sustainable'

During the 1990s, the transport policy discourse shifted due to the rising prominence of two discourses – that of sustainability and climate change – which compelled the integration of the environment into many previously resistant areas. Also important in transportation was the increasingly prominent debunking of 'predict and provide' policy; the idea that policymakers should respond to congestion by building more roads to meet the demand. 'New realists' like Phil Goodwin (who co-wrote an influential report for the UK's Rees Jeffreys Road Fund) argued that congestion could never be solved by adding motor vehicle capacity (Goodwin et al. 1991). Instead, demand management policies should be followed. In the UK, the thread of this approach can be traced back to the 'Buchanan report' in 1963 (Buchanan 1964), which, contrary to popular understanding, argued that *either* traditional towns had to be totally redesigned around the car *or* car use needed to be restricted.

By the mid-1990s, many policymakers were at least paying lip service to new realist ideas. At the EU level, the voluntary agreement to cut carbon emissions was negotiated with European carmakers amidst rhetoric on the need for the industry to become greener. Following the failure of the agreement to cut emissions substantially, a decade later the EU imposed mandatory, albeit weakened, regulation (Aldred and Tepe 2011). The threat of climate change meant that ideas about greening transport became imbued with rhetorical urgency. And, in a double-edged development, policy rhetoric moved away from the local and the politics of local threats (such as injury or local air pollution) to the global stage (seen as appropriate to climate change).

Yet, increasingly, policymakers saw changing local travel behaviour as less politically challenging than, for example, restricting air travel (barely tackled, despite tax breaks on kerosene and place-based subsidies for low-cost carriers) and long-distance freight (requiring confrontation with business and deeply implicated in the EU vision of a Europe without the friction of distance). Hence, for diverse reasons, a politics of the sustainable city that acts in response to climate change has developed. This is reflected in the UK government's White Paper on Local Sustainable Transport, *Creating Growth, Cutting Carbon*, which argues: 'It is the short-distance local trip where the biggest opportunity exists for people to make sustainable travel choices' (Department for Transport 2011: 5).

Banister's 'Sustainable Mobility Paradigm'

One of the key proponents of sustainable transport is David Banister, based at the Oxford University Centre for the Environment. Banister summarizes his approach in the 'Sustainable Mobilities Paradigm', in which he questions two of the underlying principles of conventional transport planning: (1) that travel is a derived demand; and (2) that travel is a cost minimization (Banister 2008: 73). Banister suggests instead that the activity of travelling may be valued for its own sake and that reliability is more important, for most, than the minimization of travel time. However, he argues that current approaches to transport planning fail to incorporate this as 'many of the methods used cannot handle travel as a valued activity or travel time reliability' (Banister 2008: 74).

Banister adapts a table from Marshall (2001) to contrast the 'conventional' and 'sustainable' approaches (see Table 16.1). The Sustainable Mobilities Paradigm implies a different modelling

Table 16.1 Contrasting approaches to transport planning (reproduced with permission from Banister 2009)

The conventional approach: Transport planning and engineering	An alternative approach: Sustainable mobility
Physical dimensions	Social dimensions
Mobility	Accessibility
Traffic focus, particularly on the car	People focus, either in (or on) a vehicle or on foot
Large in scale	Local in scale
Street as a road	Street as a space
Motorized transport	All modes of transport often in a hierarchy with pedestrian and cyclist at the top and car users at the bottom
Forecasting traffic	Visioning on cities
Modelling approaches	Scenario development and modelling
Economic evaluation	Multicriteria analysis to take account of environmental and social concerns
Travel as a derived demand	Travel as a valued activity as well as a derived demand
Demand based	Management based
Speeding up traffic	Slowing movement down
Travel time minimization	Reasonable travel times and travel time reliability
Segregation of people and traffic	Integration of people and traffic

methodology and a different approach to street design, suggesting that a reconceptualized street space can successfully integrate people and traffic, albeit with car users often at the bottom of the street hierarchy. Importantly, the sustainable approach follows prominent policy paradigms in focusing on the local and city-level mobility, stressing the local against the large in scale.[2] Relocalizing environment discourse undoubtedly has positive effects (Slocum 2004). On the other hand, a localized focus may potentially downgrade the city's role as a hub within broader unsustainable networks; for example, through the frequent use of air transport by some citizens or through long-distance energy-intensive commuting by car or by high-speed rail.

Banister (2008: 74) argues that his approach does not seek:

> to prohibit the use of the car, as this would be both difficult to achieve and it would be seen as being against notions of freedom and choice. The intention is to design cities of such quality and at a suitable scale that people would not need to have a car.

The difficult question is how one gets to a city where the car is not needed, given that a city designed around the car makes the vehicle seem indispensable – because the infrastructure makes walking, cycling and using public transport unpleasant, unsafe and/or slow. Banister suggests that, while public acceptability is the key barrier to change, it must be achieved through public participation throughout, phasing in controversial changes. This deliberative and rational approach is in contrast to the work of ecocentric writers such as Mayer Hillman, who has argued that politicians need to make urgent decisions even if these go against public opinion (Rowlatt 2010).

Current debates around sustainable transport

Banister's development of a sustainable transport approach has led him and others to explore new modelling techniques; for example, in the VIBAT series of projects and in the UK Visions 2030.

These approaches seek to move beyond traditional transport modelling and envision what a society based on different mobility principles would look like. In Visions 2030, the project team chose a number of different generic localities (such as a Victorian street, edge of town estate, etc.) and created static and moving images representing what these places might look like under different transportation scenarios. VIBAT UK (Visioning and Backcasting for Active Travel for the UK) starts with the aim of reducing transport emissions and then examines policy pathways to get there, rather than starting with policies and assessing their environmental impact. These approaches bring utopianism and broader social questions into modelling science and have been flagged by Dennis and Urry (2009) as sociologically interesting.

These debates are also shifting to incorporate public health perspectives. By focusing on the benefits of walking and cycling, this has partly counteracted the tendency of policymakers to promote technological solutions as ostensibly least politically problematic. In the UK, this was inaugurated by the British Medical Association's 1992 publication *Cycling: Towards Health and Safety,* marking a shift from a medical approach focusing solely on cycling as a 'risky' activity. More broadly, the World Health Organization's Healthy Cities Program helped create coalitions focused on promoting the 'healthy city' across the world. Traditional transport modelling has included injuries as a transport-related cost, but not the public health benefits of active modes of transport. In contrast, public health writers have argued that the public health benefits of increased walking and cycling must be taken into account when modelling policy effects (Woodcock et al. 2009).

Just as there are different interpretations of the new mobilities paradigm, there are tensions among sustainable transport proponents. For example, banning heavy goods vehicles from town centres might have a positive effect on local liveability and on walking and cycling rates. However, in terms of transport emissions, this might be outweighed by increased greenhouse gas emissions resulting from lower vehicle efficiency. This raises the issue of how sustainability should be defined and what role concerns of justice, ethics and equality should play (Aldred 2011). Issues that arguably deserve more focus within both fields are discussed here. Where, for example, Banister recommends pricing mechanisms, this may have a disproportionate impact on lower income groups in places such as many cities in the United States where car dependence is high and public transport scarce. When health and well-being are taken into account as well as environmental quality, inequalities may be complex and multifaceted.

Finally, the tendency of 'sustainable transport' to be used as an inaccurate and essentializing shorthand for 'sustainable *modes* of transport' has been questioned. Banister (2008) argued that it is important to take a holistic view. Where trips by sustainable modes increase, there may be rebound effects. Journeys may not replace car trips, as in London's bike hire scheme where many bicycle journeys replace public transport and walking trips, while others may be additional journeys taken for their own sake (in line with the Sustainable Mobilities Paradigm's perspective on journey time). This does not mean such schemes are a bad idea, but that their use should be scrutinized and the different potential benefits, such as increased public acceptability of cycling, examined critically (see also Whitelegg 2009 for a critique of the assumption that high-speed rail is environmentally friendly).

What can mobilities offer sustainable transport?

Common themes

Both Banister's sustainable transport paradigm and the new mobilities paradigm have critiqued traditional transport studies. For sustainable transport writers, the critique is predominantly environmental (and their methods frequently quantitative), whereas the NMP has more in common with the sociological critique of rationality and quantitative reason (and often more environmentally damaging modes are examined with little or no comment on environmental issues). However, there are similarities. Both approaches tend to treat transport as embedded in other social systems; hence Banister's emphasis on the importance of looking holistically and considering alternatives to physical movement and the impact of other policy areas on travel choices. In both, travel is meaningful not only in its own terms (not just seen as a time cost), but also for its interrelation with other areas of life. This implies a systemic analysis that does not take travel demand as a given but asks how it might be changed.

It is worth noting that some writers cannot easily be allocated to one or another field. Peter Freund and George Martin (1993, 2007), for example, are two sociologists who draw on elements from both approaches. Writing before the rise of the mobilities paradigm, they analysed and critiqued the conditions that 'auto space' places upon its users, paying close attention to interactions between identity and inequality (Freund and Martin 1993). Similarly, Winfried Wolf's *Car Mania* (1996) embodies both a call for more sustainable transport systems and a political economy analysis of the car system. Many transport geographers (for example, Shaw and Hesse 2010) are also interested in both questions of sustainability and the challenge posed by the new mobilities paradigm.

Additional foci

One key strength the mobilities literature can add to work on sustainable transport is its focus on identities and issues around normalization and social exclusion. Some of the most interesting chapters in Daniel Miller's *Car Cultures* (2001) discuss how marginalized groups attempt to reclaim automobility for themselves. Similarly, current work has discussed the problems associated with assuming a cycling identity within a low-cycling context, particularly for certain groups (see, for example, Aldred in press b; Steinbach et al. 2011). Perhaps underrepresented – as frequently happens in social science where the focus is often on the marginal – is a critical analysis of decisions made by people whose consumption choices contribute most to climate change and other environmental problems. However, this would be very much within the scope of the new mobilities paradigm.

The focus on the network may have additional benefits for sustainable transport. It can help bring into focus journeys that may disappear if they fall outside the city framework – for example, two-way long-distance journeys along European road networks before food arrives at a city supermarket. The mobilities paradigm potentially provides a critical angle on claims to sustainability made by policymakers and politicians, through, for example, bringing in issues of class, race, exclusion and gentrification in the production of the sustainable city. Sociologists and geographers have long critiqued the use of concepts such as 'the environment', as have writers from within the environmental movement itself (see, for example, Lohmann 2010). This too can enrich the study of sustainable transport.

Finally, the new mobilities paradigm can contribute to the development of methods for studying transport and mobility. Banister's comment that travel time should be seen as meaningful implies the development of new modelling frameworks and the utility of incorporating qualitative approaches focusing precisely on the meaning of transport. Within the new mobilities paradigm, ethnographers have developed the ethnography of 'the passenger', analysing the multiple things that people do while travelling. David Bissell (2010) also discussed the temporary and splintered communities created on public transport. Ethical and citizenship dimensions may be of particular interest, while the rise of bicycle hire and car share schemes may also provide further material for analysing 'the public' as it relates to transportation.

Rachel Aldred

Disagreements

I have outlined differences in emphasis (such as methodology and disciplinary orientation) in this chapter. However, the main disagreement between the two approaches relates to normative positioning. This can be briefly explored in terms of attitudes towards the car, where it has so far been most salient. Miller's (2001) view, endorsed by some but not all mobilities writers, is that cars should be seen not as having essential (negative) properties but rather meanings dependent upon the contexts and cultures within which they are used. Understandably, this anthropological aloofness can seem frustrating for those who start with a normative presumption in favour of sustainability or greening policy (however defined). Miller's take on mobilities directs attention towards cultural specificity, rather than *the* car culture, as more environmentally inclined writers might see it.

However, there are different ways of articulating a cultural approach to the car. The mobilities approach attempts to avoid essentializing the car (or other transport objects), which epistemologically would mean putting the object before the system. Yet this could be compatible with a holistic approach promoted by sustainable transport writers, looking at how transport modes are used and how they fit in with urban transport more broadly. It depends on how the interaction between system and object is formulated. Miller (2001: 17) states:

> If the car is understood to be as much a product of its particular cultural context as a force then it follows that prior to an analysis of that larger cultural environment we cannot presume as to what a car might be.

This statement is (perhaps deliberately) ambiguous: does the car as an object not have any essential properties? Is it purely determined by the particular cultural context? Or does it have properties that exert force on such contexts? For example, how specific are those often dominant ideologies of automobility associated with power, strength, money and danger (Paterson 2007)? From where are they derived?

Perhaps the answer lies in how cars, as particular types of physical objects, are mobilized within similar political and social contexts. In countries with unequal income distribution alongside large-scale private ownership and use of motor vehicles, access to cars is unequal and will remain so, albeit with minor differences. Socially excluded groups (including women, the poor, disabled people, older people, and children) are less likely to have primary car access (and, in the case of children, driving is forbidden). Therefore, it is not surprising that meanings of cars are often tied up with inequality, power and status. Given the 'normal' experience of car crashes, it would be surprising if cars were not associated with danger. Perhaps this is how one can marry culturalism and contingency with a sustainable transport perspective.

Areas for further development

I have already indicated some aspects of a sustainable mobilities research agenda. However, there are other areas where the NMP might be employed to develop research on sustainable transport. These could include the sociology of mobilities movements, which has been somewhat neglected by the mobilities literature, but which could contribute to understanding the diversity of mobility regimes or constellations. This might include, for example, a comparative analysis of the role of movements in co-producing regionally or nationally specific mobility regimes, including studies of how motorist and cyclist organizations have affected the development of policy in different countries. This aspect has thus far been relatively underdeveloped.

A mobilities approach to freight would also represent a fruitful area for further research. Currently, most work within the mobilities paradigm deals with the movement of people rather than things (see Jespersen and Drewes 2005 for an exception). In this area, there could be useful crossovers with work done on the sociology of food. Movements of other 'things' might include the study of how water is transported and used, bringing into focus cultural meanings of water and wastewater, drawing on Elizabeth Shove's (2004) work on cleanliness and convenience. Broader questions about how we conceptualize regimes or constellations of mobility could lead to more in-depth theorization of different levels or moments within such assemblages, and how these change over time. Finally, the development of a sociological analysis of transport modelling could be another area of exploration. Given the shift from policy modelling to visioning, it could contribute to the development of a theoretically informed and self-critical mixed-methods approach to sustainable transportation.

Conclusion

This chapter has briefly outlined some key components of what has become known as the new mobilities paradigm, paying particular attention to the work of John Urry. It has argued that the focus on networks, mobilities, identities and systems has synergies with work on sustainable transport. It has also provided additional foci and suggested areas for further development. The key disagreement between the two approaches is in terms of normativity. However, I have suggested that a normative approach that resists essentializing the individual transport object is possible and could draw together the two approaches. Finally, I would suggest that sustainable transport can remind new mobilities scholars of the importance of environmental questions, helping to guard against a tendency in some (but by no means all) of the mobilities literature to celebrate mobilities uncritically or to downgrade environmental considerations.

Notes

1 In this chapter I focus on the work of Urry because he has been so influential, in particular his book *Mobilities*. I am nevertheless aware that other work by Urry may formulate the approach differently – for example, with reference to chaos theory.
2 There is an interesting tension between the city as local and the city as large-scale. Also, cities are becoming part of reconstituted conglomerations, as with the Øresund/Öresund Region straddling Denmark and Sweden.

References

Aldred, R. (2010) 'On the Outside? Constructing Cycling Citizenship', *Social and Cultural Geography*, 11, 1, pp. 35–52.
Aldred, R. (2011) 'Roads to Freedom', *Red Pepper*, July 2011. Available at www.redpepper.org.uk/roads-to-freedom/. Accessed 14 January 2013.
Aldred, R. (in press a) 'The Commute', in P. Adey, D. Bissell, K. Hannam, P. Merriman and M. Sheller (eds), *The Routledge Handbook of Mobilities*. Routledge: London.
Aldred, R. (in press b) 'Incompetent, or Too Competent? Negotiating Everyday Cycling Identities in a Motor Dominated Society'. Forthcoming in *Mobilities*.
Aldred, R. and Jungnickel, K. (2012) 'Constructing Mobile Places Between "Leisure" and "Transport": A Case Study of Two Group Cycle Rides', *Sociology*. 15 March 2012, doi: 10.1177/0038038511428752, http://soc.sagepub.com/content/early/2012/03/14/0038038511428752.abstract
Aldred, R. and Tepe, D. (2011) 'Framing Scrappage in Germany and the UK: From Climate Discourse to Recession Talk?' *Journal of Transport Geography*, 19, 6, pp. 1563–1569.
Banister, D. (2005) *Unsustainable Transport: City Transport in the New Century*. London: Routledge.

Banister, D. (2008) 'The Sustainable Mobility Paradigm', *Transport Policy*, 15, 2, pp. 73–80.
Banister, D. (2009) 'Recycling Cities: Urban Planning and Spatial Form', keynote paper for presentation at the *Velo-City 2009 Conference*, Brussels, 12 May. Available at www.velo-city2009.com/assets/files/paper-Banister-Recycling-Cities-plenary1.pdf. Accessed 21 January 2013.
Barker, J. (2009) 'Driven to Distraction? Children's Experiences of Car Travel', *Mobilities*, 4, 1, pp. 59–76.
Bissell, D. (2010) 'Passenger Mobilities: Affective Atmospheres and the Sociality of Public Transport', *Environment and Planning D: Society and Space*, 28, pp. 270–289.
Böhm, S., Jones, C., Land, C. and Paterson, M. (2006) 'Impossibilities of Automobility', pp. 3–16 in S. Böhm, C. Jones, C. Land, and M. Paterson (eds), *Against Automobility*. Oxford, UK: Blackwell.
Bonham J. (2006) 'Transport: Disciplining the Body that Travels', pp. 57–74 in S. Böhm, C. Jones, C. Land, and M. Paterson (eds), *Against Automobility*. Oxford, UK: Blackwell.
British Medical Association (1992) *Cycling: Towards Health and Safety*. Oxford, UK: Oxford University Press.
Buchanan, C. (1964) *Traffic in Towns*, abridged ed. London: Penguin.
Büscher, M., Urry, J. and Witchger, K. (eds) (2011) *Mobile Methods*. Abingdon, UK: Routledge.
Butcher, M. (2011) 'Cultures of Commuting: The Mobile Negotiation of Space and Subjectivity on Delhi's Metro', *Mobilities*, 6, 2, pp. 237–254.
Castells, M. (1996) *The Network Society*. Oxford, UK: Blackwell.
Cresswell, T. (2010) 'Towards A Politics of Mobility', *Environment and Planning D: Society and Space*, 28, pp. 17–31.
Cresswell, T. (2011) 'Mobilities 1: Catching Up', *Progress in Human Geography*, 35, 4, pp. 550–558.
Dant, T. and Martin, P. (2001) 'By Car: Carrying Modern Society', pp. 143–158 in J. Gronow and A. Warde (eds), *Ordinary Consumption*. London: Routledge.
de Certeau, M. (2001) *The Practice of Everyday Life*. Berkeley, CA: University of California Press.
Dennis, K. and Urry, J. (2009) *After the Car*. Cambridge, UK: Polity Press.
Department for Transport (2011) *Creating Growth, Cutting Carbon: Making Sustainable Local Transport Happen*, London: Department for Transport.
Edwards. P., Roberts, I., Green, J. and Lutchmun S. (2006) 'Deaths from Injury in Children and Employment Status in Family: Analysis of Trends in Class Specific Death Rates', *British Medical Journal*, 333, 7559, pp. 119–122.
Fincham, B., McGuinness, M. and Murray, L. (2010) *Mobile Methodologies*. Basingstoke, UK: Palgrave Macmillan.
Freund, P. and Martin, G. (1993) *The Ecology of the Automobile*. Montreal: Black Rose Books.
Freund, P. and Martin, G. (2007) 'Hyperautomobility, the Social Organization of Space, and Health', *Mobilities*, 2, 1, pp. 37–49.
Gartman, D. (2004) 'Three Ages of the Automobile: The Cultural Logics of the Car', *Theory, Culture and Society*, 21, 4–5, pp. 169–195.
Garvey, P. (2001) 'Drinking, Driving and Daring in Norway', pp. 133–152 in D. Miller (ed.), *Car Cultures*. Oxford, UK: Berg.
Gilroy, P. (2001) 'Driving While Black', pp. 81–104 in D. Miller (ed.), *Car Cultures*. Oxford, UK: Berg.
Golbuff, L. and Aldred, R. (2011) *Cycling Policy in the UK: A Historical and Thematic Overview*. London: University of East London.
Goodwin, P., Hallett, S., Kenny, F. and Stokes, G. (1991) *Transport: The New Realism*. Oxford, UK: Transport Studies Unit, University of Oxford.
Gorz, A. (1979) *Ecology as Politics*. London: Pluto Press.
Green, J., Steinbach, R. and Datta, J. (2012) 'The Travelling Citizen: Emergent Discourses of Moral Mobility in a Study of Cycling in London', *Sociology*, 46, 2, pp. 272–289.
Illich, I. (1974) *Ecology and Equity*. London: Calder and Boyers.
Jensen, O. (2006) '"Facework", Flow and the City: Simmel, Goffman and Mobility in the Contemporary City', *Mobilities*, 1, 2, pp. 143–165.
Jensen, O. (2010) 'Negotiation in Motion: Unpacking a Geography of Mobility', *Space and Culture*, 13, 4, pp. 389–402.
Jespersen, P. and Drewes, L. (2005) 'Involving Freight Transport Actors in Production of Knowledge: Experience with Future Workshop Methodology', pp. 89–104 in T. Thomsen, L. Nielsen and H. Gudmundsson (eds), *Social Perspectives on Mobility*. Aldershot, UK: Ashgate.
Jungnickel, K. (2005) *73 Urban Journeys: Different Ways of Looking at the No. 73 London Routemaster Bus*. London, Studio INCITE, Goldsmiths, University of London, UK. Available at www.73urbanjourneys.com. Accessed 30 March 2012.

Kaufmann, V. (2010) 'Mobile Social Science: Creating a Dialogue among the Sociologies', *British Journal of Sociology*, 61, s1, pp. 367–372.

Koshar, R. (2004) 'Cars and Nations: Anglo-German Perspectives on Automobility Between the World Wars', *Theory, Culture, and Society*, 21, 4–5, pp. 121–144.

Laurier, E., Lorimer, H., Brown, B., Jones, O., Juhlin, O., Noble, A., Perry, M., Pica, D., Sormani, P., Strebel, I., Swan, L., Taylor, A., Watts, L. and Weilenmann, A. (2008) 'Driving and "Passengering": Notes on the Ordinary Organization of Car Travel', *Mobilities*, 3, 1, pp. 1–23.

Lefebvre, H. (2008) *Critique of Everyday Life*, vol. 2, *Foundations for a Sociology of the Everyday*. London: Verso.

Lohmann, L. (2010) 'Neoliberalism and the Calculable World: The Rise of Carbon Trading', pp. 77–93 in K. Birch and V. Myhknenko (eds), *The Rise and Fall of Neoliberalism: The Collapse of an Economic Order?* London: Zed.

Marshall, S (2001) 'The Challenge of Sustainable Transport", pp. 131–147 in A. Layard, S. Davoud and S. Batty (eds), *Planning for a Sustainable Future*. Spon Press: London.

Merriman, P. (2007) *Driving Spaces: A Cultural–Historical Geography of England's M1 Motorway*. Oxford, UK: Blackwell.

Miller, D. (2001) *Car Cultures*. Oxford, UK: Berg.

Paterson, M. (2007) *Automobile Politics: Ecology and Cultural Political Economy*. Cambridge, UK: Cambridge University Press.

Pooley, C., Turnbull, J., and Adams, M. (2005) *A Mobile Century? Changes in Everyday Mobility in Britain in the Twentieth Century*. Aldershot, UK: Ashgate.

Rajan, S. (2006) *The Enigma of Automobility: Democratic Politics and Pollution Control*. Pittsburgh, PN: University of Pittsburgh Press.

Richardson, T. and Jensen, A. (2003) 'Linking Discourse and Space: Towards a Cultural Sociology of Space in Analysing Spatial Policy Discourses', *Urban Studies*, 40, 1, pp. 7–22.

Roberts, I. (2007) 'Formula One and Global Road Safety', *Journal of the Royal Society of Medicine*, 100, 8, pp. 360–362.

Rowlatt, J. (2010) *Doomed by Democracy?* Transcript, BBC Radio 4. Available at http://news.bbc.co.uk/nol/shared/spl/hi/programmes/analysis/transcripts/24_05_10.txt. Accessed 19 March 2012.

Shaw, J. and Hesse, M. (2010) 'Transport, Geography and the "New" Mobilities', *Transactions of the Institute of British Geographers*, 35, 3, pp. 305–312.

Shove, E. (2004) *Comfort, Cleanliness and Convenience: The Social Organization of Normality*. Oxford, UK: Berg.

Skinner, D. and Rosen, P. (2007) 'Hell is Other Cyclists: Rethinking Transport and Identity', pp. 83–96 in D. Horton, P. Rosen and P. Cox (eds), *Cycling and Society*. Aldershot, UK: Ashgate.

Slocum, R. (2004) 'Polar Bears and Energy-Efficient Lightbulbs: Strategies to Bring Climate Change Home', *Environment and Planning D: Society and Space*, 22, pp. 413–438.

Sports Council with Cycling Council of Great Britain (1976) *Cycling: A New Deal*. London: Sport Council/Cycling Council of Great Britain.

Steinbach, R., Green, J., Datta, J. and Edwards, P. (2011) 'Cycling and the City: A Case Study of How Gendered, Ethnic and Class Identities Can Shape Healthy Transport Choices', *Social Science and Medicine*, 72, 7, pp. 1123–1130.

Urry, J. (2000) *Sociology Beyond Societies: Mobilities for the Twenty-First Century*. London: Routledge.

Urry, J. (2006) 'The System of Automobility', *Theory, Culture and Society*, 21, 4–5, pp. 25–39.

Urry, J. (2007) *Mobilities*. Cambridge, UK: Polity.

Urry, J. (2008) 'Climate Change, Travel and Complex Futures', *British Journal of Sociology*, 59, 2, pp. 261–279.

Urry, J. (2010) 'Consuming the Planet to Excess', *Theory, Culture and Society*, 27, 2–3, pp. 191–212.

Watts, L. (2008) 'The Art and Craft of Train Travel', *Social and Cultural Geography*, 9, 6, pp. 711–726.

Whitelegg, J. (2009) 'On the Wrong Track: Why High-Speed Trains Are Not Such a Green Alternative', *Guardian*, 29 April.

Williams, R. (1961) *The Long Revolution*. New York, NY: Columbia University Press.

Wolf, W. (1996) *Car Mania*. London: Pluto.

Woodcock, J., Edwards, P., Tonne, C., Armstrong, B., Ashiru, O., Banister, D., Beevers, S., Chalabi, Z., Chowdhury, Z., Cohen, A., Franco, O., Haines, A., Hickman, R., Lindsay, G., Mittal, I., Mohan, D., Tiwari, G., Woodward, A. and Roberts, I. (2009) 'Public Health Benefits of Strategies to Reduce Greenhouse-Gas Emissions: Urban Land Transport', *Lancet*, 374, 9705, pp. 1930–1943.

Part IV
Risk, uncertainty and social learning

17
Towards a socio-ecological foundation for environmental risk research

Ortwin Renn

At first glance, risk research appears to be a topic primarily for the engineering and natural sciences, which can be enriched, at best, by social science studies. However, the study of the physical consequences of human impact cannot only be related to the effects of human interventions on nature and society, but must start with the investigation of the social and cultural causes that have triggered these interventions in the first place. Humans make decisions based on mental models of what they would like to accomplish. These models are framed by social and cultural aspirations, values and norms (Adams 2003; Bell 2012; Bostrom et al. 1992). Whoever and whatever is at risk, or is affected by human interventions, has been subject to some kind of preceding decisions by another individual or a group, as long as the focus is on human-induced risks rather than natural hazards. But even with natural hazards, human decision makers are responsible for their own and others' exposure to these hazards or for failing to assure adequate protection (Liu et al. 2007). Hence, the reasons behind human interventions that lead to risks and the consequences of these interventions are both relevant topics of interdisciplinary risk research. Based on this broader risk perspective, three areas of overlap can be identified that link the biophysical with the cultural world (Mathies and Homburg 2001; Renn et al. 2007):

- basic human needs and demands with respect to resources from the natural environment (search for resources, energy, land, habitats);
- the consequences of human interventions for natural cycles, processes and structures, including biological changes in and among humans (such as health risks, pandemics); and
- the feedback of these interventions on cultural self-image, social structures and social processes (cultural identity).

Undoubtedly, all three dimensions are closely and inevitably associated with insights from the natural sciences. Without sufficient knowledge of the structure and dynamics of natural systems, the estimation of anthropogenic causes and consequences of risk-taking remains speculative. But focusing on the reactions of natural systems to human intervention is not sufficient to adequately understand the relationship between the physical and the social world (Becker et al 1999; Rosa and Dietz 2010).

Interactions arise from the energetic, material or communicative exchanges between humans and the natural environment. Applied to risk, these are the actions of people leading to environmental responses and the many feedback loops in between. The need to use natural resources and the consequences of significant adaptations to differing environmental conditions furnishes the preconditions needed to explain and understand changes in the environment. In addition, people react to the changes they have triggered with a more or less customized set of behaviours, which in turn require multiple forms of adaptation to complex responses from the natural and human environments (Weber 2006).

These ongoing processes of perceived environmental change, responsive action and renewed awareness of the respective environmental implications are highly dependent on cultural factors. Human perception is based on two elements, biologically controlled signal processing and the attribution of psychological and cultural meaning to perceived signals. What do people perceive, for example, as significant damage to the environment? Where do they address their attention? To which part of their environment do they show fear or indifference? The point at which people start to act or react can rarely be derived from the objective realities of the natural environment. It becomes apparent only from the respective psychological and cultural perspective of the viewer (Hannigan 1995; Mosler 1995). The selectivity with which people perceive, classify and evaluate the vast variety of natural phenomena may partly be controlled biologically (like the 'childlike patterns' in cute animals) but are usually culturally rooted preferences or aversions. These, above all, provide a selective perception and valuation of nature (Jaeger et al. 2001).

Traditional risk research is largely characterized by a 'dual mode model', in which concepts from the technical-natural sciences are applied as normative guidance for how society should deal with threats, and in which complementary studies on risk perception and risk communication are needed to ensure that people's behaviour then matches the objective risk knowledge that science provides (Fischhoff 1995; Horlick-Jones and Sime 2004). The scientific model is based on the ideal of a value-free exploration of the physical responses to human intervention. This is supposed to offer the best safeguard that people will gain the insight necessary to change their behaviour accordingly. This assumption is problematic for two reasons (Meyer 2001). First, the choice of research topics is already culturally influenced. The postulated assumptions about correlations between behaviours, responses to the physical environment and perception are determined by preconceived cultural frames which require their own explanation and questioning. Second, the supposed chain of knowledge and management falls short. Knowledge alone does not change behaviour. Although knowledge is a prerequisite for rational action, it does not replace the necessity of having normative priorities for weighing the consequences for the biophysical environment against other positive or negative impacts on the economy and society. At the same time, the balancing of these processes itself is dependent on social preferences. For example, political scientist Aaron Wildavsky (1984) concluded that US environmental legislation and regulatory practices systematically hide the true extent of environmental impact when damage is distributed widely and evenly, while pollution that affects socially prominent groups is targeted by the media and triggers more regulatory action. Similarly, the concentration of environmental health research on cancer may be based less on objective risk potential than on the societal preoccupation with cancer and its associated imagery (uncontrolled growths, lingering death, etc.) (Brown and Goble 1990).

In conclusion, it must be noted that risk research has to combine both components: the consequences of human behaviour for the natural and socio-cultural environments as well as the reflexive perception, assessment and evaluation of human behaviour when experiencing changes in the natural environment (Rip 2006). This therefore requires risk research that involves these two components systematically, giving priority to the interaction between physical changes and their mental representations and making it the major subject of research (Becker 2003). Here, the socio-ecological approach seems to be particularly suitable.

Risk research at the intersection of nature and culture

To move the socio-ecological approach closer to an integrative concept of risk research, some basic considerations are required. First, an analytic separation between 'nature' and 'environment' should be assumed (Mohr 1995). The environment is understood here as nature which has been designed and transformed by humans for human purposes (Kasperson et al. 1995). In addition to the natural environment, there is also a social and cultural environment which, in turn, interacts with the natural environment and gives rise to mental models that people associate with this relationship. The 'natural' refers to phenomena that exist without the intervention of humans and their effect and force (Böhme 2002). In contemporary times, the intense interaction between nature and culture has led to the development of a hybrid human-designed nature (Becker 2003: 183–186). Human cultures use parts of the natural phenomena (such as raw materials or renewable resources) to create designed environments. Humans have agency for changing the environment in order to meet their needs and desires and to create symbolic expressions of their cultural identity (e.g. French gardens as opposed to English landscape gardens). In addition, humanity transcends its concrete environment in space and time through speech, writing, science and art. Every environment, including the natural environment, is always a product of perception and, therefore, socially and culturally mediated (Scholz 2011).

Only anthropogenically altered environments could provide the conditions for populations to grow beyond the level of a hunting and gathering society (less than one person per square kilometre) while, simultaneously, giving rise to individually anchored ethics (providing the principal possibility of equal living and development opportunities for each individual) (Kasperson et al. 1995; Mohr 1995). The creation of 'artificial' human-modified environments does not mean the fall of humanity but is, rather, an anthropological necessity for a species that is capable of rational and ethical behaviour.

The opportunity to anticipate and shape natural and cultural environments presupposes the existence of concepts and ideas about causes and effects. Both the generation of consequential knowledge and the formation of intersubjective criteria for evaluating situations and options form the main evolutionary advantage of human beings. The functionalization of nature into a productive environment has its price, however. On one side, competitors for the same resources (especially land and crops) are systematically suppressed and threatened with extinction. On the other hand, through the 'artificial metabolism' of production and consumption, resources are consumed and wastes released into the environment which, in turn, affect natural processes (mostly negatively). No one doubts that human activity inevitably demands that price, unless one would desire to revive the original conditions of hunter-gatherer societies. The question for good environmental policy, therefore, is: to what degree may human beings suppress their competitors, and put nature under stress, to ensure the long-term survival of humankind under humane conditions (this is the idea of sustainability) and, at the same time, conserve the vitality of the natural environment as much as possible (Ott and Döring 2004; WBGU 2011)?

The decision regarding at what level of intervention the potential negative impacts outweigh the positive expectations is the key question of risk management. What risks to individuals and society are acceptable in comparison with the benefits expected from taking that risk? Such a balancing process requires two things: knowledge of the consequences of the interventions (scientific and technical risk research) and knowledge of the desirability and ethical justifiability of standards in order to find the 'right' measure of balance (Academy of Sciences in Berlin 1998; Renn 2008). The social and cultural sciences cannot determine this 'right' level and it cannot be derived from systematic knowledge. However, the social sciences can offer *catalytic assistance* in structuring the process of finding the adequate balance between intervention and conservation (Jaeger et al. 2001). The sociologist Ulrich Beck has expressed this as follows:

> But no expert can ever answer the question: How do we want to live? What are people willing to accept and what not, this does not follow from any technical or environmental hazard diagnosis. This question must rather be turned into a global conversation of cultures on the subject. Right here a second, cultural studies perspective takes aim. It says: the scale and urgency of ecological crisis fluctuates with intra- and inter-cultural perception and evaluation.
>
> *(Beck 1996: 119, translated by Jason Eilers)*

In conclusion, integrative risk research should follow five key objectives (see Becker et al. 1999; Dunlap et al. 1994; Renn et al. 2007):

- to gain systematic understanding of the processes of knowledge generation and value formation regarding human interventions into nature and society, and, from these findings, contribute to a better understanding of the *human–environment–nature relationship* and the cultural patterns for risk selection;
- to gain better knowledge of processes and procedures, that shape or enlighten the *social discourse(s)* about the right balance between intervention and conservation and, consequently, about ethically justifiable degrees of interventions into the natural environment on the basis of comprehensible and politically legitimate criteria;
- to investigate institutional processes and organizational structures that review, revise and regulate *individual and collective risk management decisions*;
- to identify patterns of technology development and change with the goal of identifying and understanding *potential impacts of these technologies* on humans and the environment; and
- to investigate *obstacles and barriers*, but also opportunities and incentives, which affect risk-related behaviour at both individual and collective levels, and to systematically explore and develop constructive suggestions for *resolving risk induced conflicts*.

Constructivist and realist approaches to risk research

Which approach would be in a position to be integrated into a consistent framework regarding these five aspects of risk research? From my point of view, the social-ecological perspective is particularly well suited (overview in Becker et al. 1999; Renn et al. 2007). Here the social-ecological perspective should be understood as the approach that ecological and social processes are closely interrelated and can only be understood within their mutual relationship, and as forming a unified phenomenon (Jahn and Sons 2001). A definition is given by Becker (2003: 171, translated by Jason Eilers):

> Social Ecology is the science of the relation of men to their natural and social environment. In socio-ecological research, the shapes and design possibilities of these relationships are examined in a cross-disciplinary perspective. The aim of the research is to generate knowledge for social action plans to ensure the future viability of society and its natural resources.

What is perceived as a consequence of risk in terms of future viability for nature, humans and the environment does not derive from nature itself, but from selection and reception processes that are culturally shaped. At the same time, interventions have consequences for nature and society regardless of the intent of the action and they will take place independently of human perception.

On this subject, my approach to socio-ecological research differs from the dominant social science perspective of postmodern constructivism, which assumes that human knowledge is based fundamentally on cultural attribution (overviews in Bell 2012; Berger and Luckmann 2004; Bradbury 1989; Rosa 1998; Scholz 2011). The constructivist school posits that human claims about reality are social constructions about the state of the world made on the basis of internalized worldviews and cultural traditions. These can be made internally consistent and authentic, yet they do not claim to adhere to isomorphism (i.e. a realistic representation of reality), nor even to homomorphism (i.e. a convergence of knowledge towards reality) (Buttel and Taylor 1992; Hillgartner 1992; Renn et al. 2007; Seiderberg 1984). These mental images of realty are not arbitrary but, rather than arising from the externally given nature of the observed objects, the shapes of these images are formed through social interaction and experiences within the framework of language and culture. Constructivists relate these considerations not only to causal models of human behaviour, but also to cause–effect chains within the natural environment (since the acknowledgement of these external processes depends on observation by humans). As an example of a decidedly constructivist view of the problem, we refer here to sociologist Niklas Luhmann (1986: 63, translated by Jason Eilers), who writes:

> Humans or fish may die, swimming in lakes or rivers may cause disease, there may be no more oil coming from the pump and the average temperature may go down as well: as long as it is not communicated, this has no impact on society. Society is indeed environmentally sensitive, but an operationally closed system. It observes only through communication. It cannot help but to communicate meaningfully and to regulate these communications by communication itself. Only it can, therefore, endanger itself.

In terms of risk, Sheila Jasanoff (1999: 150) has characterized the constructivist approach as follows:

> I have suggested that the social sciences have deeply altered our understanding of what 'risk' means – from something real and physical if hard to measure, and accessible only to experts, to something constructed out of history and experience by experts and laypeople alike. Risk in this sense is culturally embedded in texture and meaning that vary from one social grouping to another. Trying to assess risk is therefore necessarily a social and political exercise, even when the methods employed are the seemingly technical routines of quantitative risk assessments ... Environmental regulation calls for a more open-ended process, with multiple access points for dissenting views and unorthodox perspectives.

In contrast to the constructivist camp, *realists* believe that by applying methodological rules of empirical investigations, one is able to produce at least approximations between the actual inherent qualities of objects and their descriptions by knowledge holders (Dunlap and Catton 1994a; Freudenburg and Gramling 1989; Rosa 1998, 2008). This approximation arises primarily as a result of an ongoing review process of hypotheses testing (chain of prediction–intervention–observation), through an organized scientific process. Realists see people's behaviour as being influenced by objectively given changes in the natural environment and believe they can distinguish between physical changes or stresses on the one hand and social constructions of these changes on the other. Again, a quote can be used to illustrate the argument:

> If our discipline is going to make substantial contributions to understanding the social causes and consequences of global environmental change, we must adopt a truly ecological perspective that sensitizes us to the role that our species plays in the global ecosystem … We must develop a full-blown 'ecological sociology' that studies the complex interdependencies between human societies and the ecosystems (from local to global) in which we live.
>
> (Dunlap and Catton 1994b: 23)

The argument between constructivists and realists still spurs debate about the theoretical location of risk research and the socio-ecological perspective (Bell 2012; Horlick-Jones and Sime 2004). Is ecological risk a construct of the social processing of communicable content (and thus a product of self-harm)? Or is ecological risk an outside threat to humanity, provable by scientific knowledge, caused or at least affected by human behaviour and which can, therefore, be changed by interventions into nature and the environment?

Neither extreme of naive realism or radical constructivism can be rigorously and logically justified. Both radical relativistic solipsism (all knowledge claims have the same validity) and naive realism lead to denial of the dependence of human knowledge on cultural mediation and interpretation (Giddens 1978; Renn 2008; Rosa 1998; Searle 1995). But even if one is inclined to a moderate hybrid as the only viable alternative justification, consideration of these extremes is useful in pre-structuring the visions as well as the methods of socio-ecological research.

On this issue, I tend to a position that is described in the literature as moderate realism. It is assumed that natural control systems exist independently of our cognition and affect (Renn 2008: 3; Renn et al. 2007: 132). In addition, the appropriate knowledge tools are available to help test our notions of reality by intervening in the natural system and measuring its responses, at least within certain limits (the so-called contextual constraints such as time, location and intervening variables). We can expect that images that we create to understand our interventions into the natural world get increasingly more accurate, the more we learn to interpret the signals that the natural world sends out to our senses (or machines that substitute for them). One of the priorities of knowledge-creating institutions is to set up selection rules which allow a *rapprochement* between knowledge and real-world phenomena.

This process of aligning observations with theoretical insights occurs against the background that our knowledge is never a true representation of the real world outside of us. Our observations are always selective and culturally mediated through our senses and the use of language. Physical objects, as real as they may be appear, are reaching people's minds only through knowledge selection and sensory perception. The objective world is not directly accessible to us. We receive physical signals (such as light reflections), not meanings. So we do not see tables or cars but only shapes and colours. We then attribute meaning to these shapes and colours. We are able to communicate about the meaning of these shapes and, through interventions in the environment, create new signals (such as in experiments), which in turn we interpret by means of what our senses and sensing machines can grasp. Both the individual perception of environmental signals, as well as the communication of these signals, assume that we are capable of capturing basic relationships between us and the natural world and that processes of attribution and classification are language tools that provide us with a homomorphic image of the world around us.

As long as the destruction of forests or the death of fish is not communicated, as the quote above by Niklas Luhmann attests, that fact does not exist for society. That, however, does not mean that fish do not die when they are unobserved or that the climate does not change when people deny this. If one claims that reality only exists when we observe it, the line between knowledge and metaphysics blurs.

Constructivists underestimate the ability of knowledge-generating systems for critical self-observation and cross-system communication based on systematic observation of external signals and the analysis of feedback signals as a result of targeted interventions into the objective world (Redclift 1999a). The ability of humans to approximate reality by creating knowledge is based on the principle of anticipating responses of interventions and measuring the signals that are reflected from the targeted natural system.

Without doubt, many knowledge producers are trapped in their own thinking and may be immunized against the claims of other social systems, as some sociologists claim (e.g. Beck 1986; Luhmann 1986). Knowledge elites can get bogged down or lose sight of observable realities. This does not change the principal insight that social institutions of knowledge generation have the ability to align knowledge and reality with one another, regardless of whether they actually do so in every individual case.

The overarching societal concern of risk research is to predict potential damage as best as possible and then to prevent or mitigate it. Legions of toxicologists, epidemiologists, safety analysts and many others are charged with the task of distinguishing real from imaginary or non-perceived threats to human health and the environment. Apart from the fact that no scientist likes to be confronted with the claim that his or her findings are nothing but social constructions, it also does an injustice to scientists if one disqualifies their findings as products of social desirability or group-specific interests (Renn 2001).

Although pluralizing expertise is very popular within the social science literature on risk, it is counterproductive for effective risk governance. In fact, since consequential knowledge of risk decisions is associated with uncertainty and thus covers a wide range of legitimate claims to truth, it is necessary to separate methodologically reliable knowledge from mere conjecture and speculation. If the boundaries between scientific knowledge and mere conjecture or anecdotal knowledge blur, any fear of risk, no matter how absurd, has a quasi-scientifically based justification. The determination of the scope of methodically secure knowledge should be done by science itself, since only there are the methodological rules and verification procedures in place to make sense of and resolve competing truth claims (Redclift 1999a).

Moderate realism does not advocate an uncritical acceptance of scientific findings, but it does avoid the tendency to relativize natural science research. Validity claims for causal knowledge cannot be accomplished by epistemic self-referential systems of justification but, rather, can be made by recognizing the legitimacy of continuous flows of interventions and measurements within the framework of recognized rules of theoretical deduction and empirical induction. Only under this condition can a truly interdisciplinary discourse about complex issues be successfully conducted (see also Mittelstraß 2003; Rosa 1998; Weingarten 2003).

Characteristics of a socio-ecological approach to risk research

A moderate realist approach based on all these considerations is a first essential characteristic of social-ecological risk research. As the term 'social-ecological' suggests, it deals with the integration of social processes of perception and action with the effects and repercussions of human action on the natural environment (Becker et al. 1999; Meyer 2001; Renn et al. 2007). It is assumed that these effects and reactions are real and act independently of knowledge, observation and perception. Taking a moderate realistic approach provides us with the confidence that our knowledge of these relationships over time is approximating the changes that are occurring in the environment, without losing sight of the fact that our knowledge is limited by our physical facilities (including the related tools) and our dependence on language.

The main thesis of the social-ecological approach is that knowledge can only advance in the area of risk and environmental research by reflecting on the interaction between the social processes of knowledge generation and signal recognition and their interpretation based on language, research tools and cultural contexts. This task of establishing relationships between human observers and non-human environments by means of interventions thus stands at the beginning of the socio-ecological research process. Social-ecological risk research is self-reflexive and self-critical of its knowledge claims. This allows one to look beyond the confines of relativism and allows selected interactions between nature, environment and culture to be intersubjectively validated (similar to Becker et al. 1999; Dunlap and Catton 1994b; Gallopin et al. 2001).

The second essential characteristic of the social-ecological approach is the combination of knowledge and values (IRGC 2005). In addition to cultural selection and interpretation regarding risk selection and perception, knowledge needs to be supplemented by a culture of values and moral principles in order to derive meaningful and responsible risk decisions. Decision making includes two components: assessment of likely consequences of each decision option (the knowledge component), and evaluation of these consequences according to personal or collective desirability and ethical acceptability (the value component).

Classic risk research has been adopting a value-free approach. Research questions are normally phrased as 'What if ...?'. By not abiding to one scenario, many researchers felt that they were able to represent the plurality of societal values. This approach, however, blends out the fact that each of these 'if–then' questions represents but one of many possible scenarios and that a selection must be made. Furthermore, even the most scientific risk assessment will be limited to potential consequences that researchers think will or should matter to society (Vecchione 2011).

The new perspective of social ecology can no longer rely on neutrality. Valuations not only creep into the factual process of consideration of what could be researched, but are already incorporated into the design of assessment processes (Stirling 2006: 237). The catalytic function of science, part of the five requirements mentioned in the second section above, is based on normative foundations, such as that all those who could be affected should also be considered in the analysis or that the factual findings of appraisal processes should be made available to all parties (Redclift 1999b). That these normative standards must be made explicit is understood here as self-evident.

The third component of social-ecological research includes the methodological approach often referred to in the literature as transdisciplinary (Becker 1999; Mittelstraß 2003; Pohl and Hirsch Hadorn 2006; Scheringer et al. 2005; Scholz 2011). This term includes, despite all the differences in respective positions, the following key aspects:

a. methodological approach based on using multiple methods employed by different disciplines or methods that bridge different disciplines;
b. a problem-oriented, not phenomenon-oriented, approach;
c. a close connection between theoretical analysis and practical implications; and
d. direct involvement of individuals and groups affected by the respective problem in the research process (recursive participation).

These four aspects are addressed further, below:

a. The socio-ecological approach places particular emphasis on bridging research methods, capable of capturing the relationships between social and physical factors (Becker et al. 1999).

For example, one can take the results of medical tests on blood samples and relate them to eating habits as a means to investigate nutritional behaviour (Zwick 2011). Transdisciplinary methodology is not dependent on the development of innovative interdisciplinary research instruments. In many cases, an intelligent mix of related (proven) disciplinary methods is quite sufficient.

b. The problem-oriented approach bears on the insight of the constructivist–realist debate that humans cannot study phenomena as they exist in reality but can only construct relationships between humans and objects by receiving and interpreting signals from the external world. This process is interest-driven in its broadest sense (Beck 2009). Curiosity could also serve as an interest. Within the social-ecological tradition, the interest is given by the acknowledgement of problems and the quest for problem resolutions. Problems are thus defined socially and culturally. They include the perception of a condition felt to be unsatisfactory and they point to the perceived need for developing intervention options that are supposed to help improve this condition. Rarely, aside from curiosity, do people have interest in a single phenomenon. Rather, they want to gain knowledge in order, first, to understand a situation that they define as unsatisfactory (i.e. experiencing a problem) and, second, to work out possible solutions. Knowledge is, as Helga Nowotny (1999: 71) noted, 'made to measure, in response to the specifications that must be worked out in concrete cases'. Problems usually involve several interrelated phenomena, which are often subject to different disciplines. If we take up the example of nutrition again, one could call the problem 'Adipositas' (obesity). To understand this problem, and even more so in order to develop effective solutions, you have to relate medical, physiological, psychological, social and cultural components of dietary habits (Zwick 2011). The fixation on a single causal relationship, something like caloric intake and consumption, describes only one side of the problem and, looked at in isolation, does not help in the development of solutions.

c. The problem-oriented approach is connected with practical applications (Becker et al. 2001; Redclift 1999b). Social-ecological research will not only generate knowledge that improves our understanding of problems, but also point to solutions to these problems and test these in reality. Practical application is always related to the quality of knowledge. Practical relevance should not be misused as an excuse for inadequate research quality. Transdisciplinary science needs adequate and discriminatory quality criteria for selection and generation of knowledge (Scholz 2011). However, the question in transdisciplinary research is how precise this generated knowledge must be to initiate or justify practical action. Knowledge for change does not need to be perfect or absolutely certain. Just think about climate research in public discourse, where an accurate forecast is always asked for but which the current state of research cannot supply. However, one can hardly derive from the quest for perfect knowledge the demand to postpone precautionary climate policy (Stirling 2006). Therefore, a relevant degree of knowledge and confidence must be strived for, enabling a reasoned choice from a variety of options. It may well be the case that a high degree of uncertainty remains which cannot be resolved by further research (Stirling 2003). Nevertheless, a rational decision about alternative courses of action can be derived even under conditions of high uncertainty. Accuracy, for its own sake, which may well have its place within the framework of disciplinary phenomenon research, is unnecessary in a transdisciplinary approach. A tolerable uncertainty in the results also allows use of qualitative and semi-quantitative methods which are often particularly suitable for analysis of relationships between natural and cultural phenomena (Renn et al. 2007: 138).

d. The last controversial feature of social-ecological research in the literature is the participatory approach (Redclift 1999b; Renn 2008; Scholz 2011; Welp and Stoll-Kleemann 2006). Especially when it comes to problem selection, to localizing knowledge bases and to developing and evaluating options for socio-ecological approaches, it is recommended that those who are affected by the problem directly and indirectly should participate in the research project.

The definition of the problem, the question of which values to include and the selection of methods used to consider and evaluate solutions, all presuppose normative assumptions that need to be aligned to the preferences and values of those who will be affected. These requirements cannot be derived legitimately from research itself, but must result from social discourse. In this respect, transdisciplinary research is also conditional on the involvement of affected persons and groups in the process of knowledge discovery and evaluation.

The danger of the participatory approach comes from the blending of factual knowledge, normative orientation, values and interests (Gethmann 2001). Starting from the roots of radical constructivism, this mix is not seen as a problem but rather as an inevitable research condition (Gross 2001). If one follows the path of moderate realism, however, an analytic differentiation into various categories of knowledge and evaluation is essential (Renn 2010). For socio-ecological research, a division into three categories of knowledge has proven useful (Daschkeit 2006; Pohl and Hirsch Hadorn 2006):

- *Knowledge for orientation*: What normative assumptions will we accept as benchmarks for our own actions, and what does that imply for the specific risk to be managed?
- *Knowledge for system understanding*: What are the connections between the variables that define the problem and its context? What is the current situation? What are the factors that determine the present situation? What actions are possible to change that situation?
- *Knowledge for transformation*: What measures are appropriate to achieve the desired objectives given the knowledge for orientation (goals) and the knowledge for understanding (causal web)? How can we reach these objectives under conditions of effectiveness, efficiency and acceptability?

The combination of systemic, orientative and transformational knowledge is directed towards a *rapprochement* between the models representing reality (knowledge) and the real manifestations of reality in form of repercussions. Using some vague notion of desirability as a selection mechanism of knowledge should be avoided and an organized scepticism must strictly be adhered to, examining the validity and reliability of research results (Rosa 1998).

However, when it comes to framing problems, interpreting results and judging the acceptability of solutions, subjective values are essential for creating important benchmarks for the research process. Aspects such as selecting problems to be investigated, deciding which dimensions are classified as socially relevant, as well as balancing the arguments for and against a course of action require subjective input. The big challenge is to shape the necessary integrative research process so that, on the one hand, the analytical separation of states of knowledge and values is maintained but, on the other, a procedural 'locking in' of analysis and evaluation is ascertained. For the actual research process, this means there must be differing participatory principles and procedures in order to assess the implicit requirements of the three different categories of knowledge. Knowledge for orientation requires value input from all affected populations. Knowledge for systems understanding requires input from all who are able to contribute to the analysis of risks and opportunities of each decision option. And knowledge for transition requires the input of all those who are needed to translate insights into collective action (see the ladder of participation in Renn 2008: 280).

The creation of system knowledge should stay limited to the participatory process of collecting relevant (factual) knowledge (IRGC 2005: 49ff.). Regarding the question of orientative knowledge, the rules of normative generalization ability must be respected. As for knowledge for transformation, sound experiential knowledge of effectiveness should be included in addition to systematically exploring causal relationships.

Selection and evaluation of options are, however, dependent on discursive practical experience in which all value preferences and interests in the debate need to be fairly represented (Scholz 2011). The capacity of all participants to exchange arguments in a practical and communicative way, ensuring a debate based less on power than on argumentation, and allowing for sufficient time for substantive deliberations, determines the quality of the process (Hagendijk and Irwin 2006; Horlick-Jones et al. 2007; Webler 1999). The trick then is in the intelligent networking and integration of the two basic elements, knowledge generation and option evaluation. This is one of the key challenges for transdisciplinary research in the sense of meeting the catalytic function of social-ecological science.

Summary and outlook

What do these abstract considerations mean for a socio-ecological perspective on risk? A socio-ecological foundation for environmental risk research must be based on three requirements:

1. integration of systemic, orientative and transformational knowledge as a prerequisite for a problem being recognized and described as closely as possible to its real status in nature and society, and reflecting both its complex cause-and-effect chains and normative embedding in society;
2. procedural integration of knowledge generation and selection of individual and collective evaluation processes, without causing an impermissible mixing of perception and interest during the assessment and evaluation phases of the process; and
3. synthesis of a theoretical claim to provide a robust explanation of a problem with a practical claim to its effective handling, leading to a competent, socially fair and ethically acceptable outcome.

It is neither possible nor desirable for societies to insure against all environmental risks, for risks are associated with opportunities. Aaron Wildavsky (1990: 120) has described this dilemma with the apt phrase, 'No risk is the highest risk of all.' For this reason, it is necessary to develop appropriate risk management strategies based on clear criteria and to implement an effective and efficient form of risk reduction through a *judicious combination of policy instruments* (WBGU 1999). The main objective is to strike the right balance between *too little and too much precaution* (Renn 2008: 179).

According to the analysis above, a socio-ecological approach is both suitable and promising for meeting this objective of finding the most appropriate balance. The socio-ecological approach is interdisciplinary, problem-oriented, practical and participatory. Within the transdisciplinary realm, an integration of systemic, orientative and transformational knowledge provides the basis for a rational and problem-oriented approach to the risk of environmental hazards. Particularly important is the combination of expertise and normative knowledge to determine the acceptability of risks. The necessary basis for this orientative knowledge must be derived directly from civil society, and solutions must then withstand review by an open societal discourse.

More than ever, discursive processes should be at the centre of a rational, integrative and preventive risk policy. A discourse without a systematic knowledge base remains only superficial, and a discourse that hides the normative quality of the courses of action being considered helps immorality break through. In this respect, all discursive processes must be measured by how they have conducted the integration of systemic, orientative and transformational knowledge. The goal of this discourse is ultimately to strive for sustainable development through human intervention for all concerned and to come to terms with environmental hazards.

References

Academy of Sciences in Berlin (1998) *Environmental Standards: Scientific Foundations and Rational Procedures of Regulation with Emphasis on Radiological Risk Management*. Dordrecht/Boston, MA: Kluwer.

Adams, M. (2003) 'The Reflexive Self and Culture', *British Journal of Sociology*, 54, pp. 221–238.

Beck, U. (1986) *Risikogesellschaft*. Frankfurt: Suhrkamp.

Beck, U. (1996) 'Weltrisikogesellschaft, Weltöffentlichkeit und globale Subpolitik. Ökologische Fragen im Bezugsrahmen fabrizierter Unsicherheiten', pp. 119–147 in A. Diekmann and C.C. Jaeger (eds), *Umweltsoziologie, Special Volume of the Kölner Zeitschrift für Soziologie und Sozialpsychologie*. Opladen: Westdeutscher Verlag.

Beck, U. (2009) 'World Risk Society as Cosmopolitan Society: Ecological Questions in a Framework of Manufactured Uncertainties', pp. 47–82 in E.A. Rosa, A. Diekmann, T. Dietz and C.C. Jaeger (eds), *Human Footprints on the Global Environment: Threats to Sustainability*. Cambridge, MA: MIT Press.

Becker, E. (1999) 'Fostering Transdisciplinary Research into Sustainability in the Age of Globalization: A Short Political Epilogue', pp. 284–288 in E. Becker and T. Jahn (eds), *Sustainability and the Social Sciences*. London: Zed.

Becker, E. (2003) 'Soziale Ökologie: Konturen und Konzepte einer neuen Wissenschaft', pp. 165–196 in G. Matschonat and A. Gerber (eds), *Wissenschaftstheoretische Perspektiven für die Umweltwissenschaften*. Weikersheim: Margraf.

Becker, E., Jahn, T., Hummel, D., Stiess, I. and Wehling, P. (2001) 'Sustainability: A Cross-Disciplinary Concept for Social-Ecological Transformations', pp. 147–152 in J.T. Klein, W. Grossenbacher-Mansuy, R. Häberli, A. Bill, R.W. Scholz and M. Welti (eds), *Transdisciplinarity: Joint Problem Solving among Science, Technology, and Society*. Basel: Birkhäuser.

Becker, E., Jahn, T. and Stiess, I. (1999) 'Exploring Uncommon Ground: Sustainability and the Social Sciences', pp. 1–22 in E. Becker and T. Jahn (eds), *Sustainability and the Social Sciences*. London: Zed.

Bell. M. (2012) *An Invitation to Environmental Sociology*, 4th ed. Thousand Oaks, CA: Sage.

Berger, P.L. and Luckmann, T. (2004) *Die gesellschaftliche Konstruktion der Wirklichkeit*, 20th ed. Frankfurt: Fischer.

Böhme, G. (2002) *Die Natur vor uns. Naturphilosophie in pragmatischer Hinsicht*, vol. 33. Baden Baden: Die Graue Edition.

Bostrom, A., Fischhoff, B. and Morgan, M.G. (1992) 'Characterizing Mental Models of Hazardous Processes: A Methodology and an Application to Radon', *Journal of Social Issues*, 48, pp. 85–100.

Bradbury, J.A. (1989) 'The Policy Implications of Differing Concepts of Risk', *Science, Technology, and Human Values*, 14, pp. 80–99.

Brown, H. and Goble, R. (1990) 'The Role of Scientists in Risk Assessment', *Risk: Issues in Health and Safety*, 6, pp. 283–311.

Buttel, F.H. and Taylor, P.J. (1992) 'Environmental Sociology and Global Environmental Change: A Critical Assessment', *Society and Natural Resources*, 5, pp. 211–230.

Daschkeit, A. (2006) 'Von der naturwissenschaftlichen Umweltforschung zur Nachhaltigkeitswissenschaft?' *GAIA*, 15, 1, pp. 37–43.

Dunlap, R.E. and Catton, W.R. (1994a) 'Struggling with Human Exemptionalism: The Rise, Decline and Revitalization of Environmental Sociology', *American Sociologist*, 25, pp. 5–30.

Dunlap, R.E. and Catton, W.R. (1994b) 'Environmental Sociology: Development, Current Status, Probable Future', pp. 11–32 in W.V. D'Antonio, M. Sasaki and Y. Yonebayashi (eds), *Society and the Quality of Social Life*. New Brunswick, NJ: Transaction Publishers.

Dunlap, R.E., Lutzenhiser, L.A. and Rosa, E.A. (1994) 'Understanding Environmental Problems: A Sociological Perspective', pp. 27–49 in B. Bürgenmeier (ed.), *Economy, Environment, and Technology: A Socio-Economic Approach*. Armonk, NY: Sharpe.

Fischhoff, B. (1995) 'Risk Perception and Communication Unplugged: Twenty Years of Process', *Risk Analysis*, 15, 2, pp. 137–145.

Freudenburg, W.R. and Gramling, R. (1989) 'The Emergence of Environmental Sociology', *Sociological Inquiry*, 59, pp. 439–452.

Gallopin, G.C., Funtowicz, S., O'Connor, M. and Ravetz, J.R. (2001) 'Science for the 21st Century: From Social Contract to the Scientific Core', *International Journal of Social Science*, 168, pp. 219–229.

Gethmann, C.F. (2001) 'Participatory Technology Assessment: Some Critical Questions', pp. 3–14 in M. Decker (ed.), *Interdisciplinarity in Technology Assessment: Implementation and its Chances and Limits*. Heidelberg/Berlin: Springer.

Giddens, A. (1978) 'Positivism and its Critics', pp. 237–286 in T. Bottomore and R. Nisbet (eds), *A History of Sociological Analysis*. New York, NY: Basic Books.

Gross, M. (2001) *Die Natur der Gesellschaft – eine Geschichte der Umweltsoziologie*. Weinheim/München: Juventa.

Hagendijk, R. and Irwin, A. (2006) 'Public Deliberation and Governance: Engaging with Science and Technology in Contemporary Europe', *Minerva*, 44, pp. 167–184.

Hannigan, J.A. (1995) *Environmental Sociology: A Social Constructivist Perspective*. New York, NY: Routledge.

Hillgartner, S. (1992) 'The Social Construction of Risk Objects: Or, How to Pry Open Networks of Risk', pp. 39–53 in J.F. Short and L. Clarke (eds), *Organizations, Uncertainties, and Risk*. Boulder, CO: Westview.

Horlick-Jones, T., Rowe, G. and Walls, J. (2007) 'Citizen Engagement Processes as Information Systems: The Role of Knowledge and the Concept of Translation Quality', *Public Understanding of Science*, 16, pp. 259–278.

Horlick-Jones, T. and Sime, J. (2004) 'Living on the Border: Knowledge, Risk and Transdisciplinarity', *Futures*, 36, pp. 441–456.

IRGC (International Risk Governance Council) (2005) *White Paper on Risk Governance. Towards an Integrative Approach*. Author: O. Renn with Annexes by P. Graham. Geneva: International Risk Governance Council.

Jaeger, C.C., Renn, O., Rosa, E.A. and Webler, T. (2001) *Risk, Uncertainty and Rational Action*. London: Earthscan.

Jahn, T. and Sons, E. (2001) 'Der neue Förderschwerpunkt sozial-ökologische Forschung des BMBF', *TA-Datenbank-Nachrichten* 4, pp. 90–97.

Jasanoff, S. (1999) 'The Songlines of Risk', *Environmental Values*, 8, 2, pp. 135–152.

Kasperson, R.E., Kasperson, J.X., Turner, B.L., Dow, K. and Meyer, W.B. (1995) 'Critical Environmental Regions: Concepts, Distinctions, and Issues', pp. 1–41 in J.X. Kasperson, R.E. Kasperson and B.L. Turner (eds), *Regions at Risk: Comparisons of Threatened Environments*. Tokyo/New York/Paris: United Nations University Press.

Liu, J., Dietz, T., Carpenter, S.R., Alberti, M., Folke, C., Moran, E., Pell, A.N., Deadman, P., Kratz, T., Lubchenco, J., Ostrom, E., Ouyang, Z., Provencher, W., Redman, C.L., Schneider, S.H. and Taylor, W.W. (2007) 'Complexity of Coupled Human and Natural Systems', *Science*, 317, pp. 1513–1516.

Luhmann, N. (1986) *Ökologische Kommunikation. Kann die moderne Gesellschaft Sich auf ökologische Gefährdungen einstellen?* Opladen: Westdeutscher Verlag.

Mathies, E. and Homburg, A. (2001) 'Umweltpsychologie', pp. 95–124 in F. Müller-Rommel and H. Meyer (eds), *Studium der Umweltwissenschaften. Sozialwissenschaften*. Heidelberg/Berlin: Springer.

Meyer, H. (2001) 'Quo Vadis? Perspektiven der Sozialwissenschaftlichen Umweltforschung', pp. 153–168 in F. Müller-Rommel and H. Meyer (eds), *Studium der Umweltwissenschaften – Sozialwissenschaften*. Heidelberg/Berlin: Springer.

Mittelstraß, J. (2003) 'Von der Einheit der Wissenschaft zur Transdisziplinarität des Wissens' pp. 13–27 in G. Matschonat and A. Gerber (eds), *Wissenschaftstheoretische Perspektiven für die Umweltwissenschaften*. Weikersheim: Margraf.

Mohr, H. (1995) *Qualitatives Wachstum. Losung für die Zukunft*. Stuttgart/Wien: Weitbrecht.

Mosler, H.J. (1995) 'Umweltprobleme: Eine sozialwissenschaftliche Perspektive mit naturwissenschaftlichem Bezug', pp. 77–86 in U. Fuhrer (ed.), *Ökologisches Handeln als sozialer Prozeß*. Basel: Birkhäuser.

Nowotny, H. (1999) *Es ist so. Es Könnte auch anders sein*. Frankfurt: Edition Suhrkamp.

Ott, K. and Döring, R. (2004) *Theorie und Praxis starker Nachhaltigkeit*. Marburg: Metropolis.

Pohl, C. and Hirsch Hadorn, G. (2006) *Gestaltungsprinzipien für die transdisziplinäre Forschung – ein Beitrag des td-net*. Munich: Ökom Verlag.

Redclift, M. (1999a) 'Sustainability and Sociology: Northern Preoccupations', pp. 19–73 in E. Becker and T. Jahn (eds), *Sustainability and the Social Sciences*. London: Zed.

Redclift, M. (1999b) 'Dance with the Wolves? Sustainability and the Social Sciences', pp. 267–273 in E. Becker and T. Jahn (eds), *Sustainability and the Social Sciences*. London: Zed.

Renn, O. (2001) 'Science and Technology Studies: Experts and Expertise', pp. 13647–13654 in N.J. Smelser and P.B. Baltes (eds), *International Encyclopedia of the Social and Behavioral Sciences*, vol. 20. Cambridge, MA: Cambridge University Press.

Renn, O. (2008) *Risk Governance: Coping with Uncertainty in a Complex World*. London: Earthscan.

Renn, O. (2010) 'The Contribution of Different Types of Knowledge Towards Understanding, Sharing and Communicating Risk Concepts', *Catalan Journal of Communication and Cultural Studies*, 2, 2, pp. 177–195.

Renn, O., Schweizer, P.J., Dreyer, M. and Klinke, A. (2007) *Risiko. Über den gesellschaftlichen Umgang mit Unsicherheit*. Munich: Oekom.

Rip, A. (2006) 'A Co-evolutionary Approach to Reflexive Governance – And Its Ironies', pp. 82–100 in J.P. Voß, D. Bauknecht and R. Kamp (eds), *Reflexive Governance for Sustainable Development*. Cheltenham, UK: Edward Elgar.

Rosa, E.A. (1998) 'Metatheoretical Foundations for Post-Normal Risk', *Journal of Risk Research*, 1, 1, pp. 15–44.

Rosa, E.A. (2008) 'White, Black and Grey: Critical Dialogue with the IRGC's Framework for Risk Governance', pp. 112–165 in O. Renn and K. Walker (eds), *The IRGC Risk Governance Framework: Concepts and Practice*. Heidelberg/New York, NY: Springer.

Rosa, E.A. and Dietz, T. (2010) 'Global Transformation: Passage to a New Ecological Era', pp. 1–45 in E.A. Rosa, A. Diekmann, T. Dietz, and C.C. Jaeger (eds), *Human Footprints on the Global Environment*. Cambridge, MA: MIT Press.

Scheringer, M., Valsangiacomo, A., Hirsch Hadorn, G., Pohl, C. and Ulbrich Zürni, S. (2005) 'Transdiziplinäre Umweltforschung: Eine Typologie', *GAIA*, 14, 2, pp. 192–195.

Scholz, W.W. (2011) *Environmental Literacy in Science and Society: From Knowledge to Decisions*. Cambridge, MA: Cambridge University Press.

Searle, J.R. (1995) *The Construction of Social Reality*. New York, NY: Free Press.

Seiderberg, P.C. (1984) *The Politics of Meaning: Power and Explanation in the Construction of Social Reality*. Tucson, AR: University of Arizona Press.

Stirling, A. (2003) 'Risk, Uncertainty and Precaution: Some Instrumental Implications from the Social Sciences', pp. 33–76 in F. Berkhout, M. Leach and I. Scoones (eds), *Negotiating Change*. Cheltenham, UK: Edward Elgar.

Stirling, A. (2006) 'Precaution, Foresight and Sustainability: Reflection and Reflexivity in the Governance of Science and Technology', pp. 225–272 in J.P. Voß, D. Bauknecht and R. Kamp (eds), *Reflexive Governance for Sustainable Development*. Cheltenham, UK: Edward Elgar.

Vecchione, E. (2011) 'Science for the Environment: Examining the Allocation of the Burden of Uncertainty', *European Journal of Risk Regulation*, 2, 2, pp. 227–239.

WBGU (German Advisory Council on Global Change) (1999) *Welt im Wandel. Strategien zur bewältigung globaler Umweltrisiken*. Jahresgutachten 1998. Heidelberg/Berlin: Springer.

WBGU (German Advisory Council on Global Change) (2011) *World in Transition: A Social Contract for Sustainability*. Berlin: WBGU.

Weber, K.M. (2006) 'Foresight and Adaptive Planning as Complementary Elements in Anticipatory Policy Making: A Conceptual and Methodological Approach', pp. 189–221 in J.P. Voß, D. Bauknecht and R. Kamp (eds), *Reflexive Governance for Sustainable Development*. Cheltenham, UK: Edward Elgar.

Webler, T. (1999) 'The Craft and Theory of Public Participation: A Dialectical Process', *Risk Research*, 2, 2, pp. 55–71.

Weingarten, M. (2003) 'Von der Beherrschung der Natur zur Strukturierung gesellschaftlicher Naturverhältnisse: Philosophische Grundlagen der Umweltwissenschaften', pp. 127–144 in G. Matschonat and A. Gerber (eds), *Wissenschaftstheoretische Perspektiven für die Umweltwissenschaften*. Weikersheim: Margraf.

Welp, M. and Stoll-Kleemann, S. (2006) 'Integrative Theory of Reflexive Dialogues', pp. 43–78 in S. Stoll-Kleemann and M. Welp (eds), *Stakeholder Dialogues in Natural Resources Management: Theory and Practice*. Heidelberg/Berlin: Springer.

Wildavsky, A (1984) 'Die Suche nach einer fehlerlosen Risikominderungsstrategie', pp. 224–233 in S. Lange (ed.), *Ermittlung und Bewertung industrieller Risiken*. Heidelberg/Berlin: Springer.

Wildavsky, A. (1990) 'No Risk Is the Highest Risk of All', pp. 120–127 in T.S. Glickman and M. Gough (eds), *Readings in Risk*. Washington, DC: Resources for the Future.

Zwick, M.M. (2011) 'Die Ursachen der Adipositas im Kindes- und Jugendalter in der modernen Gesellschaft', pp. 71–90 in M.M. Zwick, J. Deuschle and O. Renn (eds), *Übergewicht und Adipositas bei Kindern und Jugendlichen*. Wiesbaden: VS Verlag.

18
Uncertainty and claims of uncertainty as impediments to risk management[1]

Raymond Murphy

If uncertainty were absolute, then no leader, no institution and no population could be held accountable for preventing harm, mitigating environmental problems or decreasing vulnerability to hazards. There would be no reason to change practices because we simply would not know of any threat. The concept of uncertainty is frequently used in this absolute sense as a rhetorical device to excuse bad decisions: decisions not to invest in protection against natural hazards; decisions not to impose higher standards on polluting industries; decisions not to pay short-term costs to avoid long-term problems. This chapter argues that uncertainty must not be reduced to a politically convenient rhetorical device. At first sight, uncertainty appears to be a well-understood concept, but it is not that simple. It is necessary to examine more precisely what it is we are uncertain about and to capture the nuances of uncertainty. For example, there is uncertainty about when, where and how you will die, but not about the fact that you will die. John Maynard Keynes distinguished uncertainty from both certainty and scientifically based calculations of probability, such as could be calculated for roulette. His nuanced conception also captured degrees of uncertainty: 'the expectation of life is only slightly uncertain. Even the weather is only moderately uncertain' (Keynes 1937: 213–214). For him, uncertainty did not necessarily indicate the complete absence of knowledge about threats. Nor, according to Keynes, did uncertainty excuse inaction. Indeed, he argued that 'the necessity for action compels us ... to do our best to overlook [the] awkward fact' that the future cannot be known with absolute certainty. This chapter explores the nuances of uncertainty and knowledge. It goes on to examine how new uncertainties and risks are being created by modern societies. Finally, it investigates whether learning has occurred to minimize risks or whether claims of uncertainty are paralysing the management of risks.

The nuances of uncertainty

Keynes (1937) conceptualized uncertainty as the absence of sufficient knowledge with which to calculate risk. This resonates with what Ortwin Renn (2008: 91), following David Resnik (2003: 332), refers to as 'decisions under ignorance'; that is, decision-contexts characterized by a lack of knowledge about the probability and/or consequences of a harmful event (see also Renn and Klinke 2012). I argue, however, that risks must be analysed in broader terms than simply calculations.

Uncertainty and risk assessment are not the simple, straightforward concepts they appear to be. There are different components, forms and degrees of knowledge, of calculability and of uncertainty. Hence, for complex problems like climate change, scientists use the expression 'preponderance of the evidence' (Schneider 2009) to guide thinking about risk in order to avoid the false dichotomy that science results in either accurate probabilistic calculations of frequency and impact of hazards or it yields nothing. The preponderance of scientific evidence is particularly significant for low frequency, high impact events that constitute an important type of risk. Even for these events which do not occur with sufficient frequency to make precise probability calculations, or that are based on an oncoming discontinuity that invalidates simple extrapolation, there is often partial knowledge available, which implies that uncertainty is also partial. Thus there is a range of possibilities from total uncertainty (we simply do not know) to certainty. In between those poles, there are various degrees and components of scientific probability calculations as well as indicative scientific knowledge (preponderance of the evidence).

Total uncertainty

Threats quite clearly exist under situations of complete ignorance if ignorance is taken to mean, as proposed by Renn (2008), the inability to calculate the likelihood and magnitude of risk before the actual event. We learn that risks were present after they have been actualized into harm. In these cases, experience provides a significant basis on which to learn lessons about past and future risks regardless of whether we had sufficient knowledge with which to calculate probabilities. To use my own study of the 1998 ice storm disaster in northeastern North America as an example, there was insufficient meteorological data at that time on which to predict an extreme weather event of the magnitude experienced, even as a worst-case scenario (Murphy 2010a). Close to total ignorance of the imminent intensity, duration and scope of the freezing rain prevailed in forecasts prior to its arrival. There was an elevated chance of harm, but even meteorologists and climatologists did not have the means to calculate it. The conclusion of the main technical inquiry into the disaster was entitled *Confronting the Unforeseeable* (CST 1999). Risk existed, but it could not be foreseen with the existing models, tools and evidence of scientists. Even at this pole, however, risk management can learn from surprises to build more robust constructions; to improve resilience so that society can return to its normal state; to reinforce monitoring and all-hazards emergency organizations; to recognize that even worst-case scenarios can be underestimations; and thus, most of all, to expect the unforeseeable (see Murphy 2009). Retrospective disaster studies typically end with a list of lessons to be learned, but all too often inconvenient lessons are ignored as the disaster recedes with the passage of time.

There is an important lesson here for social scientists: risk should not be reduced to its socially constructed estimate, not even its scientific assessment. Human awareness of risk, even in scientific form, is limited and fallible. Near the pole of total uncertainty, decision makers can be legitimately excused for being surprised. But most disasters and environmental calamities do not occur at the pole of total uncertainty.

Certainty and probability

At the opposite pole is certainty, where the outcomes of decisions are completely known (Renn 2008). Where knowledge is complete, it does not make sense to speak of risk because we are dealing with the certainty of harm rather than the chance of harm. The certainty of harm typically leads societies and individuals to take measures to avoid it. For example, the tides in the Bay of Fundy (on the Atlantic coast of North America) are higher than most tsunamis and just as powerful. But they are so regular and certain that they have become integrated into human expectations such that institutions and individuals take them into account and there is rarely any harm done.

Many potentially harmful events are less regular than tides in the Bay of Fundy, but they may, nevertheless, be predicted with various degrees of confidence. Sufficient knowledge exists to provide the basis for the reduction of risk and the prevention of harm (if only the knowledge is acted upon). One type of risk consists of high frequency events, which provide probabilistic knowledge concerning the chances of harm and what it will be. Typical examples are automobile accidents and house fires. Even though it is uncertain who the victims will be and how and when harm will occur, extensive data-sets of empirical evidence enable probabilistic assessments of risk and aggregate conclusions to be made by actuaries that are very useful to institutions like insurance companies and governments. Preventive measures are implemented, such as mandatory smoke detectors and speed limits, as well as post-accident compensation, medical care, etc.

Indicative knowledge

One form of partial knowledge is indicative scientific knowledge that, even if not definitive, provides a solid basis for social policies and risk governance. Seismologists have acquired knowledge about the grinding of tectonic plates such that they can predict whether an earthquake is likely in that area. With present understanding and evidence, however, they cannot predict its force, timing or the location of its epicentre. Nevertheless this knowledge, albeit incomplete, provides a good basis on which emergency management and political decision makers can act. The latter are then confronted by the issue of whether to reduce risk by enacting building standards and reinforcing existing buildings or whether to save money in the present by running the risk. Partial knowledge also exists concerning floods, hurricanes, cyclones, storm surges and tsunamis. There has been an accumulation of scientific evidence about these dynamics. Hence, there is a sufficient knowledge base for learning to implement preparedness measures such as the restoration of wetlands and barrier reefs, construction of dykes, restrictions on development in risky areas such as flood plains, tsunami early warning systems, and emergency preparedness measures. But these are expensive and there is still uncertainty about when and where these hazards will strike and their magnitude. Preparedness measures thus run up against economic pressures to minimize present expenses even if doing so means maximizing vulnerability and accepting risk. For example, there was good scientific evidence that constructing the Mississippi River-Gulf Outlet Canal (also known as the MRGO Canal) in New Orleans would exacerbate risk if a hurricane struck the city. But the cost–benefit analyses of developers and local politicians did not take this into account, to the detriment of the city when risk was actualized into harm (Freudenburg et al. 2009). The same is true for floods: there is good evidence that building on flood plains is dangerous, but there are powerful economic interests to do so. In each of these cases, there is a strong scientific foundation of knowledge for prevention and risk reduction in spite of the remaining uncertainties. Similarly there are many scientific studies, including those of the Intergovernmental Panel on Climate Change (IPCC), that provide a firm scientific foundation on which to form conclusions about global warming. We do know that a high carbon economy brings with it serious risks. There is not total uncertainty concerning climate change, even though there remain many uncertainties about specific harms, impacts, locations and timing. Indicative knowledge is admittedly partial, but it would be misleading and potentially harmful to confuse it with uncertainty. It is a solid basis for learning in order to improve governance when confronted by risks and uncertainties.

The fallibility of scientific risk assessment and lay knowledge

There are two other elements which make the issue of uncertainties even more complex. First, since scientific knowledge is fallible, there is always the possibility that scientific understanding and the preponderance of the evidence will change. What the present evidence now shows as being safe may prove to be risky in the future, and vice versa. Hence, there is residual uncertainty even with the best of science (see also Lockie and Measham 2012). Furthermore, scientific scepticism implies a willingness to challenge any hypothesis, particularly those having to do with forecasts of the future. But effective risk management does not allow this to reject best practices that require following the preponderance of the best available evidence concerning risk rather than dismissing it. Best practices necessitate the change of social practices that indicative knowledge suggests will likely lead to harm.

Second, scientific knowledge is not the only way to distinguish safety from risk (although for some matters it is the only way, such as the depletion of the ozone layer, the risk of earthquakes, etc.). There is also non-scientific, lay knowledge. This is usually based on experience over a lifetime, or several lifetimes if knowledge is passed down from generation to generation. Lay knowledge about risk can be very useful. Take global warming, for example: local residents can discern the regression of glaciers by using the memories of their parents or sometimes their own memory. The Inuit can see the melting of the Arctic ice cap in the last few years and compare it to the past. But there are various versions of lay knowledge and some can be misleading. It is often founded on vested economic interests and current ways of life and, hence, strongly affected by a present bias. For example, when current climate is beneficial, it is difficult for the population and leaders to recognize the risk of it changing for the worse for future generations because of the level of greenhouse gas emissions that forms the basis of present prosperity. Lay knowledge, as well as many time series of economists, are based on extrapolations from the present and recent past and are not valid for situations involving discontinuities, which is precisely the risk with global climate change and other environmental changes.

Uncertainty as ideology

Most cases do not involve uncertainty in an absolute sense, but the word is frequently used that way ideologically. For example, prevention, preparation and response for emergencies in the USA are the responsibility of the Federal Emergency Measures Agency and, ultimately, of the President. After Hurricane Katrina devastated New Orleans, American President George W. Bush said on television: 'I don't think anyone anticipated the breach of the levees' (Shane and Lipton 2005). But scientific studies had clearly indicated the risks of breachable levees and other vulnerabilities if New Orleans were struck by a Category 4 or 5 hurricane. The probability of a hurricane striking New Orleans had been calculated well before Katrina arrived and it was not low. The likelihood of harm because vulnerability had been aggravated by the construction of canals and the destruction of wetlands had frequently been pointed out (Freudenburg et al. 2009). Based on his analysis of the destruction of wetlands by oil extraction and the construction of canals that locals called 'hurricane alleys', a meteorologist predicted three months in advance of Katrina precisely what would happen (Freudenburg et al. 2009). Despite the fact that the vulnerability of the city was well known, the budget to reinforce the levees was reduced. Even though meteorologists forecasted an active hurricane season months before Hurricane Katrina struck and predicted its path and power days in advance, evacuation preparations were inadequate for the infirm, the elderly and the poor. Calling the hazard unanticipated amounts to playing the uncertainty card to avoid responsibility for risk management. President Bush was not the only leader to use that wordplay. The claim of uncertainty has often been utilized after disasters to justify lack of preparation for a supposedly unpredictable 'Act of God' or an assumed '1 in 1000-year event' (Murphy 2010a, 2010b). Claiming uncertainty is the ultimate excuse for recklessness and/or incompetence.

The claim of uncertainty about the risk of environmental problems is also frequently used by vested interests that strive to continue business as usual. Uncertainties were claimed by the chlorofluorocarbon (CFC) industry about the risks of CFCs depleting the ozone layer, by the automotive industry about the risks of leaded petrol harming children, etc. The rhetorical tactic of highlighting uncertainty has similarly been used to avoid social change needed to mitigate global warming. Those that advocate full speed ahead with fossil fuels contend that climate change 'may exist only in computer simulations' (Ray and Guzzo 1993: 27). Deniers ask how climatologists can forecast warming for the entire planet in fifty years when meteorologists are often wrong in predicting next week's local weather. Business journals and tabloid media specialize in drawing attention to any uncertainties they can find about global warming. The email controversy at the University of East Anglia was widely used to deny anthropogenic global warming and was depicted as 'Climategate' to suggest a criminal conspiracy by climatologists similar to Watergate. Three subsequent inquiries in Britain demonstrated that the science of global warming was solid, that it was not undermined by the email controversy and that uncertainty generated by the email controversy was unwarranted. But the conclusions of these inquiries were not given the same prominence in the media as the initial email controversy, so Climategate sticks in the mind as a sound bite and is still believed in some quarters to have invalidated the science of global warming. Climategate constituted rhetorically manufactured uncertainty, not scientifically based uncertainty. All these examples show that uncertainty has its attractions for those seeking to evade responsibility for making risk and to render risk acceptable to the population. The climatologist Stephen Schneider (2009: 259) concludes that the 'framing of the climate problem as "unproved", "lacking a consensus", and "too uncertain for preventive policy" has been advanced strategically by the defenders of the status quo' to avoid social change that would reduce the risk of harmful anthropogenic climate change. Highlighting uncertainties does not, however, always succeed in persuading leaders and the population to accept risk. In the cases of CFCs depleting the ozone layer and pollution from leaded petrol, states forged ahead to ensure safety despite residual uncertainties and claims of uncertainty by the CFC and automotive industries respectively.

The absolute conception of uncertainty as situations in which there is no scientific basis to indicate risk does not apply to most cases, as illustrated by Hurricane Katrina and global warming. Although knowledge is incomplete, there is a scientific basis for knowing about risk which decision makers dismiss. In both those cases, talk about uncertainty has been deployed as a rhetorical device to justify inaction. The 'ostrich defence', whereby leaders hide their heads in the sand and claim ignorance, has been rejected by the courts in litigation against directors of reckless companies like WorldCom and Enron. Similarly, claims of uncertainty are dubious as a defence when there has been a refusal to act on the available scientific evidence about the risks of environmental problems like climate change (Schneider 2009) or other possible disasters, even in the face of uncertainties that remain (see Murphy 2009: 327). The use of uncertainty in risk management as an ideology to legitimate failure to reduce vulnerabilities and to mitigate environmental problems like global warming shows that the concept of uncertainty badly needs clarification. Institutional failures that increase vulnerability by malign neglect in cases where there is evidence of risk despite some uncertainties must be distinguished from unavoidable calamities where there genuinely is no scientific evidence to indicate risk.

The making and unmaking of risks and uncertainties

There has been impressive social change in the past 250 years, particularly the development of science and technology, market globalization, increases in organizational efficiencies, etc. These have resulted in remarkable benefits, not the least of which is improved life expectancy, which, in turn, has led to a rapid increase in world population (presently seven billion and counting). The sharp rise in consumption by a growing population has, however, created risks and uncertainties on a planet that has hitherto been in a steady state of nature's dynamics, which are beneficial for human development, but is finite in terms of resources and waste sinks. Social change is producing environmental change such as deforestation, habitat loss for other species and hence biodiversity loss, depletion of non-renewable resources for future humans, failure to renew renewable ones, the despoliation of land, oceans, and the atmosphere as waste dumps, an increasingly carbonized atmosphere, global warming and many more. These threaten to turn back against contemporary societies. How can such risks be managed?[2]

Risk management by the market

What is a resource? Economists have followed Erich Zimmerman (1951: 15), who claimed that 'resources are not, they become'. Science, technology and the market transform what was not a resource into one. Julian Simon (1981) contends that human reason is the ultimate resource, because when a resource becomes scarce, market forces of supply and demand increase its price, which gives a profit-seeking incentive to search for more resources and find substitutes. He has faith that more of the resource or substitutes will indeed be found by human reason in a timely manner. Hence price increases of resources are temporary, and resources tend to become less expensive over time and, therefore, more abundant according to the premise that lower price indicates less scarcity. Bjørn Lomborg (2001, 2007) adds that, since wealthy societies can defend themselves more readily against environmental risks than poor ones, those risks can be better diminished by pursuing wealth than by policies to reduce risks. Economics, which used to be called 'the dismal science', has become 'the euphoric science' to these practitioners. Indeed, the market is the institution that will fulfil the cornucopian dream of bringing wealth and safety to all market societies. The path to sustainability is presented as the pursuit of wealth.

This perspective is, however, unconvincing. First, resources become, it is true, but they also degrade, disappear or simply become too expensive and/or risky to retrieve. Second, the belief that substitutes for non-renewable resources will always be found when needed is nothing more than blind faith. Third, this perspective glosses over the fact that the proposed solution is what caused modern environmental risks in the first place. They have been mainly created by technological development steered by market forces in pursuit of profit — for example, fossil fuel emissions produced by the hydrocarbon and automotive industries. Fourth, it presumes that societies can adapt to the environmental damage caused by market forces. Although adaptation is clearly needed, dismissing prevention constitutes a risk-maximizing and uncertainty-maximizing approach. For safety, it requires certitude that no tipping points into irreversible harm will be reached. Such certainty does not exist.

Risk management by reconfiguring the market through a state–market partnership

The economist Mark Jaccard (2005, 2009) notes that economic development since the Industrial Revolution has been based on the energy of fossil fuels, and argues that they will continue to be combusted to supply energy in both wealthy and developing countries because they are abundant, energy dense, flexible, easy to use and relatively inexpensive. As conventional sources of oil become exhausted – which has already occurred in the USA – the price of oil increases, which gives incentives to extract oil from tar sands, shale and deepwater locations, and to liquefy natural gas and coal. Hence, according to Jaccard, there is little risk of the exhaustion of fossil fuels or even of oil in the foreseeable future. Unlike Simon (1981), however, Jaccard is concerned about the risks of environmental problems attendant upon the pursuit of economic growth, particularly global warming because of the combustion of fossil fuels. He accepts the indicative evidence of the IPCC and other scientific institutions that carbonizing the atmosphere generates high impact risk as well as uncertainties such as possible tipping points into irreversible vicious circles and runaway harmful climatic changes. Thus, he argues that fossil fuels need to be made sustainable, which means making them clean, including coal. Jacaard (2009: 130) contends that to reduce the risk of global warming:

> humanity must impose a charge on the emissions of carbon and other greenhouse gases, and that this charge must grow gradually over the coming decades to be a very significant cost factor in determining energy supply-and-demand investment decisions.

He emphasizes that this should have been implemented on a global scale in the 1980s, when the Scandinavian countries first put it into practice. The price increase for the consumer would not be too high, because 'the carbon price will slow down the consumption of oil and therefore the progression from lower-cost oil to higher-cost oil and its higher-cost substitutes' (Jacaard 2009: 130). The charge on carbon would be imposed by the state either in the form of a carbon tax or a cap on carbon emissions, and would result in market incentives to reduce emissions. Hence, this proposal consists of a partnership between the institutions of the state and the market. Risk management through such forms of a carbon charge is, however, encountering fierce resistance.

An ignored, long-term risk and its management

Geologists come to dramatically different conclusions about the abundance of usable energy, particularly oil and its fossil fuel substitutes, and hence the risk of their exhaustion. J. David Hughes (2009, 2010) agrees with Jaccard (2009) that societies will likely be dependent on fossil fuel energy for a long time because of problems of scale: renewable energy is currently a small fraction of the primary energy sources in all countries, and replacing an enormous amount of fossil fuels presents major technical and economic problems as well as opposition by vested interests of many types. He also agrees that the planet is huge even if it is finite, so absolute physical exhaustion of fossil fuels is unlikely even if demand increases sharply. But getting energy from usable fossil fuels is another matter.

'Usable' is based on a fundamental feature of human interaction with nature: it takes energy to get energy. When the Saudi Ghawar oil field was brought online in 1951, for each oil-equivalent barrel of energy used for extraction, 100 barrels of oil gushed out. That energy return on investment, or net energy, has been decreasing ever since. For example, one oil-equivalent barrel of energy now yields only four barrels of oil extracted and upgraded from the Alberta tar sands. Furthermore, that does not include the energy expended to build and transport the heavy equipment used to extract oil from sand. Even at Ghawar, oil now has to be coaxed out with seven million barrels of water per day (Hughes 2009: 68). To liberally mix metaphors, once the low-hanging fruit is picked, one has to run faster on the treadmill to extract an identical amount of oil. Energy sources are increasingly becoming energy sinks. When the energy used in extraction comes close to the energy extracted, it no longer makes sense to remove the fossil fuel from the ground.

To employ an analogy: there may be oil, coal, and other forms of energy on the moon but extracting and retrieving it for use on Earth would require more energy than the moon's oil or coal delivers. Hence, it would be a net loss of energy to use it. The same applies for vast amounts of energy sources on Earth that are difficult to get at and widely dispersed. As large, easily extractable deposits become exhausted, only small, widely dispersed deposits remain, which creates technical and economic problems for their exploitation. Calculations of the amount of energy reserves in the ground are misleading if they do not subtract the energy needed to extract and upgrade them.

New unconventional sources of oil yield additional energy and seemingly attenuate the risk of depletion, but they perversely aggravate the risk of exhausting the supply of usable oil more quickly because so much energy must be consumed in extraction. This is not only true of oil, which has unique physical properties as a dense source of energy in liquid form, but also of other fossil fuels like coal. Combustion of coal produces far more pollution and greenhouse gases than oil does, and it is not in a form (liquid) that makes it relatively easy to transport and use. To give it improved physical properties, there is much talk about liquefaction of coal, capture and storage of carbon dioxide from coal-fired electricity plants and tar sands oil extraction facilities, and clean coal. But all these transformations require energy. Making coal as usable as oil will deplete energy reserves more quickly. Transforming the physical properties of coal to make it liquid and clean will also change its economic properties by making it less cheap. The risk of energy depletion, pollution and emissions does not only consist of the energy combusted to run automobiles, planes, heat and air-condition buildings, etc., but also the energy consumed to extract, upgrade and make energy less risky.

Hughes (2009: 63) refers to the twin demons of depletion of easily extractible energy and increasing demand as the 'energy sustainability dilemma'. For example, the future of oil is one that is extracted from tar sands and shale; from deepwater drilling and areas like the Arctic made accessible by global warming. But these only give short-term relief.

> If we consider that the world burns thirty-one billion barrels of oil each year, the Brazil find [i.e. deep-sea oil wells off Brazil's coast] represents sixteen months of world oil consumption and the entire circum-Arctic region perhaps three years.
>
> *(Hughes 2009: 60)*

Furthermore, these unconventional sources of energy create new risks of damaging the environment. This has been clearly documented for oil extracted from tar sands, which has resulted in greater deforestation, water contamination and greenhouse gas emissions than conventional oil (Kunzig 2009). It has been seen with the environmental damage suffered in the Gulf of Mexico from deepwater oil drilling. Risk is particularly high for extraction activities in fragile areas made accessible by global warming, such as Arctic drilling, mining and transportation of dangerous commodities. The new technology of hydraulically fracturing shale with chemicals and water at high pressures has resulted in a temporarily plentiful supply of natural gas and tight oil in North America, but it has also contaminated groundwater and provoked earthquakes. These technological developments to retrieve more energy have yielded immediate benefits but have exacerbated long-term risks and uncertainties, such as the risk of depletion of usable energy and of anthropogenic global warming. Hughes (2009) advocates energy conservation and energy efficiencies as the only effective way to mitigate these twin risks, but fails to specify how this will be institutionalized.

Conclusion: much to learn in dealing with uncertainties and risk

Curiously, the economist Jaccard (2005) has greater faith in science to find technical ways to extract usable and safe energy than the scientist Hughes (2009). But there is no certainty that technological solutions will appear when needed. They are sometimes found, but often are not. They can be hoped for, but not counted on to reduce risk and solve problems. Risk management based on their assumed development and implementation is highly uncertain.

The main uncertainty is sociopolitical; specifically, whether society will mitigate risks or exacerbate them. A quarter of a century after the Scandinavian countries began a charge on carbon, the most notable fact is that it has not been started in most countries. Whether in the form of a carbon tax or cap-and-trade mechanism, it has not been implemented in the main fossil fuel-consuming countries and there is massive resistance to it. The same is true for regulations that would diminish emissions significantly. The tabloid and business media are filled with Chicken Little apocalyptic commentary that the sky will fall and the economy ruined if a price is put on carbon, so politicians fear to implement such a policy. If a gradually increasing carbon price were implemented tomorrow in all countries, it would take decades before it rose sufficiently to moderate fossil fuel consumption, and more decades before it resulted in a stabilization of global warming. There is no sign that a significant price on carbon will be implemented globally in the foreseeable future. Hence, fossil fuel emissions and the wastage of lower-cost conventional oil accelerate, and so do the change to higher-cost, dirtier and dangerous unconventional oil and its substitutes. Carbon will have been transferred from the ground to the atmosphere before policies are enacted to decrease greenhouse gas emissions, much like closing the barn door after the horse has escaped. And carbon dioxide molecules remain in the atmosphere for almost a century. Jaccard's (2009: 130) carbon charge precondition for the use of fossil fuels has not been implemented. The state and the market have failed to reconfigure the incentive structure in order to minimize risk. His argument leads to the conclusion that environmental risks are presently being mismanaged by the institutions of most societies.

Conservation of energy and its more efficient use, as advocated by Hughes (2009), are not being practised either. The option of leaving the oil and gas in the tar sands, shale, deep seas and Arctic until environmentally safer technologies of extraction are developed has been rejected. Since none of the proposed solutions of Jaccard and Hughes (carbon pricing, conservation, efficiencies) are being practised on the scale needed to mitigate the conjoint risks of resource exhaustion and global warming, where will fossil fuels used by humans come from in double the period of pumping the Saudi Ghawar oil field 60 years from now; in double the industrial period 250 years in the future; and more so in double the calendar period 2,000 years hence? Where will the massive sinks be for anthropogenic greenhouse gas emissions? Viewed in anything but a short-term timeframe, risk is high and uncertainty is absolute. This would not be the case if energy conservation were practised, energy efficiencies enhanced and renewable energies used.

Climate change is a particularly interesting social problem, because: (1) the very successes of science, technology, the market and organizational efficiency are increasing risk and uncertainty; yet (2) the science of risk assessment is being ignored because it brings troubling news about modern prosperity and lifestyles. The paradox of the public understanding of science is that it consists of a blind faith that science will bring solutions to environmental problems in a timely manner, yet it dismisses science when it indicates the risk of harm and suggests that social change is needed. For example, it is assumed that science will bring solutions on demand to the risks of global warming and resource depletion, even though these wishful conjectures about the future lie in the realm where a conception of uncertainty as the total absence of rational scientific knowledge applies. On the other hand, the scientific warnings about these two risks are based on solid indicative knowledge but are relegated to, at most, back-of-the-mind worries that have not resulted in changes to social practices and the improved management of these risks. The claim of uncertainty about the risk of global warming and of the exhaustion of usable energy is central to the acceptance of risk, with no distinction made between total and partial uncertainty, between ignorance and valuable indicative knowledge. Although safety can never be ensured in a total sense, it can be enhanced if total and partial uncertainty is differentiated and if partial indicative knowledge is used to mitigate the risk of environmental problems and disasters rather than being dismissed as uncertain.

Notes

1 I would like to thank the Social Sciences and Humanities Research Council of Canada for a grant that supported this research.
2 For a detailed empirical investigation of the institutions and their leaders immediately involved in the management of a specific disaster and disaster preparedness, see Murphy (2009).

References

CST (Commission Scientifique et Technique), chargée d'analyser les événements relatifs à la tempête de verglas (1999) *Pour affronter l'imprévisible*. Québec: Gouvernement du Québec.
Freudenburg, W., Gramling, R., Laska, S. and Erikson, K. (2009) *Catastrophe in the Making: The Engineering of Katrina and the Disasters of Tomorrow*. Washington, DC: Island Press.
Hughes, J.D. (2009) 'The Energy Issue: A More Urgent Problem than Climate Change', pp. 58–95 in T. Homer-Dixon (ed.), *Carbon Shift: How the Twin Crises of Oil Depletion and Climate Change Will Define the Future*. Toronto: Random House of Canada.
Hughes, J.D. (2010) 'Hydrocarbons in North America', pp. 211–228 in R. Heinberg and D. Lerch (eds), *The Post Carbon Reader: Managing the 21st Century's Sustainability Crises*. Healdsburg, CA: Watershed Media.
Jaccard, M. (2005) *Sustainable Fossil Fuels*. Cambridge, UK: Cambridge University Press.
Jaccard, M. (2009) 'Peak Oil and Market Feedbacks: Chicken Little vs Dr Pangloss', pp. 96–131 in T. Homer-Dixon (ed.), *Carbon Shift: How the Twin Crises of Oil Depletion and Climate Change Will Define the Future*. Toronto: Random House of Canada.
Keynes, J. M. (1937) 'The General Theory of Employment', *Quarterly Journal of Economics*, 51, 2, pp. 209–223.
Kunzig, R. (2009) 'The Canadian Oil Boom: Scraping Bottom', *National Geographic*, 215, 3, pp. 34–59.
Lockie, S. and Measham, T. (2012) 'Social Perspectives on Risk and Uncertaint: Reconciling the Spectacular and Mundane', pp. 1–13 in T. Measham and S. Lockie (eds), *Risk and Social Theory in Environmental Management*. Canberra, ACT: CSIRO Publishing.
Lomborg, B. (2001) *The Skeptical Environmentalist*. Cambridge, UK: Cambridge University Press.
Lomborg, B. (2007) *Cool It*. New York, NY: Alfred A. Knopf.
Murphy, R. (2009) *Leadership in Disaster: Learning for a Future With Global Climate Change*. Montreal/Kingston: McGill-Queen's University Press.
Murphy, R. (2010a) 'Extract of Leadership in Disaster', *Bulletin of the American Meteorological Association*, 91, 3, pp. 390–391.
Murphy, R. (2010b) 'Environmental Hazards and Human Disasters', pp. 276–291 in G. Woodgate and M. Redclift (eds), *The International Handbook of Environmental Sociology*, 2nd ed. London: Edward Elgar.
Ray, D. and Guzzo, L. (1993) *Environmental Overkill*. New York, NY: HarperCollins.
Renn, O. (2008) *Risk Governance: Coping with Uncertainty in a Complex World*. London: Earthscan.
Renn, O. and Kinke, A. (2012) 'Complexity, Uncertainty and Ambiguity in Inclusive Risk Governance', pp. 59–76 in T. Measham and S. Lockie (eds), *Risk and Social Theory in Environmental Management*. Canberra, ACT: CSIRO Publishing.
Resnik, D. (2003) 'Is the Precautionary Principle Unscientific?' *Studies in History and Philosophy of Biological and Biomedical Science*, 34, 2, pp. 329–344.
Schneider, S. (2009) *Science as a Contact Sport: Inside the Battle to Save Earth's Climate*. Washington, DC: National Geographic Society.
Shane, S. and Lipton, E. (2005) 'Government saw Flood Risk but not Levee Failure', *New York Times*, 2 September, National page.
Simon, J. (1981) *The Ultimate Resource*. Princeton, NJ: Princeton University Press.
Zimmerman, E. (1951) *World Resources and Industries*, revised ed. New York, NY: Harper and Brothers.

19

Transboundary risk governance
Co-constructing environmental issues and political solutions

Rolf Lidskog

During the last four decades, a number of environmental problems have been conceptualized by the international community as transboundary problems that pose widespread threats to human well-being and which thus warrant global concerted action. This globalization of environmental problems is constituted through social processes in which different actors struggle to put various environmental issues on the international political agenda. Some actors may try to advance claims about the specifically transboundary character of a problem and to thus change the opportunity structure for political action. Others try to prevent an issue from becoming perceived as global or to erode the perceived seriousness of an issue that is already on the international agenda. This is because the regulation of environmental problems goes to the heart of domestic policies concerning, for example, energy policy, transport policy and employment policy. There are numerous interests beyond the explicitly environmental evoked by these policy domains and a variety of views as to whether certain environmental issues ought to be handled as matters of international priority or as the sole responsibility of domestic politics.

As with any political issue, it is those who speak for or claim to represent environmental issues who are the main causal agents in the political process. We must scrutinize why certain changes in nature are conceptualized as environmental problems, thereby gaining political priority, while others are not for at least two reasons. First, the environment has no voice of its own in contrast to categories of people (workers have labour movements, women their feminist movements, elderly people their pensioners' organizations, etc.). Changes in nature are always in need of human ombudsmen in order to attract political and public attention. However, it is likely that there is competition among human agents to represent nature and conflict over what this representation really means. Second, environmental problems are to a large extent discovered and mediated through science. Currently, it seems virtually impossible for an environmental condition to be successfully transformed into a political problem without scientific support in the form of data and analysis. However, science does not only discover and diagnose environmental problems. It also suggests solutions and proposes pathways for political action. Sometimes this is done explicitly, where science consciously formulates political advice and recommendations and tries to influence policymaking through different strategies. At other times, its proposals are more implicit in the way nature is framed, diagnosed and measured, thereby implying the direction in which the solution can be found.

At the same time, the relationship between science and policy is not simple: some environmental problems become politically prioritized despite considerable scientific uncertainty (for example, forest decline in the 1980s), while scientific evidence supports other environmental problems that have low priorities in practice (for example, the detrimental effect of many air pollutants on materials, including cultural heritage). Furthermore, science and policymaking are not discrete activities related through a one-way process, starting from the scientific discovery of a problem, to politically formulated environmental goals and ending in governmental strategies for implementing these goals. Rather, the relationship between science and policy is one of reciprocity, whereby science and policy intermingle in a complex and dynamic process where environmental issues and political solutions are co-constructed.

This does not imply that nature plays no role for environmental regulation, but only that nature is always in need of interpretation. When analysing environmental policymaking and regulation, actors' understanding of an issue cannot be disregarded. It is their interpretation and evaluation of an issue – which includes values as well as knowledge – that motivates their actions. Obviously, this interpretation does not take place in a vacuum. It is triggered by something. Dramatic environmental catastrophes, for example, call for new interpretations of the situation and social responses to it. The 2011 Tohoku earthquake in Japan that resulted in more than 15,000 deaths and the nuclear disaster at the Fukushima nuclear power plant changed many people's understanding of both the vulnerability of society and the safety of nuclear power. In some cases, this change in understanding is triggered by the direct experience of the environmental change, such as those living in the Tohoku region. But, for many, it occurs through mediated information about the catastrophe through news coverage and social media.

This chapter explores how transboundary environmental issues are made governable. In particular, it is concerned with the dynamic character of environmental regulation as various actors strive to influence regulatory processes and to negotiate, and renegotiate, the boundaries between society, science and nature. Earlier demarcations are transgressed and new ones configured. Changes in nature, scientific understanding and political organizations influence each other, and often do so in unforeseeable ways. The European regulation of transboundary air pollution is used as case for this discussion. This environmental problem has a fairly long regulatory history, not least in Europe, where the Convention on Long-Range Transboundary Air Pollution (CLRTAP) was established in the late 1970s. CLRTAP is one of the oldest international environmental conventions regulating transboundary pollutants. Also, it is seen as one of the most successful cases of transboundary environmental regulation.

This chapter consists of four main sections. The first elaborates on how environmental problems are co-produced. The second section describes the development of the regulation of transboundary air pollution in Europe, in particular how science and policy co-produced the diagnosis, but also the remedy. The third section analyses the processes that led to this regulation. It emphasizes that, in order to solve a transboundary issue, that issue needs to be territorially anchored in order to reduce its complexity and attach responsibility to certain actors. The concluding section stresses the provisional character of the new boundaries created through regulatory processes. The construction of regulatory objects enables regulation, but these objects are always open to renegotiation and revision, thus dissolving existing regulations and developing new ones.

Science, policy and the environment

Environmental regulation does not only concern politics; it also concerns science. Even though lay understandings and various other kinds of knowledge – fuelled by media coverage and public discussion – are important in putting an environmental problem on the political agenda, these understandings and claims are commonly based on scientific knowledge. In most cases, it is through measurement instruments, monitoring stations, experiments, data collection, mapping and modelling that environmental problems become objects for political intervention (Asdal 2007; Latour 1987).

The production and use of scientific knowledge are widely recognized as essential elements of environmental standards-setting (Jasanoff and Wynne 1998). By defining the criteria by which environmental degradation and health risks are assumed to be avoided or mitigated, a standard determines what are considered 'good' and 'bad' environments; for example, clean and unclean air, anthropogenic or naturally caused climate change, and toxic or non-toxic soil. Examples of such environmental standards include the Red List of Threatened Species, which evaluates the risk of extinction of species and subspecies; carbon dioxide equivalents (CDE), which describe how much global warming a given type and amount of greenhouse gas may cause; and the concepts of 'critical levels' and 'critical loads', which evaluate the environmental effects of airborne pollutants. Thus, how an environmental problem is represented and measured – provided this representation is seen as scientifically sound and the political community deems it valid – has great implications for its regulation (Lidskog and Sundqvist 2011).

Instruments such as monitoring technologies and mapping activities do not simply measure the objective parameters of environmental processes; they are also important in shaping recognition and conceptualization of environmental problems. There are no well-defined environmental problems 'out there', ready to be acted on. Instead, a decisive part of the regulatory process is to define the object at stake, and do so in a way that makes it governable. But, to make it governable, it is not sufficient to characterize a problem's causes, effects and remedies. Actors responsible for changing behaviour and actions to be taken also need to be identified. A proposed solution with no attachment to specific actors rarely works in practice.

Even if science is pivotal for defining issues, developing measures and proposing remedies, this does not take place in a social vacuum. Scientists themselves take into account political considerations when designing policy-relevant knowledge, and powers other than scientific ones also influence and shape how measurement develops. Often, on the surface, science and policy are presented as separate activities (Hilgartner 2000; Latour 1998). In practice, however, there is a dynamic interdependence between them, in which science and policy are co-produced in the sense that policy influences the production and stabilization of knowledge, while knowledge simultaneously supports and justifies that policy (Jasanoff 2004). Thus, constructing science-based policy and policy-relevant sciences goes hand in hand, even if they may be presented as separate activities. It is misleading to see scientific knowledge that supports policy processes as produced at an objective distance from political activities, only thereafter being communicated to political actors. Instead, science and policy co-construct a space populated by governable entities.

That science and policy together construct environmental issues does not mean that it takes place without any input from processes in nature. We do not have a science of environmental objects and a politics of human subjects – a world of mute nature and speaking humans (Latour 2011). Neither does nature directly speak for itself. What we have is a nature that always has to speak through something or someone – be it an experiment, a measure, a standard, a scientist or a local resident. It is misleading to conceptualize reality as divided into an external world of things and facts, on the one hand, and an internal world of understandings, definitions and meanings, on the other. Instead of drawing borders between nature and culture, things and humans, material and social, real and constructed, we have social networks and practices that draw together seemingly different entities (Latour 1993, 2004). It is the dynamic interplay between different factors – society, nature and science – that make up reality.

Thus, to make transboundary risks governable, the complexity of environmental issues must be reduced and responsibility attached to specific actors. The next section shows how a specific transboundary risk — air pollution — was constructed as an object of international regulation. In particular, it will illustrate how science and policy co-constructed both the environmental problem of transboundary air pollution as well as its solution.

Regulating transboundary air pollution

In the 1960s, it was believed that the pollutants from tall smokestacks were naturally diluted to a harmless level when released into the atmosphere. Even though there was little doubt as to the acid output from power plants based on fossil fuels, the idea of long-range air pollution was highly controversial among both researchers and politicians. By the end of the 1960s, however, Scandinavian researchers demonstrated that, contrary to earlier belief, air pollutants could travel several thousand kilometres before deposition and damage occurred (Lundgren 1998). The direct link between sulphur emissions in the United Kingdom and continental Europe and the acidification of Scandinavian lakes was established, and the burning of fossil fuels for transport, heating and power generation became not only a local, but also a transnational problem. Slowly, an environmental consciousness emerged, encompassing an understanding in which the effects of local sources were seen to have a new temporal–spatial pattern: delayed in time and extensively spread out in space. Due to the spatial characteristics of long-range air pollution, initiatives for multilateral political cooperation were undertaken. In Europe, it was discovered that transboundary air pollutants constituted a threat to a common European atmosphere (Lundgren 1998; Sundqvist 2011; VanDeveer 2004).

Because of the transboundary nature of the problem, no single country was able to solve its acid rain problems without the help of adjacent acid-producing countries, making international cooperation necessary. But the discovery also fostered competition and conflict between nation states, as the damages incurred by importing acidifying substances and the benefits reaped from exporting them were unevenly distributed among European countries. Hence, some countries ended up as net sufferers, while others were net beneficiaries. This meant that the polluters did not directly share the benefits of abatement, thus creating a negative incentive structure for international regulation of transboundary air pollution. This was reinforced by the fact that the costs of the abatement strategies differed considerably from nation to nation because of different production structures. Moreover, the benefits produced by abatement policies also differed country to country. Finally, there were scientific disputes concerning the character of the acid rain. Even if there was little doubt as to the acid output of fossil fuel-burning plants, there was no firmly established knowledge about the relationship between the emissions and their ecological consequences (Lidskog and Sundqvist 2011).

Claims concerning acid rain were made in a climate of scientific uncertainty and political misunderstanding, making it a technically complex and politically charged environmental controversy. In the early 1970s, coal-burning energy producers, the coal-mining industry, the car manufacturing and other industries, along with some political parties, contested the claims, arguing that natural factors such as volcanic eruptions, rotting vegetation and forest fires were more culpable than emissions from fossil fuel combustion. Also, West Germany and Great Britain — which were large net producers of transboundary air pollution — campaigned against international regulation (Boehmer-Christiansen 2000; Hajer 1995; Sprinz and Wahl 2000).

By providing scientific arguments, researchers helped proponents of regulation to force opponents to change their strategies from contesting scientific knowledge claims to emphasizing the cost-efficiency of proposed reductions. Thus, from the late 1960s to the early 1980s, acid rain was transformed from an esoteric branch of scientific research in certain specialized fields of atmospheric chemistry and ecology into a household term. In a process involving scientific development and political initiative, an international agreement – the Convention on Long-Range Transboundary Air Pollution – was finally adopted in 1979. This was the first multilateral, legally binding instrument addressing atmospheric environmental issues, with the aim of reducing emissions of airborne cross-border pollutants within the United Nations Economic Commission for Europe region. It was later followed by protocols regulating specific substances. Thirty-three parties, including thirty European nation states, the United States, Canada and the European Commission, signed the Convention (Munton et al. 1999: 167). Today, fifty-one parties have ratified the Convention.

Initially, the Convention developed protocols that were based on flat-rate reduction of pollutants – i.e. the same emission cuts for all countries. But 1994 marked a new era in the work of the CLRTAP regime, with a second generation of protocols that focused on varying national reduction rates based on the approach of cost-effectiveness and critical loads – i.e. effects in relation to what nature can withstand. The latest protocol, signed in 1999, was a multi-pollutant/multi-effect protocol that regulates four types of compounds – sulphur dioxide, nitrogen oxides, ammonia and volatile organic compounds – that affect human health, natural ecosystems, materials and crops through acidification, eutrophication and ground-level ozone. It is considered to be one of the most sophisticated and most scientifically based conventions ever signed, as it assesses effects rather than emissions, which is a delicate process that has more stringent requirements of scientific knowledge (Thompson 1999).

Co-producing science and policy

Through a series of workshops, mostly organized by CLRTAP, the key concepts of critical load and critical level were gradually incorporated into abatement strategies. They are defined as follows:

> 'Critical load' means a quantitative estimate of an exposure to one or more pollutants below which significant harmful effects on specific sensitive elements of the environment do not occur, according to present knowledge.
>
> *(UNECE 1999: Article 1)*

> 'Critical levels' mean concentrations of pollutants in the atmosphere above which direct adverse effects on receptors, such as human beings, plants, ecosystems or materials may occur, according to present knowledge.
>
> *(UNECE 1999: Article 1)*

Disagreements were initially articulated due to scientific uncertainty and gaps in knowledge, but consensus over the usefulness of these concepts were gradually reached amongst the researchers involved (Sundqvist et al. 2002). Once established, the concepts were further developed and underpinned by scientific work that identified many uncertainties and caused some controversies. Despite this, the critical load/level concepts have played a central role in consensus formation among scientists and politicians in developing coherent regulation of transboundary air pollution. Making such an assessment implies broad monitoring programmes for collecting data on environmental effects, as well as integrated assessment modelling that incorporates the perspectives and findings of various disciplines. This enables the evaluation of environmental impacts of alternative abatement strategies, thereby assisting negotiators and decision makers in regulating transboundary air pollution.

The first interactive computer model that used the critical loads approach and cost-effectiveness to suggest optimized abatement strategies came in the form of RAINS (the Regional Acidification Information System), developed by the International Institute for Applied Systems Analysis (IIASA) in 1984. It was developed in cooperation with researchers from a range of disciplinary and national affiliations, most of them connected to the CLRTAP. The RAINS model has been able to bring heterogeneous scientific practices together to speak with a single voice in a scientifically defensible way, as well as communicate results to policymakers (Sundqvist et al. 2002).

In the 1970s, several monitoring stations were set up in European countries and emission data was exchanged (Sundqvist et al. 2002). The EMEP (the Cooperative Programme for the Monitoring and Evaluation of Long-Range Transmission of Air Pollutants in Europe) was established in 1978. This enabled the development, support and coordination of national monitoring programmes, thereby facilitating the evaluation of protocols on emission reductions. Today, it encompasses some hundred monitoring stations in Europe, measuring sulphur dioxide, nitrogen dioxides and ground-level ozone. The EMEP was eventually integrated into the CLRTAP and became a mechanism for producing, distributing and exchanging standardized emission data, which is key to the evaluation of the effects of various abatement strategies (Lidskog and Pleijel 2011; Munton et al. 1999; Wettestad 2002). It also fostered the development of uniform abatement strategies and shared understandings and modelling of transboundary air pollution flows.

However, simply producing quantitative estimates of critical levels for various pollutants and ecosystems is of no help to policymakers when negotiating relevant, cost-effective abatement strategies. They need help interpreting the data to make sense of them, so that it can guide policymakers in developing ways to abate air pollution. In response to this, the CLRTAP's *Mapping Manual* was developed as a tool for modelling and mapping critical levels and exceedances of these levels, which later extended to dynamic modelling of acidification. It standardized the methods for deriving data with which to assess effects and risk, and became an important means for making the effects of air pollution understandable and meaningful to politicians and negotiators.

The concept of critical loads, together with the monitoring programme EMEP, the integrated assessment model RAINS, and the mapping manual, have been important not only in facilitating communication between scientists and policymakers, but also in mobilizing policy-relevant scientific research and shaping science-based policy. These science-based tools not only helped influence the way the problem was understood, but also their solutions. Hence, the environmental problem and its possible remedies were co-constructed.

Spatiality, complexity and responsibility

As shown through the brief presentation of the history of regulation of transboundary air pollution, the regulatory objects are not stable or complete entities ready to be governed by political actors. Neither are they first made governable by science and, thereafter, distributed to political actors. Instead, science and policy interacted and negotiated with each other in order to create regulatory objects. Three tasks were crucial in order to make the transboundary issue of air pollution governable: (1) territorial anchoring; (2) reduction in complexity of the issue; and (3) identification of relevant actors to involve in the regulation.

Transboundary problems and spatial solutions

In a world organized through geographical borders, issues have to be related to these borders. Thus, to become politically manageable, an issue must be territorially anchored, i.e. have a spatial identity (Lidskog et al. 2011). Through shaping the spatial identity of a problem, a certain environmental issue can be handled as a matter of international priority, as the sole responsibility of domestic politics or as a local problem to be solved by municipalities (Czarniawska and Joerges 1996). Through different kinds of scale-making exercises, actors try to relate issues to geographical borders and thereby construct regulatory objects. Thus, to define an issue as transboundary does not make space meaningless but, on the contrary, calls for clear spatial demarcations.

Earlier, air pollution was believed to be exported to a global common – the air. However, with the discovery that airborne pollutants will ultimately end in deposition on land, creating environmental damages to specific national territories, a different kind of environmental problem was discovered. This discovery of the transboundary character of air pollution did not result in the dissolution of spatial boundaries. Instead, it meant the transgression of earlier demarcations between local, national and international environmental problems, and the establishment of new ones. The EMEP revealed the geography of net exporters and importers. A political geography of winners and losers was constructed, and new and different incentives for international cooperation were thus established. The regulatory solutions were geographically differentiated, which was then reflected in the differentiated demands on emission reduction and cost-effective abatement strategies. Thus, the transboundary issue of air pollution was made governable by anchoring it territorially.

Reducing complexity by constructing risk

Environmental problems are increasingly understood as complex and ambiguous phenomena associated with considerable uncertainty (Beck 1992; Irwin and Michael 2003). To make such vague and multifarious phenomena manageable, their complexity needs to be reduced and the uncertainties associated with them eliminated or at least made manageable. A central way to reduce complexity is to conceptualize an issue in terms of risk. This means to anticipate potential harm, to average these events over time and space and to use relative frequencies (observed or modelled) as a means to specify probabilities (Jaeger et al. 2001). Also, boundaries are drawn between what are acceptable and unacceptable risks, and systems are developed to make sure that risks are within acceptable levels (Hood et al. 2001). Conceptualizing an issue as a risk opens space for action by making the future the object of calculation, deliberation and decision making. In this sense, risk calculations enrol futures and shape policy formulations (Wynne 1996).

Monitoring activities and critical levels used to identify and measure air pollution were combined with risk assessment tools such as mapping technologies and scenario building in order to analyse the risk different geographical areas had for exceeding critical levels. Through these risk assessments, geographical consequences of current emissions, as well as prognosticated future emissions, were constructed. A better picture of the geographical locations at risk, as well as the source areas of these risks, started to emerge, which then became the object of political action. Through scientific practices and political negotiations, accepted levels of emissions and exposure were constructed and abatement strategies developed. Thus, governable entities in need of regulation were defined in order to avoid the risk for detrimental environmental consequences.

Creating responsibility by ascribing capabilities

The construction of a spatial identity and reduction of uncertainty is necessary, but not sufficient, to govern a transboundary issue. There is also a need to identify which actors are best suited to take action and to be responsible for developing and implementing rules (Winter 2006). As seen above, constructing a spatial identity for and reducing the complexity of an issue involves drawing boundaries and making demarcations. These boundaries influence what tasks, mandates, responsibilities and identities are ascribed to various actors, resulting in certain actors being seen as central to the regulatory work and others as peripheral (Lidskog et al. 2009).

From the very beginning, the governable entity of transboundary air pollution was presented as an object for nation states to govern. It was clear that it was other actors – such as the coal industry – that needed to reduce their emissions. But nation states were heavily dependent on these industries. Environmental organizations, the media and citizens played an important role in initiating the process of regulation and mobilizing support for it, but they had very little say in the design of the regulation. To a large extent, the negotiation and regulatory work took place behind the closed doors of policymakers and environmental scientists. Even the CLRTAP has been, since its inception, heavily dependent on economic support from individual nation states (Hettelingh et al. 2004).

That nation states are seen as central actors does not imply that other actors have no obligation. It only means that the solution to the problem of transboundary pollution was transformed into a national problem, where the nation has to take responsibility for reducing its national emissions (which were different in scope, due to the geographically differentiated emissions the Convention demanded). In this case, many states successfully reduced their emissions on a large scale through a combination of economic restructuring, technical development and political measures.

Conclusion

This chapter has discussed how transboundary environmental issues are rendered governable. It has argued that all environmental problems call for social practices that make the problem visible, institutional mechanisms that disseminate knowledge about it and strategies that mobilize support for regulation. The reason for this is that environmental issues do not exist as regulatory objects ready to be governed. The construction of regulatory objects enables regulation, but these objects are always open for renegotiation and revision, implying that new boundaries are created in regulatory processes. Regulating transboundary issues never takes place in a vacuum. Actors struggle to construct environmental problems, both as governable and as ungovernable entities. Consequently, there is co-production of the governable space, a space in which changes in nature, scientific understandings and political organizations influence each other in potentially unforeseeable ways. The chapter has identified three dimensions of particular relevance when transforming transboundary environmental issues into governable objects: (1) the spatial demarcations that anchor issues territorially; (2) the reduction of complexity of vague and multifarious issues through the language of risk; and (3) the constitution of certain actors as responsible for handling the issue.

On the surface, national negotiators struggle to influence (or hinder) the development of international agreements. Beyond that, however, there are environmental monitoring and modelling, scientific assessment, environmental organizations, transnational companies and other practices and organizations that heavily influence the development of international regulations. What may be seen as static and solid regulation, steered by formal agreements and signed by states, is in fact the result of dynamic processes in which actors struggle to disseminate their understanding of the main causes of pollution and relevant remedies throughout society.

The political handling of transboundary air pollution is – as in the case of the international regulation of substances that deplete the ozone layer – seen by many policy evaluators as a regulatory success, especially in contrast with many other transboundary environmental issues. Emissions of several airborne substances, not least sulphur, have been radically reduced. It is important to note, however, that this regulation was heavily supported by other factors such as economic restructuring (of electricity and coal-mining industries) and technological development (for example, low-sulphur fossil fuels and desulphurization equipment). Furthermore, growing political complexity, new scientific findings, lack of public support, and competition with other transboundary problems – both environmental and non-environmental – are seen as major obstacles to advancing policy development in this area. The organizational context has recently changed due to the stronger political role of the European Union and its air policy for all member states. Scientific advances have also led to a more complex and dynamic view of ecosystems. Additionally, there has been a gradual shift in focus from aquatic and terrestrial ecosystems towards human health. This shift means that epistemic networks, established concepts and developed models may become obsolete or, at least, less relevant. Moreover, the issue of public trust has become a central concern of regulatory work. Without public legitimacy, decision makers and experts may not succeed in implementing further abatement strategies. The CLRTAP has hitherto paid very limited attention to this issue. Thus, what has been a successful regulation may rapidly become obsolete and in need of renegotiation.

References

Asdal, K. (2007) 'Enacting Things Through Numbers: Taking Nature into Accounting', *Geoforum*, 39, 1, pp. 123–132.
Beck, U. (1992) *Risk Society: Towards a New Modernity*. London: Sage.
Boehmer-Christiansen, S. (2000) 'The British Case: Overcompliance by Luck or Policy?', pp. 279–312 in A. Underdal and K. Hanf (eds), *International Environmental Agreements and Domestic Politics: The Case of Acid Rain*. Aldershot, UK: Ashgate.
Czarniawska, B. and Joerges, B. (1996) 'Travels of Ideas', pp. 13–48 in B. Czarniawska and G. Sevón (eds), *Translating Organizational Change*. Berlin: de Gruyter.
Hajer, M. (1995) 'Acid Rain in Great Britain Environmental Discourse and the Hidden Politics of Institutional Practice', pp. 145–164 in F. Fisher and M. Black (eds), *Greening Environmental Policy. The Politics of Sustainable Future*. London: Paul Shapman Publishing.
Hettelingh, J.-P., Bull, K., Chrast, R., Gregor, H.-D., Grennfelt, P. and Mill, W. (2004) 'Air Pollution Effects Drive Abatement Strategies', pp. 59–84 in J. Sliggers and W. Kakebeeke (eds), *Clearing the Air: 25 Years of the Convention on Long-Range Transboundary Air Pollution*. New York, NY: United Nations Economic Commission for Europe.
Hilgartner, S. (2000) *Science on Stage: Expert Advice as Public Drama*. Stanford, CA: Stanford University Press.
Hood, C., Rothstein, H. and Baldwin, R. (2001) *The Government of Risk: Understanding Risk Regulation Regimes*. Oxford, UK: Oxford University Press.
Irwin, A. and Michael, M. (2003) *Science, Social Theory and Public Knowledge*. Maidenhead, UK: Open University Press.
Jaeger, C., Renn, O., Rosa, E.A. and Webler, T. (2001) *Risk, Uncertainty and Rational Action*. London: Earthscan.
Jasanoff, S. (ed.) (2004) *States of Knowledge: The Co-Production of Science and Social Order*. London: Routledge.
Jasanoff, S. and Wynne, B. (1998) 'Science and Decision Making', pp. 1–87 in S. Rayner and E.L. Malone (eds), *Human Choice and Climate Change*, vol. 1, *The Societal Framework*. Columbus, OH: Battelle Press.
Latour, B. (1987) *Science in Action: How to Follow Scientists and Engineers Through Society*. Cambridge, MA: Harvard University Press.
Latour, B. (1993) *We Have Never Been Modern*. New York, NY: Harvester Wheatsheaf.
Latour, B. (1998) 'From the World of Science to the World of Research?' *Science*, 280, 5361, pp. 208–209.
Latour, B. (2004) *Politics of Nature: How to Bring the Sciences into Democracy*. Cambridge, MA: Harvard University Press.
Latour, B. (2011) 'From Multiculturalism to Multinaturalism: What Rules of Method for the New Socio-Scientific Experiments? *Nature and Culture*, 6, 1, pp. 1–17.
Lidskog, R. and Pleijel, H. (2011) 'Co-Producing Policy-Relevant Science and Science-Based Policy: The Case of Regulating Ground-Level Ozone', pp. 223–250 in R. Lidskog and G. Sundqvist (eds), *Governing the Air: The Dynamics of Science, Policy, and Citizen Interaction*. Cambridge, MA: MIT Press.
Lidskog, R., Soneryd, L. and Uggla, Y. (2009) *Transboundary Risk Governance*. London: Earthscan.

Lidskog, R. and Sundqvist, G. (2011) 'The Science–Policy–Citizen Dynamics in International Environmental Governance', pp. 323–360 in R. Lidskog and G. Sundqvist (eds), *Governing the Air. The Dynamics of Science, Policy, and Citizen Interaction*. Cambridge, MA: MIT Press.

Lidskog, R., Uggla, Y. and Soneryd, L. (2011) 'Making Transboundary Risks Governable: Reducing Complexity, Constructing Identities and Ascribing Capabilities', *Ambio: A Journal of the Human Environment*, 40, 2, pp. 111–120.

Lundgren, L.J. (1998) *Acid Rain on the Agenda: A Picture of a Chain of Events in Sweden, 1966–1968*. Lund: Lund University Press.

Munton, D., Sooros, M., Nikitina, E. and Levy, M.A. (1999) 'Acid Rain in Europe and North America', pp. 155–248 in O.R. Young (ed.), *The Effectiveness of International Environmental Regimes: Causal Connection and Behavioural Mechanisms*. Cambridge, MA: MIT Press.

Sprinz, D.F. and Wahl, A. (2000) 'Reversing (Inter)national Policy: Germany's Response to Transboundary Air Pollution', pp. 139–166 in A. Underdal and K. Hanf (eds), *International Environmental Agreements and Domestic Politics. The Case of Acid Rain*. Aldershot, UK: Ashgate.

Sundqvist, G. (2011) 'Fewer Boundaries and Less Certainty', pp. 195–221 in R. Lidskog and G. Sundqvist (eds), *Governing the Air: The Dynamics of Science, Policy and Citizen Interaction*. Cambridge, MA: MIT Press.

Sundqvist, G., Lettel, M. and Lidskog, R. (2002) 'Science and Policy in Air Pollution Abatement Strategies', *Environmental Science and Policy*, 5, 2, pp. 147–156.

Thompson, J. (1999) *Introduction to the Ministerial Declaration on Long-Range Transboundary Air Pollution*. Gothenburg, 30 November.

UNECE (United Nations Economic Commission for Europe) (1999) *Protocol to the 1979 Convention on Long-Range Transboundary Air Pollution to Abate Acidification, Eutrophication and Ground-Level Ozone*. Available at www.unece.org/fileadmin/DAM/env/lrtap/full%20text/1999%20Multi.e.pdf. Accessed 21 January 2013.

VanDeveer, S.D. (2004) 'Ordering Environments: Regions in European International Environmental Cooperation', pp. 309–334 in S. Jasanoff and M.L. Martello (eds), *Earthly Politics: Local and Global in Environmental Governance*. Cambridge, MA: MIT Press.

Wettestad, J. (2002) *Clearing the Air: European Advances in Tackling Acid Rain and Atmospheric Pollution*. Aldershot, UK: Ashgate.

Winter, G. (ed.) (2006) *Multilevel Governance of Global Environmental Change: Perspectives from Science, Sociology and the Law*. Cambridge, UK: Cambridge University Press.

Wynne, B. (1996) 'SSK's Identity Parade: Signing Up, Off-and-On', *Social Studies of Science*, 26, 2, pp. 357–391.

20

The role of professionals in managing technological hazards

The Montara blowout

Jan Hayes

In 2011, three of the five largest companies in the world (based on revenue) were publicly owned oil companies – Royal Dutch Shell, ExxonMobil and BP (Fortune n.d.). Another three – Chevron, Total and ConocoPhillips – made the top twenty. The oil industry's safety record is more or less consistent with broad industrial sector performance, but the environmental and social impacts of accidents, when they do occur, can be substantial. The history of the offshore oil and gas sector, in particular, has been punctuated by a series of major disasters. In 1980, the Alexander Kielland drilling rig capsized off the coast of Norway, causing 123 fatalities. The Ocean Ranger drilling rig sank off the Canadian coast in 1982 with 84 fatalities, and when the Piper Alpha production platform off the UK coast was destroyed by fire in 1987, 167 lives were lost. The 2010 blowout from the Deepwater Horizon drilling rig, working under contract to BP in the Gulf of Mexico, demonstrated yet again the hazards of offshore oil. In this case, the cloud of hydrocarbons flowing from the uncontrolled well ignited immediately, causing the deaths of eleven crew members and numerous injuries. The subsequent oil spill caused major environmental damage in areas of intense human activity. The financial cost to BP and its shareholders has been significant and is ongoing. The rig itself, destroyed in the disaster, cost $350 million to build in 2002 (National Commission on the BP Deepwater Horizon Oil Spill and Offshore Drilling 2011: 2). Following the accident, BP established a $20 billion compensation fund and, as of August 2011, had paid out over $7 billion to individuals, businesses and government impacted by the oil spill (BP n.d.). Litigation is expected to continue for decades.

These and other incidents have led to significant social science research interest in the social factors behind major industrial accidents. This research has two main strands. Cognitive psychology has been used as a framework to study the behaviour of individuals in the field, focusing on unsafe acts and the development of organizational interventions designed to foster behaviour modification among field personnel. Of more interest here is the research that conceptualizes accidents as linked to organizations, rather than to individuals. In this way of considering accident causality, field personnel are typically seen as passive actors whose actions are determined by the organizational context in which they work. Further, that organizational context is created by senior managers. In this chapter, it will be argued that research has largely ignored issues of conflict and power within organizations and, as a result, has failed to acknowledge the role of senior technical professionals in ongoing safety assurance. A specific case, a blowout in the Montara oil field off the northwest coast of Australia in 2009, is used to illustrate the role of both managers and technical professionals in accident causation. This has important implications for ongoing accident prevention.

Preventing disasters: the social science view of organizations

Starting with Barry Turner's work in the late 1970s (Turner and Pidgeon 1997), closely followed by analysis of the circumstances surrounding Three Mile Island (La Porte 1981; Perrow 1982), social scientists from various fields (sociology, management, organizational psychology) have taken an increasing interest in the role of organizations, rather than individuals, in industrial safety. One common approach is to study organizations that have experienced accidents, in order to identify failings. Alternatively, other studies focus on organizations that perform well, in an attempt to explain why this might be the case.

Focusing first on disaster analysis, a significant body of analysis has been produced regarding the causes of specific disasters in complex socio-technical systems. These include the loss of the space shuttles Challenger (Vaughan 1996) and Columbia (Starbuck and Farjoun 2005), the loss of US Black Hawk helicopters (Snook 2000), the explosion and fire at Exxon's Longford (Australia) gas plant (Hopkins 2000), the BP Texas City Refinery fire (Hopkins 2008) and the Montara blowout (Hayes 2012a). This research has led to generalized models regarding the organizational causes of accidents, the most well known of which is James Reason's (1997) Swiss cheese model. According to this model, system defences that are functionally designed to prevent any given hazard leading to an accident are understood to be imperfect in practice (like holes in Swiss cheese). An accident occurs when the holes in the slices of cheese line up and provide an accident trajectory through all the defences. In this model of system failure, unsafe acts by people in the field are seen to be caused by workplace factors such as competency, rostering, control room design, task design, etc. Workplace factors, in turn, are seen to be caused by overriding organizational factors such as budgets, safety priorities, leadership, etc. In this way, the performance of all components in the system is ultimately linked to management decision making.

Organizational psychologists, by contrast, tend to conceptualize 'safety culture' (or 'safety climate') as the overarching determinant of safety behaviour. Many theorists have noted inconsistencies in the way this term is used and the implications of this inconsistency for actually improving safety performance (Antonsen 2009; Haukelid 2008; Høivik et al. 2009; Hopkins 2005; Klein et al. 1995; Ocasio 2005; Pidgeon 2010; Silbey 2009). Two key assumptions (usually unstated) underpin much of the research on safety culture. First, culture is driven by top management; and, second, culture is as a consequence essentially uniform across an organization. Schein's (1992) much-quoted definition of culture – 'the way we do things around here' – has come to imply a far more uniform organizational landscape than was originally intended. Antonsen (2009), in particular, has highlighted the role of conflict and power and the extent to which these issues are ignored or even hidden in recent safety culture discussions.

Perrow (1999: 378), also, has criticized research that emphasizes safety culture at the expense of questions of power. Based on his analysis of, in the first instance, the partial meltdown at the Three Mile Island nuclear power plant in 1979, Perrow argues that the potential for accidents depends on two interconnected factors: (1) the technical complexity of the technology; and (2) the degree of coupling within the technological system (that is, the extent to which parts of the system are interconnected and hence the speed with which problems can propagate and compound). He characterizes accidents in complex, tightly coupled systems as 'normal', in the sense that such accidents cannot be prevented or controlled and are, therefore, to be expected. If it is true that organizations based around complex and tightly coupled technologies can never be managed or operated to prevent accidents, then it follows that such technologies should never be approved unless governments and communities are willing to accept the consequences of their failure. However, the concept of the 'normal accident' can be criticized on several fronts. First, closer investigation of every accident Perrow has applied this concept to reveals that the accident was not inevitable but was caused by specific management failures (Hopkins, 2012: 159ff.). Second, the concept of the normal accident fails itself to deal with questions of power or any of the other social factors that would typically be associated with sociological analysis. Third, and as a consequence, the concept provides very little that is of use for improving risk management at an organizational level.

Other authors have posed even broader structuralist explanations for industrial disasters. In their analysis of the Deepwater Horizon incident, for example, Freudenburg and Gramling (2011) highlight what they see as a series of failings on the part of the US government to exert appropriate control on safety and environmental matters in the resources sector, stretching back over decades. In their view, this is a direct result of energy policy which promotes high oil consumption from US domestic supply, to the point where 'we may have "promoted" the societal significance of our technology up to, and perhaps beyond, the point where it can actually do what we expect it to do' (Freudenburg and Gramling 2011: 20). Similar to Perrow, they argue that serious accidents in the offshore oil industry are essentially inevitable.

In parallel with disaster-focused research, a complementary strand of social science research has addressed successful organizations and how they have achieved high levels of safety performance. Researchers taking this 'high reliability' view favour an ethnographic or sociological approach to what are known as 'normal operations studies' (Allard-Poesi 2005; Bourrier 1998, 2011; Hayes 2009; Hopkins 2009; La Porte 1996; La Porte and Consolini 1991; Roberts 1994; Rochlin 1999; Schulman 1996; Weick 1987, 1993; Weick and Sutcliffe 2001). Many organizational studies describe the qualities of specific high performing organizations, but the only integrated theory which draws all these strands together is Weick and Sutcliffe's (2001) High Reliability Organizations (HRO) theory. These researchers claim that high functioning, or high reliability, organizations have the ability to detect and respond to system variations due to their state of mindfulness about their operations. In turn, mindfulness is fostered by five qualities, which are:

- *preoccupation with failure*: seeking out small faults in the system and using those to improve performance;
- *reluctance to simplify*: valuing diversity of views and resisting the temptation to jump to quick conclusions;
- *sensitivity to operations*: valuing experienced operating people who have a nuanced system understanding;
- *commitment to resilience*: using layers of protection, valuing redundancy in equipment and people; and
- *deference to expertise*: placing appropriate value on the advice of technical experts in decision making.

Unlike the work based in disasters, this area of research does not address management directly. The qualities listed above are seen as desirable, but this theoretical approach makes no claim regarding uniformity of behaviour across an organization. The chapter will now turn to an analysis of the Montara blowout, drawing on both accident causation and HRO models.

Jan Hayes

The Montara blowout

The incident

The continental shelf off the northwest coast of Australia holds rich reserves of oil and gas, with hydrocarbon exploration and production activities in the area stretching back over decades. On 21 August 2009, those in charge of drilling the Montara H1 oil and gas well lost control of the well, resulting in uncontrolled flow of hydrocarbons to the environment. This type of event is known in the oil industry as a blowout. The incident has been the subject of a statutory inquiry (Borthwick 2010).

At the time of the blowout, the facilities in the area (250 kilometre offshore) were the West Atlas drilling rig, with a crew of sixty-nine people operating over the small unmanned Montara Wellhead Platform, and Java Constructor, a construction barge with 174 people on board. As a hydrocarbon cloud engulfed the facilities, the vessel rapidly moved away from the drilling rig and the drilling rig personnel were evacuated. Oil and gas flowed for more than ten weeks before it ignited. The resulting fire continued for a further two days before the flowing fluids were brought under control on 3 November 2009 and the fire extinguished.

No one was injured or killed as a result of this incident. It has to be said that this was more good luck than good management and that, had the blowout ignited immediately, the result could have been similar to the Deepwater Horizon incident. At the time of the Montara blowout, the construction vessel Java Constructor was close by. If the gas cloud had ignited immediately, it is possible that the crew of the vessel (as well as those on the drilling rig itself) could have been directly impacted. Apart from the loss of the drilling rig and the platform due to the fire, the physical consequences of the Montara event included environmental pollution and, even in this area, luck has played a part. Due to the light nature of the escaping fluids and the remote location of the well, the majority of the hydrocarbon weathered away, leading to relatively little impact on the Australian coast or coastal marine life when compared with events such as the Deepwater Horizon blowout off the US coast (Borthwick 2010: 26). However, the Commission of Inquiry did note that impacts have been felt as far away as West Timor and that the lack of baseline data and the slow response in putting a monitoring plan in place mean that the full extent of the impact of the Montara spill will never actually be known.

Blowouts are a well-known hazard in the offshore oil and gas industry. There were thirty-nine such incidents on the US Outer Continental Shelf in the period 1992 to 2006 (Izon et al. 2007). The Australian offshore industry had also experienced six blowouts prior to Montara (with the immediately previous one being in 1984) (Borthwick 2010). Given the potential for disaster, and because of the large sums of money involved, drilling and well construction activities in this global industry are tightly controlled, both within operating companies and by regulation. Despite this, all well control systems failed at Montara.

Barriers to prevent flow from the well

At the time of the incident, the H1 well had been drilled, and two sections of concentric pipe known as casing strings had been put in place in the well. The outer casing ran from the surface to a depth of 1,640 metres and the inner casing ran inside it to a depth of approximately 3,800 metres. This was the depth at which hydrocarbons were expected to be present. The detailed sequence of events that led to the blowout starts with the cementing of what is known as a casing shoe at the bottom of the inner casing. This device is simply a one-way valve designed to prevent flow of hydrocarbons up the casing to the surface.

Table 20.1 Summary of well control barriers

Barrier	Status
Cement shoe	• Errors during installation resulted in a 'wet shoe' with impaired functionality • Test results that indicated this were ignored
Inner pressure containing corrosion cap (PCCC)	• Not designed for blowout prevention but included in well design partly for that purpose • Removed for operational reasons just before the blowout
Outer pressure-containing corrosion cap (PCCC)	• Not designed for blowout prevention but included in well design partly for that purpose • Never installed
Fluids in the well bore	• Could have provided a barrier, but weight not sufficient for that purpose and level not monitored

The operating company's Well Construction Standards required two proven barriers to uncontrolled flow to be in place when drilling activity was over. The primary pressure-containing barrier should have been the cemented casing shoe at the end of the inner casing. Based on the pressure and flow profiles measured and recorded during the cementing operation, it is apparent that the integrity of the cement was never proven and that the outcome was a 'wet shoe' with the cement contaminated by drilling fluids and/or hydrocarbons. Such contamination meant that the device could not be relied upon to prevent hydrocarbon flow up the well. The secondary barriers that should have been in place (according to the final well design) were two pressure-containing corrosion caps (PCCCs). The well design called for these to be installed on the top of both the inner string and the outer string. At the time of the blowout, neither of these caps was in place.

The well control barriers that were included in the well design and their status at the time of the blowout are shown in Table 20.1. The choices made in the months leading up to the blowout introduced a series of latent failures into the system of multiple barriers, negating the entire design philosophy and leaving the system in a very vulnerable state. Under these circumstances, uncontrolled flow of hydrocarbons to the surface became inevitable. In Perrow's (1999) terms, offshore drilling is moderately complex, but the long gestation period for this accident and the evidence that was available to be discovered during that period regarding the potential for failure mean that it cannot reasonably be called a tightly coupled system. The accident cannot reasonably be classified as 'normal' and hence we must look further for organizational explanations.

Lack of management attention to system integrity

One of the most startling events leading up to the blowout was the failure by the crew offshore to install the PCCC on the outer string, followed by their reporting to onshore management that such a cap had indeed been installed. This action potentially put the lives of the crew at risk, but they apparently did not see their actions in those terms. It is difficult to escape the conclusion that someone (indeed, probably several people) knew that the cap had not been installed and yet reported to onshore management that it was in place. This can perhaps best be considered as an indication of the relationship between offshore and onshore personnel within the organization. The onshore management team appear to have taken a 'hands-off' approach regarding work done offshore. In his statement to the Commission of Inquiry, the Drilling Superintendent stated that:

> if there was an issue with a forward plan that could not be resolved offshore [the Senior Drilling Supervisor] would call me to discuss the issue … The plans were not normally sent to the [onshore] office for review unless there was an issue that could not be resolved offshore … if the Senior Drilling Supervisor … needed additional expertise from onshore staff, he would telephone me.
>
> (Wilson 2009a: 33)

In other words, written plans for work were sent offshore (such as the drilling plan that required installation of the PCCCs), but the attitude of onshore management was to assume that offshore personnel were competent and operating in accordance with those plans without ever conducting any checks as to whether this was actually the case. The Drilling Superintendent also stated, in relation to the reports on cementing operations he was sent every day, that he had no reason to check the reports in detail and that when he 'reviewed the [daily drilling report] to see if there was any obvious errors or issues. There were none' (Wilson 2009a: 185). As the Commission of Inquiry report points out (Borthwick 2010: 3.123), the role of the Drilling Superintendent was the day-to-day supervision of activities offshore, and this involved much more than simply looking for obvious errors in written summaries of the work undertaken, especially work involving a critical safety function such as cementing the casing shoe. The Drilling Superintendent indicated again his overall attitude to supervision when he said:

> as there were no indications or reasons … to think that the wells were not suspended per the [drilling plan] and subsequent change control, there was no reason to conduct any form of audit to check that all work that was thought to be performed had in fact been completed.
>
> (Wilson 2009a: 266(c))

In fact, there was evidence available, particularly in relation to the state of the cemented casing shoe, if only he had looked. But checking on the technical integrity of completed work was not a task that management apparently saw as their responsibility. An email sent by the Drilling Superintendent to a wide range of people (including on-rig personnel, the overseas parent company managers and government representatives) five days before the blowout provides an insight into management priorities:

> Whilst we have been busy drilling some our guys have been working offline to suspend Montara H3ST-1 and Montara H1. Both wells will be fully suspended by the end of the day. This has saved us about 12-18 hours of rig time by being able to do this activity offline – a job well done.
>
> (Wilson 2009b)

Rig time is expensive. This email praised the offshore crew for saving time and hence money. In these circumstances, it is not difficult to imagine a scenario in which the PCCC was found at the last minute to be unserviceable, and offshore personnel thus decided to proceed and suspend the well anyway without the additional barrier in place. It is perhaps significant that the Daily Drilling Report from the offshore team to onshore management that first reports the fictional installation of the PCCC was sent the day after the above general note was circulated.

By sending such a report, the offshore personnel were simply confirming for onshore management what they wanted to hear.

Critical activities were conducted with no supervision, and managers put cost savings before safety. These were key failings on the part of the organization as a whole. The Reason model suggests that the unsafe acts by field personnel in failing to install the PCCC were a direct result of the lack of priority given to integrity issues by senior management. A safety-culture view of these events would also highlight similar aspects of the circumstances of the blowout, typically couched as lack of safety leadership. These are important points, but a more nuanced view highlights other factors, as described in the following section.

Separation of engineering integrity and operations functions

Statements by the Drilling Superintendent quoted in the previous section indicate the large measure of technical discretion given to offshore personnel on specialist engineering matters such as cementing operations. This is an important point. In addition to the lack of management, there was also a lack of any engineering integrity function. This means that offshore operational personnel were left to decide for themselves whether or not specialist engineering input was required. As is clear from the details of the cementing activity (Hayes 2012a), the personnel responsible for conducting these activities offshore had serious gaps in their technical ability to understand what was happening and to know at what point they needed to seek professional advice.

This situation appears to have had its roots in the roles previously held by the various individuals involved. The Drilling Superintendent had previously (prior to the operational drilling phase) held the role of Senior Drilling Engineer (Wilson 2009a: 12) In that role, he had been personally responsible for (amongst other things) the design of the wells. This was a primarily technical, rather than managerial, role. Once the drilling programme moved into the operational phase, he took on the management role of Drilling Superintendent, with all the offshore drilling crew reporting to him, including both Senior Drilling Supervisors and all the Drilling Supervisors (Wilson 2009a: 16). This is a significant change in role from technical expert in well design to manager of the implementation. Sensemaking theory (Weick 2001; Weick et al. 2005) reminds us that the cues we see depend critically on our perception of our role (or organizational identity). In this case, there was perhaps an intention on the part of the organization that the Drilling Superintendent maintain some technical oversight, but he seems to have had a different understanding of the new role. He emphasized his reliance on being told, rather than seeking out information for himself, when he said:

> Although there was appropriate communication between [the Senior Drilling Supervisor] and me on 7 March 2009 there was information that I consider, with the benefit of hindsight, could have been given to me so that I would be better able to make decisions about what needed to be done ... [Further] my post incident analysis indicates that the 9 5/8" casing shoe most probably did not form an adequate primary tested barrier however on the day with the information supplied to me, I had no reason to suspect that it was not an adequate barrier.
>
> *(Wilson 2009a: 259)*

This significant confusion over the role of individuals with both managerial and professional responsibilities was also seen in the decisions regarding use of PCCCs. Partway through the drilling of the well, the written drilling plan was changed to include PCCCs despite the fact that this type of device was not listed as a well control barrier in the company's Well Construction Standards. One key factor in assessing the suitability of these devices is the ability to test the seal provided once they are installed. This is a technical issue that was addressed by the Drilling Superintendent in a professional, rather than managerial, capacity. He described in evidence how he had sent the manufacturer's instructions offshore so that:

> the caps could be installed as per the manufacturer's instructions. I assumed that those instructions would call for an in situ pressure test after installation and I did not note prior to sending out the manufacturer's instructions that they themselves did not call for the PCCCs to be pressure tested once installed.
>
> *(Wilson 2009a: 198)*

In fact, the Operating and Service Manual for the PCCCs explicitly stated that they were not designed to operate as barriers against blowout and were only meant to be used on a well that had been otherwise secured. It seems likely that the Drilling Superintendent did not read any of the material provided by the manufacturer for this critical safety device (Borthwick 2010: 3.215).

Another individual holding a combined technical and managerial position showed similar reluctance to get involved in ensuring that technical standards were met in the field. Some months later, the decision was made (for operational reasons) to remove, and then not reinstall, the inner PCCC in the hours immediately prior to the blowout. In his statement, the Well Construction Manager (who was on the facility at the time) said that, based on the other barriers that he erroneously thought were in place, 'there was no compelling reason to re-install the 9 5/8" corrosion cap' (Duncan 2009: 251). Contrary to this, he told the Inquiry during the hearing that he had expected that the cap would be reinstalled once the cleaning work was complete and that when he discovered that it had not been installed he did not insist, on the basis that 'he did not want to give the impression to personnel on the rig that he was trying to teach them how to do their jobs' (Borthwick 2010: 3.187). It was the Well Construction Manager's role both to ensure that work was done in accordance with plans (and he apparently planned that the cap would be reinstated) and also to ensure the technical robustness of activities undertaken. To leave a well barrier uninstalled for such a reason is again an abrogation of both managerial and professional responsibilities.

In summary, as work on the well progressed, critical technical decisions were left to those offshore who then adopted an attitude of trial and error learning to fill the gaps in their knowledge of normal drilling practice. Those onshore had apparently adopted an attitude that 'No news is good news.' These are both highly problematic, and such attitudes can be contrasted with those that HRO researchers tell us are necessary to avoid accidents, such as preoccupation with failure, sensitivity to operations and commitment to expertise (Weick and Sutcliffe 2001).

It must be emphasized that this is not simply a case of requiring relevant people to have sufficient technical knowledge. It is common for incident investigations to focus on lack of technical knowledge, without considering why such knowledge may or may not be applied appropriately in any given case. The literature on professionalism gives some further insights that move the focus away from simply knowledge. The role of professionals focuses on both knowing and also, critically, on doing. Whilst the terms 'professional' and 'professions' 'are implicated in a fairly tangled and unruly web of usage' (Carr 2011: 97), there is broad agreement that, in addition to expert technical knowledge, the characteristics of professionals include being bound by a code of ethical conduct in addition to technical and/or commercial standards (Carr 2011; Friedson 2001; Middlehurst and Kennie 1997) and being able to exercise experienced judgement in specific cases, rather than relying completely on application of general rules (Dunne 2011). In the case of the Montara blowout, if senior company personnel had acted as professional drilling engineers they would have felt an ethical responsibility towards the safety of the facility and its crew and, therefore, would have endeavoured to ensure that technical standards were maintained so that the well was always sufficiently under control.

Leaders in high reliability organizations have the ability to see the potential for disaster despite their success. This has strong echoes in the literature on professionalism and professional wisdom. Swinton (2011: 157) claims that 'the development of professional wisdom is an act of the imagination' and, further, that 'you can only imagine what you have been taught to imagine'. Those working on and responsible for Montara were apparently lacking the ability to imagine that a blowout could occur and that their collective actions would directly impact the potential for such a catastrophe.

Implications for safety and future research directions

The Montara case study described above has shown that a richer picture of the social relationships within organizations, including the role of key technical professionals, in addition to the influence of senior management, provides a better understanding of accident causation in this case. This insight is not limited to the Montara accident analysis. Hopkins' (2008) work on the Texas City refinery fire highlighted that some of the contributing technical issues were known to technical specialists on the site, but that they had insufficient power within the organization for their concerns to be acted upon in an environment of strict cost control. He has advocated that organizational structures in high hazard organizations should ensure that technical specialists have sufficient organizational 'voice'. Similarly, the report into the Columbia space shuttle disaster discussed attributes of organizational design related to the power and authority of technical specialists that could be expected to prevent such incidents from occurring again. In particular, the report called for (amongst other things) 'a robust and independent programme technical authority that has complete control over specifications and requirements, and waivers to them' (CAIB 2003).

Other work on high functioning organizations has also emphasized the extent to which senior operations personnel have occupational identities as both employees and as professionals (Hayes 2009, 2012b). In organizations with a strong management focus on safety these two identities rarely conflict directly and yet professional qualities of experience, judgement, valuing understanding of complex systems and a sense of public trust are relied upon to make appropriate decisions in potentially uncertain situations. Managers must (be seen to) be in control and yet 'professional wisdom ... must incorporate a recognition of the limits of one's understanding and power – and thus avoidance of the hubris attendant on the absence of such recognition' (Dunne 2011: 23).

In the past two decades, it has become increasingly common for organizations to outsource specialist functions, essentially leaving a core of staff to manage the work of external professionals. This decreases the formal authority of professionals and leaves them with the power to provide advice at best. In many industries, including the offshore oil and gas industry, a 'career of achievement' (i.e. moving to a more senior position based on professional skills and knowledge) is seen as less desirable than a 'career of advancement' (i.e. progressing up the hierarchy of the organization chart) (Zabusky and Barley 1996). As described above, professional behaviour requires both knowing and acting. Structural isolation of professionals leaves them potentially unable to act. The nexus between knowing and acting is a potential fruitful area of research. Of course, in the case of Montara, technical and managerial roles were merged. It could be argued that separation of the two functions is potentially a more effective way to ensure that a professional perspective on operations is maintained. This is also an organizational design issue worthy of further investigation.

It is beyond the scope of this chapter to address why the regulatory regime in place in the offshore oil and gas industry failed to prevent the Montara incident. The debate over the merits of prescriptive versus risk-based regulation is extensive (see, for example, Ayres and Braithwaite 1992; Baldwin and Black 2008; Hale et al. 2002; Paterson 2000) and has been revisited in the USA since the Deepwater Horizon incident. Whilst regulators themselves are often well aware of the social science research on accident causation and prevention described above, with the exception of Norway national policymakers have generally proven to be more reluctant to focus regulatory attention on these matters with discussion mired in questions over use of models of organizational culture (Hayes in press). The work of the International Regulators Forum and national regulatory agencies in this area merit further attention.

Another potentially interesting area for research is the role of both industry lobby groups and professional associations in setting technical standards. Historically, professional societies such as the American Society of Mechanical Engineers (ASME) played a key role in standards development. In his article 'To Protect and Serve', Varrasi (2009: 28) describes ASME standards committees as 'a tradition of public service and professional responsibility that dates to the late nineteenth century industrial age'. Anecdotally, industry-based bodies are now more often playing a leading role and driving preparation of standards in parallel with other government lobbying activities. Bodies such as the American Petroleum Institute (API), the International Association of Drilling Contractors (IADC) and Oil and Gas UK (an industry trade association) are heavily involved in driving changes to technical standards as a result of the Deepwater Horizon incident. This blurring of roles at an institutional level again emphasizes the need for further research to clarify questions of power and control over key professional values.

The existence of a conflict between professionalism and managerialism has a long history in academic publications on the theory of work (see Causer and Jones 1996 for a summary). As early as the 1960s, authors have highlighted two potential sources of conflict: first, that professionals may see themselves as more closely aligned to public service than the organization for which they work; and, second, that managers may wish to constrain the independence and judgement of the professionals under their direction by imposition of bureaucratic rules. These are important issues in management of safety in complex socio-technical systems. Despite this, current models of organizational safety fail to adequately acknowledge the leadership role of professionals. Investigations into several recent accidents, including the Montara blowout, have highlighted the dangers of inadequate professional input into operations. Further research is needed into the impacts of current industry trends that are further downgrading the role of technical professionals.

References

Allard-Poesi, F. (2005) 'The Paradox of Sensemaking In Organizational Analysis', *Organization*, 12, pp. 169–196.
Antonsen, S. (2009) 'Safety Culture and the Issue of Power', *Safety Science*, 47, 2, pp. 183–191.
Ayres, I. and Braithwaite, J. (1992) *Responsive Regulation: Transcending the Deregulation Debate*. New York, NY: Oxford University Press.
Baldwin, R. and Black, J. (2008) 'Really Responsive Regulation', *Modern Law Review*, 71, pp. 59–94.
Borthwick, D. (2010) *Report of the Montara Commission of Inquiry*. Canberra, ACT: Commonwealth of Australia.
Bourrier, M. (1998) 'Elements for Designing a Self-Correcting Organization: Examples from Nuclear Power Plants', pp. 133–147 in A.R. Hale and M. Baram (eds), *Safety Management: The Challenge of Change*. Oxford: Pergamon.

Bourrier, M. (2011) 'The Legacy of the High Reliability Organization Project', *Journal of Contingencies and Crisis Management*, 19, 1, pp. 9–13.
BP (n.d.) *Claims Information*. Available at www.bp.com/sectiongenericarticle.do?categoryid=9036580and contentid=7067577. Accessed 19 November 2011.
CAIB (Columbia Accident Investigation Board) (2003) *Report of the Columbia Accident Investigation Board*, vol. 1. Washington, DC: Columbia Accident Investigation Board.
Carr, D. (2011) 'Virtue, Character and Emotion in People Professions: Towards a Virtue Ethics of Interpersonal Professional Conduct', pp. 97–110 in L. Bondi, D. Carr, C. Clark and C. Clegg (eds), *Towards Professional Wisdom: Practical Deliberation in the People Professions*. Farnham, UK: Ashgate.
Causer, G. and Jones, C. (1996) 'One of Them or One of Us? The Ambiguities of the Professional as Manager', pp. 91–112 in R. Fincham (ed.), *New Relationships In the Organised Professions*. Aldershot, UK: Ashgate.
Duncan, C.N. (2009) *Statutory Declaration of C.N. Duncan*. Available at www.montarainquiry.gov.au/exhibits.html. Accessed 22 January 2013.
Dunne, J. (2011) '"Professional Wisdom" in "Practice"', pp. 13–26 in L. Bondi, D. Carr, C. Clark and C. Clegg (eds), *Towards Professional Wisdom: Practical Deliberation in the People Professions*. Farnham, UK: Ashgate.
Fortune (n.d.) *Global 500*. Available at http://money.cnn.com/magazines/fortune/global500/2011/. Accessed 19 November 2011.
Freudenburg, W.R. and Gramling, R. (2011) *Blowout in the Gulf: The BP Oil Spill Disaster and the Future of Oil in America*. Cambridge, MA: MIT Press.
Friedson, E. (2001) *Professionalism: The Third Logic*. Chicago, IL: University of Chicago Press.
Hale, A., Goossens, L. and Van De Poel, I. (2002) 'Oil and Gas Industry Regulation: From Detailed Technical Inspection to Assessment of Safety Management', pp. 79–107 in B. Kirwan, A. Hale and A. Hopkins (eds), *Changing Regulation: Controlling Risks in Society*. Oxford, UK: Pergamon.
Haukelid, K. (2008) 'Theories of (Safety) Culture Revisited: An Anthropological Approach', *Safety Science*, 46, 3, pp. 413–426.
Hayes, J. (2009) 'Operational Decision-Making', pp. 149–178 in A. Hopkins (ed.), *Learning From High Reliability Organisations*. Sydney, NSW: CCH.
Hayes, J. (2012a) 'Operator Competence and Capacity: Lessons from the Montara Blowout', *Safety Science*, 50, 3, pp. 563–574.
Hayes, J. (2012b) 'Use of Safety Barriers in Operational Safety Decision Making', *Safety Science*, 50, 3. pp. 424–432.
Hayes, J. (in press) 'A New Direction in Offshore Safety Regulation', in M. Baram, P. Lindoe and O. Renn (eds), *Governing Risk in Offshore Oil and Gas Operations*. Cambridge, UK: Cambridge University Press.
Høivik, D., Moen, B.E., Mearns, K. and Haukelid, K. (2009) 'An Explorative Study of Health, Safety and Environment Culture in a Norwegian Petroleum Company', *Safety Science*, 47, 7, pp. 992–1001.
Hopkins, A. (2000) *Lessons from Longford: The Esso Gas Plant Explosion*. Sydney, NSW: CCH.
Hopkins, A. (2005) *Safety, Culture and Risk: The Organisational Causes of Disasters*. Sydney, NSW: CCH.
Hopkins, A. (2008) *Failure to Learn: The BP Texas City Refinery Disaster*. Sydney, NSW: CCH.
Hopkins, A. (ed.) (2009) *Learning from High Reliability Organisations*. Sydney, NSW: CCH.
Hopkins, A. (2012) *Disastrous Decisions: The Human and Organisational Causes of the Gulf of Mexico Blowout*. Sydney, NSW: CCH.
Izon, D., Danenberger, E.P. and Mayes, M. (2007) 'Absence of Fatalities in Blowouts Encouraging in MMS Study of OCS Incidents 1992–2006', *Drilling Contractor*, July/August, pp. 84–90.
Klein, R.L., Bigley, G.A. and Roberts, K.H. (1995) 'Organizational Culture in High Reliability Organizations: An Extension', *Human Relations*, 48, pp. 771–793.
La Porte, T.R. (1981) 'On the Design and Management of Nearly Error-Free Organizational Control Systems', pp. 185–200 in D.L. Sills, C.P. Wolf and V.B. Shelanki (eds), *Accident at Three Mile Island: The Human Dimension*. Boulder, CO: Westview Press.
La Porte, T.R. (1996) 'High Reliability Organizations: Unlikely, Demanding and at Risk', *Journal of Contingencies and Crisis Management*, 4, pp. 60–71.
La Porte, T.R. and Consolini, P.M. (1991) 'Working in Practice but not in Theory: Theoretical Challenges of "High-Reliability Organizations"', *Journal of Public Administration Research and Theory*, 1, pp. 19–48.
Middlehurst, R. and Kennie, T. (1997) 'Leading Professionals: Towards New Concepts of Professionalism', pp. 50–68 in J. Broadbent, M. Dietrich and J. Roberts (eds), *The End of the Professions? The Restructuring of Professional Work*. London: Routledge.

National Commission on the BP Deepwater Horizon Oil Spill and Offshore Drilling (2011) *Deep Water: The Gulf Oil Disaster and the Future of Offshore Drilling. Report to the President*. Washington, DC: National Commission on the BP Deepwater Horizon Oil Spill and Offshore Drilling.

Ocasio, W. (2005) 'The Opacity of Risk: Language and the Culture of Safety in NASA's Space Shuttle Program', pp. 101–121 in W.H. Starbuck and M. Farjoun (eds), *Organization At the Limit: Lessons from the Columbia Disaster*. Malden, MA: Blackwell.

Paterson, J. (2000) *Behind the Mask: Regulating Health and Safety in Britain's Offshore Oil and Gas Industry*. Aldershot, UK: Ashgate.

Perrow, C. (1982) 'The President's Commission and the Normal Accident', pp. 173–184 in D.L. Sills, C.P. Wolf and V.B. Shelanki (eds), *Accident at Three Mile Island: The Human Dimension*. Boulder, CO: Westview Press.

Perrow, C. (1999) *Normal Accidents: Living with High-Risk Technologies*. Princetown, NJ: Princetown University Press.

Pidgeon, N. (2010) 'Systems Thinking, Culture of Reliability and Safety', *Civil Engineering and Environmental Systems*, 27, 3, pp. 211–217.

Reason, J. (1997) *Managing the Risks of Organizational Accidents*. Aldershot, UK: Ashgate.

Roberts, K.H. (ed.) (1994) *New Challenges to Understanding Organizations*. New York, NY: Macmillan.

Rochlin, G.I. (1999) 'Safe Operation as a Social Construct', *Ergonomics*, 42, pp. 1549–1560.

Schein, E. (1992) *Organizational Culture and Leadership*. San Francisco, CA: Jossey-Bass.

Schulman, P.R. (1996) 'Heroes, Organizations and High Reliability', *Journal of Contingencies and Crisis Management*, 4, pp. 72–82.

Silbey, S. (2009) 'Taming Prometheus: Talk About Safety and Culture', *Annual Review of Sociology*, 35, pp. 341–369.

Snook, S.A. (2000) *Friendly Fire: The Accidental Shootdown of US Black Hawks Over Northern Iraq*. Princeton, NJ: Princeton University Press.

Starbuck, W.H. and Farjoun, M. (eds) (2005) *Organization at the Limit: Lessons from the Columbia Disaster*. Oxford, UK: Blackwell.

Swinton, J. (2011) 'The Wisdom of L'Arche and the Practices of Care: Disability, Professional Wisdom and Encounter-in-Community', pp. 153–168 in L. Bondi, D. Carr, C. Clark and C. Clegg (eds), *Towards Professional Wisdom: Practical Deliberation in the People Professions*. Farnham, UK: Ashgate.

Turner, B.A. and Pidgeon, N.F. (1997) *Man-Made Disasters*. Oxford, UK: Butterworth.

Varrasi, J. (2009) 'To Protect and Serve', *Mechanical Engineering*, 131, 6, pp. 28–33.

Vaughan, D. (1996) *The Challenger Launch Decision: Risky Technology, Culture and Deviance at NASA*. Chicago, IL: University of Chicago Press.

Weick, K.E. (1987) 'Organizational Culture as a Source of High Reliability', *California Management Review*, 29, pp. 112–127.

Weick, K.E. (1993) 'The Collapse of Sensemaking in Organizations: The Mann Gulch Disaster', *Administrative Science Quarterly*, 38, pp. 628–652.

Weick, K.E. (ed.) (2001) *Making Sense of the Organization*. Oxford, UK: Blackwell Business.

Weick, K.E. and Sutcliffe, K.M. (2001) *Managing the Unexpected: Assuring High Performance in an Age of Complexity*. San Francisco, CA: Jossey-Bass.

Weick, K.E., Sutcliffe, K.M. and Obstfeld, D. (2005) 'Organizing and the Process of Sensemaking', *Organization Science*, 16, pp. 409–421.

Wilson, C.A. (2009a) *Statutory Declaration of C.A. Wilson*. Available at www.montarainquiry.gov.au/exhibits.html. Accessed 22 January 2013.

Wilson, C.A. (2009b) *Email: Montara Platform Wells Morning Update*. 16 April 2009. Available at www.montarainquiry.gov.au/exhibits.html. Accessed 22 January 2013.

Zabusky, S. and Barley, S. (1996) 'Redefining Success: Ethnographic Observations in the Careers of Technicians', pp. 185–214 in P. Osterman (ed.), *Broken Ladders: Managerial Careers in the New Economy*. New York, NY: Oxford University Press.

21
Social learning to cope with global environmental change and unsustainability[1]

J. David Tàbara

In the current context of global environmental change and increasing unsustainability, social learning can be understood as the cultural and structural processes through which human societies reframe their worldviews and establish new patterns of interaction with biophysical systems to better cope with the unintended, unwanted, cumulative and often unexpected collective consequences of social action. Social learning can be conceived both as the *process*, as well as the *outcome*, of institutional transformation and creation in response to human-induced environmental change. On the one hand, social learning occurs within a given *social-ecological system of reference*, which includes both the human and biophysical conditions in which learning takes place. On the other hand, learning materializes as a complex and combined result of various *dynamics and mechanisms* whereby new frames, capacities and goals reorient collective behaviours towards desired conditions. However, the final outcomes of social learning cannot be fully anticipated. Indeed, we cannot know what we will know and what eventually will result from our knowledge, mostly because this depends on the processes we will choose for learning which, at present, are still unknown.

Yet there is not a single or universal definition of social learning. Distinct notions, theories and categorizations of learning exist in economics, psychology and policy sciences. The various approaches tend to assess the 'evidence' for learning based on different grounds. Cognitive and cultural learning perspectives tend to emphasize changes in understandings, assumptions, worldviews and motives, while relational and structural learning approaches tend to stress the creation of new capacities, the redistribution of responsibilities and the creation of long-term patterns of interaction through the transformation or constitution of new institutions. Some approaches also tend to emphasize the role of *processes* like the existence of opportunities for participation, the types of leadership, and modes of facilitation. Others concentrate their attention on *results*, like those of a cognitive, relational or institutional nature. The literature has also distinguished between *learning* and *knowing*, and some of the most promising research efforts today are oriented toward understanding the relationships between knowledge-building, learning and transformative change for sustainability (Blackmore 2007; Jennex 2008; Keen et al. 2005; KLSC 2011; Mostert et al. 2008; Pahl-Wostl 2009; Pahl-Wostl et al. 2008; Reed et al. 2010; SLG 2001; Siebenhüner and Heinrichs 2010; Wals 2007).

This chapter takes an integrative approach. Specifically, the focus here is on *sustainability learning* (Tàbara and Pahl-Wostl 2007), conceived as the complex structural process whereby changes both in worldviews and in the quality of social-ecological interactions emerge in ways that lead to more robust and resilient ways for humans to relate to the natural world. A transition to sustainability demands profound change in understandings, interpretative frameworks and broader cultural values and of beliefs, just as it demands transformations in practices, institutions and the social structures that regulate and coordinate individual actions and behaviours.

Multiple dimensions, multiple narratives

Given the vast literature devoted to formal theories and models of learning, the focus of the present contribution is limited to the examination of issues of global environmental change and (un)sustainability. I try to move away from any attempt to offer a simplified model or quick recipe for learning that claims to be able to support global sustainability. Similarly, I do not intend to provide an aseptic analysis, free of value judgements. Sustainability is both a descriptive and a normative notion. Normative statements cannot, therefore, be avoided. The approach here is nuanced and reflexive, based on a series of 'vignettes' that aim at grasping, albeit incompletely, the complexity of global environmental change, sustainability and social learning.

Adapt or modify?

In the present age of the Anthropocene, of the Great Acceleration, and of growing intertwined planetary boundaries (Crutzen 2002; Rockström et al. 2009; Steffen et al. 2007) individuals and societies are inevitably required to adapt to a human-altered biophysical environment which has never existed before. In this novel, complex, techno-social-ecological global *system of systems*, many new uncertainties and challenges emerge. The speed and scope of the changes make it difficult for individuals and organizations to reflect upon, react and transform their patterns of behaviour in time. A very common response to this situation is to intensify modification of the 'external' world (instead of ourselves) with further interventions and technologies in an attempt to 'control nature' (Murphy 2009) and whatever is perceived to be outside the most immediate human sphere of action. However, further modifications of the natural world may exacerbate the complexity and interconnections of global social-ecological systems in ways that create greater systemic risks, unwanted consequences and uncertainties, thus posing greater difficulties for future adaptation. Under these conditions, it appears that humankind is inexorably destined to both modify and adapt to a largely unknown and unknowable environment. Decisions on how much human resources and ingenuity should be devoted to one task or the other – or to both – must be taken. Notwithstanding, these decisions cannot depend on one single type of 'objective' knowledge or scientific discipline, as they are inevitably entangled in many normative judgements. Sustainability requires not only being aware of such complexities and uncertainties, but above all developing the necessary normative patterns of knowledge creation and collective behaviour that render serious consideration to these decisions. A precautionary attitude is required to minimize the irreversible loss of options for learning and renewal necessary to guarantee the viability, diversity and quality in functioning of social-ecological systems in the long term. In this process, ethics must be recognized as a central form of collective knowledge and an indispensable component of sustainability learning.

Sustainable developments

One of the greatest complications that we face in the conceptualization and application of the notion of sustainability is that it almost inevitably comes associated with the word 'development'

(Kates et al. 2005). But as Mary Midgley (2011) remarks, it is high time that we rethought the colonial idea of 'development' (in singular); that is, as the inexorable linear fate which all human societies must follow – like caterpillars turning into butterflies. Both sustainability and development are usually understood as socially desirable goals, but the combination of the two may necessitate a reframing of both. The evolution of human societies, with all their diversity and complexities, goes through multiple phases of renewal, destruction and reconstruction.[2] Different social-ecological dynamics may take different trajectories, adopt different patterns of organization, and may behave within different domains or scales in synergetic, independent or contradictory manners. What is then required is the quest for *diverse and open paths of multiple developments,* which embrace the complexity and variety of social-ecological systems and dynamics to ensure the preservation and/or restoration of the necessary, diverse conditions that make both sustainability and development possible in the many contexts in which human action takes place.

Revealing and interfacing processes

Sustainability is a language on its own, with its own vocabulary and grammar, which fundamentally can only be learned and fully understood by practising it. For this new 'language of motives' (Mills 1940) to trickle down, it requires communities of practice (Wenger 1998) as well as of individuals and organizations experienced in specific procedural and integrative competences (Martens et al. 2010; Robinson 2008). To substantiate and validate such 'narratives of hope' embedded in the discourse of sustainability, the creation of spaces for diverse and reasoned interaction between different stakeholders, with the aim of assessing risks, problems and proposing viable solutions, is especially urgent. Improving sustainability can be thought essentially as a question of revealing, understanding and improving social-ecological *processes* in the many spheres in which human action takes place. In this sense, focusing on processes and moving our attention from the question of 'What is the problem?' to 'Who is the problem?' may help improve equity and facilitate the distribution of responsibilities regarding unsustainable behaviours.[3] However, revealing the unsustainability of social practices and institutional arrangements and proposing feasible options for alternative paths of development calls for remarkable personal, transdisciplinary and professional competences. These are not often taught or promoted in science, education or policy, mostly because they do not belong exclusively to any one of them but to all at the same time. Sustainability science has been proposed as one avenue to compensate this deficiency. The new generation of 'sustainability entrepreneurs' needs a full suite of competencies to facilitate transformation and learning, such as: aptitude to translate complexity among different audiences; capacity to integrate different sources of knowledge; and the abilities to lead and create sustainability partnerships and to bridge organizations across different scales and domains of action (Berkes 2002, 2009; Berkes and Folke 1998; Cash et al. 2003, 2006; de la Vega et al. 2009; Jäger 2011; Van Kerkhoff and Lebel 2006; Weichselgartner and Kasperson 2010).

Embracing complexity

In technologically intense societies, the provision of food, health, mobility, communication, education and entertainment is made possible thanks to the existence of highly complex systems, the size, nature and dynamics of which we may not be fully aware (Castellani and Hafferty 2009; Urry 2005). The long-term viability of these systems depends on them having a *structural design* (S), which continually allows for renewal and innovation. This can be achieved by injecting new flows of *energy and resources* (E), which allow the kinetic functioning of the system, by improving the robustness and quality of *information and knowledge systems* (I) to adapt to changing conditions and by containing the potential negative and intertwined *systemic effects and changes* (C) that the actual functioning of all these systems provoke (the *SEIC model*, Tàbara and Pahl-Wostl 2007). Thus, social-ecological systems are complex systems characterized by many clusters of practices and agents, some partly independent but also mainly interconnected. Agents operating within these systems maintain many simultaneous interdependencies with many other agents and parts of the system. At the same time, such social-ecological interactions respond to forces that produce recursive, cumulative feedbacks on different scales. This tends to create non-linear dynamics where sudden bifurcations, multiple trajectories and regime shifts are possible. The complex nature of these systems is further exacerbated by the fact that different agents may see not only different parts of the system but different systems – depending on the issues at stake and how they relate to a particular subset of agents within the system. A robust definition of social-ecological systems dynamics valid to deal with global environmental change and sustainability is an arduous task and a complex co-production learning process in itself. Hence, in a complex system, the type of learning processes needed to move towards sustainability cannot be fully anticipated. The dynamics and outcomes of learning are largely unpredictable, diverse and subject to multiple contingencies and feedbacks. Embracing such complexities is a first step towards acquiring a more precautionary and open attitude fundamental in the design of long-term viable systems. Individuals living in the increasingly hyper-connected global societies should be bestowed with the necessary capacities, flexibility and degrees of freedom to become aware, adapt to and anticipate the multiple sustainability challenges inherent in the large-scale complex systems integrated into the functioning of the large-scale complex systems in which they develop their daily activities.

Managing common problems

There are numerous uncertainties concerning whether global human societies can be sustained, for how long or under what conditions. On the one hand, in sustainability, not only are the means largely unknown, but also the goals: knowledge-building processes aimed at supporting sustainability are not only goal-driven activities but, mostly, goal-searching ones (Jäger 2011). However, uncertainties are not complete. A wealth of knowledge and practices about what works to improve sustainability is already available. Innovative pathways to prevent negative impacts on the global environment and help restore some of the basic functions of global social-ecological systems have been identified and assessed – so the problem is not of that of lack knowledge but mainly of collective action. For example, collective action theory, as elaborated by Elinor Ostrom (2009, 2010), provides crucial insights into the learning mechanisms that operate in the case of Common Pool Resources as well as in global environmental change. Ostrom underlines the crucial role of polycentric institutional arrangements that allow greater opportunities for experimentation and feedback between different scales of action. While there are no panaceas for managing resources in a sustainable way, some key variables that have been identified include size and productivity of the system, predictability of its dynamics, resource unit mobility, number of users, role of leadership, existing social capital and norms, knowledge of the social-ecological system, importance of resources for users, and the type of collective-choice rules (Ostrom 2007, 2009, 2010).

Learning what not to do

Learning what not to do is a necessary and often neglected component in the improvement of sustainability, in particular for the integrated sustainability governance of complex social-ecological systems. In liberal democracies, such framing is often seen as a threat to the freedom of individuals and corporations rather than as a guarantee for the long-term viability and quality of the systems in which we operate. Understanding the constraints and opportunities for bringing such framing into science, education and policy discussions, while avoiding simple ideological discussions, is of paramount importance in sustainability learning. There are many examples in which communities and organizations have learned to regulate what *should not be done* to prevent negative consequences of their actions on the systems in which they intervene. Nevertheless, in times of fast emergence and convergence of new technologies (with huge potential for unexpected global environmental change like synthetic biology) and multiple cross-scale interactions, learning what not to do is a particularly challenging task. In order to harness such complex and cascading developments, a high degree of social and institutional innovation is needed. The governance systems that are designed to address such challenges will have to be highly integrative and follow approaches that are sufficiently open, independent and resourceful so as to anticipate and reorient possible unwanted futures in time to avert unwanted consequences. New institutional arrangements should encourage the production of socially and ecologically robust assessments, as well as democratic decision making about which development pathways are to be followed, instead of technologies and vested interests creating path dependencies which have never been publicly deliberated.

Reframing knowledge: an example

Increasing the time and resources spent reflecting on a given complex environmental issue may not lead to the reduction of associated uncertainties. In a related research project carried out in Europe in the mid-1990s, we qualitatively explored how lay people's perceptions of global warming might change following participation in informed deliberation on the issue (Kasemir et al. 2003). One team monitored the learning that occurred within a series of heterogeneously composed meetings in the Barcelona metropolitan area. Thirty-eight participants, divided into five groups, were exposed to a variety of expert and non-expert sources of information. To our surprise, and after long debates that went on for five sessions of two and a half hours each, we found that uncertainties regarding the potential causes and impacts of climate change did not diminish but increased. In fact, many participants declared they had learned a lot about the problem of global warming and, in particular, about how much they did not know about it. At the same time, the number of participants willing to take preventative action also increased (Tàbara 2005). This 'precautionary learning' appears to be a recurrent outcome of many learning processes related to global environmental change and sustainability. A first step in this direction is recognizing the limitations of our knowledge, followed by increased consensus about the need to avoid the worst outcomes derived from our ignorance. Sustainability learning is not simply about 'knowing more', but also about learning about what we don't know, about how we can collectively generate different types of knowledge, and about implementing multiple learning processes able to harness the unexpected and unwanted effects that may arise from our partial and inadequate knowledge and views.

Tools of mislearning and relearning

Models are central tools in the (self-)representation of social-ecological systems. In a similar fashion to games and metaphors, models provide powerful means to convey and communicate complexity. Global dynamics of stocks and flows of information, resources and pollution are very much affected by some of the models created by the social sciences which, in turn, have become major ideological instruments for the re-creation of the world. Tools used in economics, such as the general equilibrium macro-economic models so pervasively used to inform public policies, increasingly are being called into question in the face of mounting intertwined crises.

Not only do many of these models have great limitations in explaining and anticipating ongoing instabilities and turbulences, but also many of their assumptions about human behaviour cannot be tested empirically.[4] An alternative set of tools for explaining, educating and communicating economic behaviours and complex dynamics is now required. In particular, new tools are needed that move away from the pervasive teaching and exaltation of 'perfect competition' towards taking into account the intrinsic complexity of economic interactions and considering the required social-ecological conditions and best strategies for 'perfect global cooperation'. Learning to cooperate, and to understand the conditions for cooperation with 'different others' in concrete social-ecological contexts as a focus of the new economic models, appears to be a necessary task in the present times of cascading global risks and uncertainties.[5]

New learning, identities and power arrangements

Learning leads to the creation of new identities in the same way that the emergence of new identities tends to create new demands for learning. Identities define what is in and out in each person's systems of reference and, in this sense, identities define what is included and excluded from the cognitive and perceptual boundaries in which social action takes place. The creation of identities is parallel to the creation of meaning. Identities and meaning are both relational products in the interaction between individuals and agents in the contexts where people operate. The penetration of the sustainability worldview in science, education and policy may be leading to a broad identity shift in which the motives determining social action may take into account a more extended set of references in time, space and the subjects and objects under moral consideration. In this broader cognitive and moral space, the new references may include not only the rights of present generations but also those of future generations; not solely the rights of those belonging to our own communities and nations but the rights of all people in the world; and not just the rights of humans but also extending respect, compassion and moral consideration to non-human beings and the social-ecological processes necessary to maintain 'life in all diversity' (as stated in the first principle of the Earth Charter). In the current context of global change, the co-evolution of identities with social learning is likely to be accompanied by the development of new rules that allow for novel forms of social-ecological interaction better suited to deal with common problems of unsustainability. As new identities emerge from learning, individuals and organizations may adapt their multiple roles to the new conditions and boundaries of the situation. In some cases, as in the formation of environmental movements, the creation of new identities is a necessary condition – albeit not a sufficient one – for the mobilization of resources and for agents to engage in sustainability transitions (Ilhan 2009; Tàbara and Ilhan 2008). In other cases, obstacles to learning have risen from various resistances to transform present identities that benefit particular interests and the incumbent regime. This is why social learning needs to be understood as a political process, and as a process that is likely to become even more political in the face of escalating dangerous global environmental trends and indeterminacies.

A world of men … and women

As noted a century ago by Georg Simmel (1911: 234), 'our culture, with the exception of few domains, is mostly a masculine one'. Access to learning and advancements in the literacy of women has yielded massive changes not only in reproductive patterns but also in improvement of standards of living, reduction of violence, and the setting of innovative development strategies all over the world. In some cases, women have been found not only to be more vulnerable to environmental change (Goldsworthy 2010) but also faster in taking adaptive responses than men

(Ongoro and Ogara 2012; see also Shiva 1989). However, in our present times, most decisions related to the (over-)use and (over-)exploitation of natural resources still appear to be mainly taken by men. This pattern is characteristic not only of traditional or tribal societies but is reproduced in many of the opulent societies where most decision boards of key corporations and political parties are still mostly dominated by men. While the influence of women on corporate environmental performance is an issue that requires further research (Terjesen et al. 2009), the exclusion of women's perspectives from such decisions is both a blatant expression of inequality and a situation that prevents the emergence and consolidation of more open and diverse spaces and institutions conducive to reflectivity and social learning. Although it has been several decades since the environmental and feminist movements pointed out the need for women to have a greater role in environmental and resource management decisions (e.g. Rio Declaration and Agenda 21), substantive progress so far has been scant. It would be naive to believe that a world exclusively governed by women would just lead us to sustainability. However, when it comes to living sustainably and to reframing not only the means but also the goals of human development, it is hard to believe that men on their own will ever be able to carry out this extraordinary challenge and to learn what is needed to be learned. A multiplicity of diverse, oftentimes suppressed, perspectives must be truly respected and taken into account.

Our body, our closest environment

Our biophysical body is a primary means of experience and interaction with the external, natural environment. A close examination of the composition of our body reveals that the environment lives as much in our body as our body lives in the environment (Turner 2008; Watts 1970). Our body is a complex system, in some ways separate, but in other ways connected, in time, space and to the rest of chemical elements and living organisms. What happens – or we desire to happen – to our body is not only a manifestation of changes occurring to larger biophysical systems, but also a source of changes which are likely to affect them, too. As such, the body is as much an indicator of global environmental changes as it is a major channel through which we interpret the scope and quality of these changes. Being the closest biophysical environment to our consciousness, our attitudes to our body reflect our attitudes to the natural world, bringing in particular practices and responses. The type of food we eat (and discard in the process), the ideals and aspirations for a perfect or never-ageing body, and the lifestyles we follow, say a lot about the type of daily interactions we maintain with the environment. Social trends such as vegetarianism, slow food or green consumerism may be adaptive responses in the attempt to transform such interactions in a meaningful manner at the very individual, physical levels and to the space where particular experiences can be felt and different options discussed (see Hinton and Goodman 2010). While environmental change has often been understood as processes occurring 'out there', either in another place, time or to other people, increasingly the negative impacts on health, reproduction and quality of life are making people realize that changes are happening here and now, and to our bodies. The new global situation not only poses large threats for human well-being but also gives rise to myriad opportunities for a new social reflectivity on global change and to rethink the role of emerging technologies and consumption.

Meaning, knowledge and ignorance

In relation to environmental change, one can easily see that, while there are many types of knowledge valid to support sustainability, there is only one type of ignorance: that which disregards or denies the need to do something about it and impedes or fails to anticipate and implement the necessary and efficient measures to cope with it. Nevertheless, in order for individuals and organizations to engage in collective environmental and social action, it has to make sense doing so. Meaning is needed to make sense of the world around us, to transform information into knowledge and to mobilize the resources to deal with the mounting negative environmental trends of today. New information technologies and the greater mobility of humans and goods makes people less dependent on local conditions for obtaining resources and knowledge – while making them increasingly interdependent on the external and global systems. These multiple, interlinked processes erode many of the original collective meanings that gave sense to traditional practices in the use of natural resources. Such traditional collective understandings of humans' ultimate dependence on the biophysical environment may now be crucial in the transition towards sustainability. Whilst the destruction of meanings is relatively easy, their reconstruction can be very difficult – mostly because the social-ecological conditions which gave rise to them no longer exist (Otero et al. 2013). This destruction of meaning is reinforced by a dominant exemptionalist view about knowledge, now commonplace in science, education and policy (Dunlap 2002). That view accepts and promotes a 'placeless' universal knowledge, the validity and robustness of which is not checked against identifiable existing social-ecological systems of reference. Therefore, a key learning challenge of contemporary societies is how to manage both knowledge in an open, plural and socially-ecologically coupled mode, and ignorance, the latter too often associated with particular power arrangements and the pervasive idea that abstract, detached knowledge should prevail over socially-ecologically coupled kinds of knowledge (Tàbara and Chabay 2013). In the present conditions, where many processes of destruction and production of knowledge occur at the same time, meaning has become one of the scarcest resources to support transitions towards sustainability.

Harmonizing social learning processes

In contemporary complex societies, many different learning processes occur at the same time. Learning processes may be complementary and many kinds of collective intelligences are needed for different purposes. Likewise, they may be in competition with other collective learning processes, with learning processes in other domains and/or processes at other levels of action. Moreover, learning processes are distributed unevenly among society. Existing inequalities in social structure remove opportunities for suppressed voices to engage in social learning processes and, therefore, to participate constructively in the building of diverse paths for sustainable *developments*. In order to contribute to sustainability, the emergence of global kinds of knowledge need to respect and pay attention to local sources of expertise and judgement (Hulme 2010). The harmonization and improvement of equity of the multiple processes of systems learning and knowing that need to be developed to cope with global environmental change constitutes a meta-learning process in itself. Environmental and sustainability issues often are framed as problems of 'not knowing enough' and as problems that we would overcome if only we knew 'more' (the knowledge-gap-filling model). However, the challenge for sustainability may be approached in a different way: as the task of selecting, integrating and generating qualitatively different types of knowledge to stimulate the emergence of a diversity of learning processes suited for a variety of problems, purposes and contexts. In many cases, the lack of substantive action does not have to do with the existence of 'incomplete knowledge' but must be found in other psychological, social and political factors. These include processes of sensemaking, ways of framing issues, the lack of adequate incentives, and the unwillingness or inability to generate options and deploy resources to distribute responsibilities in accordance with the new social-ecological situations (Tàbara et al. 2010).

Limits and opportunities for learning

There are, however, many limits to learning. While on the one hand social learning should empower individuals and societies to deal with common problems, on the other learning also leads to new knowledge needs about how to deal the global challenges of unsustainability. Science and technology also display this ambivalent character, being as much part of the problem as the solution. The growing speed of flows of information (but not necessarily in the quality of those flows) poses new constraints on existing institutions, now largely unfit to govern complex dynamics in sustainable ways. In some cases, such as global warming, it will take decades until mitigation measures can produce significant effects on the climatic system, if they are successful at all. In this situation, learning to adapt cannot be avoided (Hinkel et al. 2010). In this 'learning race against time' means not only that we need to learn new ends and means to reorient our actions, but also that we need to implement the attendant measures, practices and institutions in sufficient time ahead to prevent large-scale social-ecological disruptions. Other cognitive and cultural limits to learning and to sustainability need to be overcome. These come from bounded and corporate rationalities and the institutional arrangements created around them (e.g. from property regimes to market mechanisms). The kinds of learning processes, encouraged and implemented in different contexts and organizations, depend on the type of perceptions held about the nature of problems that need to be addressed. As Caldwell (1997) suggests, environmental problems can be conceived as incidental, operational or systemic in character, with explanations and proposed remedies varying according to these different understandings (see Table 21.1).

Many opportunities for social learning towards sustainability remain. Social learning can be facilitated by specific strategies and interventions in education, science and policy. While education at all levels will play a central role, not all educational efforts will encourage the type of systems thinking and acting necessary to boost a transition towards sustainability – nor will the results of such education necessarily materialize in time. The information and communication technology (ICT) revolution is posing phenomenal challenges as well as opportunities to traditional forms of

Table 21.1 Interpretations of environmental impairment (adapted from Caldwell, 1997)

Perceived causes	Explanations	Remedies
I. Incidental Harmful behaviours occurring in the normal course of human activities	*Errors in judgement* Dereliction, ignorance, carelessness, alcohol and drug abuse	*Exhortation* Ad hoc responses, clean-up campaigns, indoctrination, education and penalties
II. Operational Misdirected policy, flawed programme planning and execution, and bureaucratic intransigence	*Ineffective management* Insufficient or incorrect information, poor morale or operating procedures, avarice and corruption	*Correction* Improved procedures, impact assessments, independent review of proposals, standards, enforcement and incentives
III. Systemic Impairment inherent in technology economic systems; unsustainable and exploitive economic practice	*Built-in hazards* Narrowly focused policies failing to assess full dimensions of environmental consequences; policies based on unwarranted assumptions	*Reorientation* Basic changes in beliefs and behaviour systems; redesigning institutions and development of alternative technologies, elimination of harmful products and procedures

education and public engagement in decision making, both of which are becoming largely obsolete. Yet, ICT may also facilitate new forms of interactions and social-ecological awareness. The development of concrete practices and procedures that contribute to new forms of socially and ecologically coupled knowledge could be a first step in this direction. This, in turn, may require a plurality of hybrid learning processes able to articulate and integrate a diversity of information and knowledge systems, sources and representations – scientific and artistic, local and universal, formal and informal, male and female, young and old – as well as chances for learning from non-human beings and a variety of forms of biophysical systems organization.

Conclusion

The conflation of social and ecological trends into what has been labelled the 'Great Acceleration' is exerting massive pressures on the ways both individuals and institutions operate. Much of the anxiety of present times derives from the need to reorient individuals' undertakings to a largely unknown New World, now mostly driven by the unintended, cumulative and often unwanted consequences of our collective actions. With multiple technologies amplifying the environmental effects of these actions, the current situation requires a profound problematization of the assumptions, values and beliefs used to build our judgements about the future ('the future is no longer what it used to be'). In these turbulent times (Folke and Rockström 2009; Milbrath 1989), many aspirations and ideals are being reformulated, many political boundaries become permeable, inequalities cannot be contained, and simple solutions are not an option. In this fluid context, sustainability learning emerges as a complex process of re-creating our identities, accepting the limits of our knowledge and setting limits to our actions in an age where apparently everything is attainable. As such, sustainability learning can be conceived as the complex process in which more extended and informed meanings are searched for in the creation of new *patterns of social-ecological interaction* that can make the long-term viability of human life on Earth feasible under high standards in the quality of life.

In the face of such large complexities, social learning cannot be avoided. On the one hand, the viability of global social-ecological systems depends on developing open, adaptable and anticipatory governance structures in which agents can become actively engaged in continuous reconfiguration and renewal. This requires flexible and multiple arrangements that preserve social-ecological diversity, allow for sufficient degrees of freedom and avoid the dominance of individual agents who prevent the learning, transformation and adaptation of others. Polycentric designs that encourage multiple cross-scale learning feedback, and bestow greater responsibilities on individuals in managing their resources, appear to be well suited to cope with accelerated global environmental change and unsustainability. In this respect, the restoration or re-creation of multiple social-ecological feedbacks for humans to learn from their aggregated actions and emergent dysfunctions on global and local ecosystems is essential.

But this will not be an easy journey. The institutionalization of practices that can contribute to new forms of social-ecological reflectivity for sustainability is likely to be increasingly contested. For some, embedding reflective practices in education, science and public communication is of paramount importance to anticipate catastrophe and avoid encroaching on impending thresholds of unsustainability. For others, reflectivity is perceived as a danger to their freedom and identities as well as a threat to economic and political interests to whom the present ecological illiteracy serves. Opportunities for social learning, global restoration and renewal of social-ecological systems are there. But at the end, as Freud (1989/1930; see also Tàbara and Giner 2004) asked: Who will win, *Eros* or *Thanatos*? Love or self-destruction? We simply cannot say.

Notes

1 This contribution has benefited from discussions with Carlo Jäger and Ilan Chabay within the EU project of the Global Systems Dynamics and Policy (www.gsdp.eu), the new IHDP initiative on 'Knowledge, Learning, and Societal Change' (www.proclim.ch/4dcgi/klsc/en/media?1913), as well as with Jill Jäger in the context of the EU project, 'Vision RD4SD' (www.visionRD4SD.eu). Responsibility for any biases and errors is the author's. Opinions presented in this chapter do not necessary express those of the cited projects.
2 In the resilience perspective, such evolution has been conceptualized under the ideas of the adaptive cycles and 'panarchy' where the sustainability of social-ecological systems can be analysed according multiple stages of exploitation, conservation, release and reorganization (Gotts 2007; Gunderson and Holling 2002; Holling 2001).
3 Albeit a bit offensive, it has been noted that one very direct but efficient way to convey this message to a variety of scientific, political and public communication audiences is by the use of the expression 'It's the process, stupid!' (rephrasing the famous Bill Clinton 1992 presidential campaign). Focusing on processes rather than only on outcomes can reveal many inconsistencies and unscrupulous behaviours with regard to sustainability, which now go untapped in economics, politics and the organization of present societies.
4 Including those related to the existence of a representative agent, complete or homogeneous information or the tendency towards market clearance. For an introduction, see Beinhocker (2007).
5 Current developments in Agent Based Models (ABM) and complexity economics appear to be promising in this regard (see www.gsdp.eu).

References

Beinhocker, E.D. (2007) *The Origin of Wealth. Evolution, Complexity and the Radical Remaking of Economics*. London: Random House Business Books.
Berkes, F. (2002) 'Cross-Scale Institutional Linkages: Perspectives From the Bottom Up', pp. 293–322 in E. Ostrom, T. Dietz, N. Dolsak, P. C. Stern, A.S. Stonich and E.U. Weber (eds), *The Drama of the Commons*. Washington, DC: National Academy Press.
Berkes, F. (2009) 'Evolution of Co-Management: Role of Knowledge Generation, Bridging Organisations and Social Learning', *Journal of Environmental Management*, 90, pp. 1692–1702.
Berkes, F. and Folke, C. (eds) (1998) *Linking Social and Ecological Systems: Management Practices and Social Mechanisms for Building Resilience*. Cambridge, UK: Cambridge University Press.
Blackmore, C. (2007) 'What Kinds of Knowledge, Knowing and Learning are Required for Addressing Resource Dilemmas? A Theoretical Overview', *Environmental Science and Policy*, 10, 6, pp. 512–525.
Caldwell, L.K. (1997) 'Environment as a Problem for Policy', pp. 1–17 in L.K. Caldwell and R.V. Barlett (eds), *Environmental Policy: Transnational Issues and National Trends*. London: Quorum Books.
Cash, D.W., Adger, W., Berkes, F., Garden, P., Lebel, L., Olsson, P., Pritchard, L. and Young, O. (2006) 'Scale and Cross-Scale Dynamics: Governance and Information in a Multi-Level World', *Ecology and Society*, 11, 2, art. 8.
Cash, D.W, Clark, W.C., Alcock, F., Dickson, N.M., Eckley, N., Guston, D.H., Jäger, J. and Mitchell, R.B. (2003) 'Knowledge Systems for Sustainable Development', *Proceedings of the National Academy of Sciences of the United States of America*, 100, 14, pp. 8086–8091.
Castellani, B. and Hafferty, F. (2009) *Sociology and Complexity Science: A New Field of Inquiry*. Heidelberg: Springer.
Crutzen, P.J. (2002) 'Geology of Mankind: The Anthropocene', *Nature*, 415, p. 23.
de la Vega-Leinert, A.C., Stoll-Kleemann, S. and O'Riordan, T. (2009) 'Sustainability Science Partnerships in Concept and in Practice: A Guide to a New Curriculum from a European Perspective', *Geographical Research*, 47, 4, pp. 351–361.
Dunlap, R.E. (2002) 'Paradigms, Theories and Environmental Sociology', pp. 329–350 in R.E. Dunlap, F.H. Buttel, P. Dickens and A. Gijswijt (eds), *Sociological Theory and the Environment: Classical Foundations, Contemporary Insights*. Boulder, CO: Rowman and Littlefield.
Folke, C. and Rockström, J. (2009) 'Turbulent Times', *Global Environmental Change*, 19, pp. 1–3.
Freud, S. (1989 [1930]) *Civilization and its Discontents*. New York and London: Norton.
Goldsworthy, H. (2010) 'Women, Global Environmental Change, and Human Security', pp. 215–236 in R.A. Matthew, J. Barnett, B. McDonald and K.L. O'Brien (eds), *Global Environmental Change and Human Security*. Cambridge, MA: MIT Press.

Gotts, N.M. (2007) 'Resilience, Panarchy and World-Systems Analysis', *Ecology and Society*, 12, 1, art. 24.

Gunderson, L.H. and Holling, C.S. (eds) (2002) *Panarchy: Understanding Transformations in Human and Natural Systems*. Washington, DC: Island Press.

Hinkel, J., Bisaro, S., Downing, T., Hofmann, M.E., Lonsdale, K., McEvoy, D. and Tàbara, J.D. (2010) 'Learning to Adapt: Re-Framing Climate Change Adaptation', pp. 113–134 in M. Hulme and H. Neufeldt (eds), *Making Climate Change Work for Us*. Cambridge, UK: Cambridge University Press.

Hinton, E.D. and Goodman, M.K. (2010) 'Sustainable Consumption: Developments, Considerations and New Directions', pp. 245–261 in M.R. Redclift and G. Woodgate (eds), *The International Handbook of Environmental Sociology*, 2nd ed. Cheltenham, UK: Edward Elgar.

Holling, C.S. (2001) 'Understanding the Complexity of Economic, Ecological, and Social Systems', *Ecosystems*, 4, pp. 390–405.

Hulme. M. (2010) 'Problems with Making and Governing Global Kinds of Knowledge', *Global Environmental Change*, 20, 4, pp. 558–564.

Ilhan, A. (2009) *Social Movements in Sustainability Transitions: Identity, Social Learning and Power in the Spanish and Turkish Water Domains*. Ph.D. thesis submitted at the Institute of Environmental Science and Technology of the Autonomous University of Barcelona. Available at http://tdx.cat/bitstream/handle/10803/5817/ai1de1.pdf?sequence=1. Accessed 23 January 2013.

Jäger, J. (2011) 'Risks and Opportunities for Sustainability Science in Europe', pp. 185–201 in C.C. Jaeger, J.D. Tàbara and J. Jäger (eds), *Transformative Research for Sustainable Development*. Heidelberg: Springer and European Commission.

Jennex, M.E. (2008) *Current Issues in Knowledge Management*. Information Science Reference. Hershey, PA: IGI Global.

Kasemir, B.J., Jäger, J., Jaeger, C.C. and Gardner, M.T. (eds) (2003) *Public Participation in Sustainability Science: A Handbook*. Cambridge, UK: Cambridge University Press.

Kates, R., Thomas, M.P. and Anthony, A.L. (2005) 'What Is Sustainable Development?' *Environment*, 47, 3, pp. 9–21.

Keen, M., Brown. V. and Dyball, R. (2005) *Social Learning in Environmental Management: Towards a Sustainable Future*. London: Earthscan.

KLSC (Knowledge, Learning and Societal Change) (2011) *Finding Paths to a Sustainable Future: Science Plan*. International Human Dimensions Programme on Global Environmental Change (IHDP). Available at http://proclimweb.scnat.ch/portal/ressources/2070.pdf. Accessed 23 January 2013.

Martens, P., Roorda, N. and Cörvers, R. (2010) 'Sustainability, Science and Higher Education: The Need for New Paradigms', *Sustainability*, 2, 5, pp. 294–303.

Midgley, M. (2011) 'Developmental Doubts', pp. 9–22 in C.C. Jaeger, J.D. Tàbara and J. Jäger (eds), *European Research on Sustainable Development*, vol. 1, *Transformative Science Approaches for Sustainable Development*. Heidelberg: Springer and European Commission.

Milbrath, L.W. (1989) *Envisioning a Sustainable Society: Learning our Way Out*. Albany, NY: State University of New York Press.

Mills, C.W. (1940) 'Situated Actions and Vocabularies of Motive', *American Sociological Review*, 5, pp. 904–913.

Mostert, E., Pahl-Wostl, C., Rees, Y., Searle, B., Tàbara, J.D. and Tippett, J. (2008) 'Social Learning in European River-Basin Management: Barriers and Fostering Mechanisms from 10 River Basins', *Ecology and Society*, 12, 1, art. 19.

Murphy, R. (2009) *Leadership in Disaster. Learning for a Future with Global Climate Change*. Montreal and Kingston: McGill-Queen's University Press.

Ongoro, E.B. and Ogara, W. (2012) 'Impact of Climate Change and Gender Roles in Community Adaptation: A Case Study of Pastoralists in Samburu East District, Kenya', *International Journal of Biodiversity and Conservation*, 4, 2, pp. 78–89.

Ostrom, E. (2007) 'A Diagnostic Approach for Going Beyond Panaceas', *Proceedings of the National Academy of Sciences of the United States of America*, 25, 39, pp. 15181–15187.

Ostrom, E. (2009) 'A General Framework for Analysing Sustainability of Social-Ecological Systems', *Science*, 325, pp. 419–422.

Ostrom, E. (2010) 'Polycentric Systems for Coping with Collective Action and Global Environmental Change', *Global Environmental Change*, 20, pp. 550–557.

Otero, I., Boada, M. and Tàbara, J.D. (2013) 'Social-Ecological Heritage and the Conservation of Mediterranean Landscapes Under Global Change: A Case Study in Olzinelles (Catalonia)', *Land Use Policy*, 30, 1, pp. 25–37.

Pahl-Wostl, C., Tàbara, J. D., Bouwen, R., Craps, M., Dewulf, A., Mostert, E., Ridder, D. and Taillieu, T. (2008) 'The Importance of Social Learning and Culture for Sustainable Resources Management', *Ecological Economics*, 64, 3, pp. 484–495.

Pahl-Wostl, C. (2009) 'A Conceptual Framework for Analysing Capacity and Multi-Level Learning Processes in Resource Governance Regimes', *Global Environmental Change*, 19, pp. 354–365.

Reed, M.S., Evely, A.C., Cundill, G., Fazey, I., Glass, J., Laing, A., Newig, J., Parrish, B., Prell, C., Raymond, C. and Stringer, L.C. (2010) 'What Is Social Learning?' *Ecology and Society*, 15, 4, resp. 1.

Robinson, J. (2008) 'Being Undisciplined: Transgressions and Intersections in Academia and Beyond', *Futures*, 40, pp. 70–86.

Rockström, J., Steffen, W., Noone, K., Persson, Å., Chapin, III, F.S., Lambin, E., Lenton, T.M., Scheffer, M., Folke, C., Schelnhuber, H. Nykvist, B., De Wit, C.A., Hughes, T., van der Leeuw, S., Rodhe, H., Sörlin, S., Snyder, P.K., Costanza, R., Svedin, U., Falkenmark, M., Karlberg, L., Corell, R.W., Fabry, W.J., Hansen, J., Walker, B., Liverman, D., Richardson, K., Crutzen, P. and Foley, J. (2009) 'Planetary Boundaries: Exploring the Safe Operating Space for Humanity', *Ecology and Society*, 14, 2, art. 32.

Shiva, V. (1989) *Staying Alive: Women, Ecology and Development*. Atlantic Highlands, NJ: Zed.

Siebenhüner, B.S. and Heinrichs, H. (2010) 'Knowledge and Social Learning for Sustainable Development', pp. 185–200 in M. Gross and H. Heinrichs (eds), *Environmental Sociology. European Perspective and Interdisciplinary Challenges*. Heidelberg: Springer.

Simmel, G. (1911) *Philosophische Kultur*. Berlin: Verlag Klaus Wagenbach, Spanish translation in G. Simmel *Sobre la Aventura. Ensayos Filosóficos* (1988) Barcelona: Ed. Península.

SLG (Social Learning Group) (2001) *Learning to Manage Global Environmental Risks: A Comparative History of Social Responses to Climate Change, Ozone Depletion and Acid Rain*, vols 1 and 2. Cambridge, MA: MIT Press.

Steffen, W., Crutzen, P.J. and McNeill, J.R. (2007) 'The Anthropocene: Are Humans Now Overwhelming the Great Forces of Nature?' *Ambio*, 36, 8, pp. 614–621.

Tàbara, J.D. (2005). 'Percepció i Comunicació del Canvi Climàtic a Catalunya' ['Public Communication and Perception of Climate Change in Catalonia'], pp. 772–815 in E. Llebot (ed.), *El Canvi Climàtic a Catalunya*. Barcelona: IEC.

Tàbara, J.D. and Chabay, I. (2013) 'Coupling Human Information and Knowledge Systems with Social-Ecological Systems Change: Reframing Research, Education and Policy for Sustainability', *Environmental Science and Policy*, 28: 71–81.

Tàbara, J.D., Cots, F., Dai, X., Falaleeva, M., Flachner, Z., McEvoy, D. and Werners, S. (2009) 'Social Learning on Climate Change Among Regional Agents: Insights from China, Eastern Europe and Iberia', pp. 121–150 in W. Leal Filho and F. Mannke (eds), *Interdisciplinary Aspects of Climate Change*. Frankfurt: Peter Lang.

Tàbara, J.D., Dai, X., Jia, G., McEvoy, D., Neufeldt, H., Serra, A., Werners, S. and West, J.J. (2010) 'The Climate Learning Ladder: A Pragmatic Procedure to Support Climate Adaptation', *Environmental Policy and Governance*, 20, pp. 1–11.

Tàbara, J.D. and Giner, S. (2004) 'Diversity, Civic Virtues and Ecological Austerity', *International Review of Sociology/Revue Internationale de Sociologie*, 14, 2, pp. 262–283.

Tàbara, J.D. and Ilhan A. (2008) 'Culture as Trigger for Sustainability Transition in the Water Domain', *Regional Environmental Change*, 8, 2, pp. 59–71.

Tàbara, J.D. and Pahl-Wostl, C. (2007) 'Sustainability Learning in Natural Resource Use and Management', *Ecology and Society*, 12, 2, art. 3.

Terjesen, S., Sealy, R. and Singh, V. (2009) 'Women Directors on Corporate Boards: A Review and Research Agenda', *Corporate Governance: An International Review*, 17, 3, pp. 320–337.

Turner, B.S. (2008) *The Body and Society*, 3rd ed. London: Sage.

Urry, J. (2005) 'The Complexities of the Global', *Theory, Culture and Society*, 22, 5, pp. 235–254.

Van Kerkhoff, L. and Lebel, L. (2006) 'Linking Knowledge and Action for Sustainable Development', *Annual Review of Environment and Resources*, 31, pp. 445–477.

Wals, A.E.J. (ed.) (2007) *Social Learning: Towards a Sustainable World*. Wageningen: Wageningen Academic Publishers.

Watts, A. (1970) 'The World is Your Body', pp. 181–193 in R. Disch (ed.), *The Ecological Conscience: Values for Survival*. Upper Saddle River, NJ: Prentice-Hall.

Weichselgartner, J. and Kasperson, J. (2010) 'Barriers in the Science–Policy–Practice Interface: Toward a Knowledge–Action–System in Global Environmental Change Research', *Global Environmental Change*, 20, pp. 266–277.

Wenger, E. (1998) *Communities of Practice: Learning, Meaning and Identity*. Cambridge, UK: Cambridge University Press.

Part V
(Re)assembling social-ecological systems

22

The social-ecological co-constitution of nature through ecological restoration

Experimentally coping with inevitable ignorance and surprise

Matthias Gross

Ecological restoration as a widely recognized field of practice and research emerged in the 1980s, partly as a reaction and addition to traditional nature conservation strategies, and partly as a result of the availability of novel technologies for restoring, cleaning up and redesigning industrially degraded or contaminated ecosystems.[1] Given the inevitability of unexpected change in the course of restoring or designing a piece of land, the self-descriptions of many restoration practitioners point to the social-ecological co-constitution of nature based on attributed 'natural activities'. In the course of almost any restoration project, at certain points, actors step back and then begin to realize that they have been participating in a piece of natural land or ecosystem, and that, by doing so, they have become part of the often unexpected and sometimes little understood ecological processes involved. This realization is part of many discussions in ecological practice.

From a sociological point of view it is one thing to register this way of speaking and perceiving, but it is quite another to integrate it into a social theory of interaction and communication. Yet if the self-descriptions of ecological restoration practitioners are to be taken seriously in social theory, non-human nature must somehow gain its own voice in the theoretical language and analytical tools of sociology. This 'voice giving' requires something very much in line with Max Weber's *verstehende* sociology; that is, the attempt to take social actors' constructions of their own life worlds seriously in the construction of social theory. To put oneself as a sociological observer into the shoes of the other gives one a sense of another person's capabilities, cultural frameworks and individual values. But it certainly does not work very well for elemental forces, such as the growth of fauna and flora, storms, water or fire. This is why, in this case, a practice theory approach in the tradition developed by Ted Schatzki (2010), the modalities of actor-network theory (Latour 2005) and the 'experimental practices' of pragmatist environmental sociology (Overdevest et al. 2010) appear useful in analysing and understanding the human and natural worlds as interdependent and 'communicative' entities. The adjective 'communicative', between humans and nature, aptly describes the attitude adopted by the practitioners, ecologists or other participants when they wait to see what influence they have on nature or how nature responds. They find themselves in a very peculiar situation of double contingency, since they neither know how nature will respond to their intervention nor how they will themselves interpret these responses (Gross 2003). The management and governance of these practices then can be understood as the proceduralization of this contingency, i.e. proceduralizing the surprises and the inevitable ignorance involved in processes of ecological restoration.

This chapter takes this observation and employs some sociological ideas in the analysis of proceduralization. Among the issues for an alternative analysis of social-ecological linkages, the concepts of non-knowledge and experiment as an alternative to notions of risk and linear planning can play a crucial role. By way of connecting concepts of the unknown and the inevitable surprising effects of human interventions to natural and human-made environments, the chapter will present an experimental approach by reconstructing possibilities for acting in the face of (well-defined) ignorance and outlining some of the necessary social and ecological capacities to cope with surprising events. By so doing, this chapter contributes some sociological insight to current debates on ecological resilience and vulnerability within social-ecological systems.[2]

The idea of an experiment as a trial or a venture into the unknown is crucial for science, notwithstanding that experiments in the laboratory are characterized by detailed minutes and controlled procedures. Viewing ecological design processes outside the laboratory as experiments challenges the premises of ecological predictability and certainty. What makes the physical, technical and procedural basis for an experiment work is that it is deliberately arranged to generate surprises (Gross 2010; Latour 2011; Rheinberger 1997). The surprising effects of experiments can be seen as the driving force for producing new knowledge, since surprises help scientists become aware of their own ignorance. Similar to experiments in the laboratory, experimental activities in the real world can thus also bring surprises. But, unlike in the laboratory, they are often not welcomed since the consequences cannot be controlled easily.

In order to conceptually frame the social-ecological co-constitution of nature through experimental design processes that can help to successfully cope with inevitable ignorance and surprise, the chapter will: first, introduce some clarification on risk and non-knowledge; second, discuss how the handling of non-knowledge and surprise in ecological restoration can be incorporated into a model of real-world experimentation exemplified in the ecological restoration of ecosystems; third, these more conceptual elaborations will be followed by the example of revitalizing industrially contaminated landscapes; and finally, offer conclusions on the experimental strategies in environmental research and ecological restoration in order to scrutinize some of the benefits but also the limits of such an approach in contemporary societies.

Risk and non-knowledge

Risk is most widely understood as the probability of a harmful event multiplied by the amount of harm the event is expected to inflict. In these terms, dealing with ignorance clearly differs from taking or limiting risks, since the risk of a certain event occurring presupposes knowledge of both the character of events that may occur and the probability that they will do so. Although discussions on risk have a long tradition in many academic fields, it was Ulrich Beck's *Risk Society* (*Risikogesellschaft*) from 1986 that fostered a wide debate in mainstream sociology. Beck tried to frame a theory about a distinctive form of modernization in which new threats to health and the environment challenge those institutions responsible for managing and distributing risk. Niklas Luhmann (1993), by contrast, used the notion of risk to describe the attribution of regrettable consequences of a decision to a decision maker. Authors in the micro-sociological tradition meanwhile conceptualize risk as 'dependent on decisions which are made in an attempt to calculate and rationalize away unpredictability and uncertainty' (Jones 2007: 38). However, these broad understandings of risk beg the question as to what decisions are, in practice, not risk-based.

Many observers have pointed to limitations in the field of risk studies (e.g. Campbell and Currie 2006; Green 2009; Hubbard 2009; Wehling 2011), and some have argued that the notion of risk should be replaced with clearer alternative concepts (Dowie 1999). Risk assessments have the distinguishing characteristic of not spelling out what should be done but, at best, what should *not* be done. Certainly, there exist numerous definitions and concepts of risk that undoubtedly can be useful for research in sociology and related fields (e.g. Aven 2003; Jaeger et al. 2001; Renn 2009). However, it is argued here that in many cases when social scientists talk about risk or uncertainty it would be empirically and theoretically more feasible and meaningful to frame decisions, utterances and practices in terms of different shadings of *ignorance* (see also Gross 2010; Smithson 2010; Smithson et al. 2000; Wehling 2011).

Such a view contrasts with traditional risk assessment methods that often gloss over areas of ignorance and instead offer constricted frameworks for management and decision making. In this line of thinking, Andrew Stirling has made the point that using risk assessments when clear knowledge about probabilities and outcomes are not available would be 'irrational, unscientific and potentially misleading' (2007: 311). In turn, it can be argued that it is more often the things that are not known which are most important for decision makers and thus most pivotal for sociological analysis. This position departs from the still common view that ignorance is necessarily detrimental. It points to the merits of not knowing and – more importantly for the cases discussed here – of how ignorance can serve as a productive resource. In order to be able to act, it is suggested, actors need to agree on what is not known and to take this into account for future planning. They need to decide to act in spite of (sometimes) well-defined ignorance, or what has more recently been called 'non-knowledge'; a term used to refer to the possibility of becoming knowledgeable about one's own ignorance. Unlike the term 'ignorance', non-knowledge points to symmetry between accepted 'positive' knowledge and ignorance that is sufficiently well defined (Gross 2010). The many meanings of the term 'ignorance' – with connotations ranging from actively ignoring something, to attributions of not even knowing that something is unknown – makes 'non-knowledge' more precise as a specified form of the unknown in order to analyse decisions. While the term 'ignorance' is quite ambiguous, 'non-knowledge' refers more precisely to a specific form of the unknown. This specified form can be positively used for further planning and activity or can be rendered unimportant. 'Ignorance' as a generic term can be defined as knowledge about the limits of knowing in a certain area but also including unknown knowledge gaps that actors only realize on hindsight. The latter can be referred to as 'nescience', a different epistemic class from non-knowledge or ignorance, since no one can refer to their own current nescience, as it is not part of their conscious non-knowledge, but may guide their gut feelings and intuition (see Table 22.1 for an overview). The aim is thus to specify ignorance so that it can be used in a meaningful and constructive way.

The following sections contribute to the debate on social-ecological linkages by emphasizing the centrality of dealing successfully with the fact of knowing that you can never know what

Table 22.1 Types of ignorance (adapted from Gross 2010)

Nescience	Unknown ignorance can only be known in retrospect, but can also be things people are not aware of but in fact 'know' (e.g. intuition)
General non-knowledge (broad ignorance)	The acknowledgement that some things are unknown, but not specified enough to take action
Positive non-knowledge	Known and specified ignorance used for further planning and activity
Negative non-knowledge	Known but not necessarily specified (yet), e.g. because rendered unimportant and dangerous at this point in time

to expect in ecological restoration practices. Whereas having faith in total control and complete knowledge of ecological systems and social processes implies an ability to act only when everything is known in advance, an 'experimental approach' makes it possible to accommodate different factors in spite of knowledge gaps.

'Let the fire decide': experimenting with the unpredictable

An experiment in the most general sense can be defined as a cautiously observed venture into the unknown. An experiment is deliberately arranged to generate unexpected events so that the surprising effects derived out of the experimental setup can be used as a springboard for producing new knowledge, since surprises help scientists become aware of what they did not know. As explained above, surprises are the impulse for unexpected knowledge. The difference from the everyday life world is that the surprising abruptness in a laboratory experiment is normally welcomed.

The idea of also thinking about processes outside the laboratory as experiments goes back to Francis Bacon's reflections on the relation of the experimental method to society. Bacon was clearly one of the most influential thinkers in the formation of the worldview that distinguished the experimenter's realm from the world of objects experimented upon, privileging human (rational) beings as masters of a world to which they essentially do not belong. However, Bacon's most provocative proposal was the idea that approval of the experimental method in science would turn society itself into a large-scale experiment (Krohn 2009). In this view, modern society should give science an experimental opportunity, since the promises of gains from modern science cannot be justified by anticipatory argument, but only by practising and implementing the new method into the larger society. In other words, although moving the dangers of the laboratory experiment into the wider society can be questionable, at the same time to not do so would mean losing the opportunity for unexpected and potentially useful innovation. Extending this notion of public experimentation to ecological restoration means that unexpected events are not necessarily rendered as purely negative but often considered crucial in order to learn and to help create better and safer solutions for restoration technologies. In this sense, the interest in the experimental co-constitution of nature through restoration focuses on the positive outcomes and possibilities of coping with inevitable ignorance and surprises in human–nature interactions.

An experimental approach in the sense outlined above can be conceived of as a way of coordinating the contingent activities of diverse actors in ecological restoration processes which are pursued despite uncertainties and an acknowledged awareness of ignorance, so that projects as a whole do not have to be interrupted. In the case of chemical risk assessments, for example, Claire Waterton and Brian Wynne (2004) have observed an 'alternative policy order' in the European Environmental Agency where the acknowledgement of ignorance implies a revision of responsibilities between science, formal policy institutions and other stakeholders. As they put it, 'if it is not possible to predict the risks of this or that chemical, then the question arises: do we need it and the social purposes it is serving; and do we want the uncontrollable uncertainties which its use brings?' (Waterton and Wynne 2004: 100). As a consequence, it is not science alone but a much broader range of stakeholders that are called on to articulate values, concerns, and perhaps even aesthetic preferences, in the course of decision making.

Such an approach can be observed in ecological restoration since, as field practice, its laboratory is not a controlled setting but an ever-changing world of human actions and natural fluctuations. The practices of ecological restoration can be perceived as similar to processes of continual refiguration of natural setups in the pursuit of an intended capture of nature's agency. Ecological restoration can thus be understood as doing work *in* and *with* nature that leads to the observation that the ecosystem answers to human attempts at building and designing a piece of land and that human work and attitudes in turn adjust to nature's answers. In restoration work, there is the human capacity to 'let nature do the work'; that is, to use natural powers in order to carve out certain inventions of nature that seem to be desirable. Steven Packard (1993: 14), an early practitioner of ecological restoration, stated that 'every restorationist knows the ecosystem will respond in unpredictable ways that rise out of itself. That's precisely what we want to liberate.' Thus, restoration is a human activity in nature that is integrative and focuses on variability and ignorance as absolutely fundamental and perhaps even welcomed. Consequently, any fragmentary understanding of nature is not simply to be mourned as incomplete, but accepted as necessarily full of nescience and non-knowledge. Thus understood, ecological restoration provides a basis for what William R. Jordan III (2006: 25), who coined the term 'restoration ecology' in the mid-1980s, has termed a 'new communion with nature', which explicitly includes elements 'we find inconvenient, useless, ugly, or even dangerous, as well as those we find useful or attractive for some other reason'.

Following established convention and considering only the agency of humans as the legitimate object of sociological study would, in the case of ecological restoration, miss half of 'where the action is' (alluding to Goffman 1967; see also Brewster and Bell 2010; Gross 2003). Human agency and the effects of nature can be assessed separately but, in the practice of ecological restoration, they are too closely linked to be meaningfully separated. This view goes hand in hand with Georg Simmel's (1998: 213) interpretations of the objective character of things designed by humans that hold in their inner core 'strengths and weaknesses, components and significances, that we are completely innocent of and which take us by surprise'. An example of this is Packard's (1988) project to restore a tallgrass savanna. In order to bring some existing landstrips of prairie in the Chicago area together with the necessary old oaks, a certain technique that used the natural powers of prairie burnings was developed by a group of restoration activists and volunteers. As Packard (1988: 14) summarizes:

> We wanted to use natural forces to bring them together, however. So we relied on the fires we were using to restore and maintain the adjacent, open prairies. We let the fires blast into the brush lines as far as they would go. 'Let the fire decide' became our motto. That was the natural scheme; that's what we wanted.

In this view, the unexpected powers of the ecosystem should be unleashed to observe them carefully and, by so doing, learn to better understand the system. Reflecting on the restoration of the savanna, Packard came to the conclusion that this type of public experimentation in nature is in fact the most sustainable way to gain knowledge about ecosystems. He says:

> we learned by a trial-and-error process using hundreds of varying uncontrolled restoration experiments. If we had proceeded more systematically, we would by now either have spent a small fortune, or, using those resources available to us, we would only now be getting the results of the first experiments, all of which were failures. But using craft and intuition we have developed techniques that seem to work.
>
> *(Packard 1988: 13)*

Thus understood, restoration is pieced together and built, thought about and tried out, formulated and reformulated, always in negotiation with other people and nature. A distinctive feature of this conception of ecological restoration is that it can be rendered experimental. Like in an experiment, nature when 'in action' produces results that are not substantively attributable to any human agent or intention. In the negotiations with nature, plans and goals of restorationists that seemed to be feasible in the first place are at stake and liable to revision. The human performances and actions that accompany the restoration of a prairie are themselves products of temporarily emergent sequences of passive adaptation and active social action. The practice of ecological restoration then could also be framed by what Andrew Pickering's (1995: 21) famous metaphor calls 'a dance of agency'. Restorationists build and operate new ecosystems, but the specific outcome of, say, the configuration of a prairie, lies not in the hand of any human actor alone. Furthermore, it cannot be planned with certainty if the prairie is going to be a prairie at all.

As William Jordan (2010) knows from Curtis Prairie in Madison, Wisconsin, where sweet clover took over the whole prairie at one point, the random element of nature had to be reintroduced.[3] This goes so far as to include imitating even the random quality of natural processes in, for example, the distribution of plants or the timing of events such as floods or burns. Overlooking this random or stochastic element can cause problems in restoration projects, just as much as does leaving out a species or letting a troublesome species get out of hand. To illustrate, Jordan tells the story of sweet clover, a European native, on Curtis Prairie (see Gross 2003). In the 1980s, the ecological restorationists there had fallen into the habit of burning their prairies on a regular schedule, most areas being burned in alternate years. This, however, favoured reproduction of the sweet clover, a biennial plant that takes two years to set and seed. For this reason, the biennial burns amplified its reproduction at the expense of other species. Once this was understood, the solution was simple. As soon as the crew of restorationists broke the rhythm of the burns by shifting to an irregular burning schedule, the growth of sweet clover quickly declined and the prairie species reasserted themselves. By simply turning the work over to the unpredictable elements in nature, they gave up the idea of control. It is true that, when it comes to actually picking a burn day and time, the ecological restorationists make decisions, taking into consideration factors such as safety, convenience and cost. Yet at one level they very deliberately relinquished control by deciding not to decide, but instead turned the decision over to the surprise element in ecosystems. Thus understood, the public experiment *in* and *with* the prairie might go on for many more years, without a fixed plan and without definite sets of knowledge. This, however, is not a strategy of incompetency or a sign of premature science, but a conscious site-specific approach that is aimed at doing justice to the non-human nature found in these sites. Indeed, this is perhaps the only way to move forward in the face of inevitable ignorance and surprise. Restoration activists pick certain possibilities to design nature, and nature, in turn, moves the restorationist in a certain direction while providing a series of options to be chosen in the next step.

Ecological restoration can thus be understood as a turn-taking process between human and natural agency in the sense that there are passages of mainly human activity as well as periods of human passivity where humans wait to see how nature will react to their suggestions (e.g. a prairie fire). Here, a crucial aspect of dealing with unknowns needs to be noted. It is the intentionality of dealing with non-knowledge; that is, the question of how actors can shield their objective from the consequences of failure, enabling them to acquire resources and facilitating their ability to assert expertise in the face of unexpected events. This also raises questions about the varied ways that actors may seek to not know about certain things in the sense that they may consciously avoid knowledge from emerging in the first place (Frickel et al. 2010; Hess 2010; Kempner et al. 2011). A crucial question here is: How much do actors need to know in order to make strategic use of deliberate knowledge avoidance? On the other hand, this points to disinterestedness as well as lack of thoughtfulness and patience. In this context it may also be important to analyse how ignorance is believed to be able to be turned into extended or new knowledge. As actors in the field of restoration and management of contaminated sites know, however, many things, such as the type of contaminants in the ground, cannot be known even with the most sophisticated investigation technologies. Consequently, for a sociological observation, it is the whole process of moving back and forth between as-yet-unknown phenomena and human decision making that needs to be put into the picture.

If the example of restoring prairies in the face of different 'unpredictabilities' may appear uncontroversial, it is crucial to ask whether the sociological approach to reconstructing human interaction *with* and the morphing *of* nature proposed here can be meaningfully applied to more hazardous areas of ecosystem restoration. The next section will thus present examples from revitalization projects in contaminated sites management. It will highlight strategies applied to the unexpected appearance of contaminants during the process of cleaning up and restoring complexly contaminated industrial sites to further illustrate the co-constitution of unexpected natural powers and human activities.

Restoring and revitalizing industrially contaminated land

Estimates indicate that between 500,000 and 1.6 million contaminated sites exist in Europe (EEA 2000; Frauenstein 2010), some 20,000 of which can be classified as 'megasites' due to the complexity of soil and groundwater contamination. These megasites pose a range of technical and management challenges, including the need to cope with ignorance regarding multiple contaminant sources and plumes from previous industrial activities. Since much of the contamination on deserted brownfield sites originates from the industrial production of the early twentieth century, records of accidents and waste disposal leading to contamination are often poor. Very little documentation (archive materials or maps) of chemicals dumped into the ground exists today. Worse, buffering capacities of many soil types and the capability to filter chemicals means that contaminants are often not perceived until the damage is advanced. Furthermore, since there was a lack of awareness about contaminants until the late 1960s, many of the events can only be reconstructed from interviews with the people who lived or worked in a particular area long enough to remember some of the activities on these sites.

Given the predetermined timeframes of most clean-up and restoration projects and the limited budget available, actors today generally know that they will have to make decisions based on non-knowledge (see Bleicher and Gross 2012; Gross 2010, 2012). They know that it is useless to try to know things that cannot yet be known. To wait for complete knowledge would mean extending operations beyond project timeframes and frittering away the chance to promote new investments and further economic development. However, this means that the actors involved have to expect that surprising things can happen. As soon as construction workers remove the topsoil, it is possible they may find something that does not seem 'normal'. The actors involved then know what they did not know before (positive non-knowledge) and are able to use this knowledge as a basis for further planning and action. This positive non-knowledge can emerge from a general state of ignorance. Based on this non-knowledge, for instance, engineering companies are consulted (because now it is known what is unknown), and samples are taken to evaluate the soil. They may then conclude that an area is heavily contaminated with, for instance, liquid tar and different solvents. Commenting on such an incident in the former East Germany, a representative from the engineering company in charge of a major restoration project on contaminated land stated: 'The discovery of tar meant that the whole philosophy of the project had to be rethought. A completely new plan had to be drafted ad hoc.'[4] In a similar vein, a representative from the project management team explained:

> Construction and cleaning-up has to be understood as a continuous process. They [the workers on the ground] started to remove the material, and when they did so they found that the subsoil was totally different from what we expected. The problem is that tar, which turns to liquid at a certain temperature, makes it necessary to use another type of water barrier for the rain storage reservoir ... At times like these, we have to act first and get the permits changed later.

Actors were prepared for unexpected events and so were able to make decisions quickly and flexibly. In other words, the actors involved agreed on what was not known and took this into account for their subsequent activities; they decided to act on the basis of their (positive) non-knowledge. Positive non-knowledge, however, can be developed into new knowledge, which is fed into the next stage of the process to be subjected to further observation and assessment (Bleicher and Gross 2012).

Experts' willingness to disclose the limits of knowledge in their communication with other stakeholders can be understood as an important part of the process of implementing successful development projects. This includes changes of permits (administration), change of plans (engineering companies and research institutes) and last, but not least, flexibility in redeploying capital for remediation. One major factor that helps to generate such a strategy is an explicit agreement on shared objectives made at an early stage of restoration projects. In addition, special clauses in the contracts and permits between the actors (the so-called collateral clauses) make particular mention of the unknown. Collateral clauses are agreements between joint contractors to pool their guarantees in handling a large restoration project. The objectives are that others will also have recourse to the contractor's warranty to continue the project in the face of unexpected changes.[5]

A further important issue with regard to successfully coordinating such projects is the institutionalization of contacts and information exchange. Although this might seem obvious, consultation here also implies that all the actors involved must communicate their own non-knowledge. This is understood not in terms of a failure or sloppy investigation but, rather, as a creative way of dealing with contaminants. The approach taken to cope with this phenomenon is one that accommodates project activities despite the presence of unknown factors. Put another way, dealing with ignorance is not simply a process of trial and error and learning from failure, as failures suggest that mistakes have been made. Instead, stakeholders and actors in the restoration of industrially contaminated land take seriously the impossibility of avoiding ignorance, so that there is no target for blame or 'finger pointing'.

Many activities in contaminated sites management as presented here included ignorance and non-knowledge explicitly in their planning. This seemed to encourage the development of innovative strategies and to make full use of the potential and resources of the actors involved to achieve a common goal. The actors achieved a state of preparedness, found organizational solutions for cushioning the impacts of surprising events, and thus developed accepted strategies for coping with the unknown. Given the enormous number of unknown factors regarding the dynamics of ecosystems, acting in spite of ignorance by acknowledging non-knowledge can be seen as a crucial factor in our understanding of nature–society interactions.

Outlook: practices of experimentality

One is tempted to say that all crisis-ridden events of the early twenty-first century follow a similar pattern. The tsunamis in Indonesia in 2004 and Japan in 2011, Hurricane Katrina in 2005, as well as the financial crash of 2008 and the 2010 Gulf of Mexico oil spill, all have one thing in common – they all belong to the unforeseen. Each time such a catastrophe happens, the current institutional and cognitive horizon of expectations was out of reach. However, if we accept that this is the case and acknowledge that radical surprises such as these cannot be avoided, then we may need to start looking at everyday events that seem to follow similar patterns. These events obviously are smaller in scale but are also happening (or so it seems) on a more regular basis. After all, in everyday life everyone recognizes that many things may happen by surprise – from falling in love to unexpected weather changes. However, theoretical tools available to frame how people successfully cope with what everybody knows are still in their infancy. If ignorance and surprise is unavoidable, then it is high time to look for examples of best practice and at least some theoretical patches that can help us to better understand how to *constructively* deal with these types of events (instead of merely 'critically' pointing to the fact that they happen at all). Of course, this does not mean that we look forward to events such as hurricanes, oil spills and earthquakes. But, instead, we can hope to better understand how, on a small scale, actors are capable of successfully coping with ignorance and surprise and to delineate ways of reducing injury and damage in the future. Given the regionally definable borders of most ecological design and landscape restoration projects, conceptualizing ignorance and surprise can be acknowledged as a first step towards theory building on coping with what everybody knows.

Taken to its logical conclusion, an experimental approach as outlined above would mean a shift away from navigation and orientation built on previous experience and historical extrapolations in nature and society (e.g. risk assessments). Instead, the emphasis will be more towards prospective and temporary notions of knowns and unknowns that help to govern social-ecological relationships by experimentally coping with ignorance and surprise *in* and *with* the natural world. To further such experimental processes of searching for solutions in sociological analysis and to uncover how different shadings of non-knowledge explicitly belong to negotiation processes in human–nature interactions appears to be a promising alternative and an extension to risk-centred analysis. It also promises to deliver insights to different possibilities in coping with disturbing changes in order to strengthen modern societies' resilience capacities (Holling and Gunderson 2002) or, better yet, to go beyond resilience and to reshape socio-technical infrastructures or to create new systems alternatives. At least on a small or regional scale, it seems possible to communicate non-knowledge and to cope with surprising events without losing credibility. In the cases discussed here, actors (self-)organized to put flexible preparatory measures in place in order to guarantee the continuity of the overall process. Clean-up processes instigated to deal with such cases of 'creeping' contaminants demand fast and flexible governance structures rather than blame and finger pointing. To start from this point and to draw on the experimentality of everyday life in order to conceptualize the proceduralization of the practices of actors appears to be one of the most exciting tasks facing theory building on environment–society interactions. In sum, the production of non-knowledge in the interaction between humans and the natural world does not have to be seen as a deviation or as an obscure residual category in a world oriented towards the 'revelation' of unknown phenomena, but as a potentially fruitful everyday aspect in the social-ecological co-constitution of nature.

Notes

1 By now the literature on the development and variations of ecological restoration has become immense. For a few of the more recent books on different models and the global scale of restoration activities, see Allison (2012), Comín (2010), France (2011), Hall (2010), Jordan and Lubick (2011) or Egan et al. (2011). Information on the Society for Ecological Restoration (SER) founded in 1988 can be found on the official website at www.ser.org. For the more academic orientation of restoration, see the journal *Restoration Ecology* (since 1993). For a practitioner perspective, refer to *Ecological Restoration* (since 1981, until 1999 called *Restoration and Management Notes*). Since the early 2000s, many more journals with the word 'restoration' in their titles have been founded: e.g. the Australian journal *Ecological Management and Restoration* (since 2000). For an early attempt at reconstructing the relationship between some of the academic streams of restoration ecology and the more practical approaches, see Gross (2002).

2 Many different concepts of resilience and vulnerability can be found in the recent literature (see Hess 2011; Kuecker and Hall 2011; Kuhlicke 2010; Pelling 2011). Most authors dealing with ecological issues, however, build on the work of Holling and his colleagues (see Holling and Gunderson 2002). An experimental approach as introduced in this chapter shows some similarities to what Holling and Gunderson (2002) call 'ecosystem resilience' – the capacities of a system to experience disturbance. However, it also departs from the concept by moving away from the idea of maintaining a system's standard functions, since in many cases this could mean to defensively rely on the very mechanisms that may have created a crisis in the first place. On the conservatism inherent in the notion of resilience, see also Gross (2010: 78–79). For a general discussion on the limits of the resilience concept for the analysis of social systems, see Davidson (2010).

3 Curtis Prairie, named after the Wisconsin botanist John Curtis, is located at the University of Wisconsin Arboretum. It is sometimes referred to as the world's oldest restored prairie. Many early experiments on the use of fire in ecosystem management started there in as early as the 1930s. For a more detailed analysis and different restoration examples from the Midwest of the United States, see Gross (2003).
4 These quotes are from interviews conducted with different stakeholders and decision makers in the SAFIRA II project (*Revitalization of Contaminated Land and Groundwater at Megasites*), funded by the German Federal Ministry of Education and Research (BMBF) from 2006 to 2012.
5 This is based on what the German legal system calls *Auflagenvorbehalt* – a provision for the addition of supplementary conditions – and a *Nachtragsangebot* – the possibility for both the contractor and the customer to demand a supplementary offer or a follow-up proposal that may deviate from the original specification (see Bleicher 2012).

References

Allison, S.K. (2012) *Ecological Restoration and Environmental Change: Renewing Damaged Ecosystems*. London: Routledge.
Aven, T. (2003) *Foundations of Risk Analysis: A Knowledge and Decision-Oriented Perspective*. New York, NY: Wiley.
Beck, U. (1986) *Risikogesellschaft: Auf dem Weg in eine andere Moderne*. Frankfurt am Main: Suhrkamp.
Bleicher, A. (2012) 'Entscheiden trotz Nichtwissen: Das Beispiel der Sanierung kontaminierter Flächen', *Soziale Welt*, 63, 2, pp. 97–115.
Bleicher, A. and Gross, M. (2012) 'Confronting Ignorance: Coping with the Unknown and Surprising Events in the Remediation of Contaminated Sites', pp. 193–204 in S. Kabisch, A. Kunath, P. Schweizer-Ries and A. Steinführer (eds), *Vulnerability, Risks, and Complexity: Impacts of Global Change on Human Habitats*. Göttingen: Hogrefe.
Brewster, B.H. and Bell, M.M. (2010) 'The Environmental Goffman: Toward an Environmental Sociology of Everyday Life', *Society and Natural Resources*, 23, 1, pp. 45–57.
Campbell, S. and Currie, G. (2006) 'Against Beck: In Defence of Risk Analysis', *Philosophy of the Social Sciences*, 36, 2, pp. 149–172.
Comín, F.A. (ed.) (2010) *Ecological Restoration: A Global Challenge*. Cambridge, UK: Cambridge University Press.
Davidson, D.H. (2010) 'The Applicability of the Concept of Resilience to Social Systems: Some Sources of Optimism and Nagging Doubts', *Society and Natural Resources*, 23, 12, pp. 1135–1149.
Dowie, J. (1999) 'Against Risk', *Risk, Decision and Policy*, 4, 1, pp. 57–73.
EEA (European Environment Agency) (2000) *Down to Earth: Soil Degradation and Sustainable Development in Europe: A Challenge for the 21st Century*, Environmental Issue Report No. 16. Copenhagen: European Environment Agency.
Egan, D., Hjerpe, E.E. and Abrams, J. (eds) (2011) *Human Dimensions of Ecological Restoration: Integrating Science, Nature, and Culture*. Washington, DC: Island Press.
France, R.L. (ed.) (2011) *Restorative Redevelopment of Devastated Ecocultural Landscapes*. Boca Raton, FL: CRC Press.
Frauenstein, J. (2010) *Stand und Perspektiven des nachsorgenden Bodenschutzes*. Dessau, Germany: Umweltbundesamt.
Frickel, S., Gibbon, S., Howard, J., Kempner, J., Ottinger, G. and Hess, D.J. (2010) 'Undone Science: Charting Social Movement and Civil Society Challenges to Research Agenda Setting', *Science, Technology and Human Values*, 35, 4, pp. 444–473.
Goffman, E. (1967) *Interaction Ritual: Essays in Face-to-Face Behavior*. New York, NY: Doubleday.
Green, J. (2009) 'Is It Time for the Sociology of Health to Abandon "Risk"?' *Health, Risk and Society*, 11, 6, pp. 493–508.
Gross, M. (2002) 'New Natures and Old Science: Hands-on Practice and Academic Research in Ecological Restoration', *Science Studies*, 15, 2, pp. 17–35.
Gross, M. (2003) *Inventing Nature: Ecological Restoration by Public Experiments*. Lanham, MD: Lexington Books.
Gross, M. (2010) *Ignorance and Surprise: Science, Society, and Ecological Design*. Cambridge, MA: MIT Press.
Gross, M. (2012) '"Objective Culture" and the Development of Nonknowledge: Georg Simmel and the Reverse Side of Knowing', *Cultural Sociology*, 6, 4, pp. 422–437.
Hall, M. (ed.) (2010) *Restoration and History: The Search for a Usable Environmental Past*. London: Routledge.
Hess, D.J. (2010) 'Environmental Reform Organizations and Undone Science in the United States: Exploring the Environmental, Health, and Safety Implications of Nanotechnology', *Science as Culture*, 19, 2, pp. 181–214.

Hess, D.J. (2011) 'Sustainable Consumption and the Problem of Resilience', *Sustainability: Science, Practice and Policy*, 6, 2, pp. 26–37.
Holling, C.S. and Gunderson, L.H. (2002) 'Resilience and Adaptive Cycles', pp. 25–62 in L.H. Gunderson and C.S. Holling (eds), *Panarchy: Understanding Transformations in Human and Natural Systems*. Washington, DC: Island Press.
Hubbard, D.W. (2009) *The Failure of Risk Management*. New York, NY: Wiley.
Jaeger, C.C., Renn, O., Rosa, E.A. and Webler, T. (2001) *Risk, Uncertainty, and Rational Action*. London: Earthscan.
Jones, D.S. (2007) 'Fearing the Worst, Hoping for the Best: The Discursive Construction of Risk in Pregnancy', pp. 37–50 in S. Jones and J. Raisborough (eds), *Risks, Identities and the Everyday*. Aldershot, UK: Ashgate.
Jordan, W.R., III. (2006) 'Ecological Restoration: Carving a Niche for Humans in the Classic Landscape', *Nature and Culture*, 1, 1, pp. 22–35.
Jordan, W.R., III (2010) 'Some Reflections on Curtis Prairie and the Genesis of Ecological Restoration', *Ecological Management and Restoration*, 11, 2, pp. 99–107.
Jordan, W.R., III and Lubick, G.M. (2011) *Making Nature Whole: A History of Ecological Restoration*. Washington, DC: Island Press.
Kempner, J., Merz, J.F. and Bosk, C.L. (2011) 'Forbidden Knowledge: Public Controversy and the Production of Nonknowledge', *Sociological Forum*, 26, 3, pp. 475–499.
Krohn, W. (2009) 'Francis Bacons Literarische Experimente', pp. 33–52 in M. Gamper, M. Wernli and J. Zimmer (eds), *Es ist nun einmal zum Versuch Gekommen: Experiment und Literatur I, 1580–1790*. Göttingen: Wallstein.
Kuecker, G.D. and Hall, T.D. (2011) 'Resilience and Community in the Age of World-System Collapse', *Nature and Culture*, 6, 1, pp. 18–40.
Kuhlicke, C. (2010) 'The Dynamics of Vulnerability: Some Preliminary Thoughts about the Occurrence of "Radical Surprises" and a Case Study on the 2002 Flood (Germany)', *Natural Hazards*, 55, 3, pp. 671–688.
Latour, B. (2005) *Reassembling the Social: An Introduction to Actor-Network-Theory*. Oxford, UK: Oxford University Press.
Latour, B. (2011) 'From Multiculturalism to Multinaturalism: What Rules of Method for the New Socio-Scientific Experiments?' *Nature and Culture*, 6, 1, pp. 1–17.
Luhmann, N. (1993) *Risk: A Sociological Theory*. New York, NY: De Gruyter.
Overdevest, C., Bleicher, A. and Gross, M. (2010) 'The Experimental Turn in Environmental Sociology: Pragmatism and New Forms of Governance', pp. 279–294 in M. Gross and H. Heinrichs (eds), *Environmental Sociology: European Perspectives and Interdisciplinary Challenges*. Heidelberg: Springer.
Packard, S. (1988) 'Just a Few Oddball Species: Restoration and the Rediscovery of the Tallgrass Savanna', *Restoration and Management Notes*, 6, 1, pp. 13–20.
Packard, S. (1993) 'Restoring Oak Ecosystems', *Restoration and Management Notes*, 11, 1, pp. 5–16.
Pelling, M. (2011) *Adaptation to Climate Change: From Resilience to Transformation*. London: Routledge.
Pickering, A. (1995) *The Mangle of Practice: Time, Agency, and Science*. Chicago, IL: University of Chicago Press.
Renn, O. (2009) 'Integriertes Risikomanagement als Beitrag zu einer nachhaltigen Entwicklung', pp. 553–568 in R. Popp and E. Schüll (eds), *Zukunftsforschung und Zukunftsgestaltung: Beiträge aus Wissenschaft und Praxis*. Berlin: Springer.
Rheinberger, H.-J. (1997) *Toward a History of Epistemic Things: Synthesizing Proteins in the Test Tube*. Stanford, NJ: Stanford University Press.
Schatzki, T. (2010) 'Materiality and Social Life', *Nature and Culture*, 5, 2, pp. 123–149.
Simmel, G. (1998/1919) *Philosophische Kultur. Gesammelte Essays*. Berlin: Wagenbach.
Smithson, M. (2010) 'Understanding Uncertainty', pp. 27–48 in G. Bammer (ed.), *Dealing with Uncertainties in Policing Serious Crime*. Canberra, ACT: Australian National University Press.
Smithson, M., Bartos, T. and Takemura, K. (2000) 'Human Judgment under Sample Space Ignorance', *Risk, Decision and Policy*, 5, 2, pp. 135–150.
Stirling, A. (2007) 'Risk, Precaution and Science: Towards a More Constructive Policy Debate', *EMBO Reports*, 8, 4, pp. 309–315.
Waterton, C. and Wynne, B. (2004) 'Knowledge and Political Order in the European Environmental Agency', pp. 87–108 in S. Jasanoff (ed.), *States of Knowledge: The Co-Production of Science and Social Order*. London: Routledge.
Wehling, P. (2011) 'Vom Risikokalkül zur Governance des Nichtwissens', pp. 529–548 in M. Gross (ed.), *Handbuch Umweltsoziologie*. Wiesbaden: VS Verlag.

23
Biological invasions as cause and consequence of 'our' changing world
Social and environmental paradoxes[1]

Cécilia Claeys

Concern for the environment has arisen from claims about the depletion of living species and has, as such, raised awareness among decision makers and the general public about endangered species, thresholds of irreversibility and declining biodiversity. Yet, sitting alongside the list of endangered species published by the International Union for Conservation of Nature (IUCN), is a list of invasive species. This list identifies a major new ecological risk. Biological invasions are caused by 'alien species whose establishment and spread threaten ecosystems, habitats or species with economic or environmental harm' (Article 8(h) of the Convention on Biological Diversity). The new semantic field of biological invasion recycles and extends naturalist discourses denouncing the depletion of species. Living matter is now understood quantitatively (as biomass) and qualitatively (as endemic, or valued, species), which has complicated the notion of biodiversity.

The concept of 'invasive species' was first coined by ecologist John Elton in 1958, but only became common in scientific research in the 1980s. The 1992 Rio Declaration pointed to biological invasions as one of the primary causes of declining biodiversity. Climate change is also seen as an aggravating factor that may increase or intensify the risk of biological invasions (Burgiel and Muir 2010). However, the concept of biological invasions is not universally accepted among conservation biologists and managers of protected areas. Further, the concept and the controversies which surround it have moved beyond the purely scientific realm and generated reactions, attempts at appropriation and criticism from, inter alia, users, inhabitants, farmers and consumers.

The ways in which the concept of biological invasions is used socially shed new light on the relationship between humankind and nature. Nature is redefined as fluctuating; rare in some places, proliferating in others; different today from tomorrow; sometimes threatened, sometimes threatening; it is both predator and prey. Given this, the debate over biological invasions offers a particularly good example of the interpenetration of social and environmental change. The sociological analysis put forward in this chapter is based on qualitative and quantitative data gathered from managers, decision makers, experts and inhabitants in the South of France,[2] as well as an extensive literature review of international scientific research. The taxonomic debate (the first part of the chapter) and the socio-technical controversies related to biological invasions (the second part) challenge the dichotomous and unidirectional order of humankind's relationship with nature. Finally, the third part explains how the concept of biological invasions has rekindled and revamped the great myths of Western society in which nature and its fluctuations are at once a subject of fear and control and subject to idealization.

Taxonomies: permeable and shifting frontiers between nature and culture

The IUCN defines alien species (also called non-native, non-indigenous, foreign or exotic) as:

> a species, subspecies or lower taxon introduced outside its normal past or present distribution; it includes any part, gametes, seeds, eggs, or propagules of such species that might survive and subsequently reproduce.
>
> *(McNeely 2001: 3)*

As such, biological invasions are the result of the transgression of spatial and temporal frontiers. To identify a species as foreign or, conversely, indigenous, one must be able to answer at least two questions: 'Is it from here or from elsewhere?' and 'Since when?' Researchers and other social actors have underscored the difficulty of this task due the lack of data and the impossibility of defining a zero point.

Despite their relative diversity, naturalists' answers tend to involve 'taxonomic acrobatics' (Boltanski 1979), which shift from natural history to the history of human civilization. The European colonization of the Americas is taken as the first phase of biological invasions. The Industrial Revolution and subsequent large-scale development is seen as the second phase, and the contemporary period and the globalization of trade is understood as the third. These three periods are seen as acceleration thresholds in the introduction of species and, thus, the risk of biological invasions. The claim that the circulation of goods, people and species increased quickly starting in the fifteenth century and exponentially from the nineteenth century is based on well-founded observations. However, there is an arbitrary side to every threshold that gives a social, cultural and/or economic dimension to ostensibly objective scientific criteria. Some people equate the fifteenth century with the discovery of the Americas and the rise of great exploratory journeys and colonization. The nineteenth century is associated with the rise of modernity. The contemporary Western intellectual's perception of these periods is mired in both admiration and guilt. The image of Christopher Columbus's caravels – presented to European schoolchildren from an early age – is one of the founding myths of Western civilization, but also the premise behind its glory and its corruption. These three phases in the rise of Western civilization – colonization, industrialization and globalization – are thus connected to three degrees of severity in biological invasions and are, as such, a critique of the ravages of culture on nature. Yet, at the same time, they intertwine social and environmental change. This taxonomic shift from natural history to human history undermines the solid grounding of the nature/culture dichotomy (Claeys 2010).

Taxonomic paradoxes persist, or even increase, when we seek to define the causes of biological invasions (Woods and Veatch Moriarty 2001). The naturalists interviewed in this study tend to point out the exclusively anthropogenic origins of this process in a sort of biocentric Rousseau-ism (which argues that nature is essentially good; it is man who corrupted it). As James T. Carlton (1999: 1) in his editorial in the first issue of *Biological Invasions* writes:

> Living organisms have flowed across the face of the Earth for hundreds of millions of years ... These natural range expansions of organisms, extended over vast sweeps of geological time, stand in dramatic temporal contrast to the rapid reorganization of the biotic world

that has attended the flood of humans across the face of the Earth in the past 10,000 and more years ... At least partially as a result of this, a very large number of these invasions have caused extraordinary environmental, social, and economic changes.

From this perspective, there are biological invasions which are considered normal (i.e. understood as natural and part of the dynamic nature of ecosystems) and invasions which are considered abnormal (since they are of anthropogenic origin). Abnormal species movements, in turn, both produce undesirable environmental and social change and accelerate as a consequence of undesirable environmental and social change. Climate change, in particular, is considered likely to lead to quantitative and qualitative increases in biological invasions (Burgiel and Muir 2010).

The concept of naturalized species, however, complicates a neat differentiation between species movements deemed normal and natural, on the one hand, versus those deemed abnormal and anthropogenic, on the other. In this case, the species is indeed an introduced alien species, but its development is not considered problematic. The idea behind naturalized species is that 'good' alien species may exist. This positive qualification is based on two main criteria: (1) the species' problem-free integration into the ecosystem, which results in 'the washing away of human influence' (Hettinger 2001); and (2) its usefulness for the human species. As a botanist we interviewed explained, numerous fruits and vegetables in French kitchens are of foreign origin, and allowances have to be made. Conversely, an alien species worsens its case if, on top of affecting the ecosystems dear to naturalists, it disturbs human activity or, worse still, public health. Examples include the allergenic *Casuarina equizetifolia* (Australian pine, Coast She Oak), *Schinus terebinthifolius* (Brazilian Pepper) and *Melaleuca quinquenervia* (Niaouli, Broad-Leaved Paperbark) (Binggeli 2001).

This dualism between good and bad alien species is derived from a synthesis of biocentric reasoning (good/bad for the ecosystem) and anthropocentric reasoning (good/bad for human activity). Anthropocentric reasoning about economic consequences is constantly used in scientific research, which nonetheless takes a biocentric approach to nature (see works by McNeely 2001; Richardson et al. 2000). The implicit or explicit dualism of invasion biologists is similar to the traditional dichotomies of gardeners and farmers who distinguish between good and bad weeds and useful species and pests. In some cases, biocentric and anthropocentric logics are in conflict and lead to the redefinition of status for some species. The European brown bear and wolf are archetypes of such emblematic species: officially classified as pests for centuries they have recently been reclassified as endangered species (Bobbé 2002; Mauz 2006; Micoud 1993; Skogen et al. 2008). For other species, anthropocentric and biocentric classification systems support each other. *Myocastor coypus* (Coypu, Nutria, Nutra-Rat) is a particularly good example of the anthropo-/biocentric conflation (Mougenot and Roussel 2006). The animal, which has a beaver-like body and a rat-like tail, was introduced to Europe, particularly France, during the nineteenth century, for its fur. When the intended financial success did not materialize (it lacked the luxury good appeal and was not therefore profitable), farmed *Myocastor coypus* were set free. The animals quickly moved to their habitat of choice: wetlands, streams, rivers and canals. Their burrows, which are several metres long, weakened the riverbanks and embankments, and were blamed for accentuating flood risks in regions such as Camargue (Claeys-Mekdade 2003). While, to the great pleasure of tourists and amateur photographers, these relatively tame animals are now easily visible from the roadside, they are also listed in the Invasive Species Atlas drawn up by the Camargue Regional Natural Park (Costa 2005), because they have become a threat to the ecosystem. Of the seven reasons for their inclusion in the list, three belong to the biocentric register (biodiversity, hydraulic activity and habitat) and four to the anthropocentric register (infrastructure, agriculture, animal health and human health).

The PNRC Atlas indicates that *Myocastor coypus* is legally classified as a pest (based on four national decrees) and urges hunters to take up a gun or bow to eliminate it.

The mosquito is another example of the anthropo-/biocentric conflation that exists in both scientific and vernacular taxonomies, and the crossover between what is considered 'pest' and 'useful'. Along the French Mediterranean coast, city mosquitoes (mainly *Culex pipiens*) are of little ecological interest to naturalists. They are considered a bothersome animal species that infests sources of urban miasma (e.g. sewers, septic tanks, etc.). Mosquitoes in farmed fields (different indigenous *aedes*) belong more or less to the bygone era of insecticide-free farming, while mosquitoes in natural areas (actually the same *aedes*) abundantly reproduce in protected wetlands which are free from mosquito control programmes. Given such a spatialized taxonomic layout, naturalists gladly agree to mosquito control through insecticide use in cities, but are nonetheless opposed to mosquito control in protected wetlands, particularly in the Camargue region.

There are two complementary arguments for this. Not only would the eradication of mosquitoes affect biomass and the balance of nature by weakening the first step of the food chain, it would also pollute the ecosystem with the use of pesticides. Local residents are more or less of the same opinion. Although most would like to see mosquitoes eradicated in inhabited areas, they recognize the role they play in so-called natural areas, where calls for mosquito control are much less common. Nevertheless, this neat spatial divide is challenged by the fact that these winged creatures do not differentiate between urban, rural and natural frontiers when they are carried by the wind, sometimes over several kilometres. The 'good and useful' mosquito, which is allowed to legitimately proliferate in natural areas, can therefore leave its designated area and become associated with the anthropophilic misdeeds of the 'bad and pesky' city mosquito whose proliferation is reprehensible (Claeys-Mekdade 2003; Claeys-Mekdade and Nicolas 2009; Claeys-Mekdade and Sérandour 2009). The usefulness (albeit limited) ascribed to mosquitoes that proliferate in natural areas is based on a biocentric conception of the world. But it is also a cultural and political pretext for some territories with strong ties to their identity such as the Camargue or Morbihan (Southern Brittany) regions: 'Real' *Camarguais*, like 'true' *Bretons*, are supposedly not bothered by mosquitoes and can live with them. As an integral part of local identity, abundant mosquitoes are also the gatekeepers which keep 'foreigners' – those city dwellers and tourists who 'invade' the Camargue and Morbihan regions – at bay (Claeys-Mekdade 2003; Claeys-Mekdade and Nicolas 2009; Huneau 2008). Here, a sense of both ecological and cultural indigeneity work together against the intrusion of foreigners.

The taxonomic debate over whether the indigenous mosquito is useful or a pest took a new turn with the recent introduction and proliferation along the French Mediterranean coast of an alien mosquito species which is a vector of numerous diseases (including dengue fever and the chikungunya virus). *Aedes albopictus* (Asian Tiger Mosquito, Forest Day Mosquito) has not stopped expanding its range, first in tropical areas and, more recently, in temperate regions, notably due to international trade in retreaded tyres, which often contain stagnant water where females like to lay their eggs (www.eid-med.org). The proliferation of this mosquito, which could worsen with global warming, is a real problem for humans. This thriving alien mosquito excels in transgressing established spatial–temporal zones. Unlike indigenous mosquitoes in temperate regions that bite only at certain times of the day, *Aedes albopictus* fills its anthropophilic appetite any time of day or night. Its dormant period in winter is particularly short, leaving humans subject to a biting season nearly twice as long as the indigenous species. Finally, it has settled in regions that up to now have had very few mosquitoes due to the absence of wetlands. This is especially notable along the famous Côte d'Azur coast, where *Aedes albopictus* has been proliferating since the early 2000s (www.eid-med.org). The concern expressed by residents and other stakeholders is further exacerbated by the fact that the region has built its economy around the high quality of life promoted for residents and tourists (Claeys 2010). As a result, a health monitoring operation has been established by public authorities that mobilized the entire health sector. Public awareness campaigns are frequently held to teach residents how to limit Tiger Mosquito proliferation in their gardens, and mosquito control operations are carried out each time an individual returning from a tropical region is detected with a suspicious fever. And yet, no naturalist (to date) has identified any negative effects on indigenous ecosystems caused by this alien insect.

Controversies: science, networks and decision making

Naturalists' concerns are not always cognitively or lexically shared by other social actors, even when the invasive process has a significant social or ecological impact. In some places, there is veritable opposition to naturalists' views. Elsewhere, alliances are formed where they are least expected. The debate over managing *Larus michahellis* (Yellow-Legged Gull) is a good example of such controversy. As a Mediterranean seabird, *Larus michahellis* was, until recently, listed as an endangered species. But the naturalist community started to observe an abnormal demographic increase in the species which benefited from urban growth and, more specifically, from the development of open-air landfills. The bird had become a threat to local biodiversity, notably because it attacks the young of other animal species and destroys vegetation with its overly abundant droppings. More recently, it has become a vector for new diseases acquired through its consumption of pharmaceutical products scavenged in the landfills (Clergeau 1997). In developing strategies to control the Gull population, naturalists turned to local hunting clubs. The first armed campaigns were a technical failure because *Larus michahellis* are extremely fertile and were able to reproduce rapidly to replace their fallen offspring. The campaigns were also a social failure due to protests from walkers and hikers opposed to the 'bloodthirsty' pastime of hunting. Residents, further, perceive these birds to be emblematic of the area and some feed the seabirds, drawing them to their windowsills and balconies for their elegance (Gramaglia 2010).

Residents elsewhere have demanded population control for certain species considered to be invasive. For example, on Gotland Island in the southeast of Sweden, an urticaria- (hives-) inducing caterpillar *Thaumetopoea pinivora* (Northern Pine Processionary Moth) is a problem for residents, who have demanded its eradication. And yet, no action against the insect has been taken because decision makers and experts have had trouble making the right technical-scientific decisions and getting the required economic investment (Lidskog 2010). In this respect, although economic arguments are often used to condemn biological invasions, they are also frequently used against naturalists. Indeed, economic profit is one of the driving forces behind the voluntary introduction of alien species, and the lack of economic interest may explain why decision makers are reluctant to finance programmes to combat invasive species (McNeely 2001).

The case of two plant species – *Cortaderia selloana* (Pampas Grass) and *Baccharis halimifolia* (Eastern Baccharis, Groundsel Bush, Consumption Weed etc) – considered invasive by naturalists is a good example of the tension that exists between naturalist and economic concerns. Introduced to Europe for their attractiveness and hardiness as ornamental plants in gardens, these two species are also used by municipalities in landscaping green spaces and roundabouts. However, *Baccharis halimifolia* thrives in wetlands, which are also prime hunting and fishing grounds, turning these open spaces into closed areas and making them inaccessible. The inconvenience they cause to hunters and fishing groups, who are already restricted by regulations limiting the times and places they are allowed to pursue these activities, directly influences opinions about the plant. The argument employed in support for the eradication of the plant species flows from a now classic attempt at preserving the heritage and greening the reputation of hunting and fishing (Barthélémy 2006; Fabiani 1984). Among other things, the cultural and political demands of French hunters, expressed notably through the right-wing *Chasse, Pêche, Nature et Tradition* (Hunting, Fishing, Nature and Tradition) political party, reveal an identity-based attachment to a territory. Hunters and fishers are thus ideologically predisposed to accepting the naturalist dichotomy ('indigenous' vs 'exotic') and have developed strategies to combat the two species with scientific and technical support from natural area managers. As opposed to the above-mentioned Gull example, there is no risk of being called bloodthirsty in this case.

The invasive adjective attributed by naturalists and their ad hoc allies – hunters and fishers – to these two plant species is not, however, shared by most inhabitants and green space managers. Some people do not hesitate to criticize scientists' opinions, either scoffing at or radically opposing them. Others accept the invasive adjective, which is sometimes new to them, but believe the problem is quite secondary when put in perspective (Claeys 2010). Everyday resistance to the idea of invasive alien species is supported by the fact that such plants are often thought of as typically Mediterranean. This is particularly the case with *Cortaderia selloana*, which is selected by amateur and professional gardeners for its supposed typicality despite having a common name which betrays its origins in the Pampas region of South America. Whilst naturalists wage a (sometimes gruelling) battle against these two invasive alien plants, the catalogues and websites of major horticulture brands promote their merits. When asked about the status of indigenous and exotic plants, a regional and national representative from the horticultural sector said:

> Today, if you will, plants have become global. They travel everywhere. You would be extremely surprised if, when you bought a small plant, it could tell you the route it had travelled. It may have left Sicily to go to Holland, only to return to Paris and then get sent to Marseilles. That is entirely possible. Plants travel everywhere.

The indigeneity of a plant is taken into account when it can be used as a selling point. The horticulturists and green space managers interviewed mentioned that Mediterranean gardens are the current trend. Yet, critics of invasive alien plants should not yet declare victory because, while the names 'Provençal Garden' and 'Mediterranean Garden' entail a limited list of local species, they are conversely synonymous with the dreams and imaginations of professionals and their clients, who are quite happy to evade botanical law. Furthermore, some avant-garde landscapers are outrightly opposed to combating invasive alien species. The 2000s were particularly marked by the landscaper movement, which challenged the former frontiers between 'natural' and 'domestic', 'useful plant' and 'weed' and which advocated a laissez-faire attitude towards plants. The book *Mauvaise Herbe! (Weed!)* set the tone by demanding the renaming of weeds (Pigeat and Paye-Moissinac 2003), as did *Mauvaises Herbes: utilitaires du jardin (Useful Weeds for the Garden)* by Deschamps and Maroussy (2004). Pierre Vignes (2004) has also dedicated his work to the 'splendour and harmony of free plants'. The most virulent attacks have most certainly come from Gilles Clément's *Éloge des vagabondes (A Eulogy for Vagabonds)*, in which he denounces the 'fundamentalism' and 'fraud' of invasion biologists in particular, and of conservation biology in general, and opposes a 'blindly conservative attitude' (Clément 2002: 160; translated here).

The strongest disagreements are often between those with similar interests. Indeed, ecologist-managers and gardener-landscapers are both the descendants of scholarly acclimatization societies from centuries ago. The scientific community was once a central player in promoting voluntary introductions. Until a few decades ago, the French National Society for the Protection of Nature (SNPN) was called the Society for the Acclimatization and Protection of Nature. Originally called the Imperial Zoological Acclimatization Society, it was founded in 1854 by

Isidore Geoffroy Saint-Hilaire, a professor at the National Museum of Natural History (Laissus 1995). For some biologists, this heritage is seen as a mistake of youth made by their predecessors that they want to correct by combating the spread of alien species. Conversely, for some landscapers, this heritage represents the basis for 'planetary gardens' (Clément 1999), widening the plant palette available for their unfettered creations.

The debates are ongoing and growing in complexity. Recently, criticisms have emerged from within the biological sciences community itself. Blaming the overdramatization of a story that is always presented in subjective terms of good versus evil, the botanist David Pearman (2007) argues that, first, a majority of alien species do not cause any trouble, and second, the spread of some native species is also harmful to ecosystems. Thus, he contends that the main problem is not about alien versus native species but, rather, the management of the countryside shifting from overmanagement to undermanagement, and vice versa.

Further, one cannot ignore the economic dimension of the alien species debate, which often precedes other concerns. As previously highlighted, the economic issue can be as much an ally as an enemy of invasive species managers. Thus, whatever the taxonomies and the disputes about their definitions and frontiers, the economic dualism balancing costs and benefits can at any time override socio-technical controversies about a socio-natural subject. Increasingly rewarded for their economic rationalism, nature managers and ecologists try to fight with the weapons of their enemy, acting in some ways in the spirit of what has been defined as 'ecological modernization' (Mol 2002). Indeed, economic costs have now been included in official definitions of invasive species and the cost–benefit calculation integrated into biological modelling, which allows European experts to advocate policy action to combat invasive alien species. In a report to the European Commission, the Institute for European Environmental Policy argued that policy action on invasive species 'is likely to bring more benefits (e.g. avoided costs) than it is estimated to cost' adding that the costs 'are foreseen to diminish over time' and that 'EU-level measures … can help to reduce costs for EU-27 as a whole' (Shine et al. 2010). Thus, if according to scholars such as Alain Touraine (1971), Anthony Giddens (1991) and Ulrich Beck (1992), the social change embodied in environmental thinking is associated with the dawn of deep-rooted cultural change which gives rise to a new postmodern era, for now it rather resembles a sequence of jolts unable to completely shake the predominant speculation-based capitalist economic system.

Cosmogony: fear, control and sublimation

Words are never neutral, not even those which name scientific concepts. The biological sciences, which use many analogies, fall victim to a number of inextricable lexical ambiguities in a context of 'terminological rioting' (Lévêque 2001). The problem is amplified by the fact that monitoring the neutrality of science is not enough to free it from its ideological attachments. Indeed, it is arguably futile even to try and distinguish between these two fields (Claeys 2005; Lévêque et al. 2010). The vocabulary of the biological sciences in general, and of invasion biology in particular, deploys war-like metaphors and ancestral imagery to fuel new fears about nature's deregulation (Claeys and Sirost 2010). The concepts of 'proliferation', 'parasite', 'invasive' and 'hybrid' are all pejoratively depicted figures in what Brendon Larson (2005) calls the 'war of the roses'.

The disorganized abundance of species is metaphorically associated with the disorder of human societies and its bustling crowds. This fear of disorder translates into a fear of losing control. That is one of the first paradoxes in managing nature. The expression itself is contradictory: if nature is that which is not dependent on humans, then any form of managerial intervention will de facto denature the ecosystem involved, leading to what Raphael Larrère (1994b) called the 'wildly artificial'. This is exacerbated by attempts to combat biological invasions, driving the defenders of biocentrism towards heightened interventionism. Protecting nature becomes creative – or even recreational[3] – based on an imaginary climax community (Drouin 1991). Some natural area managers are aware of this paradox, but have trouble finding a way out of what can become a vicious circle of naturalist intervention. As one manager explained:

Manager: For me, a good area is one which functions the most naturally possible ... But the problem is that we don't have enough hindsight.
Interviewer: Could it be a matter of patience?
Manager: Yes, but as managers, you know, we're also paid; it's our job, if you know what I mean. All nature reserve conservationists and such are also evaluated on their ability to save or preserve biodiversity, which is increasingly threatened not only by invasive species but also by other stuff.
Interviewer: Is it hard for a manager to say I'll leave it alone and it will sort itself out?
Manager: Indeed. Personally, I can't do it.

In becoming nature's gardener, managers are confronted with a vicious circle of interventionism; attempting to heal the wounds inflicted on nature by their own species through yet more human intervention, the consequences of which can never be fully understood in advance. Attempts to break out of this incessant spiral have periodically been made. In the South of France, managers of the protected Marais du Vigueirat wetlands carried out an experiment of this type. Confronted with the spread of *Ludwigia spp* (Water Primrose) – a genus of invasive aquatic plants – they stopped grubbing and let the plants flourish. After an initial period of proliferation, it reached a saturation point where eutrophication started to occur and the plant began to suffer the negative effects of its own invasion, thus freeing up the space it had conquered to the detriment of other species. Talking about this experiment, the person in charge of the site expressed his feelings, worries, doubts and then his satisfaction. He nonetheless underscored the difficulty and the risk involved in taking such a laissez-faire approach to endangered species on a large scale.

In such situations, managers are confronted with two major challenges: accepting their inability to control everything and accepting the mismatch between different timescales. Indeed, nature's longer timescale – whose deceptive immutability makes sudden changes impossible for the human eye to see – does not fit well with anthropic time and its political and economic constraints. This mismatch between timescales is reinforced by the assumption that stability, or at least the anthropocentric illusion of stability, corresponds to an ideal state of nature and the famous climax community from which biologists can never actually escape despite their efforts at conceptual rewording (Drouin 1991). In dreaming of a state of natural balance, scientists, immersed in the illusion of naturalist atheism, have never totally escaped from Judeo-Christian guilt, the avatar of original sin and of paradise lost. That said, the 'distinction between harming a natural system and changing it' (Throop 2000: 181; translated here) is not always clearly established. It is, at times, the same naturalists who condemn invasive alien species and who support the creation of ecological corridors, thereby confining some species while prompting others to spread.

Mistaking nature for the Garden of Eden and then letting oneself get carried away in a spiral of interventionism is testimony to a fear of social and environmental change and hence of losing control. Separating the desire for control from the quest for power (and vice versa) is only one step. As Brazilian researcher Lorelai Brilhante Kury stated in an interesting examination of the history of the Paris Museum: 'From the end of the *Ancien Régime* to the first decades of the 19th century, mastering nature through science was fundamental to the exercise of power' (Kury 2001: 5; translated here). Could humanity's irrepressible desire for control over nature, in its biocentric and postmodern version, be an even more advanced form of what Michel Foucault (2004) has called biopower: control over living species and their biological mechanisms as the ultimate means of exercising political power? Anthropologist Françoise Dubost (1997: 23; translated here) asks whether:

> Protecting what is wild, cultivating what is wild [should be understood as] nature's return or [the] subterfuge [of nature] at its epitome? The last step in the domestication of nature or the rules of a new game to protect the freedom of living species?

Resistance to biological change is nourished by, among other things, the idealization of a certain past. This imagined past is that of a pre-modern era in which humanity knew how to live in harmony with nature (Larrère 1994a). Modernity is thence cast as the final stage in a long process which distanced humanity from the Garden of Eden and, in turn, ecosystems from the state of climax community. However, as social and environmental history has shown, the supposed harmony between our ancestors and nature has largely been over-interpreted (Peretti 1998).

Fear of change and the idealization of a partially imagined past, combined with a semantic field filled with warlike connotations unique to the management of biological invasions, has led some social scientists to question the possible consequences of what they have termed 'biological nativism', which 'raises troubling scientific, political and moral issues' (Peretti 1998: 183). Mark Sagoff (1999) and Banu Subramaniam (2001) note disturbing parallels between the warlike vocabulary of invasions and the highly gendered character of discourses opposing human migration. To both the biological and human 'invader' are ascribed a range of:

> undesirable characteristics includ[ing] sexual robustness, uncontrolled fecundity, low parental involvement with young, tolerance for 'degraded' or squalid conditions, aggressiveness, predatory behavior.
>
> *(Sagoff 1999: 18)*

This xenophobic temptation (Claeys-Mekdade 2003) is not an offshoot problem. It is inherent to the very principle of protecting nature. As Jonah Peretti (1998) notes, the words 'native' and 'natural' are closely linked. The Latin *nascor* is the original root for several English words including native, natural, nation, and natality. When, amongst other things, taxonomies explicitly or implicitly rely on the dualism of good versus bad species, morality is involved and a new shift occurs, this time from Rousseau-ist naturalism towards an unconfessed, perhaps unmentionable, anthropomorphism. In this respect, Gröning and Woschke-Bulmahn (1992, 2010) have eloquently described the Third Reich's attachment to eradicating exotic plants from German gardens.

Those opposed to invasive alien species have denied such xenophobic intention. Others have counter-argued with cultural relativist discourse. Along the lines of the 1992 Rio Declaration, a connection has been established between natural diversity and cultural diversity that advocates recovering vernacular knowledge as a means of protecting the environment (and vice versa). Ethnic communities and traditional tribal groups are upheld as examples and opposed to modern Western society's loss of sense of place and sense of origin (McNeely 2001). Yet this attractive argument, full of humanism, nonetheless raises new questions. In seeking to emphasize the positive connections between nature and some cultures, the defenders of socio-natural communitarianism naturalize such ethnic/tribal groups. As such, the idealization of an imagined past is projected on to the contemporary descendants of these tribal cultures. Furthermore, the naturalization of these tribal cultures is similar, albeit in a biocentric and humanistic form, to the ethnic museum approach taken to extremes in the exhibition of 'specimen' in traditional dress in Western zoos from the sixteenth to the nineteenth century and at the Universal Expositions until the late nineteenth century (Mullan and Marvin 1999). Finally, in breaking away from its colonial past, contemporary ethnology has largely exposed the subjective and socially constructed nature of concepts such as ethnic groups and tribes (Barth 1969; Douglas 1992), just as genetics has recently shown the scientifically unfounded basis of the concept of race within the human species (Templeto 2003).

Conclusion

Both in the past and at present, when humans – regardless of whether they are scientists or lay people – identify a species as native or foreign, they engage in a taxonomic exercise which is essential to their understanding of the world. It is imprecise, since the subjects at hand are as recalcitrant as frontiers are permeable. The spatial-temporal arrangements observed here are based on an incessant shifting from biological to social and from social to biological. Whether they combat or defend invasive species using biocentric or anthropocentric arguments, the different players involved appear to be actors in a strange melodrama in which they are all fighting to play the same role: nature's gardener.

Some naturalists give in to panopticism,[4] arguing that overt biocentrism has, to date, managed only to establish another form of anthropocentrism. And yet could it be otherwise? The most radical fringe place human beings at the top of the invasive species list and, in doing so, revive old eugenicist theories. Others, however, have declared a truce on accusing and criticising humans. The boldest, like Michael L. Rosenzweig (2003), who advocates 'Win–Win Ecology', have chosen to take nature in the city seriously and call for the reconciliation of nature and culture. Finally, do social actors not share the same fear facing what Steven Yearley (2005) calls 'the end or the humanization of nature', which in both cases, leads humankind to take 'control of a system that we thought was external to us', raising our 'difficulties of living after nature' (Yearley 2005)? Would this be the true central challenge of environmental and social change leading to the so-called postmodern society?

Notes

1 Translated from French to English by Jocelyne Serveau.
2 National Research Program INVA BIO supported by the French Ministry of Environment (2004–2006), European Programs Life 99 env/F/000489 (1998-1999) and Life 08/env/F/000488 (2010–2013).
3 See notably the French Ministry of Ecology's *Recréer la nature* [*Recreating nature*] research programme (2007): www.ecologie.gouv.fr.
4 This panopticism, in the Foucauldian sense of the term (Foucault 2004), which is naturalist here, would be to the ecosystem what medicine is to the body – a means for humankind to control living matter as a specific way of exercising power, called biopolitics.

References

Barth, F. (1969) *Ethnic Groups and Boundaries: The Social Organization of Culture Difference*. Oslo: Universitetsforlaget.
Barthélémy, C. (2006). 'Du Mangeur d'aloses au carpiste sportif: esquisse d'une histoire de la pêche amateur en "rance"', *Courrier de l'Environnement de l'INRA*, 53, pp. 121–128.
Beck, U. (1992) *Risk Society: Towards a New Modernity*. London: Sage.
Binggeli, P. (2001) 'The Human Dimensions of Invasive Woody Plants', pp. 145–160 in J.A. McNeely (ed.), *The Great Reshuffling: Human Dimensions of Invasive Alien Species*. Gland, Switzerland/Cambridge, UK: IUCN.
Bobbé, S. (2002) *L'Ours et le loup: essai d'anthropologie symbolique*. Paris: Fondation de la Maison des Sciences de l'Homme, Inra.
Boltanski, L. (1979) 'Taxinomies sociales et luttes de classes: la mobilisation de la "classe moyenne", l'invention des "cadres"', *Actes de le Recherche en Sciences Sociales*, 29, pp. 75–105.

Burgiel, S.W. and Muir, A.A. (2010) *Invasive Species, Climate Change and Ecosystem Based Adaptation: Addressing Multiple Drivers of Global Change*. Washington, DC/Nairobi, Kenya: Global Invasive Species Programme.

Carlton, J.T. (1999) 'Editorial: A Journal of Biological Invasions', *Biological Invasions*, 1, 1, pp. 1–1.

Claeys, C. (2010) 'Les "bonnes" et les "mauvaises" proliférantes: controverses camarguaises', *Études rurales*, 185, pp. 101–117.

Claeys, C. and Sirost, O. (2010) 'Proliférantes natures: introduction', *Études rurales*, 185, pp. 9–21.

Claeys-Mekdade, C. (2003) *Le Lien politique à l'épreuve de l'environnement. Expériences camarguaises*. Bruxelles: Peter Lang.

Claeys-Mekdade, C. (2005) 'A Sociological Analysis of Biological Invasions in Mediterranean France', pp. 209–220 in *Proceedings of the Workshop on 'Invasive Plants in Mediterranean Type Regions of the World*, Mèze, France, 25–27 May, Council of Europe, Environmental Encounters Series 59.

Claeys-Mekdade, C. and Nicolas, L. (2009) 'Le Moustique fauteur de troubles', *Ethnologie Française*, 39, 1, pp. 109–116.

Claeys-Mekdade, C. and Sérandour, J. (2009) 'Ce que le moustique nous apprend sur le dualisme anthropocentrisme/biocentrisme: Perspective interdisciplinaire sociologie/biologie', *Natures, Sciences, Sociétés*, 17, pp. 136–144.

Clément, G. (1999) *Le Jardin planétaire: Reconcilier l'homme et la nature*. Paris: Albin Michel.

Clément, G. (2002) *Éloge des vagabondes: herbes, arbres et fleurs à la conquête du monde*. Paris: NIL Éditions.

Clergeau, P. (ed.) (1997) *Oiseaux à risques en ville et en campagne. Vers une gestion intégrée*. Paris: INRA.

Costa, C. (2005) *Atlas des espèces invasives présentes sur le périmètre du Parc Régional de Camargue*. Camargue: Parc Naturel Régional de Camargue. Available at www.parc-camargue.fr. Accessed 25 January 2011.

Deschamps, L. and Marroussy, A. (2004) *Mauvaises Herbes: utilitaires du jardin*. Rennes: Éditions Ouest-France.

Douglas, M. (1992) *Risk and Blame: Essays in Cultural Theory*. London: Routledge.

Drouin, J.-M. (1991) *L'Écologie et son histoire*. Paris: Flammarion.

Dubost, F. (1997) *Les Jardins ordinaires*. Paris: L'Harmattan.

Fabiani, J.L. (1984) 'La crise de légitimité de la chasse et l'affrontement des représentations de la nature', *Actes de la Recherche en Sciences Socials*, 54, pp. 81–84.

Foucault, M. (2004) *Naissance de la biopolitique: cours au Collège de France (1978–1979)*. Paris: Le Seuil.

Giddens, A. (1991) *The Consequences of Modernity*. Cambridge, MA: Stanford University Press.

Gramaglia, C. (2010) 'Les goélands leucophée sont-ils trop nombreux ? L'émergence d'un problème public', *Études Rurales*, 185, pp. 133–148.

Gröning, G. and Wolschke-Bulmahn J. (1992) 'The Ideology of the Nature Garden: Nationalistic Trends in Garden Design in Germany During the Early Twentieth Century', *Journal of Garden History*, 12, 1, pp. 73–80.

Gröning, G. and Wolschke-Bulmahn J. (2010) 'The Myth of Plant-invaded Gardens and Landscapes', *Études Rurales*, 185, pp. 197–218.

Hettinger, N. (2001) 'Exotic Species, Naturalisation and Biological Nativism', *Environmental Values*, 10, 2, pp. 193–224.

Huneau, V. (2008) 'Étude socio-environnementale de la présence des moustiques dans l'est du Golfe du Morbihan (56, France)', *Bulletin de la Société des Sciences Naturelles de l'Ouest de la France*, 30, 4, pp. 201–215.

Kury, L. (2001) *Histoire naturelle et voyages scientifiques (1780–1830)*. Paris: L'Harmattan.

Laissus, Y. (1995) *Le Muséum National d'Histoire Naturelle*. Paris: Édition Gallimard.

Larrère, R. (1994a) 'L'Art de produire la nature une leçon de Rousseau', *Le Courrier de l'environnement de L'INRA*, 22, pp. 5–13.

Larrère, R. (1994b) 'Sauvagement artificiel', *Le Courrier de l'Environnement de l'Inra*, 21, pp. 35–37.

Larson, B.M.H. (2005) 'The War of the Roses: Demilitarizing Invasion Biology Frontiers', *Ecology and the Environment*, 3, 9, pp. 495–500.

Lévêque, C. (2001) *Écologie. De l'écosystème à la biosphère*. Paris: Dunod.

Lévêque, C., Mounolou, J.C., Pavé, A. and Schmidt-Lainé, C. (2010) 'À Propos des introductions d'espèces: écologie et idéologies', *Études Rurales*, 185, pp. 219–234.

Lidskog, R. (2010) 'Governing Moth and Man: Political Strategies to Manage Demands for Spraying', *Études Rurales*, 185, pp. 149–162.

Mauz, I. (2006) 'Introductions, réintroductions: des convergences, par-delà les différences', *Natures Sciences Sociétés*, 14, pp. 3–10.

McNeely, J.A. (ed.) (2001) *The Great Reshuffling: Human Dimensions of Invasive Alien Species*. Gland, Switzerland/Cambridge, UK: IUCN

Micoud, A. (1993) 'Vers un nouvel animal sauvage. Le Sauvage 'naturalisé vivant'', *Natures Sciences Sociétés*, 1, 3, pp. 202–210.

Mol, A. (2002) 'Ecological Modernization and the Global Economy', *Global Environnemental Politics*, 2, 2, pp. 92–115.

Mougenot, C. and Roussel, L. (2006) 'Peut-on Vivre Avec le Ragondin? Les Représentations Sociales Reliées à un Animal Envahissant', *Natures, Sciences, Sociétés*, 14, pp. 22–31.

Mullan, B. and Marvin, G. (1999) *Zoo Culture: The Book About Watching People Watch Animals*. Champaign, IL: University of Illinois Press.

Pearman, D. (2007) 'Don't Blame the Alien', *Plantsman*, 60, pp. 60–61.

Peretti, J.H. (1998) 'Nativism and Nature: Rethinking Biological Invasion', *Environmental Values*, 7, 2, pp. 183–192.

Pigeat, J.P. and Paye-Moissinac, L. (2003) *Mauvaise Herbe!* Ivry-sur-Seine: Conservatoire International des Parcs et Jardins et du Paysage, Archipel studio.

Richardson, D.M., Pyšek, P., Rejmánek, M., Barbour, M.G., Panetta, F.D. and West, C.J. (2000) 'Naturalization and Invasion of Alien Plants: Concepts and Definitions', *Diversity and Distributions*, 6, 2, pp. 93–107.

Rosenzweig, M. (2003) *Win–Win Ecology: How the Earth's Species Can Survive in the Midst of Human Enterprise*. Oxford, UK: Oxford University Press.

Sagoff, M. (1999) 'What's Wrong with Exotic Species?' *Report from the Institute for Philosophy and Public Policy*, 19, 4, pp. 16–23.

Shine, C., Kettunen, M., Genovesi, P., Essl, F., Gollasch, S., Rabitsch, W., Scalera, R., Starfinger, U. and Brink P. (2010) *Assessment to Support Continued Development of the EU Strategy to Combat Invasive Alien Species*, Final Report for the European Commission. Brussels, Belgium: Institute for European Environmental Policy.

Skogen, K., Mauz, I. and Krange, O. (2008) 'Cry Wolf! Narratives of Wolf Recovery in France and Norway', *Rural Sociology*, 73, 1, pp. 105–133.

Subramaniam, B. (2001) 'The Aliens Have Landed! Reflections on the Rhetoric of Biological Invasions', *Meridians: Feminism, Race, Transnationalism*, 2, 1, pp. 26–40.

Templeto, A.R. (2003) 'Human Races in the Context of Recent Human Evolution: A Molecular Genetic Perspective', pp. 234–357 in A. Goodman, D. Heath and S. Lindee (eds), *Genetic Nature/Culture: Anthropology and Science Beyond the Two Culture Divide*. Berkeley/Los Angeles, CA: University of California Press..

Throop, W. (2000) 'Eradicating the Aliens', pp. 179–191 in W. Throop (ed.), *Environmental Restoration: Ethics, Theory, and Practice*. Amherst, NY: Humanity Books.

Touraine, A. (1971) *The Post-Industrial Society. Tomorrow's Social History: Classes, Conflicts and Culture in the Programmed Society*. New York, NY: Random House.

Vignes, P. (2004) *Splendeur et harmonie des plantes libres*. Aix-en-Provence: Edisud.

Woods, M. and Veatch Moriarty, P. (2001) 'Strangers in a Strange Land: The Problem of Exotic Species', *Environmental Values*, 10, 2, pp. 163–191.

Yearley S. (2005), 'The "End" or the "Humanization" of Nature?' *Organization and Environnement*, 18, 2, pp. 198–201.

24
Biological resources, knowledge and property[1]

Luigi Pellizzoni

The United Nations declared 2010 as the International Year of Biodiversity, and 2011–2020 as the Decade on Biodiversity. The solemnity of such declarations is hardly surprising. Biological resources are used more and more extensively and intensively. Their protection has become a planetary question, yet the very meaning of protection becomes increasingly problematic, at the crossroads of preservation of life forms, replacement with human artefacts, and 'remoulding' according to human needs and preferences. In the meantime, policy approaches have significantly evolved from state- to market-centred, with property rights playing an increasingly salient – and contested – role.

In short, life in its richness and variety raises today novel questions of government. To make sense of these questions, the 'governmentality' approach seems more suitable than the more popular 'governance' one. The latter adopts a traditional concept of power, focusing on its transfer from state to non-state actors, from single actors to networks, and from law to market mechanisms, as triggered by economic globalization and technological innovation (Rhodes 1997). In this framework, rights in biological property expand as a pragmatic reply to issues of social complexity and technical efficiency. The governmentality perspective (Dean 1999; Foucault 2008) allows us to look more deeply into this process. Five tenets are especially relevant. First, power is not a force exerted in opposition to people's will, but a way to shape and orient their free action – to conduct their conduct. Second, government is a set of activities that express specific mentalities or forms of reasoning around problems of rule. Third, the emergence of these mentalities or rationalities is less a matter of revolutionary changes than of the 'intensification' of existing patterns and ideas:

> the lightening, saturation, becoming-more-efficient and transversal linkage of existing practices … [up to] tipping points … where the object or subject mutates into another form.
>
> *(Nealon 2008: 38–39)*

Fourth, contemporary forms of government exert power in increasingly indirect and expert-mediated ways, and are primarily focused on *bio*-power; that is, on regulating and optimizing the biological life of people. Fifth, present forms of government are dominated by neoliberal rationality.

These five tenets frame our journey into the current relationship (or reciprocal constitution) of biological resources, knowledge and property rights. The next section discusses the rationale of property-based regulation, connecting it with a prevailing economic understanding of biodiversity. The following section locates the expansion of property rights over biological nature within the neoliberal rationality of government. The fourth section of this chapter focuses on gene technologies regulation as the main locus where the reciprocal constitution of biology, technology and property unfolds. The penultimate section addresses the issue of resistance to the 'neoliberalization of nature'. The conclusion suggests some lines for further reflection.

Regulating biological resources

The 1992 UN Convention on Biological Diversity and a number of declarations, reports and studies connect biodiversity with sustainability. This means that protection and production, preservation and use of biological resources generally are understood as two sides of a same coin. Both ecology and economics refer to the *oikos* ('household'); the whole policy discourse of biodiversity implies an evaluative framework: biodiversity has to be measured in order to find how to preserve it most efficiently (Bowker 2006). In short, biological nature is increasingly addressed according to an economic rationale.

This 'intensified' economic understanding is closely connected with scientific and technological innovation. Biology involves today bigger budgets, a greater workforce, broader economic consequences, more ethical implications and greater effects on human welfare than physics (Dyson 2007). The precise role of technoscience, however, is problematic. If biodiversity is generally recognized as crucial to sustainability, there is no agreement on whether and the extent to which biological resources and ecological functions are to be regarded as irreplaceable stocks. In a 'strong' sense, sustainability means that biological resources have to be used within their threshold of reproducibility and, in the case of irreproducible resources like oil, as parsimoniously as possible. In a 'weak' sense, it is not deemed unreasonable to assume that technological artefacts can replace virtually all types of non-human-made resources, at least in the long run. However, this classic contrast, which lies behind the enduring controversy between 'deep' and 'shallow' ecology (Devall and Sessions 1985) and between 'ecological' and 'environmental' economics (Martinez-Alier 1990), does not fully capture an emergent trait in humans' relationship with the biological sphere. The interpenetration of life and artefact, nature and culture, has intensified to the point of engendering a qualitative change. Agricultural biotechnologies aimed at working out more productive varieties of plants are certainly not new. Yet, intuitively, the FlavrSavr™ tomato (the first commercialized transgenic plant, in 1994) intermingles culture and nature, technology and biology, at a different level compared with any other variety of tomatoes previously existing.

From a regulative viewpoint, biological resources raise the problem of difficult excludability and use rivalry that Garrett Hardin (1968) labelled the 'tragedy of the commons'. Tragedy because, if users have equal access and competing interests in a resource, without being compelled to take care of its maintenance, the likely result is its overexploitation and ultimate exhaustion. Hardin's argument about state sovereignty or private property as the only answers to this problem has been challenged by theoretical, experimental and experiential research on non-state and non-market institutions for the management of common-pool resources (Poteete et al. 2010). Such institutions work effectively when users belong to a network of established relationships and use the resource in similar, well-known ways, and when legal rights, traditions or physical boundaries prevent a totally open access (Baland et al. 2007; Ostrom et al. 1994). Yet these conditions are difficult to meet. Moreover, technoscientific and organizational changes entail an increase in the number of users and/or in their capacity of exploitation, which means that a growing variety of resources face problems of rivalry and prospective extinction. At the same time, the ability of public powers to regulate effectively the use of biological resources has been increasingly questioned. Growing complexity of human relations and intensified use of nature highlight the state's limits of knowledge, steering and monitoring capacity.

One is, therefore, seemingly compelled to turn to private property. Ownership and market exchange arguably represent the best incentive for people to use a resource in the most efficient – most durably productive, or productively durable – way. The notion of 'Knowledge-Based Bio-Economy' (KBBE), supported by think tanks and governmental institutions (e.g. CEC-DG Research 2004), offers a recent, influential version of this idea: economic growth and environmental sustainability, understood as increasingly eco-efficient industrial productivity, can be ensured by tightly coupling technoscientific knowledge and biological resources within a market framework. The latter, in turn, requires commodifying and privatizing both technoscientific knowledge and biological resources (e.g. in the form of patent, copyright and licences) (Birch et al. 2010). In this view, sustainability, technoscience and the market reinforce each other. Market competition stimulates innovation, which enables eco-efficiency, triggering further market competition.

Limits to this solution may arguably come from either practical or principled problems. Some resources seem difficult or even impossible to parcel (think of fisheries or the atmosphere). Others can raise ethical issues in their commodification (think of the human body) and appropriation (think of biopiracy, on which see below). Yet, are these drawbacks really insurmountable? In the neoliberal framework, the answer is 'definitely not'.

Nature, indeterminacy and neoliberal rationality

Neoliberalism is often portrayed as a project of social change, for which:

> human well-being can best be advanced by liberating individual entrepreneurial freedoms and skills within an institutional framework characterized by strong private property rights, free markets, and free trade.
>
> *(Harvey 2005: 2)*

At its most basic, then, the 'neoliberalization of nature' (Castree 2008) means the growing dominance of a specific 'green governmentality' (Rutherford 2007) – a particular way to define goals, problems, solutions, relevant knowledge, key agents and addressees of action vis-à-vis natural resources. The latter are increasingly managed through market-oriented arrangements, by offloading rights and responsibilities to private firms, civil society groups and individual citizens – with state power, in its national and transnational incarnations, providing the general rules under which markets operate (Bumpus and Liverman 2008).

Biological resources are undoubtedly central to nature's neoliberalization. The expansive phase of capitalism begun in the 1970s, under the aegis of neoliberal ideas, programmes and policies (Harvey 2005), overlaps with the growing centrality of biology as a core technoscientific engine of economic growth.

> Genes, genetically modified organisms and other products of the new biotechnology have been explicitly targeted ... as important arenas for the expansion of capital accumulation in agriculture and health sciences, not to mention other spheres.
>
> *(Heynen et al. 2007: 10)*

Bioeconomy, the technology-enabled capture of the latent value in biological processes[2] is, therefore, increasingly central to capitalist economy in both a substantive and a symbolic sense. Production, consumption and capital accumulation draw to an increasing extent on biological processes,[3] and the whole (understanding of) economy focuses on vital traits of dynamism, growing capacity of change, evolution, differentiation and enhancement. In different ways, thus, 'life sciences represent a new face and a new phase of capitalism' (Rajan 2006: 3).

Yet to what extent does this represent something novel? The privatization and commodification of land, forests and many other resources (what Marx called the 'primitive accumulation' of capital, begun with the seventeenth-century British 'enclosures') has long been advocated by liberal thinkers (John Locke stands as a central figure in this respect, providing 'crucial ideological and discursive foundations for liberal and modernist dispositions toward non-human nature'; McCarthy and Prudham 2004: 277) and promoted by liberal governments on the grounds that, since nature gains value through the application of human labour, conferring exclusive control over natural resources to those individuals who work them is both morally right and collectively beneficial. From this viewpoint, we can talk of the present as a new phase of 'accumulation by dispossession' (Harvey 2003), conceptually analogous to previous ones. Novel spheres of accumulation are being opened up in the terrain of the biological commons and recaptured through the production of scarcity under the property form. Yet the contribution of technoscience to this endeavour is increasingly crucial. As a consequence, new rights in property over biological matter increasingly build on, or intertwine with, intellectual property rights (see below). The enclosure of nature, in other words, is ever more inseparable from the enclosure of knowledge.

Moreover, while in classical liberalism, as well as in early accounts of the ecological crisis (Meadows et al. 1972), one can find widespread concerns for the material limits to economic growth, neoliberal discourse is dominated by Promethean accounts of technological and economic expansion. The case for limits to growth is reversed into a case for the growth of limits (Lemke 2003; McCarthy and Prudham 2004). This is not entirely new. Both cases have been used to simultaneously criticize and support capitalist economy. Many ecologist discourses echo Malthusian ones, despite opposite aims (questioning vs defending the existing order), while opposing the idea of objective barriers to growth is at the basis of both Marx's critique of Malthus's reification to natural givens of contingent social, political and economic relations, and of the legitimation of expansive economic programmes, as in the aftermath of the Second World War and with the neoliberal agenda. Furthermore, nature – and more specifically biological nature, since only life is able to continuously recreate itself and overcome its own limits (Prigogine and Stengers 1984) – is traditionally regarded by science and capital as a manipulable entity. Yet there *is* something novel and peculiar to the neoliberalization of nature. This lies in the way manipulation is conceived; that is, in the specific performance assigned to human action, or the specific understanding of nature's plasticity. To explore this process and its implications in more detail it is useful to start with the neoliberal understanding of the market and human agency.

Contrary to the liberal view, the neoliberal framework does not regard markets as self-regulating institutions building on humans' natural tendency to exchange. Rather, markets have to be purposefully constructed, steered and policed in order to promote humans' natural tendency to compete (Burchell 1996; Foucault 2008; Tickell and Peck 2003). Everyone is (to become) an entrepreneur of his or her own self, responsible for their own choices, committed to valorizing their own 'human capital', governing their own conduct towards the maximization of some definition of happiness and fulfilment (Feher 2009; Rose 1996) – a personal view, yet not immune from cultural pressures and indirect, discreet and penetrating advice from a variety of expertises. The distinctions, central to liberalism, between production and reproduction, public and private, professional and domestic spheres, are eroded, and the values and attachments which market calculations and labour traditionally were embedded in are increasingly subsumed by an entrepreneurial logic, a logic provided with an intrinsic proprietary character. 'The expansion of property monopolies into all human activity is a central aim of the neoliberal programme' (Zeller 2008: 88). Yet it is not that every social and existential domain is to be privatized and marketized, but that everything is to be aligned with such rationality, in this way smoothing and increasing the efficiency of individual and social dynamics of growth. Nowhere does this possibly become more evident than in the health sector. With gene technologies, 'our somatic, corporeal, neurochemical individuality becomes a field of choice, prudence and responsibility' (Rose 2007: 39–40) vis-à-vis ourselves and those who we care about, while 'ignorance, resignation and hopelessness in the face of the future is deprecated' (Rose and Novas 2006: 442).

This corresponds to a major shift in the understanding of agency. The liberal view of freedom, rationality and responsibility entails a future neither totally fixed nor totally random. Risk means a future event related to behavioural choices, the probability of which is amenable to calculation. At the same time, non-calculable uncertainty prevents humans from being prisoners of an inevitable path. Real profits, suggest economists like Keynes and Knight, stem from 'unpredictable risks', as related for example to innovation, which are the object of a few strategic decisions. From this viewpoint, the problem of biodiversity – as the value of biological resources for present or future uses (so-called option value), or even beyond any use (so-called existence value) – looks intractable. Biological life on the planet is a complex system and human intervention makes things even more complex. We know a lot, but still very little, about life (its processes, variety, interactions, relevance for our own life), which leads to irresolvable questions of how biodiversity can really be preserved. Should we privilege, for example, the most genetically different species? Should we consider whole ecosystems, rather than species? And are we sure we have a sound understanding of ecosystems?

For the neoliberal *homo economicus*, this is far less problematic. Uncertainty 'makes free' in a different, more 'intense' way. Decision making under uncertainty becomes an empowering, everyday situation. Uncertainty is premised on entrepreneurial creativity, which requires intuition, foresight, flexibility, experiential judgement, rules of thumb and so on (O'Malley 2004, 2008). In this picture, indeterminacy does not mean constraining non-determinability, but enabling non-determination. Turbulence and contingency, as produced by global trade, innovation-based competition and floating exchange rates, do not mean paralysing uncontrollability but, rather, lack of limits, room for manoeuvre, opening up of opportunities. The more unstable the world, the more manageable (Pellizzoni 2011a). In short, the basic orientation of neoliberalism is speculative: proper calculations of risk are seen as the exception, while reasoned bets over unpredictable futures are regarded as the rule.

The emergence of a promissory financial market has been crucial to the development of biotechnology research and industry (this is discussed further below). More generally, one can notice a gradual shift away from traditional terms of controversy over technoscientific uncertainty and toward new ways to articulate indeterminacy and social choice. Many conflicts can be drawn to a basic divide with regard to innovation (Pellizzoni 2010). On one side, there are those who have more to lose from false positives. Detecting problems that do not actually exist affects research, profits, national competitiveness and the health and environmental benefits of technology advancement. Hence scientists, corporations and policymakers usually support research and development (R&D) designs aimed at reducing false positives. On the other side, there are those who have more to lose from false negatives. Failing to detect problems that actually do exist affect end-users, local communities and the environment. Social groups and ecologists, therefore, typically are critical, with experimental studies focusing on exposure to single agents or conditions rather than mixtures, which are suitable to reducing false positives but may fail to grasp the multi-causal character of many biological effects.

Approaches to calculable uncertainty thus traditionally have their own political constituencies (Hammond 1996). Yet if one thinks, for example, of the controversies over the use of genetically modified organisms (GMOs) in agriculture (Levidow and Carr 2009), or over genetic testing of disease predispositions (Rose 2007), it is easy to see that what is really at stake is the indeterminacy built into innovation. What is confronted is not so much different ways to get to reliable predictions, but different visions of the future – utopian and dystopian, if one wishes. On one side, we have those who, applying what Hans Jonas (1984) has called the heuristics of fear, plead for restraint and precaution. On the other, we have those who, building on promissory anticipations, make a case for a brave new world. The appeal to 'sound science' in this framework is not so much about designing research in a way appropriate to given goals as it is about moralized orientations towards a technological future (Berkhout 2006). The ecological or moral discourse is captured in this way by the same speculative logic of its scientific, corporate or political targets (Pellizzoni 2011a). The performative role of hype, expectations, anticipations and imaginaries (Borup et al. 2006; Felt and Wynne 2007; Pollock and Williams 2010) can be observed both at the highest policy level, as with the narrative of 'converging technologies',[4] and at the level of what Rose and Novas (2006: 442) call the 'political economy of hope': the mobilization of individual citizens, patients' groups and corporate actors around the 'knowable, mutable, improvable, eminently manipulable' character of human biology, as a way to either escape from a blind destiny or a foreseen but implacable fate, or extract 'biovalue' from the appropriation, commodification and marketization of life.

Gene technologies, biological entities and private property

Biotechnology, it has been remarked, 'is a form of enterprise inextricable from contemporary capitalism' (Rajan 2006: 3). The genetic revolution began in 1973 with the invention of recombinant DNA and related technologies for cutting up and splicing DNA molecules. These technologies allow different genes and DNA sequences to be expressed in vector organisms (usually bacteria or viruses). In this way, sequences of genetic information can be transferred across barriers of species and genus – that is, horizontally instead of vertically – generating new forms of life. Compared with previous biotechnologies, concerned with the industrial-scale reproduction of standardized lifeforms, gene technologies allow the extension of a logic of flexibility and destandardization, as found in post-Fordist economy, to the sphere of life (Cooper 2008).

For the development of an entire industry, a new technology must be matched to a 'suitable' economic environment. Three institutional elements, originating in the United States, stand as crucial to the building of such environment:

- The availability of funds for investment in high risk ventures, allowed by new legislation on financial capitals. As the main market for venture securities, NASDAQ (National Association of Securities Dealers Automated Quotation, created in New York in 1971) has been pivotal to the institutionalization of a largely promissory market in life science innovation.
- The enforcement, in 1980, of the Patent and Trademark Amendment (or Bayh–Dole Act), the aim of which was to promote the patenting of publicly funded research and the private exploitation of patents by their holders (either by issuing licences to private companies, or by entering into joint ventures, or else by creating their own start-up companies). The novel and by now famed figure of the scientist-entrepreneur stems from this legislative innovation.
- The extension of patents to the biotech field. In its 1980 ruling, *Diamond* v. *Chakrabarty*, the US Supreme Court stated that a genetically modified bacterium is human-made (that is, it is a new composition of matter) and that whether an invention is alive or not is not a legitimate legal question. In short, 'anything under the sun made by man', including living matter, can be patented.

The worldwide spread of this regulatory approach was ensured by the 1994 TRIPs (Trade Related Aspects of Intellectual Property Rights) Agreement, strongly advocated by North

American pharmaceutical, software and entertainment industries. This agreement extends the scope of patentability to all commercially exploitable products and processes, including genetically modified plants and animals. Ratifying TRIPs is mandatory for joining the World Trade Organization and every signatory state is compelled to conform its legislation accordingly.

The last decades have witnessed an explosion of intellectual property monopolies in the biotechnology field. From 1990 to 2000, for example, the annual rate of increase of biotechnology patents was 15 per cent in the US and 10.5 per cent in Europe, compared to a 5 per cent overall increase of all types of patents during the same period (OECD 2002). Besides numbers, a look at the logic of biotech patenting allows a better understanding of the neoliberal rationality of government. A product patent for a genetic sequence, for example, entails regarding it as a 'composition of matter', novel in that in its isolated and purified form it is not available in nature, and the 'utility' or 'industrial applicability' of which lies in the disclosure of its function. Such disclosure basically corresponds to understanding the biochemistry of the protein a gene produces and how this leads to a specific trait of the organism. Therefore, genes are regarded as carriers of information, suitable for translation into different media. Though information, such as ideas or scientific theories, is excluded from patenting, the demonstration of some technical effect or functionality allows property rights claims. This entails disregarding the often complex connection between genes and traits – one gene may be involved in the production of many proteins and there are typically several molecular interactions, cascades and feedback loops responsible for the final phenotype (Calvert 2007).

In short, patents carve out task environments of their own, transforming indeterminacy from a drawback into a valuable resource. On one hand, any difference between living and non-living entities is erased. On the other, a living entity is considered an artefact if basic functional parameters can be controlled, thus reproduced, and a correspondence is implicitly established between matter and information, so that rights in property over information can be subsumed into rights in property over the organisms incorporating such information, and vice versa. The ambiguous status of the gene, its ontological oscillation between material and information, can be found in actual court rulings, where patents are recognized to cover either genes or whole organisms, or both.[5] This 'ontological fluidity' (Carolan 2010) emerges also from the 'substantial equivalence' argument central to commercial applications, by which, for any practical purpose, patented artefacts are indistinguishable from nature and thus do not require any specific regulation. Artefacts are thus simultaneously identical to and different from (more usable, more valuable) than natural entities.

Biotechnology patents, therefore, are more than a mere extension of a well-proven property rights-based approach to regulating innovation. They produce things the ontology of which (their character or identity) is peculiarly indefinite. Moreover, this is by no means limited to the biotechnology field, being observable elsewhere. To what extent, for example, is a nanomaterial analogous to its bulk variant? Are we in front of the same thing at a smaller scale or a different thing, since its properties can be very different? And what is the global warming potential of a greenhouse gas? Is it a conventional 'CO_2 equivalent' as used in carbon emission trading calculations, or a real atmospheric phenomenon or process in its own right? For each of these examples, nature, knowledge and property rights can hardly be regarded as interacting yet separate (material, intellectual and regulatory) entities, but reciprocally constituting ones.

All this has inevitable effects on our conceptual equipment. Take Polanyi's notion of 'fictitious commodities'. Thanks to the medium of money, he says, capitalism treats elements of nature, such as water or trees, 'as if' they were marketizable resources disembedded from any socio-cultural meaning and biophysical function (Polanyi 1944). Yet biotech patenting is more than an abstracting procedure. There is more than an 'as if' at stake. There is the actual crafting of entities that previously did not exist and that are characterized by a structural ambivalence or oscillation: equivalence–difference, materiality–virtuality, substance–information. Or take Marx's concept of labour's 'formal' and 'real' subsumption to capital which, by extension, can be applied to nature. Formal subsumption of nature, then, means that capital exploits natural resources according to the biophysical features of those resources, as with mineral extraction and traditional fisheries. Real subsumption means that industries alter the properties of nature, increasing or intensifying its productivity and consequently enhancing capital accumulation (Boyd et al. 2001). Talking of alteration of nature may be suitable to non-genetic biotechnologies, yet seems inaccurate in the case of patented genes. Here we are confronted with the creation of resources that belong to nature but, at the same time, are totally internalized within the economic process of accumulation. We can apply the notion of real subsumption only by stretching its meaning to encompass not only an enhancement of the productivity of natural resources but also a profound redefinition of their ontology according to market logic. If quantification and calculation of the 'global-warming potential' of various gases makes it possible to parcelize and marketize the atmosphere, gene patents make it possible to parcelize and marketize any biological entity.

If, according to this view, there are no material limits to the appropriation and marketization of biodiversity, perhaps limits stem from our moral feelings towards the human body or our sense of justice. Take the notion of 'benefit-sharing'. The idea is that 'participants in research deserve some form of returns, precisely because their participation is leading to lucrative products' (Hayden 2007: 731). This perspective applies, for example, to bio-prospecting: the collection of forms of life and related 'traditional knowledge', usually from developing countries, for their technoscientific-commercial potentials. According to the UN Convention on Biological Diversity, such forms of life and culture cannot be considered part of the global commons; therefore, source communities and nations are entitled – in an ethical rather than legal sense – to share the benefits, including economic value, that those forms of life and traditional knowledge generate. The principle of benefit-sharing is aimed at limiting 'biopiracy' – the commercial exploitation of biodiversity and local knowledge without permission and compensation – even though practical problems stem from specifying the form (monetary or else) and entity of returns, and their beneficiaries (individual local residents, communities, ethnic groups, national researchers, etc.). In clinical research, benefit-sharing allegedly provides emerging collective research subjects and other expressions of 'biosociality' or 'biological citizenship' – patient and 'high risk' groups, parents of diseased persons, volunteers for pharmaceutical trials, ethnic groups, even entire countries in the case of national biobanks (Rabinow 1996; Rose and Novas 2006) – with a sound reply to growing evidence that biological samples and their derivatives are lucrative forms of property for companies. Clinical research, however, raises an additional ethical challenge; namely, the need to avoid undue inducement to participate. This problem is far from theoretical, as demonstrated by traffic in human organs (Scheper-Hughes 2006). The thin line between unethical inducement and ethical benefit-sharing increases the relevance of 'what' and 'to whom' is given back, and the perceived need to avoid any legal drift (attribution of rights in the properties of one's own body and thus to the results of research).

Does benefit-sharing represent a shelter from the neoliberalization of nature? Not so much. Benefit-sharing contradicts the traditional logic of gift-giving to strangers, especially relevant with regard to the human body (think of blood donation), but lying also at the basis of traditional curiosity-driven scientific access to biological and cognitive commons. We have here an example of the colonizing power of the neoliberal rationality of government. As noted, its goal is not to commodify and marketize everything (as would happen through attribution of rights in property to body parts and connected biovalues, or to local practices over specific ecosystems or species), but to create the conditions for market exchange, which requires non-market relationships to be aligned with a market logic. Then, however the recipients and character of the benefits are defined, redistribution is privatized: 'There are no strangers in this vision, only fellow (and competing) stakeholders' (Hayden 2007: 749). The ethical framing of the issue (equity vs property as a guiding rule) is, therefore, misleading. Apparently working as an antidote to the full commodification of bodies, knowledge and biodiversity, ethics effectively acts as a legitimating mediator between science and the market. Conspicuous by its absence in the benefit-sharing narrative is the political aspect of the issue: the power relations built into the 'deal', and their distributional effects. The identification of recipient communities is exemplary of the displacement of the political. These collectivities are treated as already-existing ethical subjects, yet they are typically constituted through the encounter of clinical or local human and non-human participants with their scientific-corporate partners, a constitution that represents a classic political gesture of inclusion and exclusion. Significantly, the ontology of these collectivities is hybrid, oscillating between nature and culture, genes and tissues or habits, biological condition (disease) and social relation (care), aggregates of individuals and risk categories, the sharing of a physical place and the sharing of a role in the productive chain. An indefiniteness that, once more, does not hamper but enables purposeful action.

Resisting the neoliberalization of life

If the neoliberal rationality of government is the driving force in the reciprocal constitution of biology, knowledge and property rights, one may ask what is wrong with it and what one can do about it.

There is actually no shortage of criticisms against neoliberal politics and policies. A typical allegation is that they have been unable to deliver their promised goods. Instead of more economic efficiency and stability, wealth and well-being, freedom and autonomy, instability has grown, the distribution of wealth and health is increasingly uneven, and dependence and deprivation have spread. The superior effectiveness of market-based policy approaches is at best unproven. As regards biotechnology patents, arguments about their depressive effects on research and innovation have been advanced and contested. Patents may cover basic biological matter and processes, even if their use is not demonstrated at the time of the patent application. This broadens the need for researchers to pay royalties or make deals with patent holders, with consequent restriction of knowledge circulation and innovation (Heller and Eisenberg 1998; Nelson 2004).[6] On the other hand, the non-use of patents means a loss of opportunity in front of the high financial risks of cutting-edge R&D. In addition, patents are wasting assets, given their time limits and the general speed of innovation (Epstein and Kuhlik 2004). At least 'for the vast majority of universities patenting has been a losing financial proposition' (Lave et al. 2010: 666). There is, however, evidence that patents can be used instrumentally; that is, with no intention to apply them but just for licensing and asserting patent rights (sometimes also for defensive reasons, that is to avoid litigation or allow the holder to counter-sue), or as a way to increase the financial value of companies by raising expectations regarding their future profitability (Zeller 2008).

Alongside these unresolved issues, unease with growing privatization and appropriation of biophysical and intellectual commons is widespread. Yet repeated crises and sustained criticisms have not substantively affected the dominant policy framework. This is undoubtedly due to the major interests lying behind neoliberal policies and their capacity to evolve and adapt to changing institutional and cultural contexts and conditions (Brenner et al. 2010; Ong 2006).[7] Of no lesser relevance, however, is the fact that neoliberalism is a political project that seeks to create a social reality it maintains already exists (Lemke 2003). It develops institutional practices and rewards for expanding competitive entrepreneurship, yet simultaneously claims to present 'not an ideal, but a reality; human nature' (Read 2009: 26). This blurring of description and prescription provides neoliberal ideology with a hegemonic status (Pellizzoni and Ylönen 2012), which does not help a considered assessment of different policy approaches on a case-by-case basis. Any failure of the market, any evidence opposed to the promised increase in efficiency, simply marks the distance between a transhistorical reality and contingent flaws, constraints, oppositions and irrationalities.

The most problematic aspect of the 'neoliberalization of nature', however, possibly lies in the immunitarian drift entailed by commodification and privatization processes. Being immune means being exonerated from communal duties, having nothing in common, no obligations to others, being owners of, identical to, oneself (Esposito 2008). Dealing with biological nature is increasingly drawn to explicit or implicit (as with benefit-sharing) contractual relationships. The latter have an intrinsically immunitarian structure (Pellizzoni 2008). Commitments and connections are established within the deal and according to the deal, and end with the conclusion of the deal. The 'solipsism' of contract is strengthened by the neoliberal view of a virtually unlimited agency over an indeterminate, pliant biophysical world. Growing immunization, in this context, leads to playing more and more with indeterminacy, seen as just pertaining to an inner task environment. Yet growing exposure to risk and uncertainty also leads to growing need of immunitary protection, as testified by the securitarian drift observable in virtually any country and any policy field (Mattelart 2010). Can this spiralling movement proceed indefinitely? Nobody knows, but an eventual crash, also in the form of a self-immunitary shock, is not unlikely. For example, the machinic commodification of the body, dominated by the idea of contract and choice over proprietary detachable parts, can put virtually anyone at the 'wrong' side of the deal – not only with organ transplants, where social inequalities play a major role, but also with the 'meliorative' surgeries and 'enhancing' medications available to the more affluent. The 'real' world can bite beyond any technological construction of contingency, as the Fukushima disaster has shown at a totally different scale. In short, growing allocation of nature to a distinction internal to human manufacturing processes can be confronted with nature's most terrible rejoinder: its total indifference to the survival and well-being of humans, as individuals and as a species.

In front of this extreme but by no means implausible scenario stand a number of countervailing trends. The latter are to be assessed in their inevitable ambivalences. Resistance, according to Foucault (2008), is an effect of, stems from, builds on, power. Resistance can overhang power but can never be external to, or independent from, power.

Take the increasing relevance of knowledge in our access to, and regulation of, biology. The technicalities of environment and technoscience-related policymaking have often hollowed out public debate and marginalized lay and local concerns. However, scientific methods and languages are increasingly applied in protests. Neglected issues are brought to the fore and counter-expertises are built up, opening up new spaces of contestation (Frickel et al. 2010; McCormick 2007; Pellizzoni 2011b). Such contestations, on the other hand, are easily drawn to focus on technical issues of safety, quality and efficacy of innovation, to the detriment of political–economic issues of need, desirability and distribution; that is, those already excluded or marginalized as dismissible in front of, or acting as a disincentive to, innovation.[8]

Or take the central role of ethics in the government of biology. If today, as we have seen, we are in front of what may be called a 'politics of ethics' (Felt and Wynne 2007: 47), then this also opens new spaces of debate and contestation in both their subjects and objects. Consumers, for example, increasingly make their choices according to considerations of justice, fairness, environmental protection, animal welfare and so on, engaging in new forms of individualized collective action (Lockie 2002; Micheletti 2003). They use the market for checking the market.[9] Yet consumer-citizenship corresponds also to a neoliberalization of activism (Roff 2007), where problems are framed as people's 'wrong choices' rather than state and corporate policies, where citizen power ultimately lies in the individual's own capacity of purchase, and where markets intercept a critical demand through further diversification (Fair Trade, organic food, etc.). For some, like the emerging de-growth movement (Latouche 2006), the only actual solution is an exit from, rather than a reinvention of, consumerism. Similarly, the 'making up' of biological citizens entails cutting across the body politic and the individual identity according to expert (scientific but also legal, insurance, administrative, etc.) concepts and classifications. This opens up new areas of struggle, demand for recognition, claims to expertise and, as already remarked, collective mobilization, where 'the co-production of health and wealth is [regarded as] a profoundly ethical endeavour' (Rose and Novas 2006: 457). Yet, we have seen that this may come at the cost of forgetting or marginalizing the profoundly political role such co-production exerts.

Finally, new spaces of contestation often take the shape of hybrid forums of discussion between scientists, technology developers, policymakers, local communities, end users, affected groups, etc. The extent to which these forums represent a means for reframing highly contentious issues remains, however, an open question. Despite being usually advocated as an antidote to the technicization, privatization and individualization of policies, they can be seen as well as aligned with the neoliberal rationality of government. With critical consumerism and biological citizenship, hybrid forums contribute to a 'politics of activation' by which '"publics" in various guises are constituted, mobilized and activated' (Hobson 2009: 182). At the very least, there is a serious risk of reproducing in subtler forms traditional divides between experts and non-experts or welcome and unwelcome concerns (Felt and Wynne 2007; Wynne 2001). In many cases, moreover, deliberative arenas hide the presence of profound political divides behind the search for a generalized consensus on 'fair' solutions to technical problems. Questions of 'whether or not' or 'to the benefit of whom' tend to disappear behind questions of 'how', addressed by a public of stakeholders rather than citizens in the fullest sense of the word (Goven 2006; Urbinati 2010). The conflict over agricultural gene technologies has prompted plenty of these forums at different scales, one conspicuous case of which was the 'GM Nation?' public dialogue organized in the UK in 2003. Some commentators are damning:

> In giving the appearance of democracy, such talk actually diverts from a more adequate onslaught on deeper institutional and epistemic commitments ... Little has changed: we are simply in the old nexus of technocratic aspirations with the public construed as an obstacle to progress.
> *(Irwin 2006: 316–317)*

And yet, for others, this very experience shows that deliberative arenas can also offer an opportunity for challenging boundaries, performing different models of the public and questioning dominant expert assumptions (Levidow 2007).

Conclusion

What I have termed the reciprocal constitution of biology, technology and property is more than a pragmatic reply to emerging problems of governance. It expresses a peculiar mentality of government – a particular answer to the question of who can govern, who and what is to be governed, and how. More specifically, it expresses a peculiar understanding of human agency vis- à-vis biological nature. The hegemonic status of the neoliberal framework makes difficult what is most needed: a dispassionate discussion of the virtues and drawbacks of property rights-based regulation, in different contexts and policy fields. Dispassionate, however, does not mean solely 'technical' in the sense of apolitical. Nowhere, as with biological nature, are 'the ways in which we know and represent the world (both nature and society) inseparable from the ways we choose to live in it' (Jasanoff 2004: 2). And the way we choose to be as reflexive human beings, one might add.

Thus, it is not property rights themselves that need to be urgently addressed[10] but, rather, the proprietary character that life in all its declensions increasingly is taking. Such character, synthesized in the idea of the 'neoliberalization of nature', is the latest, most accomplished result of the short circuit between life, economy and politics that represents the metaphysical foundation of modern biological science, economic enterprise and political action. Never as today do biology, economy and politics constitute themselves by pointing beyond themselves, each to the other two realms. Countervailing tendencies, then, should be assessed according to their contributions, not only to voicing unease and dissent, but also to carving a way out of the 'flexible paradise of neoliberalism' (Lemke 2003: 64), which finds in this nexus its premise.

Any real change, as Michael Walzer (1985) has argued, shares its basic features with the biblical exodus. For the moment, the contours of such exodus are obscure. If there is no other place to go, then what can be left is a game and its rules: starting a new game with new rules, or where old rules take new meaning. An exodus, in any case, is a process, not a single revolutionary event. It involves individuals, but cannot be enacted individually. The de-growth and other significant initiatives, such as the 'transition movement',[11] point in different ways towards a (re-)establishment or strengthening of common goods and communal ties among humans, and between humans and non-human nature. In one form or another, this is likely to entail a rejection of commodification, appropriation and unlimited enhancement as driving values; a rejection grounded on crucial distinctions – for example, between manufactured and proprietary or between makeability and acceptability – and on recognition of a fundamental element of alterity and the unattainableness of life to human agency.

Notes

1 I am grateful to the editors for their careful review of the chapter. David Sonnenfeld provided useful comments, prompting a clarification of some passages.
2 For the OECD, the bioeconomy emerging in primary production, health and industry involves three elements: 'the use of advanced knowledge of genes and complex cell processes to develop new processes and products, the use of renewable biomass and efficient bioprocesses to support sustainable production, and the integration of biotechnology knowledge and applications across sectors' (OECD 2009: 7).
3 The classic Marxian view is that capital-intensive technological innovation transforms living labour into dead labour, reducing the surplus value. As a consequence, capital has always sought geospatial expansion to appropriate new biological and human resources. Then, it is the intensification of this process and the increasing difficulty in distinguishing living and dead labour, biology and technology, that characterizes the present situation.
4 'Converging technologies' is shorthand for the alleged ongoing combination of nano-bio-info-cogno technoscience towards an unprecedented 'enhancement' of industrial productivity and human biological and mental capacities. This policy narrative has proven influential on both sides of the Atlantic (e.g. Nordmann 2004; Roco and Bainbridge 2002).
5 Cf., for example, the 2004 *Monsanto Canada Inc.* v. *Percy Schmeiser case* on the use of the Roundup Ready™ canola plants, discussed in Carolan (2010).
6 Beside patents, one should also consider the growing diffusion of 'material transfer agreements' (MTAs), by which tangible research materials, such as cell lines, plasmids or reagents, are transferred between two organizations, when the recipient intends to use it for its own research purposes. By means of 'reach-through clauses', MTAs allow to lay claims upon any intellectual properties arising from research using the related tools (Lave et al. 2010).
7 In this sense, the co-existence, even within a single country, of contradictory trends towards promoting, adjusting or resisting neoliberalization (Hess 2012; Jessop 2002), does not question but rather highlights the dominant position of neoliberal ideology.
8 Examples come from Kinchy et al. (2008), who analyse the regulation of recombinant bovine growth hormone (rBGH) and the Cartagena Biosafety Protocol on the transport and use of genetically modified organisms. On a similar line, Ottinger (2010) finds that focusing on quality standards may support but also limit the scope of citizen mobilization against pollution.
9 Similarly, carbon markets are not only typical neoliberal tools in climate change policy, but also spaces of contention and diversion. NGOs such as Greenpeace regard carbon markets as open-ended political frameworks amenable to redesign along social and environmental lines, while it is possible for individuals to use their own capacity of purchase of emission permits to raise polluting industries' costs (Blok 2011, 2012).

10 Some scholars, for example, warn that giving up property monopolies, as with the Open Source movement, does not automatically entail avoiding integration in, or subjugation to, the logic of capital (Suarez-Villa 2009). On the other hand, there exist strong cases for an ecologically sensible use of property rights-based regulation, for example, within the 'ecological modernization' approach (Mol et al. 2009).
11 While the de-growth movement focuses on the downscaling of production and consumption that entails a contraction of economy, the transition movement builds on an ecosystems approach towards constructing resilient, integrated society–nature assemblages (see www.transitionnetwork.org/). There are obvious areas of overlap between the two approaches.

References

Baland, J.M., Bardhan, P. and Bowles, S. (eds) (2007) *Inequality, Cooperation, and Environmental Sustainability*. Princeton, NJ: Princeton University Press.
Berkhout, F. (2006) 'Normative Expectations in Systems Innovation', *Technology Analysis and Strategic Management*, 18, 3–4, pp. 299–311.
Birch, K., Levidow, L. and Papaioannou, T. (2010) 'Sustainable Capital? The Neoliberalization of Nature and Knowledge in the European "Knowledge-Based Bio-Economy"', *Sustainability*, 2, 9, pp. 2898–2918.
Blok, A. (2011) 'Clash of the Eco-Sciences: Carbon Marketization, Environmental NGOs and Performativity as Politics', *Economy and Society*, 40, 4, pp. 451–476.
Blok, A. (2012) 'Configuring Homo Carbonomicus: Carbon Markets, Calculative Techniques and the Green Neoliberal', pp. 187–208 in L. Pellizzoni and M. Ylönen (eds), *Neoliberalism and Technoscience: Critical Assessments*. Farnham, UK: Ashgate.
Borup, M., Brown, N., Konrad, K. and Van Lente, H. (2006) 'The Sociology of Expectations in Science and Technology', *Technology Analysis and Strategic Management*, 18, 3–4, pp. 285–298.
Bowker, J. (2006) 'Time, Money and Biodiversity', pp. 107–123 in A. Ong and S. Collier (eds), *Global Assemblages. Technology, Politics, and Ethics as Anthropological Problems*. London: Blackwell.
Boyd, W., Prudham, S. and Schurman, R. (2001) 'Industrial Dynamics and the Problem of Nature', *Society and Natural Resources*, 14, pp. 555–570.
Brenner, N., Peck, J. and Theodore, N. (2010) 'Variegated Neoliberalization: Geographies, Modalities, Pathways', *Global Networks*, 10, 2, pp. 182–222.
Bumpus, A. and Liverman, D. (2008) 'Accumulation by Decarbonization and the Governance of Carbon Offsets', *Economic Geography*, 84, 2, pp. 127–155.
Burchell, G. (1996) 'Liberal Government and Techniques of the Self', pp. 19–36 in A. Barry, T. Osborne and N. Rose (eds), *Foucault and Political Reason*. London: UCL Press.
Calvert, J. (2007) 'Patenting Genomic Objects: Genes, Genomes, Function and Information', *Science as Culture*, 16, 2, pp. 207–223.
Carolan, M. (2010) 'The Mutability of Biotechnology Patents: From Unwieldy Products of Nature to Independent "Object/s"', *Theory Culture and Society*, 27, 1, pp. 110–129.
Castree, N. (2008) 'Neoliberalising Nature: The Logics of Deregulation and Reregulation', *Environment and Planning A*, 40, pp. 131–152.
CEC-DG Research (2004) *Towards a European Knowledge-Based Bioeconomy*. Luxembourg: Office for Official Publications of the European Communities.
Cooper, M. (2008) *Life as Surplus. Biotechnology and Capitalism in the Neoliberal Era*. Seattle, WA: University of Washington Press.
Dean, M. (1999) *Governmentality*. London: Sage.
Devall, B. and Sessions, G. (1985) *Deep Ecology. Living as if Nature Mattered*. Salt Lake City, UT: Peregrine Smith.
Dyson, F. (2007) 'Our Biotech Future', *New York Review of Books*, 54, 12, 19 July. Available at www.nybooks.com/articles/20370. Accessed 25 January 2013.
Epstein, R. and Kuhlik, B. (2004) 'Is There a Biomedical Anticommons?' *Regulation*, 27, 2, pp. 54–58.
Esposito, R. (2008) *Bios: Biopolitics and Philosophy*. Minneapolis, MN: University of Minnesota Press.
Feher, M. (2009) 'Self-Appreciation; or, The Aspirations of Human Capital', *Public Culture*, 21, 1, pp. 21–41.
Felt, U. and Wynne, B. (eds) (2007) *Taking European Knowledge Society Seriously*. Report for the European Commission. Luxembourg: Office for Official Publications of the European Communities.
Foucault, M. (2008) *The Birth of Biopolitics*. London: Palgrave Macmillan.
Frickel, S., Gibbon, S., Howard, J., Kempner, J., Ottinger, G. and Hess, D. (2010) 'Undone Science: Charting Social Movement and Civil Society Challenges to Research Agenda Settings', *Science, Technology and Human Values*, 35, 4, pp. 444–473.

Goven, J. (2006) 'Dialogue, Governance and Biotechnology: Acknowledging the Context of the Conversation', *Integrated Assessment Journal*, 6, 2, pp. 99–116.
Hammond, K. (1996) *Human Judgement and Social Policy*. New York, NY: Oxford University Press.
Hardin, G. (1968) 'The Tragedy of the Commons', *Science*, 162, pp. 1243–1248.
Harvey, D. (2003) *The New Imperialism*. Oxford, UK: Oxford University Press.
Harvey, D. (2005) *A Short History of Neoliberalism*. Oxford, UK: Oxford University Press.
Hayden, C. (2007) 'Taking as Giving: Bioscience, Exchange and the Politics of Benefit-Sharing', *Social Studies of Science*, 37, 5, pp. 729–758.
Heller, M. and Eisenberg, R. (1998) 'Can Patents Deter Innovation? The Anticommons in Biomedical Research', *Science*, 280, pp. 698–701.
Hess, D. (2012) 'The Green Transition, Neoliberalism and the Technosciences', pp. 209–230 in L. Pellizzoni and M. Ylönen (eds), *Neoliberalism and Technoscience: Critical Assessments*. Farnham, UK: Ashgate.
Heynen, N., McCarthy, J., Prudham, S. and Robbins, P. (2007) 'Introduction: False Promises', pp. 1–21 in N. Heynen, J. McCarthy, S. Prudham and P. Robbins (eds), *Neoliberal Environments: False Promises and Unnatural Consequences*. London: Routledge.
Hobson, K. (2009) 'On a Governmentality Analytics of the "Deliberative Turn": Material Conditions, Rationalities and the Deliberating Subject', *Space and Polity*, 13, 3, pp. 175–191.
Irwin, A. (2006) 'The Politics of Talk: Coming to Terms with the "New" Scientific Governance', *Social Studies of Science*, 36, 2, pp. 299–320.
Jasanoff, S. (2004) 'The Idiom of Co-Production', pp. 1–12 in S. Jasanoff (ed.), *States of Knowledge: The Co-Production of Science and Social Order*. London: Routledge.
Jonas, H. (1984) *The Imperative of Responsibility: In Search of Ethics for the Technological Age*. Chicago, IL: University of Chicago Press.
Jessop, B. (2002) 'Liberalism, Neoliberalism and Urban Governance: A State-Theoretical Perspective', *Antipode*, 34, 3, pp. 452–472.
Kinchy, A., Kleinman, D. and Autry, R. (2008) 'Against Free Markets, Against Science? Regulating the Socio-Economic Effects of Biotechnology', *Rural Sociology*, 73, 2, pp. 147–179.
Latouche, S. (2006) *Le Pari de la décroissance*. Paris: Fayard.
Lave, R., Mirowski, P. and Randalls, S. (2010) 'Introduction. STS and Neoliberal Science', *Social Studies of Science*, 40, 5, pp. 659–675.
Lemke, T. (2003) 'Foucault, Governmentality and Critique', *Rethinking Marxism*, 14, 3, pp. 49–64.
Levidow, L. (2007) 'European Public Participation as Risk Governance: Enhancing Democratic Accountability for Agbiotech Policy?' *East Asian Science Technology and Society*, 1, pp. 19–51.
Levidow, L. and Carr, S. (2009) *GM Food on Trial. Testing European Democracy*. London: Routledge.
Lockie, S. (2002) 'The Invisible Mouth: Mobilizing "the Consumer" in Food Production–Consumption Networks', *Sociologia Ruralis*, 42, 4, pp. 278–293.
McCarthy, J. and Prudham, S. (2004) 'Neoliberal Nature and the Nature of Neoliberalism', *Geoforum*, 35, 3, pp. 275–283.
McCormick, S. (2007) 'Democratizing Science Movements: A New Framework for Mobilization and Contestation', *Social Studies of Science*, 37, 4, pp. 609–623.
Martinez-Alier, J. (1990) *Ecological Economics: Energy, Environment and Society*. Oxford, UK: Blackwell.
Mattelart, A. (2010) *The Globalization of Surveillance. The Origin of the Securitarian Order*. Cambridge, UK: Polity Press.
Meadows, D.H., Meadows, D.L., Randers, J. and Behrens, W. (1972) *The Limits to Growth*. New York, NY: New American Library.
Micheletti, M. (2003) *Political Virtue and Shopping*. New York, NY: Palgrave Macmillan.
Mol, A.P.J., Sonnenfeld, D.A. and Spaargaren, G. (eds) (2009) *The Ecological Modernisation Reader*. Abingdon, UK: Routledge.
Nealon, J. (2008) *Foucault Beyond Foucault: Power and its Intensification since 1984*. Stanford, CA: Stanford University Press.
Nelson, R. (2004) 'The Market Economy and the Scientific Commons', *Research Policy*, 33, 3, pp. 455–471.
Nordmann, A. (ed.) (2004) *Converging Technologies – Shaping the Future of European Societies*, High Level Expert Group 'Foresighting the New Technology Wave'. Luxembourg: Office for Official Publications of the European Communities.
O'Malley, P. (2004) *Risk, Uncertainty and Governance*. London: Glasshouse.
O'Malley, P. (2008) 'Governmentality and Risk', pp. 52–75 in J. Zinn (ed.), *Social Theories of Risk and Uncertainty*. London: Blackwell.

OECD (Organisation for Economic Cooperation and Development) (2002) *Genetic Inventions, Intellectual Property Rights and Licensing Practices. Evidence and Policies*. Paris: Organisation for Economic Cooperation and Development.
OECD (Organisation for Economic Cooperation and Development) (2009) *The Bioeconomy to 2030. Designing a Policy Agenda: Main Findings and Policy Conclusions*. Paris: Organisation for Economic Cooperation and Development.
Ong, A. (2006) *Neoliberalism as Exception*. Durham, NC: Duke University Press.
Ostrom, E., Gardner, R. and Walker, J. (1994) *Rules, Games, and Common-Pool Resources*. Ann Arbor, MI: University of Michigan Press.
Ottinger, G. (2010) 'Buckets of Resistance: Standards and the Effectiveness of Citizen Science', *Science, Technology and Human Values*, 35, 2, pp. 244–270.
Pellizzoni, L. (2008) 'The Antinomy of Accountability', pp. 210–230 in M. Boström and C. Garsten (eds), *Organizing Transnational Accountability*. Cheltenham, UK: Edward Elgar.
Pellizzoni, L. (2010) 'Environmental Knowledge and Deliberative Democracy', pp. 159–182 in M. Gross and H. Heinrichs (eds), *Environmental Sociology: European Perspectives and Interdisciplinary Challenges*. Dordrecht: Springer.
Pellizzoni, L. (2011a) 'Governing Through Disorder: Neoliberal Environmental Governance and Social Theory', *Global Environmental Change*, 21, 3, pp. 795–803.
Pellizzoni, L. (2011b) 'The Politics of Facts: Local Environmental Conflicts and Expertise', *Environmental Politics*, 20, 6, pp. 765–785.
Pellizzoni, L. and Ylönen, M. (2012) 'Hegemonic Contingencies: Neoliberalized Technoscience and Neorationality', pp. 47–74 in L. Pellizzoni and M. Ylönen (eds), *Neoliberalism and Technoscience: Critical Assessments*. Farnham, UK: Ashgate.
Polanyi, K. (1944) *The Great Transformation*. Boston, MA: Beacon Press.
Pollock, N. and Williams, R. (2010) 'The Business of Expectations: How Promissory Organizations Shape Technology and Innovation', *Social Studies of Science*, 40, 4, pp. 525–548.
Poteete, A., Janssen, M. and Ostrom, E. (2010) *Working Together: Collective Action, the Commons, and Multiple Methods in Practice*. Princeton, NJ: Princeton University Press.
Prigogine, I. and Stengers, I. (1984) *Order out of Chaos. Man's New Dialogue with Nature*. London: Heinemann.
Rabinow, P. (1996) *Essays on the Anthropology of Reason*. Princeton, NJ: Princeton University Press.
Rajan, K. (2006) *Biocapital: The Constitution of Postgenomic Life*. Durham, NC: Duke University Press.
Read, J. (2009) 'A Genealogy of Homo-Economicus: Neoliberalism and the Production of Subjectivity', *Foucault Studies*, 6, pp. 25–36.
Rhodes, R. (1997) *Understanding Governance*. Buckingham, UK: Open University Press.
Roco, M. and Bainbridge, W. (eds) (2002) *Converging Technologies for Improving Human Performance*. Arlington, VI: National Science Foundation.
Roff, R. (2007) 'Shopping for Change? Neoliberalizing Activism and the Limits to Eating Non-GMO', *Agriculture and Human Values*, 24, pp. 511–522.
Rose, N. (1996) 'Governing "Advanced" Liberal Democracies', pp. 37–64 in A. Barry, T. Osborne and N. Rose (eds), *Foucault and Political Reason*. London: UCL Press.
Rose, N. (2007) *The Politics of Life Itself*. Princeton, NJ: Princeton University Press.
Rose, N. and Novas, C. (2006) 'Biological Citizenship', pp. 439–463 in A. Ong and S. Collier (eds), *Global Assemblages: Technology, Politics, and Ethics as Anthropological Problems*. London: Blackwell.
Rutherford, S. (2007) 'Green Governmentality: Insights and Opportunities in the Study of Nature's Rule', *Progress in Human Geography*, 31, 3, pp. 291–307.
Scheper-Hughes, N. (2006) 'The Last Commodity: Post-Human Ethics and the Global Traffic in "Fresh" Organs', pp. 145–167 in A. Ong and S. Collier (eds), *Global Assemblages: Technology, Politics and Ethics as Anthropological Problems*. London: Blackwell.
Suarez-Villa, L. (2009) *Technocapitalism*. Philadelphia, PA: Temple University Press.
Tickell, A. and Peck, J. (2003) 'Making Global Rules: Globalization or Neoliberalization?' pp. 163–181 in J. Peck and H.W. Yeung (eds), *Remaking the Global Economy*. London: Sage.
Urbinati, N. (2010) 'Unpolitical Democracy', *Political Theory*, 38, 1, pp. 65–92.
Walzer, M. (1985) *Exodus and Revolution*. New York, NY: Basic Books.
Wynne, B. (2001) 'Creating Public Alienation: Expert Discourses of Risk and Ethics on GMOs', *Science as Culture*, 10, pp. 1–40.
Zeller, C. (2008) 'From the Gene to the Globe: Extracting Rents Based on Intellectual Monopolies', *Review of International Political Economy*, 15, 1, pp. 86–115.

25

Disassembling and reassembling socionatural networks

Integrated natural resource management in the Great Bear Rainforest

Justin Page

In the late nineteenth century, John Muir – founder of the Sierra Club, the first environmental organization in the USA – battled to establish California's Yosemite Valley as the first parkland set aside for preservation and public use in the USA. Muir's tireless work to protect the valley's 'holy temples' from economic development set the standard for the contemporary environmental movement which, similar to its predecessors in the 1960s and 1970s, worked to protect natural areas by excluding people. By the end of the twentieth century, however, protected areas – and the assumptions that underpin them – came under increasing scrutiny from both within and without the environmental movement.

Critics characterized this approach to nature conservation as 'fortress conservation' (Brown 2002: 6), arguing that it rests on an apolitical, neo-Malthusian assumption that environmental degradation results from unchecked human demographic growth and demand for resources. Critics argued that this view not only ignores the political and economic causes of environmental destruction, but also fails to recognize that protected areas can alienate land from local resource users, distribute the costs and benefits of conservation unequally, and generally impoverish, disenfranchise and disempower already disadvantaged groups (Robbins 2004). Moreover, environmental scholars argued that, despite the boundaries drawn around protected areas to separate nature from society, nature remains a social and material construction – from the myth of the pristine wilderness to the legal, material, political and economic infrastructure put in place to create and maintain parks (Cronon 1996; Willems-Braun 1997). From a strategic standpoint, ecologists and biologists argued that protected areas often failed to achieve their objective of preserving nature, as parks became disconnected and fragmented biological islands that were too small to sustain important species (Diamond 1975; Harris 1984; MacArthur and Wilson 1967; Saunders et al. 1991).

In response, many conservation projects no longer focus on installing boundaries between nature and people, but instead find ways to include local human populations and economies in the conservation of natural areas. Such initiatives fall under the category of 'integrated natural resource management' (INRM) (Bellamy et al. 1999; Campbell and Sayer 2003), which includes community-based natural resource management, ecosystem management and conservation and development projects that 'integrate protected areas into the economic and social context locally, regionally and nationally' (Brown 2002: 6). In place of efforts to segregate nature and society, the goal of the conservation movement is more centred on tying people, animals, ecologies and economies more closely together (for reviews of INRM initiatives, see Conley and Moote 2003; Hughes and Flintan 2001; Pavlikakis and Tsihrintzis 2000).

This chapter aims to investigate how integration is achieved in practice. This is an important goal as knowledge of the process of integration can inspire processes that may foster it. Working through a recent case of INRM in the Great Bear Rainforest (GBR) land use agreement in British Columbia, Canada, this chapter examines integration as a process of *transformation*, particularly with respect to the identities and interests of the humans and non-humans that are brought together in conservation projects. This analysis is supported by the conceptual tools of actor–network theory (ANT), a line of research that avoids conceiving of nature and society as separate domains, and instead focuses on how humans and non-humans become associated in common networks.

This analytic approach contributes to existing research paradigms deployed to study INRM, including common property theory (Berkes 1989; McCay and Acheson 1987; Ostrom 1990), social capital and social learning theories (Daniels and Walker 1996; Pahl-Wostl and Hare 2004, 2007; Pretty 2003; Pretty and Ward 2001; Schusler et al. 2003), political ecology (Blaikie and Brookfield 1987; Bryant and Bailey 1997; Peet and Watts 1996) and social constructionism (Braun 2002; Bryant 2001; Burningham and Cooper 1999; Fairhead and Leach 2003). However, it also represents a departure from these paradigms, as it is guided by very different assumptions. Generally, social science research on natural resource management focuses on the social factors thought to influence management outcomes, such as principles and models, social capital and trust, political economy, and language and representational practices. Nature, in these research programmes, is conceived of as a given biophysical reality over which social groups debate. If nature enters these debates at all, it is only in the form of cold, hard scientific facts that function to close debate rather than to foster it (see, for example, Kranjc 2002).

Natural resource conflict, from this perspective, amounts to *politics-about-nature*, in which multiple well-defined human groups with static interests debate a common biophysical reality (Latour 2004). However, situations often arise in resource management conflicts where new groups emerge and disintegrate, existing groups transform, interests shift and evolve over time, and where the underlying biophysical reality that is the object of concern is precisely at issue. In fact, situations like this may be the norm in INRM, where the *integration* of people, animals, habitats, ecosystems and economies is the central focus. Integration means 'to put together parts or elements and combine them into a whole'.[1] In this process, elements are mutually transformed as they come together to create a heterogeneous composite that did not previously exist. To assume that INRM is a purely *social* process in which social structures and forces lead to a common understanding of a *given* non-human reality (or to the domination of one view over others) is to reinforce a dualism between society and nature that INRM explicitly seeks to avoid. It also leaves no way to account for how non-humans participate in processes of integration, except in the form of cold, objective facts. Moreover, this view can lead to bad politics, a sort of 'police operation' where a priori definitions of the common world to be assembled are used as an arbiter for disputes (Latour 2004: 455).

This chapter considers the integration of humans and non-humans in the Great Bear Rainforest as a fundamentally transformative process. Based on an analysis of reports, campaign materials and media accounts, as well as in-depth interviews with environmentalists, forestry companies, First Nations[2] and local and provincial governments associated with the agreement, it examines how the identities and interests of ecosystems, people and animals shift as they become associated with one another. I consider this process to be constructive – one that produces and performs the realities in question. But this can also be destructive, where the integration of one world may require the disintegration of another world. Avoiding analytic restriction to the social sphere, I take into account the role of non-humans – particularly the grizzly bear – in this process. Finally, I examine the work of the actors involved to create their own definitions and frameworks to order the common world they compose together.

INRM and actor–network theory

Actor–network theory (ANT) – originally developed in the field of science and technology studies by Bruno Latour (1987, 1988), Michel Callon (1980, 1986) and John Law (1986) (see also Callon and Latour 1981; Callon and Law 1982) – provides a unique alternative to the politics-about-nature approach detailed above (Latour 2004). Rather than viewing nature as a given ontological domain more or less accurately disclosed by science, and politics as a conventional domain pertaining to social relations, ANT concerns itself with *ontological politics* (Mol 1999), or the politics involved in producing realities. For ANT scholars, reality is produced and performed by a wide array of human and non-human elements that come together to form a network of relations. In the words of Law (2009: 141), 'everything in the social and natural worlds [is] a continuously generated effect of the webs of relations within which they are located'. Given the underlying heterogeneity of relations, the effects that they produce are simultaneously material, discursive and collective (Latour 1993) and are viewed as an achievement of heterogeneous networks that are at once scientific and political, human and non-human, material and symbolic.

Several environmental social scientists have recognized the value of ANT to the study of environmental issues and problems (see, for example, Castree 2003; Castree and Braun 2001; Demeritt 1998; Ivakhiv 2002; Murdoch 1997, 2001; Whatmore 2002). In particular, environmental scholars suggest that ANT avoids treating nature and society as separate domains of reality (Murdoch 2001), to instead apprehend them as hybrid *socionatures* (White 2006). Instead of nature and society, from the perspective of ANT, there are heterogeneous networks that are assembled to create more or less stable outcomes. This view is particularly applicable to the question of integration in INRM, as it avoids treating humans and non-humans as different entities, but instead examines the specific processes whereby humans and non-humans become associated with one another in common networks.

This close, even 'myopic' attention to the process of association has prompted ANT theorists to recognize the need to rework conventional concepts such as power, agency and structure (Latour 2005). Association entails transformation or, in the language of ANT writers, *translation* (Callon 1986). When entities are enrolled in new networks, their identities are redefined, their interests reworked and their roles re-established. This process involves varying degrees of coercion and negotiation and thus can hardly be said to be devoid of power relations. However, the power to enrol is the power of a network (which, in turn, is dependent on the entities enrolled) rather than a group or individual with access to resources on the basis of position in a social structure. Moreover, power can be productive as well as destructive. When entities are enrolled in a network, they are able to act in new ways and to achieve interests that they did not previously possess.

Agency, as with power, is an attribute of networks rather than of individuals or groups. Elements in a network enable or afford the action of those with which they are associated, and vice versa. Since networks are comprised of human and non-human elements, this entails that non-humans are able to participate in the course of action. Agency is not considered by ANT scholars to be a capacity held by particular kinds of entities on the basis of attributes such as intentionality, but a network characteristic. When actors (or 'actants', as ANT scholars refer to them to recognize the multiple forms of figuration that actors may take) act, they do so always with and through others, and often in ways that they do not intend. Non-humans are actants when they participate in the course of action together with others, causing other elements to act in turn.

ANT writers therefore do not conceive action as the outcome of pre-existing structures and forces. If this were the case, then actors would be mere intermediaries for existing forces and no real action could take place as outcomes can be predicted well in advance (Latour 2005). This is not to suggest that actors are associated in random, haphazard or anarchic ways, but rather that their ordering derives from the actors themselves instead of pre-existing structures. Actors produce the frameworks and structures that render their networks coherent and they may do so in multiple, overlapping ways (Law 1994). It is the role of the analyst to reveal these processes.

How might an analytic approach that focuses on associations in heterogeneous networks, the translation of identities and interests, the roles of non-human actants and the ordering activities of the actors themselves be of use in the study of INRM? First, by eschewing a priori assumptions about nature and society to instead follow how the actors construct hybrid socionatural networks, ANT is able to account for how the practice of INRM works to integrate humans and non-humans in common networks. Second, since ANT views non-humans as full-blown actants, it can take account of the ways in which non-humans participate in decisions that affect their fate. Third, by accounting for the ways in which diverse elements are stitched together, ANT can account for the unexpected consequences of actions. Fourth, ANT can show how the resolution of multi-stakeholder conflicts may not necessarily derive from compromise, tradeoffs or an uneasy balance between competing interests, but through the translation and transformation of identities and interests. Finally, through the study of such translation, ANT locates power in the relation, thereby recognizing the dispersed and distributed nature of power. Such a view can be empowering for actors: by avoiding the assumption that outcomes are determined by large, impenetrable forces, actors can intervene in networks to shift relations of power.

In the following, I present an actor–network analysis of the Great Bear Rainforest. Instead of treating this outcome as the result of human politics-about-nature, I examine the processes leading to the agreement as a case of ontological politics. In particular, rather than investigating how social structures and forces influence Environmental Non-Governmental Organizations' (ENGO) efforts to police the boundary between society and nature, I examine ENGOs' efforts to disassemble a criticized socionatural network (the 'mid-coast timber supply area') and reassemble a better one (the 'Great Bear Rainforest'). Throughout, I consider how heterogeneous elements are enrolled in the emerging networks, how their identities and interests are translated, the workings of power and the strategies invented by the actors to order the networks convened into a coherent whole.

Disassembling the 'mid-coast timber supply area'

In June 1997, the government of British Columbia (BC) convened a multi-stakeholder Land and Resource Management Planning (LRMP) process for BC's central coast, which, at the time, was referred to simply as the 'mid-coast timber supply area'. The LRMP forum was premised on the idea 'of integrated resource planning at the community level', a process that would provide an 'opportunity for all interest groups, local government and First Nations to comment on how Crown land will be managed' (Government of British Columbia 1996). On the face of it, the LRMP appears to be a model of INRM, yet environmentalists (and most First Nations) rejected it and put forward three sets of concerns about the process.

First, they argued that the process – despite its appearance of openness – was in fact highly constrained and imposed narrowly predetermined parameters within which collaborative decision making could operate. Second, environmentalists felt that the format of the LRMP was very unwieldy, since it allowed any and all interest groups (from recreation groups to tourism operators and the Cattlemen's Association) to have a say, resulting in very large meetings (up to sixty participants) in which little could be accomplished. More to the point, the very nature of this format forced groups into position taking (i.e. rigid articulations of interests) that might not be taken under more informal circumstances. As one environmentalist put it, 'the way the LRMP was set up ... people took positions because in a public space like that with the Government recording it, they're not going to admit to certain things'.

Third, environmentalists believed that the interests and voices of non-humans were not adequately taken into account by the process. Given that past decisions about protected areas were generally not made on the basis of biological considerations, but as a result of the government's attempt to appease environmentalists (with areas that protected little more than 'rock and ice') while responding to the greater influence of the forestry industry with whom they formed a political compact (Marchak 1983; Wilson 1998), environmentalists chose not to participate.

In opposition to this 'cookie-cutter' process, a coalition of local and international environmental organizations worked to intervene in networks supporting status quo forestry and to replace them with alternative networks, in the process redefining the identities and interests of a variety of human and non-human elements. Following two relatively unsuccessful attempts to blockade forestry operations at Roderick Island in May 1997 and King Island in June 1997, the coalition turned their attention to the networks that lay behind forestry companies. Through a variety of methods – such as sleuthing at lumber yards and electronic tracking of forest products – environmentalists traced the transformation of trees into commodities and their movements through space to particular nodes, such as manufacturing sites, ports and lumber yards. Of course, the purpose was not just to trace the movement of forest products but to intervene in the networks they laid bare.

Activists began to target an important node in the commodity chain network: namely, retail customers of forest products. Hanging banners such as 'Don't buy rainforest destruction' on retail outlets such as Staples, IKEA and Home Depot, selling products originating in BC's forests, environmentalists raised the spectre of a consumer boycott. Companies began to capitulate, cancelling contracts and implementing more ecologically sound procurement policies, or, as ENGOs put it in a December 1998 advertisement placed in the *New York Times*, companies became environmentalists by 'saving ancient rainforests without ever chaining themselves to a tree'. In turn, forestry companies operating in BC's coastal forests were forced to change their strategies and tactics (which until then focused on countering environmentalists' claims through a PR-style campaign) and negotiate with environmentalists on ecological protection and forestry practices. The two parties agreed to a 'standstill period' in which ENGOs ceased their campaign and forestry companies ceased logging operations in contested valleys. The industry even went so far as to form a new environmentally oriented coalition of forestry companies called the Coast Forest Conservation Initiative, in order to reorient their engagement with ENGOs.

Reassembling the Great Bear Rainforest

By this point, ENGOs had successfully traced and intervened in a network supporting status quo industrial forestry. In the process, they had altered their own identities (forming a precarious coalition), the identities of retail customers of forestry products (now considered to be activists through their procurement policies), and that of forestry companies (now reinvented as a conservation initiative). However, the interests and identities of non-humans were also an issue, particularly as ENGOs simultaneously engaged in the ontological politics of crafting an alternative socionature to industrial forestry. This procedure involved a series of interlinked scientific, political and cultural processes, beginning in the early 1990s, that not only redefined coastal forests but provided them with personality and a spokesperson.

The first step involved the recategorization of the forests into a new 'biome'. ENGOs combined measures of precipitation, vegetation and location with a provisional definition of the 'coastal temperate rainforest', calculating that British Columbia (BC) has the largest expanse of undeveloped coastal temperate rainforest (approximately one quarter) remaining in the world.[3] This categorization of BC's coastal forests not only drew a scientific boundary around them, but also placed them in a wider political frame of reference (the 'rainforest') which, by the 1980s, had considerable public sentiment attached to it. According to one environmentalist, when ENGOs successfully established that BC's coastal forests were also 'rainforests', leaders in the forestry industry 'just went white' because they understood the implications of such a framing.

The coastal temperate rainforest is neither a given biophysical reality objectively discovered by scientists, nor a merely symbolic representation projected onto real forests, but a hybrid material-symbolic production. This socionature gained increasing depth as environmentalists worked through material and symbolic meanings to furnish it with a personality. The Raincoast Conservation Society, a small environmental organization spent five years doing on-the-ground lay research, collecting plants, animal tracks, photographs and stories, which they later presented in a coffee table book called *The Great Bear Rainforest: Canada's Forgotten Coast* (McAllister et al. 1997). The imagery, descriptions and stories contained in the book – each traceable back to an individual valley, estuary or animal first roughly captured in the coastal temperate rainforest category – came to revolve around the central character of the grizzly bear, 'the most profound symbol of what this ancient ecosystem is all about' (McAllister et al. 1997: 25–26). These bears, according to the book, are not only beautiful, graceful and industrious, but are ideal representatives of the coastal forests, as 'we can look at healthy grizzly populations and have confidence that the integrity of the coastal ecosystem is intact' (McAllister et al. 1997: 25). As this quotation indicates, the designation of the grizzly bear as representative of the coastal temperate rainforest is not simply symbolic. Indeed, a later conservation biology report commissioned by the ENGO coalition drew on the biological habits of grizzlies – in particular, their large territorial range – to describe a network of core protected areas and ecological corridors (Jeo et al. 1999). Serving as an umbrella species, a biological representative of fauna and flora, the grizzly bear thereby came to participate in debates over land use in coastal BC in important ways.

Composing frameworks to order networks

In its guise as representative of the coastal forests (henceforth the 'Great Bear Rainforest'), the grizzly bear became an important mediator of relations among humans. In its work to disrupt the network supporting status quo industrial forestry, the bear was deployed in a variety of forms and figurations by ENGOs, such as costumed protesters outside forestry offices, a model in a travelling rainforest bus diorama, and a 40-foot inflatable member of demonstrations held outside major retailers like Home Depot. In each case, the bear took on different figurations as the result of a network of elements. However, in its scientific figuration as an umbrella species in the Conservation Area Design (CAD), the authors attempted to sever the network and make resource management decisions based on conservation science *rather than* aesthetic, social, political or economic considerations (Jeo et al. 1999: 19). Despite this attempt at a particularly modern framing – one which orders networks into nature and society (Latour 1993) – the bear remained a mediator, connecting biological, social, political and economic processes.

When ENGOs and forestry companies negotiated over the valleys suggested by the grizzly bear for protection, their own identities and interests were transformed. For example, ENGOs agreed that key areas would be protected, but not the whole region. On the other hand, forestry companies agreed to reduce the rate of logging and continue 'sustainable forestry'. With the bear linking them, the two groups formed a single, hybrid Joint Solutions Project focused on creating innovative ways to order the relations among bears, trees and people. Originally referred to as a conservation-biology/ecosystem-based planning approach for the region, the framework gave central place to the interests of non-humans, while recognizing important connections between ecosystems and people.

Ideally, all elements in a network participate in processes that determine their relations and order them into a coherent whole (Callon et al. 2009). In practice, however, this is a highly charged process, as network elements test the extent of their representation. For example, local mayors of forest-dependent communities argued that their interests in ongoing employment were being sidelined in these processes and formed an initiative called Operation Defend to oppose changes to land use arrangements. Additionally, some environmentalists suggested that ENGOs in the Joint Solutions Project had 'sold out' and conceded too much, and some forestry companies left the coalition. Meanwhile, First Nations formed a new coalition (Turning Point, now known as Coastal First Nations) to ensure that their interests in rights and title to the land in question were adequately taken into account and ensured their active participation in the process.

Eventually, after a series of workshops and meetings, the ENGOs, forestry companies and First Nations formed a hybrid forum, the Kitasoo-Gitga'at Pilot Implementation Team (Kit-Git-Pit), where the framework could be further developed. Based in two First Nations territories, the group worked on an ecosystem-based management (EBM) framework. In this approach, non-human elements necessary for ecological integrity, such as bear dens and course woody debris, would be considered first before resource harvesting would be considered. In contrast to the CAD, however, decisions were not based solely on non-human nature: resource harvesting would take place to support forestry companies, their employees and First Nations. The novel solution that the group proposed was to create hierarchies of scale and risk and to then make decisions that cut across all these categories.

Land was partitioned into First Nations territory, landscape, watershed and site levels. Risk from human use of the land was categorized into high, moderate, low and very low.[4] The two scales were then brought together. Areas at high risk from disruption would be protected from economic activity. Notably, protected areas and working areas were not only applied to the landscape level. Rather, at each level, from entire watersheds to individual stands, land is to be designated for protection. In this way, both resource harvesting and resource protection can occur on the same land base. Risk levels are allowed to vary: at lower levels, such as watersheds and stands, risk is allowed to increase all the way to the high threshold, as long as overall risk sums to the low threshold at the territory level.

A second mechanism was created by the actors to order elements in the network. Whereas the first mechanism involved framing the forests though a hierarchy of scales and risk thresholds in order to render them compatible with economic activity while maintaining ecological integrity, the second involved reframing economic activity in order to render it compatible with healthy forests. Again, in an experimental pilot project with First Nations, the Conservation Investments and Incentives Initiative, ENGOs set out to start a new economy for the region that fostered rather than undermined ecological integrity. ENGOs and First Nations raised US$60 million from large US philanthropic foundations to invest in conservation and sustainable business development for the region. This amount was later matched by the government, culminating in a total of US$120 million in support of the new conservation economy for the coast. In this new economy, EBM forestry would continue to play a role, but would be one piece of a wider, more diverse economy. Half of the funds (US$60 million) were applied to new protected areas, providing employment to First Nations to carry out conservation studies and watchmen programmes. The other half were designated to the development of businesses that rely on ecological integrity as a condition of their profitability, such as ecotourism, non-timber forest products and shellfish aquaculture. Again, these business opportunities were directed towards local First Nations communities.

These two mechanisms – EBM and the conservation economy – reordered the reassembled elements of the network in ways that defied the ontology of the older network. There were protected areas and working areas, but they were not ordered in a tradeoff or attempted balance. Indeed, the overall level of protection (approximately 33 per cent in the final agreement) is a superficial indicator of the outcome, since the logic of protection and use was disembedded from the regional level and re-embedded in a hierarchy of scales. The areas were not designated as pristine nature and land for human development, but as hybrids. Both protected areas and working areas were socionatures. New conservancies allowed a degree of low impact economic activity for First Nations. Similarly, economic activities including industrial forestry were permitted in working areas, but conservation came first, with bear dens, stands, watersheds and landscapes left in place to ensure ecological integrity.

The conservation economy also creates socionatures, with protected areas generating economic activity, such as conservation research, and business development promoting conservation in the form of ecotourism, for example. Within this framework, the reassembled network members – First Nations, non-humans, forestry companies, ENGOs – are all transformed and reordered.

Conclusion

The Great Bear Rainforest agreement contributes to a new episode of environmental politics in nature conservation. Earlier conflicts (in Yosemite, BC and elsewhere) were often centred around environmentalists' attempt to preserve 'pristine wilderness' from industrial development. Campaigns to protect BC's central and north coasts from industrial forestry also focused on important ecological values, but they became more nuanced than the earlier conflicts. ENGOs worked to support the establishment of alternative forestry practices and to develop an alternative economy for the coast. These efforts are in line with wider developments in INRM practice, which increasingly recognizes the value of integrating people with non-humans in conservation strategies.

This chapter analysed the integration of humans and non-humans in the GBR as the result of networks of associations and consequent transformations of actors' identities and interests. This approach contrasts with conventional social scientific accounts of resource management that tend to restrict analyses to the social factors that influence resource management outcomes. Instead of adopting a politics-about-nature analysis advocated by these accounts, I chose to deploy the conceptual tools of actor–network theory to analyse the politics of socionatures in the GBR. I discussed how ENGOs intervened in networks to disassemble one socionature – the 'mid-coast timber supply area' – and the ontological politics involved in assembling an alternative network – the 'Great Bear Rainforest'. This approach has the advantage of avoiding conceptual dualisms between society and nature, which INRM projects also seek to avoid. Now let me consider some of the implications of this analysis for the practice of INRM.

First, my analysis of the GBR indicates that the integration of humans and non-humans in INRM is contingent on processes of transformation. Rather than forming well-defined interest groups with static positions, environmentalists, forestry companies, customers of forest products,

First Nations and even ecosystems and grizzly bears all shifted their identities and interests as they became associated with one another in a common network. This indicates that, for integration to be achieved, participants must be open to the transformation of their identities and interests as they become associated with others and as they work toward the production of a common world. Overly proscribed processes such as the LRMP amount to 'police actions' that result in the production of rigid identities, interests and definitions of the world. In contrast, the formation of solution spaces outside of official channels allows actors to experiment and create new options and solutions on the basis of relationally defined concerns.

Second, these processes cannot amount to a human politics-about-nature, or one in which multiple human views are progressively aligned on a given biophysical reality. For integration of humans and non-humans to occur, non-humans ought to play a part as key actants among others, and not simply as cold, hard facts. If non-humans are to have a hand in creating the common world that they are to inhabit along with humans, then they need to speak for themselves, through their associations with others, as grizzlies did with respect to their territorial habits and their ability to represent other non-humans. Moreover, for non-humans to participate in these processes, they cannot be considered as facts divorced from their social, economic and other relations, as was the focus of the CAD. Biophysical reality cannot be assumed as a given when the construction of such a reality is precisely at issue, as was the case with ENGOs' describing BC's forests as a 'coastal temperate rainforest'. The realities that emerge are neither objective discoveries nor subjective representations, but heterogeneous networks that emerge over time as more elements are added and as they increase in reality.

Third, my actor–network analysis of the GBR demonstrates that it is important to follow imbroglios all the way down and take responsibility for consequences of actions wherever they arise in the network. When integration is the goal, it is no longer possible to focus on saving wilderness from human activity and denying responsibility for the impacts of these actions in any domain other than that of ecology. Actions to protect ecological integrity will inevitably have consequences for the people who are connected to the ecologies in question. Integration requires taking responsibility for the entire heterogeneous composite under construction, rather than one or another piece of it. As ENGOs in the GBR demonstrated, this may entail activity in areas traditionally foreign to environmentalists, such as the creation of economies to support local livelihoods.

Finally, it is important to recognize that the integration of human and non-human elements into new composites entails the disintegration of other composites and the rearticulation of elements in ways that may be offensive to some members. Processes of integration are processes of power, wherein humans and non-humans become transformed and their relations redistributed. For some, such as First Nations, this process may be empowering, enabling a level of influence that was not available in other networks. For others, such as local non-indigenous communities, this process may be felt as disempowering, severing the networks in which they used to enjoy influence. For yet others, such as ENGOs and forestry companies, the process redistributes power in interesting ways. As my analysis demonstrated, power is not something possessed by particular groups on the basis of their position in a social structure, but as a function of networks. Producing the conditions conducive to integration may necessitate shifting networks of power, as ENGOs achieved through their interventions in the commodity chain and their enrolment of the industry's key ally, their retail customers. This shift in relations of power did not destroy forestry companies' influence over resource management, but prompted shifts in their identities and interests and their influence remained relational. ENGOs, forestry companies, First Nations and grizzly bears came together to work in a network to influence land use decisions. Moreover, as this case shows, the power to order relations among human and non-human entities comes not from abstract social forces but the actors themselves. In this case, it was in the form of ecosystem-based management and the conservation economy.

The integration of people, trees, animals, ecosystems and economies in INRM, demonstrated in this chapter, can benefit from processes that foster openness to transformation, the inclusion of non-humans as actants, responsibility for imbroglios and care with redistributions of power. What these processes might look like in general terms is something that future research can investigate, although it must be with the recognition that it is up to the actors themselves to design these processes. The role of research, guided by the concepts and tools of ANT, is to render these processes explicit so that they might be adopted, adapted and transformed in future projects.

Notes

1 Definition based on the *Online Etymology Dictionary*. Accessed 11 March 11 2011 from http://dictionary.reference.com/browse/integrate.
2 In Canada, aboriginal peoples are referred to as 'First Nations'.
3 Temperate rainforest was defined as 'areas between 32 and 60 degrees latitude, with the presence of vegetation (if not currently then originally in a forested condition), with at least 2000 mm (80 in) of annual rainfall' (Weigand et al. 1992: 4). 666
4 Risk is understood in relation to ecosystems' range of natural variation, recognizing that ecosystems are prone to disturbance and that human disturbance, if within the natural range, can be absorbed by ecosystems.

References

Bellamy, J.A., McDonald, G.T., Syme, G.J. and Butterworth, J.E. (1999) 'Evaluating Integrated Resource Management', *Society and Natural Resources*, 12, 4, pp. 337–353.
Berkes, F. (ed.) (1989) *Common Property Resources: Ecology and Community-Based Sustainable Development*. London/New York: Belhaven Press.
Blaikie, P.M. and Brookfield, H.C. (1987) *Land Degradation and Society*. London/New York: Methuen.
Braun, B. (2002) *The Intemperate Rainforest: Nature, Culture, and Power on Canada's West Coast*. Minneapolis, MN: University of Minnesota Press.
Brown, K. (2002) 'Innovations for Conservation and Development', *Geographical Journal*, 168, 1, pp. 6–17.
Bryant, R.L. (2001) 'Political Ecology: A Critical Agenda for Change', pp. 151–169 in N. Castree and B. Braun (eds), *Social Nature: Theory, Practice and Politics*. Malden, MA: Blackwell.
Bryant, R.L. and Bailey, S. (1997) *Third World Political Ecology*. London/New York: Routledge.
Burningham, K. and Cooper, G. (1999) 'Being Constructive: Social Constructionism and the Environment', *Sociology*, 33, 2, pp. 297–316.
Callon, M. (1980) 'Struggles and Negotiations to Define What Is Problematic and What Is Not: the Sociology of Translation', pp. 197–219 in W.R. Knorr, R. Krohn and Richard P. Whitley (eds), *The Social Process of Scientific Investigation: Sociology of the Sciences Yearbook*. Boston/London: Reidel.
Callon, M. (1986) 'Some Elements of a Sociology of Translation: Domestication of the Scallops and the Fishermen of St-Brieuc Bay', pp. 196–233 in J. Law (ed.), *Power, Action and Belief: A New Sociology of Knowledge*. London: Routledge and Kegan Paul.
Callon, M., Lascoumes, P. and Barthe, Y. (2009) *Acting in an Uncertain World: An Essay on Technical Democracy*. Cambridge, MA: MIT Press.
Callon, M. and Latour, B. (1981) 'Unscrewing the Big Leviathan: How Actors Macrostructure Reality and How Sociologists Help Them to Do So', pp. 277–303 in K.D. Knorr-Cetina and A.V. Cicourel (eds), *Advances in Social Theory and Methodology: Toward an Integration of Micro- and Macro-Sociologies*. Boston, MA: Routledge and Kegan Paul.
Callon, M. and Law, J. (1982) 'On Interests and Their Transformation: Enrollment and Counter-Enrollment', *Social Studies of Science*, 12, 4, pp. 615–625.
Campbell, B.M. and Sayer, J. (2003) *Integrated Natural Resource Management: Linking Productivity, the Environment, and Development*. Wallingford, UK: CABI Publishing in association with the Center for International Forestry Research.
Castree, N. (2003) 'Environmental Issues: Relational Ontologies and Hybrid Politics', *Progress in Human Geography*, 27, 2, pp. 203–211.

Castree, N. and Braun, B. (2001) *Social Nature: Theory, Practice and Politics*. Malden, MA: Blackwell.

Conley, A. and Moote, M.A. (2003) 'Evaluating Collaborative Natural Resource Management', *Society and Natural Resources*, 16, 5, pp. 371–386.

Cronon, W. (1996) *Uncommon Ground: Rethinking the Human Place in Nature*. New York, NY: W.W. Norton and Company.

Daniels, S.E. and Walker, G.B. (1996) 'Collaborative Learning: Improving Public Deliberation in Ecosytem-Based Management', *Environmental Impact Assessment Review*, 16, 2, pp. 71–102.

Demeritt, D. (1998) 'Science, Social Constructivism and Nature', pp. 172–192 in B. Braun and N. Castree (eds), *Remaking Reality: Nature at the Millennium*. London/New York: Routledge.

Diamond, J. (1975) 'The Island Dilemma: Lessons of Modern Biogeographic Studies for the Design of Natural Reserves', *Biological Conservation*, 7, 2, pp. 129–146.

Fairhead, J., and Leach, M. (2003) *Science, Society, and Power: Environmental Knowledge and Policy in West Africa and the Caribbean*. Cambridge/New York: Cambridge University Press.

Government of British Columbia (1996) *Backgrounder: Land and Resource Management Plans*. Available at http://archive.ilmb.gov.bc.ca/slrp/lrmp/nanaimo/cencoast/news/bkgrnd.ht. Accessed 16 March 2011.

Harris, L.D. (1984) *The Fragmented Forest: Island Biogeography Theory and the Preservation of Biotic Diversity*. Chicago, IL: University of Chicago Press.

Hughes, R. and Flintan, F. (2001) *Integrating Conservation and Development Experience: A Review and Bibliography of the ICDP Literature*. London: International Insitute for Environment and Development.

Ivakhiv, A. (2002) 'Toward a Multicultural Ecology', *Organization and Environment*, 15, 4, pp. 389–409.

Jeo, R.M., Sanjayan, M.A. and Sizemore, D. (1999) *A Conservation Area Design for the Central Coast Region of British Columbia, Canada*. Salt Lake City, UT: Round River Conservation Studies.

Kranjc, A. (2002) 'Conservation Biologists, Civic Science and the Preservation of BC Forests', *Journal of Canadian Studies*, 37, 3, pp. 219–238.

Latour, B. (1987) *Science in Action: How to Follow Scientists and Engineers Through Society*. Cambridge, MA: Harvard University Press.

Latour, B. (1988) *The Pasteurization of France*. Cambridge, MA: Harvard University Press.

Latour, B. (1993) *We Have Never Been Modern*. Cambridge, MA: Harvard University Press.

Latour, B. (2004) *Politics of Nature: How to Bring the Sciences into Democracy*. Cambridge, MA: Harvard University Press.

Latour, B. (2005) *Reassembling the Social: An Introduction to Actor-Network-Theory*. Oxford, UK: Oxford University Press.

Law, J. (1986) 'On the Methods of Long-Distance Control: Vessels, Navigation and the Portuguese Route to India', pp. 234–263 in J. Law (ed.), *Power, Action and Belief: A New Sociology of Knowledge?*. London: Routedge and Kegan Paul.

Law, J. (1994) *Organizing Modernity*. Oxford, UK: Blackwell.

Law, J. (2009) 'Actor Network Theory and Material Semiotics', pp. 141–158 in B.S. Turner (ed.), *The New Blackwell Companion to Social Theory*. Oxford, UK: Wiley-Blackwell.

McAllister, I., McAllister, K. and Young, C. (1997) *The Great Bear Rainforest: Canada's Forgotten Coast*. San Francisco, CA: Sierra Club Books.

MacArthur, R. and Wilson, E. (1967) *The Theory of Island Biogeography*. Princeton, NJ: Princeton University Press.

McCay, B.J. and Acheson, J.M. (1987) *The Question of the Commons: The Culture and Ecology of Communal Resources*. Tucson, AZ: University of Arizona Press.

Marchak, M.P. (1983) *Green Gold: The Forest Industry in British Columbia*. Vancouver: University of British Columbia Press.

Mol, A. (1999) 'Ontological Politics: A Word and Some Questions', pp. 74–89 in J. Law and J. Hassard (eds), *Actor Network Theory and After*. Malden, MA: Blackwell.

Murdoch, J. (1997) 'Inhuman/Nonhuman/Human: Actor-Network Theory and the Prospects for a Nondualistic and Symmetrical Perspective on Nature and Society', *Environment and Planning D-Society and Space*, 15, 6, pp. 731–756.

Murdoch, J. (2001) 'Ecologising Sociology: Actor-Network Theory, Co-Construction and the Problem of Human Exemptionalism', *Sociology*, 35, 1, pp. 111–133.

Ostrom, E. (1990). *Governing the Commons: The Evolution of Institutions for Collective Action*. Cambridge, UK: Cambridge University Press.

Pahl-Wostl, C., Craps, M., Dewulf, A., Mostert, E., Tabara, D. and Taillieu, T. (2007) 'Social Learning and Water Resources Management', *Ecology and Society*, 12, 2, art. 5.

Pahl-Wostl, C. and Hare, M. (2004) 'Processes of Social Learning in Integrated Resources Management', *Journal of Community and Applied Social Psychology*, 14, 3, pp. 193–206.

Pavlikakis, G.E. and Tsihrintzis, V.A. (2000) 'Ecosystem Management: A Review of a New Concept and Methodology', *Water Resources Management*, 14, 4, pp. 257–283.

Peet, R. and Watts, M. (1996) *Liberation Ecologies: Environment, Development, Social Movements*. London, New York: Routledge.

Pretty, J. (2003) 'Social Capital and the Collective Management of Resources', *Science*, 302, 5652, pp. 1912–1914.

Pretty, J. and Ward, H. (2001) 'Social Capital and the Environment', *World Development*, 29, 2, pp. 209–227.

Robbins, P. (2004) *Political Ecology: A Critical Introduction*. Malden, MA: Blackwell.

Saunders, D.A., Hobbs, R.J. and Margules, C.R. (1991) 'Biological Consequences of Ecosystem Fragmentation: A Review', *Conservation Biology*, 5, 1, pp. 18–32.

Schusler, T.M., Decker, D.J. and Pfeffer, M.J. (2003) 'Social Learning for Collaborative Natural Resource Management', *Society and Natural Resources*, 16, 4, pp. 309–326.

Weigand, J., Mitchell, A. and Morgan, D. (1992) *Coastal Temperate Rainforests: Ecological Characteristics, Stats and Distribution Worldwide*. Portland, OR: Ecotrust and Conservation International.

Whatmore, S. (2002) *Hybrid Geographies: Natures, Cultures, Spaces*. London: Sage.

White, D.F. (2006) 'A Political Sociology of Socionatures: Revisionist Manoeuvres in Environmental Sociology', *Environmental Politics*, 15, 1, pp. 59–77.

Willems-Braun, B. (1997) 'Buried Epistemologies: The Politics of Nature in (Post) Colonial British Columbia', *Annals of the Association of American Geographers*, 87, 1, pp. 3–31.

Wilson, J. (1998) *Talk and Log: Wilderness Politics in British Columbia, 1965–96*. Vancouver: University of British Columbia Press.

26
Land use tensions for the development of renewable sources of energy

Giorgio Osti

Increasing use of renewable sources of energy (RSEs) creates new pressures and tensions in relation to land use. For much of the industrial age, the dominance of fossil fuels has focused land pressures on extraction fields and on those places where energy infrastructures are built. However, novel sources of energy, including renewable sources, require more space, more surface (Nonhebel 2003). The quantity of land occupied varies according to the energy source but the net result appears to be an increase in the pressure on open, less densely populated, low productive areas. For the most part, these are rural areas. For example, the generation of wind energy on a significant scale requires open sites, preferably at some distance from houses and other human activities. Activities such as grazing or cropping can be undertaken near wind turbines, but with some limits. Offshore wind parks place constraints on shipping and fishing. Photovoltaic panels are usually located on roofs but, in some cases, they are located in fields where it becomes impossible to cultivate crops. Biogas and biodiesel need dedicated crops that enter into competition with food crops. Water for hydroelectric plants require large watersheds or, at least, the channelling of streams and rivers. Geothermal energy, by contrast, appears to have less impact on the land. However, as with fossil fuels, the associated generation and transmission infrastructure may have land use impacts in places previously unexposed to energy infrastructure.

The issue is not solely a quantitative matter of how much land is dedicated to renewable energy exploitation. The reorganization of land use for renewable energy production has at least three consequences. First, environmental balances are modified. The biodiversity of agricultural enterprises dedicated to the production of food crop, for example, may be richer than those dedicated to the production of fuel crops. The converse may also be true. Second, social well-being is modified. New opportunities for income and employment may arise for local populations, but so too may conflicts over the desirability of particular land uses, the identities of communities and so on. People tend to oppose the location of large power plants near their residences regardless of whether those plants are fed with fossil fuels or biomass. Energy self-provision from micro power plants is seen as more desirable by many. Third, new opportunities for companies and new regulatory challenges for governments arise. The new energy sources must be directed by authorities in order to assure equal and efficient exploitation. Oftentimes, energy exploitation has created negative tradeoffs between social equity and economic development.

Social analysis of renewable sources of energy can help to understand the new trends in land use. While land use changes associated with renewable energy have been concentrated, to date, on relatively marginal lands, conflict over changes is becoming more visible. As demand for renewable energy grows we can expect to see escalations in land use demand, conflict and regulation. It is argued here that, in order to understand the land use impacts of renewable sources of energy, we need to develop an appropriate analytical framework and then to apply this to specific cases. The following section will develop such a framework, which will then be applied to cases taken from rural areas, with particular attention to Italy. These are broadly representative of the situation across Western European countries.

A framework for the spatial dimension of energy

Energy is a basic factor in landscape change. The development of technology to exploit fossil fuel energy was fundamental to the growth of easy mobility, the large metropolis and the specialized agriculture we know in our times. People can live far from their workplaces and they can import food from any corner of the world. The landscape has been modified by the enormous development of cheap and rapid transport. Energy (especially oil) is the root of what Russett (1967) terms the 'mobiletic revolution'. The expansion of mobility can be seen as a chance for individual freedom; a physical reworking of the liberal revolution and its notion that every subject ought to be free to find its own right position in society. The final consequence of this vision is the deterritorialization hypothesis: individuals being so free to move that physical borders disappear (Tomlinson 1999). Barriers resist as residuals of a past that will be overcome. However, we know that deterritorialization is relative, because new spatializations continuously raise (see Lefebvre 1974).

Energy, nevertheless, continues to mould space. Energy generation and distribution shifts the urban–rural balance of modern societies, with urban deconcentration (suburbanization and sprawl) on one side and the territorial overspecialization of agriculture (monoculture) on the other. Another explanation for the spatial specialization of agriculture comes from the theory of comparative advantage (Carter and Zhong 1991). This theory is used to justify the freedom and convenience of trading food throughout the world, mirroring liberal ideological endorsement of mobility and territorial specialization and linking these to capital accumulation. Capital is the strongest force in the modern world; it has an intrinsic tendency to grow in the hands of private subjects. Land becomes a means of production subjected to the same tendency (Harvey 2005).

Energy is thus also an exceptional means of accumulation; a means so strong that it can span the different phases of accumulation of capital (Podobnik 2006). This is not an argument for technological determinism (as we find in some histories of energy discoveries); on the contrary, energy remains a means. Its ownership and the capacity to manipulate it according to production relationships determine the level, the speed and the inequality of capital accumulation.

The explanatory sequence suggested by these arguments is the following:

capital accumulation → energy manipulation → spatial change

In fossil fuel economies, new energy landscapes arise as energy intensive and/or petrochemical-based industries cluster around oil refineries and other sources of energy (Ghosn 2010). Rural landscapes more generally are reshaped as industrialized agriculture substitutes human and biological processes with fossil fuel-derived energy, fertilizer and chemical inputs. Urban landscapes are transformed by the private motor vehicle, but less directly affected by other aspects of energy supply, as most kinds of energy are distributed by underground grids (gas and electricity).

Through each of these landscapes, points of concentration are evident in refineries and power plants. The energy landscape is like a network in which multiple components are indelibly linked to larger nodules constituted by power and distribution plants (in Italian, these nodules are properly called *centrali*). The characteristics of refineries and power plants thus become crucial in the distribution of economic and political power. Concentrated energy infrastructure requires substantial investment but with the capacity to generate equally substantial profits. Energy companies provide a fundamental resource for economic and civil activity. They have been a privileged source of accumulation for generations of capitalists. And yet, complicating this picture (and its land impacts), some components of energy provision have traditionally been in the hands of the state and distribution in the hands of local authorities.

In some countries such as France, the energy industry is largely controlled by the state. State agencies or enterprises either monopolize the energy industry or control major companies through majority ownership of shares. In other countries, there are strong energy authorities that regulate market competition. The energy world is thus better represented as a diarchy of state-market than as either an entirely market or state-based system. Of course, we have national variations according to a combination of specific factors (legal tradition, internal availability of energy sources, government policies; see Helm 2004). The diarchy is easily related to institutional approaches (Jacobsson and Bergek 2004; Wolsink 2000) and to literature on different types of capitalism (Amable 2003). The question relevant to this chapter is: What kinds of energy landscape are produced by such diarchies?

The active involvement of the state in the energy sector implies a stronger distributive effort than may have been expected in an entirely private and profit-driven energy system. Large public power plants located in proximity to urban-industrial districts feed a network of energy distribution infrastructures that spread through every corner of a country. In Italy, as in other countries like the USA, considerable state efforts were devoted in the mid-twentieth century to rural electrification. The imperative was to cover the entire national space with at least basic energy services. The very idea of rurality has subsequently become associated with images of electric poles lining dusty country roads. This project of electrification was followed, in Italy, by a second wave of nation-building energy distribution with the construction of a natural gas grid through the combined action of a national monopolist and hundreds of municipal utilities.

With the intervention of the state, the energy landscape becomes more uniform. Energy infrastructure is used as an instrument of territorial balance or spatial equality. Resource extraction and power generation plants are built in peripheral areas in order to increase local employment rates. At the same time, their placement in less developed areas is a way to avoid NIMBY ('Not in my backyard') reactions in more politically organized urban communities. Power plants in underdeveloped areas are seen as a sort of 'white knight', rescuing the culturally backward and economically peripheral rural dwellers from their miserable destiny (see Hanley and Nevin 1999; Rozakis et al. 1997). However, the construction of large resource extraction and power generation infrastructure in less dense areas also means the construction of equally large energy transport infrastructure: roads, railway lines, high voltage electric lines, gas or oil pipelines, regasification terminals, ports and so on. Such landscapes, however, with their slim networks and large nodules represented by the refineries and the power plants, have been transformed more recently by: (1) the partial deregulation or liberalization of energy production and markets; and (2) the search for new, notably renewable, sources of energy.

These two processes have stimulated a proliferation of energy operators and the dissemination of power plants through space. In Italy, this has been most evident in the growth of privately owned and operated natural gas power plants. Yet it would be misleading to conclude that energy sector unbundling, or deregulation, has simply produced a new energy oligopoly. The energy arena, in fact, becomes more crowded (see Toke 2000). It may still be represented as a network, but the network is now flatter and thicker with more nodules of variable strength. Numerous enterprises are attempting to enter the energy business with considerable variation both in their size and influence and in the energy sources and technologies they deploy.[1] The energy arena also becomes populated, as a consequence, with a variety of different socio-technical systems that are only partially integrated (Foxon 2007; Walker and Cass 2007). These tendencies, and the tensions they are generating, will be illustrated through a more detailed examination of the Italian case. Particular attention will be paid to renewable sources of energy, as it is highly likely that these will come to occupy more space, especially in rural areas.

RSE diffusion in rural areas

The underlying idea here is that, in the energy landscape, rurality matters. To put this another way, the conception and organization of spaces deemed 'rural' has a significant influence on the development and dissemination of new sources of energy, renewable sources in particular. The 'rural' is a multifold concept. It may signify low population densities, economic dependence on primary production (agriculture, forestry and livestock breeding), particular ways of life, or simply the residual spaces and communities of the 'non-metropolitan'. For some, rurality thus signifies less engagement with modernity, lower levels of income and economic development and, perhaps, a more impalpable cultural lag in terms of education, entrepreneurship, self-esteem and so on.

In order to understand the influence of rurality on RSEs, this chapter distinguishes between the rural, as a specific kind of residential community, and agriculture, as an economic sector. This distinction is important because agriculture in developed countries has been subject to rapid processes of industrialization and specialization that have weakened both its social and its environmental base (see Lotti 2010; MacCannell 1988; Stofferahn 2006). Farming has become a self-contained, vertically integrated industrial sector and it has lost or degraded, in the process, many of its links to soil and to communities (Osti 2010; Ploeg 2008). The disembedding of agriculture from locally distinct agro-ecologies can be evaluated in different ways. The important point for the purposes of this chapter is that the distinction between rural communities and agriculture is both plausible and useful in framing the different forms of energy landscape associated with RSEs.

Figure 26.1 proposes a typology in which to locate RSEs in rural contexts. The two dimensions of the typology reflect the relative importance of rural communities and agriculture to distinct rural spaces. Located, then, within the four cells of the typology are various forms of renewable energy generation and use most likely to be associated, it is suggested, with the resultant landscape. The scheme is proposed to help understand which kinds of RSE are likely to be developed within distinct rural areas, and/or the ways in which rural landscapes are likely to be reshaped by RSEs. Is this theoretically derived schema then reflected in actual cases?

According to data collected in Northern Italy, there is no favourable or unfavourable bias for the diffusion of photovoltaic panels in rural areas. Rather, diffusion of photovoltaic panels through this most densely populated part of Italy seems linked to local systems with either a more advanced cultural sensibility for solar energy (South Tirol and Trentino) or proximity to small photovoltaic producers and installers (Venetia region). The broad situation in Northern Italy is thus not interpretable along any rural–urban continuum or cleavage but according to the strength of these local systems. Similar results are evident in Germany (Rode and Weber 2011). Examining some rural areas in more detail such as the axis along the River Po (traditionally dominated by rural communities and agriculture) reveals large installations of photovoltaic

Land use tensions for the development of renewable sources of energy

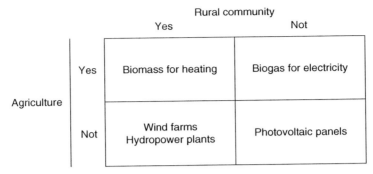

Figure 26.1 Framework of renewable energy source development in rural areas

panels (or solar farms). Big companies have bought or rented cropland for installing panel fields of more than 100 kilowatt capacity, generating tension over land use and the potential impacts on agriculture of further photovoltaic development. Domestic photovoltaic systems of 3–5 kilowatts do not generate such tension.

Wind energy development is almost exclusive to the rural areas of Southern Italy. Wind farms are concentrated on the ridges of mountain and hills, far from urban settings that are concentrated on the coast. This kind of spatial arrangement, as suggested by Figure 26.1, places wind farms in proximity to rural communities but in ways that have little to do with agriculture. Occasionally, wind turbines are placed on a farm property for legal convenience, i.e. to speed up the authorization process. Farmers also rent land for the installation of wind farms, but they do not have any further involvement in the operation of the turbines. Electricity is generated not for the use or benefit of local farmers or communities but is transferred to distant markets through high voltage grids. Wind farms are installed by multinational companies whose only aims are to supply these markets and receive subsidies. They have no interest in agriculture and their activities are almost completely detached from local farming.

At the same time, however, some indirect advantages accrue to local communities through the royalties wind companies pay to landowners. Such companies may also seek to make more direct contributions through, for example, the provision of public services (Osti 2012; see also the US case: US Department of Energy 2004). Nevertheless, land use tensions arise where wind farms are incompatible with other land uses and industries. Wind farms have been located, for example, in places where recently eco- and agro-tourism experiences have been created. Wind turbines compete with these alternative pathways of local development, the royalties or the land rent for turbines appearing more attractive to some than the less certain prospects of tourism.

Tensions also arise between wind farming and landscape and environmental protection associations. High capacity turbines (approximately 1 megawatt capacity) require towers 60–70 metres high and create, as a consequence, substantial visual impacts. Environmental associations are often divided over whether to support wind farms, for their 'clean' energy production, or to oppose wind farms, for their impacts in relation to visual amenity, noise, etc. Opposition generally is weaker in Southern Italy, where the local population is poorer. Opposition is also generally less pronounced to smaller, or micro, wind turbines, but micro systems have not spread even in the windy regions of Southern Italy (Battisti 2008; CO.SVI.G 2009). According to some experts, only large turbines are efficient (De Decker 2009). This is probably true in some respects, but it is a restricted vision of efficiency that does not consider transmission costs and local impacts. Both small and large wind turbines receive government subsidies in Italy, but it would appear that the dissemination of small systems is disproportionately affected by institutional obstacles, including the absence of certification for small turbines, high cost of wind measurements, delays in connection to the grid and fiscal limits for farmers (Misuraca 2011).

Biogas for electricity is the third form of RSE to be evaluated against the rural community-agriculture typology. Biogas is produced through the fermentation of various kinds of biomass. Such gas generally is then burned and the heat produced used to drive power turbines. There is a clear spatial pattern to the distribution of biogas power plants in Italy. Such plants are concentrated in the North, where there is greater access to the effluent from livestock farms and to corn for use as feedstock. According to CRPA (2010), in 2010 there were 16 biogas plants in Southern Italy, 10 in Central Italy and about 250 in Northern Italy.

This kind of RSE is classified as 'agriculture yes – rural community not' because the biogas-electricity process is conceived by farmers as a pure source of income for their farm. The electricity produced is sent to the grid (which is not locally controlled), while the heat produced by the turbines is dispersed or, at best, partially used on the farm. Biogas-electricity production does not strengthen relationships with the local community. On the contrary, it contributes to further isolation of agriculture from rural communities. In fact, in many cases, local populations are hostile to biogas plants (Magnani 2012), particularly when these are large (more 1 megawatt capacity), pushing farmers to locate plants as far as possible from residential areas.

Smaller biogas-electricity plants (about 250–500 kilowatt capacity), combined with pig or cattle stock-breeding, encounter less opposition from rural communities as they process animal effluent that may otherwise pose environmental, amenity and health problems. Regulatory controls on nitrate pollution also help to encourage small-scale integrated animal production and biogas-electricity operations (Carrosio 2011a). Early biogas operations were thought of as systems for collecting and treating what would otherwise be the wastes, and potential pollutants, generated by farming and related processing activities such as abattoirs. However, incentives for renewable electricity production encouraged the rethinking of biogas as a closed activity, a form of commodity production in its own right. Land use tension has arisen where, for example, fertile fields of corn are used for producing biogas rather than for feeding people or animals. A 1-megawatt biogas plant requires at least 300 hectares of corn. Animal manure is not suitable for electricity production on this scale due to its low energy power. Even a large livestock farm with thousands of pigs is not able to meet the appetite of a 1-megawatt biogas digester. Livestock farmers attempting to operate large biogas plants must intensify their operations, increasing herd size and buying the food for their animals in the market.

Demand for corn or other cereals affects farm crop ordering and rotation. It pushes farmers towards monoculture and thus undermines attempts to encourage diversification (biological and economic) through concepts such as multifunctionality. In Europe, at least, diversification has been seen as an intelligent policy response to the farm cost-price squeeze and the reduction of public subsidies for agriculture. Monocultural farming operations have limited opportunity to participate in or contribute to virtuous circles of sustainable agro-ecologies and economic diversification through agri-tourism and/or the production of higher value foods (such as traditional local foods recognized with Protected Designation of Origin or Protected Geographical Indication status). In other countries, including Germany, the situation is similar, with more pronounced land use conflict surrounding crop production for industrial-scale biodiesel production than surrounding small biogas plants (Thrän and Kaltschmitt 2007).

The last type of RSE in the typology is represented by the use of biomass, mostly wood, to produce heat and, less frequently, cooling. World statistics on the extent of biomass use are unclear, because the traditional use of wood in domestic stoves is not adequately accounted for. In any case, a great difference between the Global South and the Global North emerges. In the less developed countries, 'fuelwood and charcoal is driven primarily by growing number of rural poor' (Matthews 2004). Nevertheless, this form of heating is typical of European rural areas, too. Moreover, in recent years there has been a rediscovery of this kind of heating for reasons that deserve to be listed. First, the high cost of fossil fuels has encouraged many householders to maintain or reinstall a biomass-burning stove. Second, technological advances allow householders: (a) to install higher performance stoves that, for example, recover and burn smoke; (b) to integrate these stoves into a central heating system for the whole house; and (c) to automate these systems with electronic controls and pellet-based fuel. Third, wood stoves have lost their image as backward domestic devices. Modern systems are not only more efficient and effective; the visible flame in many stoves creates an agreeable atmosphere sought by educated and high-class people.

In our typology, biomass production for warmth is considered relevant where either agriculture or rural communities have a role. We have at least three situations when this occurs. The first is the simple wood self-provision of people living near woodland. Very often these people are not full-time farmers but simply those with opportunity and right to harvest wood. In some rural regions residents are entitled every year to a fixed quantity of wood from a common forest. Wood to burn in a stove or central heating system is cheap and easily collected. In such areas, local people have the tools for farming small plots and for collecting wood. The second situation arrives when forestry is organized in an economic way. Very often the activity is driven by farmers, according to a multifunctional scheme that includes crop cultivation, livestock breeding and forestry. Wood can be used for timber, heat or both, depending on its quality and on local processing facilities, markets, etc. In some cases, there are sawmills that produce sawdust used as fuel in special stoves. In other words, in some rural areas there is a wood chain, whose rings are forestry, timber manufacturing and building heating. In Italy, the timber manufacturing industry has become very dependent on timber imports from large producers, like Russia and Finland, but usually is located in rural areas.

The third positive combination between biomass, agriculture and rural community occurs when a wood-heating district is created. The wood fuel comes from sawmills, forestry and residuals of orchards and vineyards. Wood burning is centralized in a large boiler and hot water, then pumped through a grid into nearby buildings. Such grids can measure tens of kilometres. In the Italian (or South) Tyrol there are more than fifty heating districts. The circuit between the wood economic activity and the heating service to rural people is virtuously established. Local people are involved as input providers (wood and its byproducts) and as output receivers (heat or hot water). Farmers are the key actors, even if in a multi-dimensional fashion. In that sense, wood for heating is located in the conjunction between agriculture and rural community.

However, there are land tensions also for this apparently successful case. Heating districts were created in order to reuse abandoned forests and revitalize forestry. The more systematic exploitation of wood that followed created new jobs in forestry and heating system maintenance. But it also opened new conflicts with nature conservation activists who had seen, in the semi-abandoned woods, opportunities to enhance biodiversity. Larger heating systems, not surprisingly, have a larger spatial footprint and generate correspondingly larger impacts. For some, local biomass is not sufficient. Short rotation forestry has been introduced to some farming areas, creating competition with food and feed crops (Bringezu et al. 2012). In others, wood chips or pellets are imported from as far away as Brazil and Indonesia (GSE 2009), severing links between biomass-based energy and agriculture altogether. The local community is able to sustain its way of life but potentially exports environmental impacts through the indirect effects of energy practices on land use elsewhere (Fritsche et al. 2010; Kitzes et al. 2009). They are also exposing themselves to the risk of burning pellets that have been manufactured with recycled contaminated wood.

Giorgio Osti

The explanation of RSE diffusion in rural areas

As outlined at the beginning of this chapter, the diffusion of renewable sources of energy in rural areas may be understood from a variety of theoretical perspectives: geographical determinism, structural explanations, state-market diarchy and governance/networks. The geographical characteristics of rural areas matter, but they are not the sole determinants of RSE diffusion. For example, the diffusion of photovoltaic panels (in terms of the per capita number of plants) across Europe bears little relation to the intensity of sunlight and, therefore, the abundance of the solar resources. This is illustrated by the greater photovoltaic capacity of Germany relative to Italy. The geography of biogas plant locations, similarly, does not fully respect the distribution of land biomass capacity.

The case of biomass energy in non-European countries may appear to confirm the structural role of profit accumulation and agricultural specialization (Mol 2007). For the Italian case examined here, however, it is clear that such structuralist explanations are too rough. Particular kinds of biomass energy (bioethanol and biodiesel, in particular) used in Italy are produced in, and for, global markets dominated by multinational companies (Carrosio 2011b). However, markets for biogas and wood for burning are more spatially and socially segmented, making it more difficult for a restricted elite of capitalists to accumulate profit through their systematic exploitation. Segmentation is linked to, but not completely derived from, the action of governments (Dewald and Truffer 2009). Institutional support and public incentives have promoted an acceleration of RSE exploitation since a decision by the national government, following the model of Germany, to introduce incentives (Jacobsson and Lauber 2006). In 2005, the first real incentive for RSEs was introduced in the form of the *Conto Energia*, a 'double sustain' which offered a feed-in tariff and net-metering for solar energy (EU Directive 2001/77/CE; see also Meyer 2003). For biogas-power plants, the main incentive was introduced in 2007 and consisted of a generous all-inclusive tariff (0.28 Euro per kilowatt hour) for power plants of less than 1-megawatt capacity. For wind farms, the main incentives include both favourable prices and tradable 'green certificates' (also known as Renewable Energy Certificates), which certify that energy has been produced from allowed renewable sources. The latter were introduced before the 'renewable wave' commencing around the year 2000 and responded to a problem of regulation for electric energy produced in waste incinerators. Green certificates only apply to large power plants.

Also relevant to providers of power and gas in Italy are certificates of energy efficiency ('white certificates'), which may be exchanged between those plants that are more efficient and those that are less. For the purposes of this chapter and its concern with rural RSEs, it is important to note that these incentives do not apply to heat from wood-fuel plants – even large boilers feeding heating districts. What was identified above, therefore, as arguably the most appropriate source of energy for rural areas (wood) is the only one not to receive any national incentives!

The absence of incentives for wood-energy systems shows the limits of an institutional explanation largely based on the action of governments, more or less stimulated by the European Union. Especially for the medium and large wood-fuel heating plants, other institutions and incentives have been important. These are best represented as 'institutionalized networks' (Dawkins and Colebatch 2006; Price Boase 1996). Institutions still matter, but unlike the national system of incentives relevant to other RSEs it is local institutions that are relevant to the role of local networks. Local institutions here include municipalities and provinces with special statutes of autonomy and spatial reaches. Local networks are usually professional groups such as lumberjack and farmer associations, community cooperatives, etc. In some rural contexts, such as the aforementioned Autonomous Province of South Tyrol, a consensus between local authorities, locally placed economic interests (tourism, forestry, farming, etc.) and households produced the rapid construction of heating districts. In other parts of Italy that are not as ethnically distinct (in the Italian or South Tyrol, the majority of the population speaks German), the mechanism is similar, even if the plants are usually smaller. There is a public commitment and an enduring agreement among local operators.

These are institutionalized networks in the sense they are groupings of private and semi-private organizations to whom a public service is entrusted. This is not in itself unique, as legal frameworks exist in Italy and the EU for the contracting out of public services such as waste and water management. The de facto difference is that the service provider in wood for heating rural systems is generally a kind of not-for-profit organization such as a consumer cooperative or a professional group. The term 'kind of' is used because the borders of the not-for-profit world are not clear, despite many laws that try to define them. In this case, they are an expression of the interest of many small producers. They are a network officially acknowledged by the public institution and delegated to carry out tasks that have relevant public aspects. RSE provision and fruition could be a totally private service, but because of its environmental meaning (saving energy, reducing CO_2, local self-provision) it has a public side that is conveniently managed by a not-for-profit, semi-public organization. This quasi-public mark appears in some biogas-electricity plants as well. In such cases, the plant owner is a farmers' cooperative and its members have signed an agreement for supplying a biogas digester with a fixed amount of their crops. The network works more as a private firm than as a not-for-profit organization, but the border between the two conceptions is thin.

To sum up the interpretations of RSE development in rural areas we could say, first, that energy as a source of profit accumulation explains well the limited development in Italy of biofuels like ethanol and biodiesel. Second, the idea of a state-market energy diarchy explains the rise of many biogas-power plants, large solar farms located on arable land, and wind farms (hydropower dams before them) in marginal areas. The state-market diarchy is symbolized by the rich incentives given indiscriminately to everyone who has installed a plant supplied with renewable sources. The first generation of incentives advantaged large external investors in rural areas that were without financial resources and self-organizing capacity. Third, institutionalized networks (the RSE drive depends on grassroots organizations, acknowledged and supported by local authorities) explain better the development of wood for heating systems, mostly placed in the mountains and a variety of small RSE power plants.

Conclusions

Conclusions are necessarily provisional. With the exception of hydropower plants, large-scale diffusion of RSEs in rural areas only began a few years ago and governments are uncertain over whether to provide continuing support for them or not (Lauber and Schenner 2011). Moreover, there is a great variety of experiences that are not easily classified. It is symptomatic that for every combination of model and case illustrated by Figure 26.1 there is an exception. Examples of such exceptions include: biofuel produced by a small factory and used by the local municipality to power public transport; a wind turbine installed by another municipality for supplying electric energy to a nearby eco-village; and a wind farm promoted by a public–private company with hundreds of citizens living nearby as members (see, in other European countries, Dinica 2008; Toke 2005). A further exception – in this case, a negative example – is evident in heating districts that are managed by energy municipal companies with no involvement of local people. These companies establish purely commercial relationships with individual consumers, sometimes using heat from waste incinerators. This latter exception usually concerns towns or entire regions rather than exclusively rural areas. This allows for the identification of a final consequence of RSE diffusion in the countryside. Rurality is mainly a descriptive concept; it does not imply any aspect of self-organization of local communities. Undoubtedly, some communities are more organized in relation to the opportunity afforded by RSEs. Others are more subjected to external forces. Usually, European rural areas, because of their scattered and elderly populations, suffer more external pressures than towns and industrial districts.

In conclusion, rurality, when combined with other social categories, becomes an appropriate tool for framing the rise of renewable sources of energy. Spatial arrangements vary, but a rural pattern emerges nevertheless, at least in Italy. Valley morphology, the relative strength of farmers and a marked cultural identity are the conditions for integrated RSE development. On the contrary, open/plain spaces, elderly farmers and a weak local identity produce a countryside colonization, even if through the RSEs. These 'dark and light' results of RSEs are quite common in Western Europe (Walker et al. 2007). A possible explanation of this ambivalence can be found in the landscape, conceived as an awesome combination of space, culture and nature. When RSEs are included in a local image of space (or landscape), they have chances to be accepted and, more important, to be locally appropriated. On the contrary, when rural people have not any image of their territory, when farmers are not seen as champions of landscape, even RSEs result as a menace and easily appropriated by external forces.

To date, the European rural landscape has not been drastically changed by the rise of RSEs and land use conflicts are still small in scale. But pressure on the countryside is destined to grow as fossil fuel resources become more difficult and expensive to use. Some rural areas seem better equipped to face the coming change. They have not only the capacity to control land use at the local level but also to organize local interests, householders included. Householders or energy consumers are probably the most neglected stakeholders in RSE development. The tightening relationships between producers and consumers, evident in numerous alternative food networks, are not yet common in rural RSEs, but the path should be similar: to exchange in a more conscious and equal way the resources of each place as they are transformed into new energy landscapes. A new exchange on energy provision–consumption can help to diversify rural areas' economy and culture, overcoming their traditional flattened conformity. RSEs are now surely a source of diversification of European countryside. Even in the most colonized situations, the variety is growing and, with it, employment opportunities and services. Further analysis is needed to understand the quality of these jobs, the secondary effects on the environment and the vulnerability of rural communities that generously host the new sources of energy.

Note

1 The global governance literature captures elements of the new energy landscape (Goldthau and Witte 2010; Jegen 2009; Kemp et al. 2007), even though the emphasis is on actors' involvement and not-for-profit semi-public organizations (Horst 2008).

References

Amable, B. (2003) *The Diversity of Modern Capitalism*. Oxford, UK: Oxford University Press.

Battisti, L. (2008) 'Minieolico', pp. 263–274 in L. Pirazzi (ed.), *Nuove vie del Vento*. Rome: Franco Muzzio.

Bringezu, S., O'Brien, M. and Schütz, H. (2012) 'Beyond Biofuels: Assessing Global Land Use for Domestic Consumption of Biomass: A Conceptual and Empirical Contribution to Sustainable Management of Global Resources', *Land Use Policy*, 29, 1, pp. 224–232.

Carrosio, G. (2011a) 'Energy Production From Biogas in the Italian countryside: Policies and Organizational Models', presented to *Renewable Energy Governance in China and the EU (RenErGo) Workshop*, Wageningen

(NL), 22–23 May 2011. Available at www.wageningenur.nl/en/show/RenErGo.htm. Accessed 28 January 2013.

Carrosio, G. (2011b) *I Biocarburanti. Globalizzazione e Politiche Territoriali*. Rome: Carocci.

Carter, C.A. and Zhong, F. (1991) 'Will Market Prices Enhance Chinese Agriculture? A Test of Regional Comparative Advantage', *Western Journal of Agricultural Economics*, 16, 2, pp. 417–426.

CO.SVI.G (Consorzio per lo Sviluppo delle Aree Geotermiche) (2009) *Lo Sviluppo del Mini e del Micro Eolico Nella Regione Toscana*, Monterotondo Marittimo, Settembre, Collana CITT Formazione n. 3. Available at www.distrettoenergierinnovabili.it/der/citt/ruolo/fabbisogni-formativi-nel-settore-eolico.pdf. Accessed 28 January 2013.

CRPA (Centro Ricerche Produzioni Animali) (2010) L'Agricoltore Crede nel Biogas e i Numeri lo Confermano. *L'Informatore Agrario*, 65, 30, pp. 63–71.

Dawkins, J. and Colebatch, H.K. (2006) 'Governing Through Institutionalized Networks: The Governance of Sydney Harbour', *Land Use Policy*, 23, 3, pp. 333–343.

De Decker, K. (2009) 'Small Windmills Put to the Test', *Low-Tech Magazine*, 19 April. Available at www.lowtechmagazine.com/2009/04/small-windmills-test-results.html. Accessed 28 January 2013.

Dewald, U. and Truffer, B. (2009) 'Markets and Space in Technological Innovation Systems: Diffusion of Photovoltaic Applications in Germany', *DRUID Summer Conference*, Copenhagen Business School, 17–19 June. Available at www2.druid.dk/conferences/viewpaper.php?id=5742&cf=32. Accessed 28 January 2013.

Dinica, V. (2008) 'Initiating a Sustained Diffusion of Wind Power: The Role of Public–Private Partnerships in Spain', *Energy Policy*, 36, 9, pp. 3562–3571.

Foxon, T.J. (2007) 'Technological Lock-in and the Role of Innovation', pp. 140–152 in G. Atkinson, S. Dietz and E. Neumayer (eds), *The Handbook of Sustainable Development*. Cheltenham, UK: Edward Elgar.

Fritsche, U.R., Hennenberg, K. and Hünecke, K. (2010) *The 'iLUC Factor' as a Means to Hedge Risks of GHG Emissions from Indirect Land Use Change*. Darmstadt: Öko-Institut. Available at www.oeko.de/oekodoc/1030/2010-082-en.pdf. Accessed 28 January 2013.

Ghosn, R. (2010) *New Geographies, 2: Landscapes of Energy*. Cambridge MA: Harvard University Press.

Goldthau, A. and Witte, J.M. (eds) (2010) *Global Energy Governance: The New Rules of the Game*. Berlin: Brookings Institution Press/Global Public Policy Institute.

GSE (Gestore Servizi Energetici) (2009) *Rapporto Statistico: Biomasse*. Rome; Gestore Servizi Energetici.

Hanley, N. and Nevin, C. (1999) 'Appraising Renewable Energy Developments in Remote Communities: The Case of the North Assynt Estate, Scotland', *Energy Policy*, 27, 9, pp. 527–547.

Harvey, D. (2005) *The New Imperialism*. Oxford, UK: Oxford University Press.

Helm, D. (2004) *Energy, the State, and the Market: British Energy Policy Since 1979*. Oxford, UK: Oxford University Press.

Horst, van der D. (2008) 'Social Enterprise and Renewable Energy: Emerging Initiatives and Communities of Practice', *Social Enterprise Journal*, 4, 3, pp. 171–185.

Jacobsson, S. and Bergek, A. (2004) 'Transforming the Energy Sector: The Evolution of Technological Systems in Renewable Energy Technology', *Industrial and Corporate Change*, 13, 5, pp. 815–849.

Jacobsson, S. and Lauber, V. (2006) 'The Politics and Policy of Energy System Transformation: Explaining the German Diffusion of Renewable Energy Technology', *Energy Policy*, 34, 3, pp. 256–276.

Jegen, M. (2009) 'Swiss Energy Policy and the Challenge of European Governance', *Swiss Political Science Review*, 15, 4, pp. 577–602.

Kemp, R. Loorbach D.A., and Rotmans, J. (2007) 'Assessing the Dutch Energy Transition Policy: How Does it Deal With Dilemmas of Managing Transitions?' *Journal of Environmental Policy and Planning*, 9, 3–4, pp. 315–331.

Kitzes, J., Galli, A., Bagliani, M., Barrett, J., Dige, G. and Ede, S. (2009) 'A Research Agenda for Improving National Ecological Footprint Accounts', *Ecological Economics*, 68, 7, pp. 1991–2007.

Lauber V. and Schenner, E. (2011) 'The Struggle over Support Schemes for Renewable Electricity in the European Union: A Discursive-Institutionalist Analysis', *Environmental Politics*, 20, 4, pp. 508–527.

Lefebvre, H. (1974) *La Production de l'espace*. Paris: Anthropos.

Lotti, A. (2010) 'The Commoditization of Products and Taste: Slow Food and the Conservation of Agrobiodiversity', *Agriculture and Human Values*, 27, 1, pp. 71–83.

MacCannell, D. (1988) 'Industrial Agriculture and Rural Community Degradation', pp. 15–75 in L.E. Swanson (ed.), *Agriculture and Community Change in the US: The Congressional Research Reports*. Boulder, CO, Westview Press.

Magnani, N. (2012) 'Exploring the Local Sustainability of a Green Economy in Alpine Communities', *Mountain Research and Development*, 32, 2, pp. 109–116.
Matthews, E. (2004) 'Undying Flame: The Continuing Demand for Wood as Fuel', pp. 37–38 in R. Thornton, C. Starbird and S. Ertenberg (eds), *World Energy: Empowering the Future*. Centre for Teaching International Relations, University of Denver.
Meyer, N.I. (2003) 'European Schemes for Promoting Renewables in Liberalized Markets', *Energy Policy*, 31, 7, pp. 665–676.
Misuraca, L. (2011) 'Cosa Frena il Decollo del Minieolico in Italia', *Quale Energia*, 31, May.
Mol, A.P.J. (2007) 'Boundless Biofuels? Between Environmental Sustainability and Vulnerability', *Sociologia Ruralis*, 47, 4, pp. 297–315.
Nonhebel, S. (2003) 'Land Use Changes Induced by Increased Use of Renewable Energy Sources', pp. 187–202 in A.J. Dolman, A. Verhagen and C.A. Rovers (eds), *Global Environmental Change and Land Use*. Dordrecht: Kluwer.
Osti, G. (2010) 'The Chains of Knowledge in the Rice Production of the Po River Delta', *Rivista di Economia Agraria*, 63, 3, pp. 443–463.
Osti, G. (2012) 'Wind Energy and Rural Development in Italy', pp. 245–259 in S. Sjöblom, K. Andersson, T. Marsden and S. Skerratt (eds), *Sustainability and Short-Term Policies: Improving Governance in Spatial Policy Interventions*. London: Ashgate.
Ploeg, J.D. van der (2008) *The New Peasantries: Struggles for Autonomy and Sustainability in an Era of Empire and Globalization*. London: Earthscan.
Podobnik, B. (2006) *Global Energy Shifts: Fostering Sustainability in a Turbulent Age*. Philadelphia, PE: Temple University Press.
Price Boase, J. (1996) 'Institutions, Institutionalized Networks and Policy Choices: Health Policy in the US and Canada', *Governance: An International Journal of Policy and Administration*, 9, 3, pp. 287–310.
Rode, J. and Weber, A. (2011) 'Die Räumliche Diffusion von Photovoltaikinstallationen in Deutschland', in M. Helbich, H. Deierling and A. Zipf (eds), *Proceedings of the 19. Deutschsprachige Kolloquium für Theorie und quantitative Methoden in der Geographie*. Heidelberg: Geographische Bausteine, Selbstverlag des Geographischen Instituts, Universität Heidelberg.
Rozakis, S., Soldatos, P.G., Papadakis, G., Kyritsis, S. and Papantonis, D. (1997) 'Evaluation of an Integrated Renewable Energy System for Electricity Generation in Rural Areas', *Energy Policy*, 25, 3, pp. 337–347.
Russett, B.M. (1967) 'The Ecology of Future International Politics', *International Studies Quarterly*, 11, 1, pp. 12–31.
Stofferahn, C.W. (2006) *Industrialized Farming and Its Relationship to Community Well-Being: An Update of a 2000 Report by Linda Lobao*. Prepared for the State of North Dakota, Office of the Attorney General. Grand Forks, ND: University of North Dakota. Available at www.und.edu/org/ndrural/Lobao%20&%20Stofferahn.pdf. Accessed 28 January 2013.
Thrän, D. and Kaltschmitt, M. (2007) 'Competition: Supporting or Preventing an Increased Use of Bioenergy?', *Biotechnology Journal*, 2, 12, pp. 1514–1524.
Toke, D. (2000) 'Policy Network Creation: The Case of Energy Efficiency', *Public Administration*, 78, 4, pp. 835–854.
Toke, D. (2005) 'Community Wind Power in Europe and in the UK', *Wind Engineering*, 29, 3, pp. 301–308.
Tomlinson, J. (1999) *Globalization and Culture*. Chicago, IL: University of Chicago Press.
US Department of Energy (2004) *Wind Energy for Rural Economic Development*. DOE/GO-102004-1826. Washington DC: US Department of Energy.
Walker, G. and Cass, N. (2007) 'Carbon Reduction, "the Public" and Renewable Energy: Engaging with Socio-Technical Configurations', *Area*, 39, 4, pp. 458–469.
Walker, G., Evans, B., Hunter, S., Fay, H. and Devine-Wright, P. (2007) 'Harnessing Community Energies: Explaining and Evaluating Community-Based Localism in Renewable Energy Policy in the UK', *Global Environmental Politics*, 7, 2, pp. 64–82.
Wolsink, M. (2000) 'Wind Power and the NIMBY-Myth: Institutional Capacity and the Limited Significance of Public Support', *Renewable Energy*, 21, 1, pp. 49–64.

Index

accidents 63, 97, 129, 223, 241–3, 245, 248–50, 275
accumulation 32–3, 294–5, 299, 320–1, 326–7
acidification 234–6
acid rain 234
actions: political 58, 81; preventative 257
activists: environmental 58–9, 60, 63–6, 172, 273–4, 311, 325; human rights 59
actor-network theory (ANT) 4, 22, 192, 269, 308–10, 314
adaptation 2, 3, 6–8, 12, 18, 25, 72, 82, 87, 89, 101, 103–4, 107, 111, 113–6, 142–3, 145–9, 159, 161–2, 167, 172–3, 176, 208, 226, 254–6, 258–9, 261–3, 274
agency 4, 7, 73, 109–110, 116, 125, 128, 133–9, 144, 146, 192, 209, 272–4, 295–6, 302–3, 309
agriculture 10–1, 35, 38–9, 41, 47, 50, 56, 66, 73–4, 82–4, 86–7, 100, 282, 293–4, 296, 302, 319–20, 322–6
air pollution 8, 46–7, 129, 159, 190–1, 196, 232, 234–9
anthropology 61, 135–6, 200, 209, 288
anti-environmental movement(s) 2
Africa: sub-Saharan 25, 146; west 63
Agenda 21 259
Alberta, Canada 118, 122; tar sands 128, 227
Alier, Joan Martinez 58, 60, 293
Amazonian Indians 65
ammonia 47, 56, 235
Anthropocene 95–6, 103, 144, 254
Arctic 118, 120, 127, 143, 224, 228–9
Argentina 16, 62
Asia 16, 21, 24, 54, 118, 127, 142, 159, 161
austerity measures 37
Australia 1, 9, 16, 21, 70, 78, 110–1, 158–60, 164–6, 196, 242, 244; Montara oil field blowout 241–52; National Landcare Programme (NLP) 73–4

Barcelona 257
barriers 11, 39, 66, 75, 83, 159–60, 210, 244–5, 248, 295, 297, 320

Beck, Ullrich 18, 20, 22–3, 97–8, 121, 210, 213, 215, 237, 270, 286
Beijing 45
biodiversity 2, 10, 25, 40, 65, 71, 75–6, 78, 88, 95, 103, 173, 190, 226, 280, 282, 284, 287, 292–3, 296, 299–300, 319, 325; auctions 74; degradation 1; loss 1, 95, 104, 226
biogas 11, 319, 324, 326–7
biological invasion(s) 10, 280, 291
biology, synthetic 257
biomass 39, 280, 283, 319, 324–6
biophysical environment 3, 17, 208, 254, 259–60
blind faith 226, 229
body, the 135, 137, 259, 299, 301
Bolivia 64
BP (British Petroleum) 122, 162, 241–2
Brazil 16, 26, 64, 106, 108, 228, 325
British Columbia 11, 308, 310, 312
Brundtland report 17
bundle of services 77
Burkina Faso 64
Bush, George W. 160, 224
business studies 18
Buttel, Frederick 15, 23

Canada 11, 72, 99, 118, 122–3, 235, 241, 307–18
cap and trade 75, 113, 116
capacity building 16, 74–5, 84–5, 107
capitalism 3, 15, 17–8, 22, 31–3, 37–9, 40, 54, 60, 63, 78, 81, 85, 119–20, 123, 126, 176–7n, 181, 256, 276, 286, 294–5, 297–9, 303n, 304n, 308, 320–1, 326
carbon dioxide (CO_2) 36–8, 56, 102, 110, 157, 228–9
carbon dioxide equivalents (CDEs) 233
carbon 161–6, 223, 227–8; capture and storage 122, 124; pricing 86, 229; tax and trading 8, 111–4, 116, 176, 227, 229
Castells, Manuel 191–2
catastrophe(s); environmental 232

certification: criteria 6; standards 6
chemical enterprise 66
child labour 63
Chile 63
China 26, 45–57, 78, 106, 108, 116, 118, 120–2, 127
chlorofluorocarbons (CFCs) 225
circular economy 50
citizenship 62, 73–4, 96, 193, 199, 299, 302
civil society 19, 25–6, 39–40, 46, 51, 59, 64, 111, 116, 121, 126, 127, 179–82, 184, 217, 294
climate change 1–3, 6–9, 58, 65, 70–1, 86, 95–7, 101–16, 118, 121, 123, 125, 128–9, 133, 143–4, 146–7, 149–51, 157–69, 171, 180, 191, 196, 199, 222–25, 229, 233, 257, 280, 282, 303n; anthropogenic 1, 70, 76, 79, 97, 101, 103–4; global 2, 12, 21, 36, 78
climate governance 6, 8, 102, 104, 106–7, 157–62, 167, 170
climate justice 1, 108–9
climate modeling 6, 96, 101–4
climate policy 103, 106, 108, 162, 215
climate politics 6
climate refugees 7, 147, 150
climate regimes 6, 8, 107, 109–11, 116, 170–1
climate risk 1
coal 48, 50, 111, 118, 121–22, 124–5, 227–8, 234, 238–9
co-evolution 2–3, 12, 72, 96, 101, 258
collaboration 8, 19, 64, 160, 179–83, 185–6, 193, 310
collective behaviours 253–4
colonialism 63, 84, 119, 255, 289
commodification 60, 63, 99, 194, 294–5, 297, 299–301, 303
common pool resources 256
communities 11, 62, 65–7, 72–4, 85, 110, 127, 143, 148, 163–6, 182, 184–5, 307, 310, 322–6
community-based natural resource management 72–3
comparative methods 33
comparative research 20
competition 10, 49, 51, 55, 73, 78, 82–4, 124, 231, 234, 239, 258, 260, 294, 296, 319, 321, 325
complex systems 128, 173–4, 176, 249, 255–6
consensus 2, 64, 147–8, 181, 235
conservation 10, 60, 65, 70, 72–7, 123, 164–5, 184, 210, 228–9, 269, 280, 285, 287, 307–8, 311–4, 316, 325; soil 2, 71; water 47
conservationists 285; Northern 60, 65
constructivism 3–4
consumerism 82, 89–90, 302; green 19, 259
consumption 2, 6–7, 15, 19, 21–3, 25–6, 31–7, 40, 45, 48–50, 52–6, 64–5, 74, 81–91, 100, 114, 118–22, 125–8, 133–41, 161, 173, 175–6, 181, 190, 199, 209, 215, 226–9, 259, 294

Convention on Biological Diversity 1, 280, 293, 299
Convention on Long-Range Transboundary Air Pollution (CLRTAP) 9, 232, 235
Copenhagen 7, 106–117
corporate environmental performance 259
cost-shifting 38
critical level 233, 235–7
critical load 233, 235–6
Crosby, Alfred 59, 123, 128
cross-national research 5, 23–4, 32–3, 35–6
culture 3, 5–7, 21, 26, 63–6, 75, 85, 88–9, 95, 125–8, 133–8, 143–4, 146, 148–9, 165, 191–4, 199–200, 207–12, 214–5, 232–3, 242, 247, 253–4, 281, 283, 286, 288–9, 293, 300, 312, 321–6, 328

dams 174, 327; hydroelectric 79
data: environmental 49, 81; public health 64
debt: foreign 37–9, 109; international 38; repayment 37
Deepwater Horizon 241, 243–4, 250
deforestation 33–5, 39–40, 62, 114, 147, 172, 174, 226, 228
degradation 1, 10, 17–8, 24, 31, 33–6, 39, 52, 58, 61–2, 66, 70–1, 73–4, 76, 114, 139, 145–7, 151, 172, 226, 233, 269, 288, 307, 322
dematerialization 54
democratization 6, 7, 54, 59, 85, 96–7, 119, 158, 181–2, 186, 212, 256–7, 302,
dependency 1–2, 5, 7, 11, 32–4, 37, 39–40, 46, 60, 63, 70, 74–5, 78, 95, 97–8, 100–1, 103, 111, 118–9, 120–1, 123–4, 126, 128–9, 135, 143, 163, 173, 181, 185, 190, 194, 198, 200, 208, 211, 214–5, 217, 227, 233, 238, 242, 247, 253–4, 256–7, 260–2, 270, 286, 300, 309, 313, 322, 325, 327
desulfurization 239
developing countries 5, 31–6, 53–5, 60, 83, 85, 107–9, 111–6, 121, 133, 172, 227, 299
development 2, 5, 8–10, 12, 15, 17–20, 23–5, 31–7, 46–52, 53–6, 59, 60–1, 63–4, 66, 70–1, 78, 81, 83–8, 100–1, 103, 107, 110–1, 116, 119, 123–5, 127–8, 133–4, 139–9, 142, 147, 149, 157–9, 161–7, 170, 173, 176, 182–5, 196–7, 199, 200–1, 209–10, 215, 217, 223, 226, 229, 232, 235–6, 238–9, 241, 249–50, 254–5, 257–9, 261–2, 275–6, 277n, 281–2, 284, 296–7, 307, 313–4, 319–3, 325, 328
devolution 72–4, 79
Diamond, Jared 53, 59, 297, 307,
disasters 1–2, 62, 122, 142–5, 147, 149, 150, 222, 225, 230, 232, 241–4, 249, 301
discontinuity 6, 72, 79, 222, 224,
discursive processes 23, 72–3, 182, 217, 295, 309
displacement 7, 54, 73, 142–6, 148–51, 171

disproportionate impact 1, 32, 58, 63, 198, 324,
dominance 62, 86, 106, 116, 167, 262, 294, 319,
domination 4, 59, 143, 192, 308

Earth Summit 1, 7; 1992 1, 7, 107–8, 280, 288
eco-imperialism 6, 58–69, 87
eco-label(s) 6, 19, 75, 84, 86–8, 90
ecological change 1–5, 8–9, 11–2, 22–4, 26,
  31–3, 35–7, 40–1, 79, 86, 95–8, 100–1, 144,
  147–8, 151, 166, 172, 180, 254, 256, 260, 262
ecological economics 19, 174
ecological footprint 1, 23, 34–6, 121–2
ecological illiteracy 262
ecological Marxism 4
ecological modernization 4–5, 15–26, 45, 48,
  51–6, 96–7, 179–80, 186, 286, 304; theory 5,
  15–26, 45, 51–6, 96, 179–80
ecological rationalization 18
ecological restoration 10, 270–1, 272–5,
  277n, 269
ecologically unequal exchange 5, 32–37, 40
economic 1–2, 5, 7–8, 10–1, 16–9, 21, 23–4,
  26–7, 31–8, 45–54, 55–6, 60, 64, 66–7, 81,
  70–9, 83–6, 88, 95, 97, 101, 103, 106–7,
  109, 112–5, 119–23, 126, 128–9, 133–4, 138,
  142–7, 149, 151, 158–9, 163, 165–7, 172–5,
  177, 179, 182, 186, 194, 197, 223–4, 226–8,
  235, 238–9, 253, 257–8, 261, 262–3, 275,
  280–2, 284, 286–7, 292–7, 299, 300–1, 303,
  307, 312–5, 319, 321–2, 324–5, 327
economic actors 8, 10, 16, 18–9, 46, 53, 71, 73,
  76, 81, 83, 88, 95, 109, 126, 133, 158, 163,
  166, 172, 175, 179, 182, 186, 238, 275, 281,
  284, 292, 297, 313–5, 325
eco-socialism 3, 22
ecosystems 1, 4, 10–1, 70, 75, 77–8, 101, 103,
  143, 149, 173, 212, 235–6, 239, 262, 269–70,
  273–4, 276, 280, 282, 284, 286, 288, 296,
  299, 304n, 308, 313, 315–6; aquatic 239,
  287; global 10, 70, 173, 212, 262; local 10,
  262; terrestrial 239
eco-taxes 19, 74–6
eco-tourism 82–4, 314
Ecuador 64
education 73, 77, 85, 88, 125, 139n, 165, 171,
  255, 257–8, 260–2, 322,
efficiency 50, 53–4, 56n, 75, 77–8, 110–1, 119,
  127–8, 133–4, 138, 159, 163, 165–6, 174, 186,
  198, 216, 229, 234, 292, 294, 296, 300–1,
  323, 326
electronics industry 65
emergency management 223
emissions 36–8, 40, 45, 47–8, 50, 55, 56n,
  70–1, 75–6, 78, 86, 89, 102–4, 107–8, 110–2,
  109, 113–5, 118, 180, 121–2, 127, 149, 157–9,
  160–3, 165–6, 171–2, 190, 196, 198, 224,
  226–7, 228–9, 234–8, 298; permits 70, 78

empowerment 73, 85, 90, 261, 296, 307, 310,
  313, 315
endangered species 60, 280, 282, 284, 287
energy: conservation 52, 165, 228–9;
  consumption 7, 48, 50, 56n, 125–7, 133–41,
  328; depletion 228; efficiency 110, 127–8,
  138, 159, 163, 165, 326; policy 138, 231,
  243; renewable 11, 47, 53, 110, 124, 127,
  163, 226–7, 229, 319–20, 322–8;
  sustainability dilemma 83, 228
engineering 1, 79, 98, 176, 197, 207, 247,
  249, 275
entrepreneurs 71, 73, 96, 255, 294–7, 300, 322
environmental auditing 19
environmental awareness 51–2, 59
environmental capacity 16, 18
environmental crisis 8, 11, 15–6, 170, 175–6
environmental consciousness 65
environmental deterioration 15, 21–2, 24, 49,
  54, 147
environmental devastation 15, 121–2, 224
environmental displacement 7, 143–51
environmental flows 5, 25, 180,
environmental governance 6, 8, 18–20, 46, 50,
  52, 53, 70–80, 101–2, 104, 122, 170, 176, 179,
  181–3, 186
environmental hazards 9, 65, 217
environmental health 64, 208
environmental governance 46–8
environmental justice (injustice) 5–6, 58–69,
  97, 127
environmental law 64
environmental management 19, 56, 75
environmental mobilization(s) 6, 58, 66
environmental migration 142–151
environmental protection 45, 47–9, 50–1, 56,
  60, 75, 112, 158, 179, 323
environmental racism 59, 61
environmental reform 15, 17–25, 46, 51–6
environmental sociology 15–6, 22, 25, 35, 46,
  56, 124, 128, 269
environmentalism 60
environments 2–3, 17, 49, 55, 103, 142–4, 146,
  148, 151, 172, 208–9, 214, 233, 270, 298;
  biophysical 3, 17, 208, 254, 259–60
epistemic networks 239
equity 5, 54–6, 58, 67, 78, 89, 107–9, 119, 167,
  182, 186, 248, 255, 260, 300, 319,
Escobar, Arturo 60
ethics 1, 75, 81, 84, 88, 150, 198–9, 209–10,
  214, 217, 249, 254, 293–4, 299–302
ethnic minorities 58
European 15–6, 26n, 52, 59–60, 64, 78, 86–7,
  109, 135, 137–8, 145, 160, 170, 179–180, 190,
  196, 199, 232, 234–6, 239, 272, 274, 281–2,
  286, 289n, 320, 325–8; Commission 235,
  239, 286

European Union 86, 78, 170, 190, 239, 326
eutrophication 235, 287
exemptionalism 26n, 260
externalization 32, 35, 76
extraction 1, 5, 33, 35, 38–40, 60, 75, 77–8, 119, 121–2, 173, 224, 227–9, 297, 299, 319, 321

farmers 64, 77, 83, 150, 280, 282, 323–8,
fertilizers 39, 41n, 177n, 320,
financial crisis (2007) 45
financial services 39
First Nations 11, 310, 313–316n
fish 10, 81, 83, 85, 87, 135, 211–2, 284–5, 294, 299, 314, 319
food 1, 6, 25, 39, 49, 58–60, 64, 81, 83, 86–7, 100, 103, 118, 136–7, 143, 148, 165, 173, 184, 199, 201, 255, 259, 283, 301, 319–20, 324–5, 328; consumption 25; production 25, 39, 64, 86, 165, 319; security 1, 6, 58–9, 64, 86, 103, 118, 184; slow 259
Foreign direct investment (FDI) 32, 33, 37–40
forests 2, 10–1, 33–5, 38–40, 56n, 60, 62, 65–6, 81–2, 84–7, 110, 114, 147, 172, 174, 186n, 212, 226, 228, 232, 234, 283, 295, 307–8, 310–5, 322, 325, 327
forest dwellers 65–6
forest fires 234
fossil fuels 7, 38, 113–4, 118–9, 121–4, 127–9, 158, 190, 225–7, 228–9, 234, 239, 319–20, 325, 328
Foucault, Michel 71, 73, 288, 301
freight 8, 191, 195–6, 201
Freud, Sigmund 262
Freudenburg, Willam 121–2, 126–7, 179–80, 211, 223–4, 243

Garnaut, Ross 70–1, 78
genes 10, 294, 297–300, 302
gene technologies 10, 296–7,
genetics 1, 10–1, 49, 98, 289, 294, 296–7, 298–9, 303n
genetic engineering 1, 98
genetically modified crops 49
geography 2, 16, 18, 22, 24, 61, 167n, 175, 182, 193, 237, 326; political 237
Germany 16, 18, 112, 129, 234, 275, 322, 324, 326
Giddens, Anthony 18, 22–3, 97, 101, 121, 179, 212, 286
global commons 299
global economy 25, 70, 175
global governance 8, 37, 107, 170, 172–3, 328n; multi-scalar 8, 170, 172, 175–6
global networks 1, 25, 65
global warming 2, 97, 100, 110, 116, 173, 223–9, 233, 257, 261, 283, 299

globalization 17, 19–20, 25, 33, 37, 54–5, 67, 96–7, 175, 192, 226, 231, 281, 292
government 2, 5–6, 11, 15–6, 18, 20, 37, 45–56, 64, 70–2, 96, 110–2, 127–8, 149, 158–62, 171–2, 174–5, 181–6, 196, 223, 232, 243, 246, 250, 292–5, 298, 302, 310–1, 313, 319, 321, 323, 326–7
Great Bear Rainforest 11, 307–16
Great Britain 38, 40, 50, 55, 70, 76, 78–9
greenhouse gas (GHG) 37–8, 40, 50, 55, 70–1, 76, 78–9, 101–4, 106–7, 110–1, 113, 118–9, 121–2, 127, 129, 144, 146, 150–1, 158, 161, 166, 172, 198, 224, 227–9, 233, 298
greening 48, 53, 196, 200
greenwashing 53, 89
grizzly bear 11, 309, 312–3, 315
Guha, Ramachandra 60
Gulf of Mexico 228, 241, 276
Guyana 61

habitat loss 226
Harvey, David 21, 23, 63, 96–8, 144, 158, 192, 294–5, 320
hazardous waste 40, 61, 65
hierarchical roles 9
holistic 12, 55, 198–200
Honduras 60, 146
Huber, Joseph 16, 52
human geography 2, 16–8, 22
human resources 254
human rights 6, 9, 62–3, 66, 150
human well-being 10, 31, 37, 40, 102, 231, 259, 294
hydraulic fracturing (hydrofracturing) 228
hydroelectric 11, 63, 319
hyperglobalization 25

idealism 3
identity 10–1, 60, 64–5, 126, 138, 193–4, 199, 201, 207, 209, 237–8, 247, 249, 258, 262, 283, 285, 298, 302, 308–11, 314–5, 319, 328
ideology 46, 50, 52, 56, 86, 126, 136, 159, 175, 224–5, 300, 303n
incentive structure 229, 234,
indigenous 1, 6, 10, 59–60, 62–3, 65–6, 85, 110, 116, 281, 283–5, 315; knowledge 62–3; movements 63, 65; people(s) 65, 85; rights 1, 6, 59, 62, 85
industrial 1, 10, 15, 17–8, 25, 32–4, 38, 40, 47, 49–50, 52, 55–56n, 59, 61, 64, 90, 95–100, 111, 114, 119, 121, 123–4, 129, 143, 173, 176, 226, 229, 241–3, 250, 269–70, 275–6, 281, 294, 297–8, 303n, 311–2, 314, 319, 321–2, 324, 328; accidents 241; pollution 40, 49, 59; society 1, 17–8, 52, 97–8, 100
industrialization 52, 95, 98–100, 107, 281, 322

inequality 2–3, 5, 11, 21–2, 25, 31–2, 49, 61, 65, 171, 175–6, 190, 192–3, 198–200, 259–60, 262, 301, 320
information communication technology (ICT) 261
infrastructure 1, 11, 25, 27, 33, 38, 47, 49, 120–1, 129, 138, 157, 161–2, 164, 166, 172–3, 183, 192, 197, 277, 282, 307, 319, 321
injustice 54, 61, 63, 107, 213
innovation 6–8, 10–1, 17–8, 20, 35, 71, 74, 78, 83, 98–100, 106–7, 109–11, 114, 116, 124, 129, 157, 163–6, 172, 174, 191, 255, 257, 272, 296–8, 292–4, 300–1, 303n
intellectual property 10, 63, 295, 297–8
interest groups 11, 310–1
Intergovernmental Panel on Climate Change (IPCC) 96, 143, 223
international division of labour 31
International Monetary Fund (IMF) 33
international trade 5, 40, 100, 283
Inuit 224
investment 5, 32–3, 37–40, 49, 53, 74–5, 101, 111, 113–4, 119–20, 125, 174–6, 227, 275, 284, 297, 313, 321

Janicke, Martin 16, 18
Japan 1, 16, 78, 108, 121, 129, 137–8, 150, 164, 180, 232, 276
Jasanoff, Sheila 211, 233, 302
justice 1, 5–7, 40n, 55, 58–67, 82, 85, 97, 107–9, 127, 167, 176, 190, 198, 274, 299, 301; activists 58, 66; distributional 61–2; environmental 5, 58–67, 97, 127; gender 61; intergenerational 61; procedural 62; social 40n, 55, 59–60, 63, 66, 82, 85, 190

Katrina, Hurricane 146, 142, 151, 224–5, 276
Keynes, John Maynard 72, 221, 296
Korea 16, 78
Kyoto Protocol 71, 78, 107–15, 121, 160, 166, 171–3

labour 11, 31, 38, 50, 63–4, 66, 78, 96, 123, 143, 181, 194, 231, 295, 299, 303n; markets 11, 78; unions 63
lakes 135, 211, 234
land conflicts 66
land rights 1
land use 11, 110, 181, 308, 312–3, 315, 319–20, 323–5, 328
landholders 73–4
language 3–4, 45, 108, 151, 211–2, 214, 238, 255, 269, 301, 308–9
Latin America 16, 24, 59, 159, 161
Latour, Bruno 1, 4, 22, 96, 100, 102, 233, 269–70, 308–10, 312
leadership 9, 11, 108, 127, 160–1, 163, 171, 173, 181, 183, 242, 247, 250, 253

legitimacy 7, 65, 73, 83–4, 87, 98, 109–10, 119, 213, 239
lifestyle 5, 26, 55, 126–7, 146, 164, 229, 259
livelihood 7, 11, 55, 58, 60–1, 64, 67, 119, 122, 146–9, 315
living conditions 63–4
local government(s) 48, 50–1, 110, 158–9, 171–2, 310
logging 39, 311, 313
Luhmann, Niklas 26n, 211–3, 270

Malthusianism, neo- 22–4, 70–80, 307
management 4, 6, 9, 11, 19–20, 50, 56, 72–3, 75, 77, 82, 84–6, 97, 119, 158, 173, 175–6, 181–2, 184, 196–7, 208–10, 217, 221–7, 229, 242–7, 249–50, 259, 261, 270–1, 274–8n, 286, 288, 293, 307–8, 310, 312–6 327; environmental 19, 56, 75; natural resource 11, 72–3, 77, 181, 307–8; risk 9, 97, 210, 217, 221–7, 229
manufacturing 36, 38, 190, 234, 301, 311, 325
mapping 120, 233, 236–7
Mapuche Indians 63
market 5–6, 8, 10–1, 18, 20, 25–6, 33, 35–6, 39, 46, 51–4, 63–4, 70–9, 81–5, 87–9, 102, 104, 113–5, 120, 122, 128, 131, 134, 144, 174–6, 180, 182, 186, 226–7, 229, 261, 263n, 292–7, 299, 300–1, 303n, 321, 323–7; dynamics 18, 46, 52–4; failure 70–1, 74–6; mechanisms 71, 114–5, 261, 292; reform 74–5
market-based environmental governance 70–1, 76–8
market-based (policy) instruments (MBIs) 70–1, 74–6
marketization 19, 297, 299
Marxism 4, 17–8, 22–3
material 1–3, 7, 10, 32–3, 37–9, 46, 49–50, 52, 54–5, 59, 64, 67, 96, 119, 122–4, 126, 128, 133–9, 163–4, 167n, 173–4, 184, 192–3, 199, 208–9, 232–3, 235, 248, 275, 295, 298–9, 303n, 307–9, 312; exchange(s) 32, 208; flow(s) 1, 49, 55; resources 67
materialism 3–4, 10, 99
media 1, 62, 65–6, 81, 122, 150, 175, 208, 225, 229, 232, 238, 298, 308
methodological symmetry 3
methodology 149, 159, 193, 197, 200
Mexico 64, 66, 106, 118
mining 38–9, 49, 63, 81, 121, 228, 234, 239
mitigation 1, 7, 70, 75–6, 79, 101, 103–4, 107–9, 111, 113, 115–6, 121, 129, 148–50, 159, 161–2, 167, 172, 176, 261
mobility 7–8, 123, 126, 137–8, 191, 193–5, 197–201, 255–6, 260, 320; paradigm 8, 191, 198–9, 201
models 6, 9, 19–20, 32, 35–6, 60, 63, 66, 85, 95, 101–4, 158, 163–4, 166, 185, 191, 196–9,

Index

201, 207–8, 233, 235–6, 242–3, 247, 250, 254, 256–8, 260, 270, 286, 310, 312, 326–7; dynamic 236; mental 207, 209
modernization 4–5, 15–9, 20–6, 45–6, 48, 51–6, 96–7, 99–100, 179–180, 186, 270, 286, 304n
modernized mixtures 25, 27n
monitoring 48–9, 77–8, 149, 158, 222, 233, 235–8, 244, 284, 286, 294
moral principles 214
multidisciplinary 11–2
multilateral agreements 6, 51, 106
municipal government 64, 158–9, 161, 171

nation states 8, 20, 52, 54–5, 102, 111, 116, 157, 173, 179, 190, 235, 238
native peoples 65
natural capital 32
natural disasters 1–2, 142, 149
natural resources 1–2, 6, 11, 21, 31–5, 39, 46–7, 52–3, 60, 64, 72–3, 77–9, 86, 90, 143, 181–2, 208, 210, 259–60, 294–5, 299, 307–8
natural sciences 2, 23, 172, 207–8, 213
negotiation 7, 11, 62, 65, 102, 107–9, 111–6, 171–2, 237–8, 273, 309, 277
neoliberalism 26n, 70–5, 77, 79, 294, 296, 300, 303
neoliberalization 71–4, 79, 293–5, 299–303n; designer 72, 74, 79; rollback 72–4, 78–9; rollout 72–4, 79
Neo-Marxism 18, 22–3
network 1, 4, 8–10, 19, 22, 25, 55, 65–7, 72, 87, 96, 110, 124, 128, 158–9, 160–2, 163–4, 167n, 170–2, 174–6, 180–2, 186, 191–2, 195, 197, 199, 201, 233, 239, 269, 292–3, 284, 307–9, 311–5, 321–2, 326–7; effect(s) 4; epistemic 239; society 25
New Orleans 224, 146, 223
New York City 8, 161, 179, 183–6
New Zealand 1, 16, 21, 78, 113
Nigeria 60, 118–20
nitrogen oxide (NOX) 235
non-governmental organizations (NGOs) 39, 171, 179, 181–2, 310
non-linear dynamics 256
North America 16, 21, 83, 118, 120–1, 135, 138–139n, 142, 145, 158, 164–5, 167n, 222, 228
Norway 118, 137–9n, 241, 250
not in my backyard (NIMBY) 11, 321
nuclear 1, 97–9, 121, 125, 127, 129, 232, 242

occupational illness 63
oil 9, 60, 64, 112, 118–28, 165, 195, 211, 227–9, 241–4, 249–50, 276–7, 293, 320–1
offshore oil and gas 241, 244, 229–50

Organisation for Economic Co-operation and Development (OECD) 16, 24, 26, 35, 37, 109, 133, 298, 303n
Ostrom, Elinor 182, 256, 293, 308
Ozone 173, 224–5, 235–6, 238

parks 2, 183, 185, 307, 319
participatory approach 216,
patents 10, 294, 297–300, 303n
path dependency 173, 257
payments for ecosystem services (PES) 74, 76–7
peak oil 9, 118, 123, 128, 165
penalties 78, 261
Pepper, David 16
perception 3, 65, 127, 135, 194, 208–15, 217, 247, 257, 261, 281
Perrow, Charles 242–3, 245
pesticides 1, 39, 41n, 86, 283
photovoltaic panels 11, 319, 322, 326
planetary boundaries 254,
political discourse 52, 58, 73, 128, 145
political economy 59, 119, 151, 160–1, 199, 297, 308
political modernization 18, 46
political science 16, 18, 22–3
politicization 102
pollutants 34, 49, 56n, 77, 232–7, 324
pollution 1, 9, 34, 38, 40, 46–50, 53–4, 59, 62, 70–2, 77–8, 122, 129, 159, 173–4, 190–1, 196, 208, 225, 228, 232, 234–8, 244, 257, 303n, 324; air 9, 46–7, 129, 159, 190–1, 196, 232, 234–8; halo hypothesis 38; industrial 47, 49, 59; reduction 48; taxes 70; water 34, 40, 46, 173
polycentric 182, 256, 262
postmodernism 3, 23, 211, 286, 288–9
power 4–5, 9, 21–2, 31–3, 35, 37, 55, 60–1, 63–4, 73, 76, 90, 96, 98–100, 106–7, 110, 116, 119, 121, 123–5, 127–9, 133, 139, 163, 165, 175–6, 181–2, 186, 194, 200, 217, 224, 232–4, 241–3, 249–50, 258, 260, 273, 275, 287–9n, 292–4, 299–301, 309–10, 315–6, 319, 321, 324, 326–7; relations 32, 55, 61, 63–4, 194, 300, 309; poverty 165
precaution 9, 76, 102, 112, 215, 217, 254, 256–7, 297
precautionary principle 76, 102, 112
pricing 74, 198, 229
privatization 19, 72, 295, 300–2
production 2–5, 19, 21–2, 25, 31–3, 35–6, 38–41n, 47, 49–50, 52, 54–5, 60, 64–5, 70, 74–7, 86, 88, 100, 106, 118, 123–4, 127–9, 135, 144, 146, 151, 161, 164–5, 175–7n, 190, 192–3, 199, 209, 233–4, 241, 244, 257, 260, 274–5, 277, 293–5, 297–8, 302–4n, 312, 315, 319–25; consumption chains 19; sustainable 25, 75–6, 303n

property 6, 10-1, 63, 73-7, 79, 111, 261, 292-5, 297-300, 302, 304n, 308, 323; intellectual 10, 63, 295, 297-8; regimes 261; rights 6, 11, 63, 73-5, 77, 79, 111, 292-5, 297-8, 300, 302, 303n
protectionism 8, 170-1, 176
protests 59, 63-4, 66, 284, 301
psychology 242, 253
public-private partnerships 73, 163, 180-1

radical ecology 22-3
railway 38, 139n, 197-8, 321
rationality 18, 26, 71-3, 76, 79, 134, 143, 199, 261, 292-4, 296, 298-302
Rawls, John 61
realism 3, 5, 9, 128, 196, 210-3, 215-6
reasoning 9, 180, 282, 292; technical 9
recycling 39, 52-3, 65, 165, 174-5, 180,
Redclift, Michael 175, 213-5
reflective practices 262
reflexive modernization 18
reform: environmental 15, 17-25, 46, 51-6; market 74-5
Regional Acidification Information System (RAINS) 236
regulatory; dumping 73; objects 232, 236-8
relativism 212, 214, 288
Resnik, David 221
resource access rights 8, 58
restoration 9-10, 78, 183, 223, 255-6, 262, 269-79
rhetoric 73, 82, 160, 196, 221, 225
Rio Declaration 259, 280, 288
Rio Earth Summit (1992) 1, 18
risk 1, 3, 6-9, 18-9, 22, 33, 49, 54, 61, 79, 84, 96-100, 102, 121, 127, 121-2, 129, 145-6, 148, 198, 207-240, 243, 250, 254-5, 258, 270-2, 277, 280, 287, 296-7, 299-302, 313, 316n
rural 11, 45-6, 49-50, 54, 63, 73, 121, 146, 173, 177n, 283, 319-26, 323-4, 327-8; economies 11, 173
rural land degradation 73
rural land uses 11

sacred sites 64
safety 1, 49, 64, 103, 198, 213, 224-6, 230, 232, 241-3, 246-50, 274, 301
Saudi Arabia 137; Ghawar oil field 227, 229
scale 4-11, 15, 25, 35, 38-9, 54, 66, 70-1, 74, 77, 81, 83-4, 86, 97-8, 100, 102, 109, 110-1, 114-5, 119, 123, 157-8, 161, 170-1, 174-6, 180-2, 186, 194, 197, 200-1n, 210, 227, 229, 237-8, 255-7, 261-2, 272, 276-7, 281, 287, 297-8, 301-2, 313-4, 319, 324, 327-8
Scandinavia 227, 229, 234
Schlosberg, David 22

Schnaiberg, Allan 66
Schneider, Stephen 225
science and technology studies 22, 309
sea level rise 70, 143, 151
self-interest 52, 79, 81, 143
Sen, Amartya 62-3, 139n
Shanghai 45, 161
Simmel, Georg 258, 273
small farmers 64
small producers 63-4, 327
smog 47, 121
social action 157, 210, 253, 258, 260, 274
social class 2-3, 26, 61, 86, 96, 164-5, 194, 199, 271, 325
social conflict 55, 124, 128
social construction 3, 104, 211-13, 308
social-ecological 1-12, 63, 77, 95-105, 107, 144-5, 147-8, 150, 174, 207-220, 253-8, 260-3, 267-330
social ecology 210, 214
social equity 54, 56, 319
social inequality 3, 5, 21-2, 301
social learning 3, 8-10, 138-9, 253-65, 308
social metabolism 3
social movements 1-2, 26n, 52, 54, 66, 127, 179
social networks 8, 186, 233
social practices 7, 19, 96, 133-41, 191, 224, 229, 238, 255
social reflectivity 5, 259
social research 2, 11
social resources 11
social sciences 1-12, 15-18, 20-21, 24, 40, 53, 99, 118-9, 121-2, 128-9, 134-6, 148, 192-3, 195, 199, 207, 210-1, 213, 222, 242-3, 250, 257, 271, 288, 308-9, 314
social structures 126, 134, 207, 254, 260, 308-10, 315
social theory 3, 5, 16-8, 20, 24, 96, 129, 269
social welfare 72, 78, 85
socio-environmental development 5
socio-natural 150, 286, 288
socio-technical systems 7, 24, 27n, 99-100, 133, 164, 192, 195, 242, 250, 277, 280, 286, 322
sociology 2, 4, 11, 15-9, 22, 25-6n, 32, 34-6, 39, 46, 49, 51-2, 55-6, 96, 104, 118-32, 135, 179, 191-3, 195, 198-201, 210-3, 242-3, 269-71, 273-5, 277, 280
soil 2, 39, 59, 62, 71, 77, 144, 147, 190, 233, 275, 322
South Africa 64, 106, 108, 166
Sovereignty 8, 51, 134, 176, 293
spatial demarcation(s) 237-8
species 1, 4, 10, 60, 78, 98, 173, 209, 212, 226, 233, 274, 280-9, 296-7, 299, 301, 307, 312
spillover effects 38
stakeholders 11, 85-7, 89, 181, 255, 272, 276, 284, 300, 302, 310, 328

Stern, Nicholas 70
structural adjustment programmes 37–8
structuralism 17, 194, 243, 326
subjectivity 4,
sulfur dioxide (SO2) 45, 47
superindustrialization 52
surprise 10, 16, 97, 102, 222, 257, 269–70, 272–4, 276–7
sustainability 5–6, 9–10, 22, 45–6, 52, 55, 64, 79, 81–9, 98, 110, 133, 135, 139, 158–9, 171, 174–6, 185, 196, 198–200, 209, 226, 228, 253–9, 260–2, 263n, 293–4
sustainable 2, 6, 8–9, 12, 19, 23–6, 34, 36, 40, 45–6, 50–1, 54, 60, 64, 66, 74–6, 81–4, 86–90, 11, 119, 133, 138–9, 160, 172, 176n–7n, 190–1, 194–201, 217, 227, 254, 256, 260–1, 273, 313, 324; consumption 19, 81–2, 89–90, 139; development 2, 9, 45–6, 50, 83–4, 86, 88, 111, 217, 254; growth 45; production 75–6, 303n; transport 8, 190–1, 194–201
symbolism 3, 64–5, 81, 96, 124, 128, 209, 294, 309, 312
systems theory 22

technological determinism 21, 135, 320
technologies of agency 73
technologies of performance 73
technology studies 19, 22, 134, 309
telecommunications 39
territory 60, 64–5, 232, 236–8, 283, 285, 312–3, 315, 320–1, 328
Three Mile Island 242
threshold effects 6, 79, 98, 102
time 2, 8, 12, 16, 19–21, 23–4, 26, 35–7, 39–40, 50–1, 53–4, 64, 73, 76–8, 84, 86, 89, 95–101, 104, 107–11, 113, 115–6, 119, 121, 124–5, 129, 134, 136–7, 139, 142, 149–50, 158–9, 162, 173–4, 177n, 180, 184, 196–7, 198–9, 201, 208–9, 211–13, 217, 222, 224, 226–7, 232, 234, 237, 244–6, 248, 254–61, 271–2, 274, 276, 281, 283–4, 286–8, 293, 296, 299–300, 308, 310, 315, 321, 323, 325
tipping points 96, 226–7, 292
Tohoku earthquake and tsunami 1
tradable permits 77–8
transboundary 5, 9, 181, 231–2, 234–9
transdisciplinary 214–7, 255
transformative change 253
transnational 6, 25, 33, 37–9, 63–7, 102, 110, 119, 158, 160–1, 170, 234, 238, 294; -ization 58; advocacy networks 65–7
transport 8, 118–20, 136–8, 158–9, 173, 190–201, 227–8, 231, 234, 303n, 320–1, 327
treadmill of destruction 35–6

treadmill of production 22, 35–6
triad network model 19
tropical conservation 65
trust 7, 19, 22, 84, 88–9, 98, 126, 184, 239, 249, 308

uncertainty 3, 6, 8–9, 64, 66, 76, 79, 97, 102, 106, 112, 115, 145, 147, 160, 213, 215, 221–6, 229–30, 232, 234–5, 237, 270–1, 296, 301
unemployment 45, 62, 120
unintended consequences 97
United Kingdom 234
United Nations 71, 88, 106, 109, 145, 171, 235, 292; Framework Convention on Climate Change (UNFCCC) 106, 171
United States 36, 59, 109, 123, 146, 150, 180–1, 198, 235, 278, 297
urban 3, 8, 25, 45–7, 49–50, 54, 61, 63, 121, 129, 146, 157–67, 170–6, 181, 183–4, 200, 283–4, 320–3; rural pollution transfers; environmental change 54
urbanism 8, 157, 160, 162–4, 166
Urry, John 22, 191, 201

valuation 19, 90, 208, 214
volatile organic compounds (VOC) 235
vulnerability 1, 61, 100, 103, 143–4, 149–50, 157, 165–6, 223–5, 232, 270, 277n, 328

waste 32–3, 36–7, 40, 46–7, 49, 52, 54, 56n, 61, 65, 121, 157–8, 163, 173–4, 176 226, 275, 326–7
water 34, 38, 40, 46–7, 49, 56n, 71–2, 75–8, 88, 122, 137, 143, 164, 166, 173, 175, 201, 228, 269, 275, 283, 287, 298, 319, 325, 327
water pollution 34, 40, 46, 173
watershed 73, 76–7, 313–4, 319
Weber, Max 269
well-being 2, 10, 31, 36–8, 40, 59–63, 70, 76, 102, 120, 122, 125, 184, 198, 231, 259, 294, 300–1, 319
wetland 223–4, 282–4, 287
Wildavsky, Aaron 208, 217
wilderness protection 60
wind farms 11, 323, 326–7
women's associations 63
women's literacy 258
World Bank 33–4
world economy 20, 31–2, 39, 122
world-systems 5, 22, 31–4, 37–40
worldview 22, 126, 211, 253–4, 258, 272

Zambia 64